Irena Kostova

General and Inorganic Chemistry

Also of interest

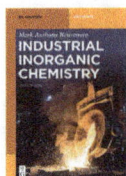

Industrial Inorganic Chemistry
Mark Anthony Benvenuto, 2024
ISBN 978-3-11-132944-4, e-ISBN (PDF) 978-3-11-132951-2

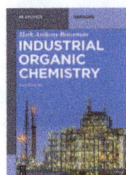

Industrial Organic Chemistry
2nd Edition
Mark Anthony Benvenuto, 2024
ISBN 978-3-11-132991-8, e-ISBN 978-3-11-133035-8

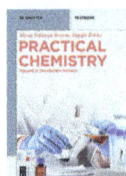

Practical Chemistry.
Transition Metals
Mesay Solomon Tesema and Digafie Zeleke, 2024
ISBN 978-3-11-157384-7, e-ISBN 978-3-11-157434-9

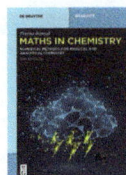

Maths in Chemistry.
Numerical Methods for Physical and Analytical Chemistry
2nd Edition
Prerna Bansal, 2024
ISBN 978-3-11-133392-2, e-ISBN 978-3-11-133444-8

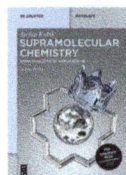

Supramolecular Chemistry.
From Concepts to Applications
2nd Edition
Stefan Kubik
ISBN 978-3-11-131507-2, e-ISBN 978-3-11-131517-1

Irena Kostova

General and Inorganic Chemistry

——

Theoretical Foundations, Main Group Elements,
Transition Metals, Lanthanides, and Actinides

DE GRUYTER

Author
Prof. Irena Kostova
Department of Chemistry
Faculty of Pharmacy
Medical University
1000 Sofia
Bulgaria
irenakostova@yahoo.com

ISBN 978-3-11-171222-2
e-ISBN (PDF) 978-3-11-171224-6
e-ISBN (EPUB) 978-3-11-171254-3

Library of Congress Control Number: 2025931334

Bibliographic information published by the Deutsche Nationalbibliothek
The Deutsche Nationalbibliothek lists this publication in the Deutsche Nationalbibliografie;
detailed bibliographic data are available on the Internet at http://dnb.dnb.de.

© 2025 Walter de Gruyter GmbH, Berlin/Boston, Genthiner Straße 13, 10785 Berlin
Cover image: MF3d/E+/Getty Images
Typesetting: Integra Software Services Pvt. Ltd.

www.degruyter.com
Questions about General Product Safety Regulation:
productsafety@degruyterbrill.com

Preface

This textbook is intended for students studying pharmacy and medicine. It covers the program included in the general and inorganic chemistry course. Divided into two parts, each part further divided into sections and subsections to facilitate the interrelationship of topics, this textbook will enable researchers, professors, and students to find the wide range of topics, including the most cutting-edge topics in general and inorganic chemistry. The content of the textbook is consistent with the curriculum. In the discipline "general and inorganic chemistry," the fundamental principles and general directions of chemistry are considered, and in the second part on the chemistry of the elements, the general physical and chemical properties of the elements and their compounds with an emphasis on their biological role are presented.

The first part of the textbook presents basic theoretical topics such as the structure of the atom, periodic table and law, chemical bonding, and complex compounds. The following are the topics related to the course of chemical processes (chemical thermodynamics, chemical kinetics, catalysis, chemical equilibrium, redox processes, and physicochemical analysis), as well as topics studying solutions (disperse systems, electrolyte solutions, and colloidal solutions). This part would give students systematic theoretical and practical knowledge in the field of general chemistry with an emphasis on biochemical processes.

The second part of the textbook is dedicated to chemical elements. It is built on the concept of interconnection "place in the periodic table – chemical properties – biological role of chemical elements and their compounds" and is adapted to the needs of pharmaceutical practice. It includes an analysis of the sources and preparations of the elements, their common compounds, their physical and chemical properties, and their applications. Attention is specifically focused on the role and influence of chemical elements and their compounds on biological systems and mainly on the human body. Students are expected to build the necessary thinking and skills to apply this knowledge in their professional realization.

The compulsory course in general and inorganic chemistry is in line with the modern requirements for in-depth fundamental knowledge and practical skills in the training of pharmacy and medical students. At the same time, the students pursuing MSc in chemical engineering and other professional studies will also find the book extremely useful. The objective is to provide the students with comprehensive treatment of the subject on modern lines.

https://doi.org/10.1515/9783111712246-202

Contents

Section B: **Chemical elements**

Section B.I: **Main groups of the periodic table**

Section B.II: **Secondary groups of the periodic table**

Section A: **Theoretical foundations**

Chapter 1
Introduction

In the new millennium, the world is facing new and extreme challenges, so making a solid foundation in chemical principles is becoming increasingly important. Scientific disciplines are becoming more integrated and interdependent, and the separation of knowledge in chemistry, physics, biology, environmental sciences, geology, etc. is almost fading. Of all scientific disciplines, chemistry is perhaps the most broadly related to other fields of knowledge. It is essential for understanding much of the natural world and this serves as the foundation of many fields, including astronomy, geology, biology, and medicine. The practical application of chemistry is applicable in varied aspects of life.

The development of chemistry can be divided into several periods. The first, the so-called old period, is characterized by the accumulation of knowledge through random observations. The alchemical idea was substantiated by Aristotle through his doctrine of the transformation of elements and the desire to transform base metals into precious ones. In the first half of the fifteenth century, it was believed that diseases were caused by certain substances. The fights against these substances were carried out using other known compounds, marking the beginning of the era of medicinal chemistry. It was during this time that chemistry emerged as an independent science. After the introduction of the concept of chemical elements by Robert Boyle in 1661, the field of chemistry was formally defined and evolved into a modern science. During this period, many scientists made attempts to explain various phenomena, distinguishing existing knowledge of known inorganic substances, and to summarize individual observations into theories. Until then, chemistry has been largely a purely descriptive science. The next period began with the works of John Dalton, and his attempts to explain chemical laws through the knowledge of atomic theory led to the development of the atomic-molecular theory. This was the first scientific theory in chemistry, approved at a Congress of Chemists in Karlsruhe in 1861. It systematized all existing knowledge in the field of chemistry and confirmed the basic chemical concepts. The discovery of the periodic law and the creation of the periodic table by D. I. Mendeleev are considered the greatest achievements in chemistry, which gave a serious impetus to the development of the chemistry of the elements and the study of the internal structure of the atom. In the first half of the nineteenth century, the composition of a large number of inorganic and organic compounds was established. This accumulation of facts led to important discoveries that were a prerequisite for chemistry to focus on theoretical principles. It is now clear that describing the facts in the chemistry of elements and their compounds is insufficient and that these facts must be linked to fundamental theoretical principles wherever possible.

The history of chemistry begins with inorganic compounds. Logically, it is in inorganic chemistry that many fundamental concepts and theoretical principles arise,

https://doi.org/10.1515/9783111712246-001

which contribute to the overall development of general chemistry. The discipline of general and inorganic chemistry, as a branch of science, is related to the properties, structure, and composition of substances, as well as the interactions between them and regularities in their change in accordance with theoretical laws. Inorganic chemistry studies the synthesis, properties, and reactions of more than 100 known chemical elements and of the extremely diverse compounds they form. Science and technology of the twenty-first century increasingly rely on natural and synthetic materials, many of which are inorganic compounds. At present, there is a growing focus on the synthesis of new materials, understanding the relationships between properties and structures of compounds, and studying the mechanisms of their chemical reactions. As a result, inorganic chemistry is no longer viewed as an isolated academic discipline, but rather as an important part of basic scientific knowledge with applications in various fields of science and everyday life. Inorganic chemistry plays a crucial role in our daily lives. Industrially, it involves the production of all the chemicals essential to our economy. Environmental chemistry largely encompasses the inorganic chemistry of the atmosphere, water, and soil. Recently, special attention has been given to the biological role of chemical elements and their compounds. Fields such as medicine, pharmacology, nutrition, and toxicology focus specifically on how chemicals interact with the chemical components of the living organisms. Knowledge of chemistry supports the understanding of biological and natural processes, geochemical, pharmaceutical, and biochemical concepts. The goal of any chemistry textbook is to address the increasingly close relationship between different disciplines and demonstrate the contemporary importance of chemistry as a separate discipline.

Chapter 2
Structure of the atom

All matter is composed of atoms, which are the smallest parts of an element that can take place in chemical reactions. An atom is composed of two regions: the nucleus, which is in the center of the atom and contains protons and neutrons, and the outer region of the atom, which holds its electrons in orbit around the nucleus. Protons and neutrons have approximately the same mass, about 1.67×10^{-24} g, defined as one atomic mass unit (amu) or one Dalton. Each electron has a negative charge (−1) equal to the positive charge of a proton (+1), 1.602×10^{-19} C. Neutrons are uncharged particles. The protons have a positive charge and the neutrons have a neutral charge, thus the nucleus has an overall positive charge. Negatively charged electrons are found in orbitals around the nucleus.

2.1 Early models of the atom

2.1.1 Thomson model

Thomson proposed that an atom may be regarded as a sphere carrying positive charge due to protons and in which negatively charged electrons are embedded. *Drawback of Thomson model:* This model was able to explain the overall neutrality of the atom, but it could not satisfactorily explain the results of scattering experiments carried out by Rutherford in 1911.

2.1.2 Rutherford's α-particle scattering experiment

In 1911, Rutherford performed some scattering experiments in which he bombarded thin foils of metals like gold, silver, platinum or copper with a beam of fast-moving α-particles.

Whenever α-particles struck the screen, a tiny flash of light was produced at that point. From these experiments, he made the following observations: (i) most of the α-particles passed through the foil without undergoing any deflection; (ii) a few α-particles underwent deflection through small angles; (iii) very few mere deflected back, that is, through an angle of nearly 180°. From these observations, Rutherford drew the following conclusions: (i) since most of the α-particles passed through the foil without undergoing any deflection, there must be sufficient empty space within the atom; (ii) a small fraction of α-particles was deflected by small angles. The positive charge has to be concentrated in a very small volume that repelled and deflected a few positively charged α-particles. This very small portion of the atom was called

https://doi.org/10.1515/9783111712246-002

nucleus; (iii) the volume of nucleus is very small as compared to total volume of atom; (iv) the nucleus is surrounded by electrons that move around the nucleus with a very high speed in circular paths called orbits; and (v) electrons and nucleus are held together by electrostatic forces of attraction.

2.1.2.1 Drawbacks of Rutherford model

Rutherford's model cannot explain the stability of atom if the motion of electrons is described on the basis of classical mechanics and electromagnetic theory. Rutherford's model does not give any idea about distribution of electrons around the nucleus and about their energies (Figure 2.1).

However, scientists still had many unanswered questions: If the electrons are orbiting the nucleus, why don't they fall into the nucleus as predicted by classical physics? If all atoms are composed of these same particles why do different atoms have different chemical properties? How is the internal structure of the atom related to the discrete emission lines produced by excited elements? Bohr addressed these questions using a seemingly simple assumption: what if some aspects of atomic structure, such as electron orbits and energies, could only take on certain values?

2.1.3 Developments leading to the Bohr's model

Two developments played a major role in the formulation of Bohr's model of atom: (i) dual character of the electromagnetic radiation, which means that radiations possess both wave and particle properties, and (ii) experimental results on atomic spectra, which can be explained by quantized electronic energy levels in atoms.

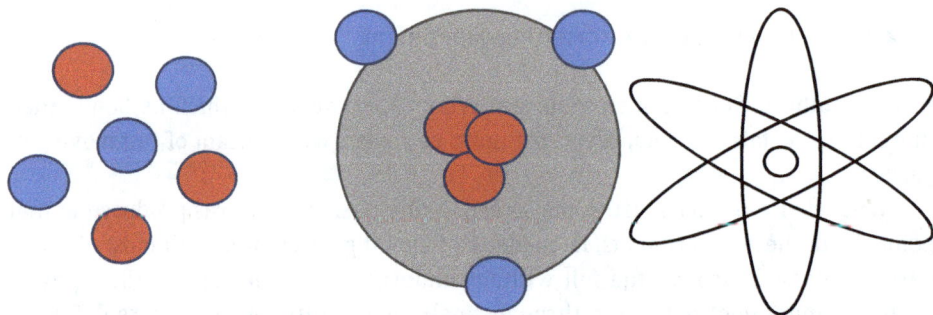

Figure 2.1: Atomic models of Thomson (1897), Rutherford (1911), and Bohr (1913).

The Bohr model shows the atom as a central nucleus containing protons and neutrons with the electrons in circular orbitals at specific distances from the nucleus. These orbits form electron shells or energy levels, which are a way of visualizing the number of electrons in the various shells.

Bohr postulated that electrons in the atom move in what he called *energy states*. These energy states are "stationary" (of fixed energy, not fixed position) at different distances from the nucleus. He further reasoned that electrons can undergo transitions from one energy state to another. Bohr called these transitions quantum jumps. He agreed with Planck's assertion that quantum chunks of radiation are emitted when such a quantum jump occurs from a higher to a lower energy state.

2.2 Quantum mechanical approach

In 1900, Max Planck postulated that the energy of a radiated quantum of energy is proportional to the frequency of radiation: $E \sim v$. With the constant of proportionality h (Planck's constant, $h = 6.63 \times 10^{-34}$ J.s):

$$E = hv$$

Five years later, Einstein proposed not only that material energy is quantized but also the light itself exists as quantum corpuscles or photons.

The simplest Bohr's model of hydrogen is based on the nonclassical assumption that electrons travel in specific orbits around the nucleus. Bohr explained the hydrogen spectrum in terms of electrons absorbing and emitting photons to change energy levels, where the photon energy is $\Delta E = hv$. Bohr's model does not work for systems with more than one electron.

2.2.1 Quantization and photons

By the early 1900s, scientists were aware that some phenomena occurred in a discrete, as opposed to continuous, manner. Max Planck and Einstein had theorized that electromagnetic radiation not only behaves like a wave, but also like particles called *photons*. Planck studied the electromagnetic radiation emitted by heated objects, and he proposed that the emitted electromagnetic radiation was "quantized" since the energy of light could only have values given by the following equation:

$$E_{photon} = n.h.v$$

where n is a positive integer, h is Planck's constant, and v is the frequency of the light, which has units of 1/s. As a consequence, the emitted electromagnetic radiation must have energies that are multiples of hv. Einstein used Planck's results. When something is *quantized*, it means that only specific values are allowed.

2.2.2 Atomic line spectra

Atomic line spectra are an example of quantization. When an element or ion is heated by a flame or excited by electric current, the excited atoms emit light of a characteristic color. For the relatively simple case of H atom, the wavelengths of some emission lines could even be fitted to mathematical equations. The equations did not explain why the hydrogen atom emitted those particular wavelengths of light. Prior to Bohr's model of the hydrogen atom, scientists were unclear of the reason behind the quantization of atomic emission spectra.

2.2.3 Bohr's model of the hydrogen atom: quantization of electronic structure

Bohr suggested that perhaps the electrons could only orbit the nucleus in specific orbits or *shells* with a fixed radius. When an atom absorbs energy, an electron is promoted to a higher energy level. When it falls back, light of energy equal to the difference in energy between the shells is emitted from the atom. Only shells with a radius given by the equation would be allowed, and the electron could not exist in between the shells. Mathematically, the allowed values of the atomic radius as $r(n) = n^2 \cdot r(1)$, where n is a positive integer, and $r(1)$ is the *Bohr radius*, the smallest allowed radius for hydrogen. He found that Bohr radius $= r(1) = 0.529 \times 10^{-10}$ m. By keeping the electrons in circular, quantized orbits around the positively charged nucleus, Bohr was able to calculate the energy of an electron in the nth energy level of hydrogen, where the lowest possible energy or *ground state energy* of a hydrogen electron $E(1)$ is 13.6 eV.

The energy is always going to be a negative number, and the ground state, $n = 1$, has the most negative value. This is because the energy of an electron in orbit is relative to the energy of an electron that has been completely separated from its nucleus, $n = \infty$, which is defined to have an energy of 0 eV. The electron in orbit around the nucleus is more stable than an electron that is infinitely far away from its nucleus.

According to the Bohr's model each orbit is associated with definite energy and therefore these are known as energy levels or shells. These are numbered as 1, 2, 3, 4, ... or K, L, M, N, ... Only those energy orbits are permitted for the electron in which angular momentum of the electron is a whole number multiple of $h/2\pi$. Angular momentum of electron:

$$m.v.r = nh/2\pi$$

where $n = 1, 2, 3, 4$, etc., m is the mass of the electron, v is the velocity of the revolving electron, r is the radius of the orbit, h is Planck's constant, and n is an integer. As long as electron is present in a particular orbit, it neither absorbs nor loses its energy, therefore, remains constant. When energy is supplied to an electron, it absorbs energy only in fixed amounts as quanta and jumps to higher energy state away from the nucleus known as *excited state*. The excited state is unstable, the electron may jump back to the lower energy state and in doing so it emits the same amount of energy:

$$\Delta E = E_2 - E_1$$

The Bohr model worked for explaining the H atom and other single electron systems such as He$^+$. The theory could not explain the atomic spectra of the atoms containing more than one electron or multielectron atoms. Furthermore, the Bohr model had no way of explaining why some lines are more intense than others or why some spectral lines split into multiple lines in the presence of a magnetic field – the Zeeman effect. In the following decades, work by scientists such as *Schrödinger* showed that electrons can be thought of as behaving like *waves* and behaving as *particles*. This means that it is not possible to know both a given electron's position in space and its velocity at the same time, stated in *Heisenberg's uncertainty principle*. The uncertainty principle contradicts Bohr's idea of electrons existing in specific orbits with a known velocity and radius. Instead, we can only calculate *probabilities* of finding electrons in a particular region of space around the nucleus.

Bohr's theory failed to explain the ability of atoms to form molecule formed by chemical bonds. Classical physics is not sufficient to explain all phenomena on an atomic level. Bohr was the first to recognize this by incorporating the idea of quantization into the electronic structure of H atom, and he was able to thereby explain the stability of an atom and the emission spectra of H. The modern quantum mechanical model can do that.

2.2.4 Dual behavior of matter (de Broglie equation)

In 1924, de Broglie proposed that matter, like radiation, should also exhibit dual behavior, that is, both particle-like and wave-like properties. This means that like photons, electrons also have momentum as well as wavelength. The de Broglie wavelength associated with particle is related to the Planck's constant (h), mass (m), and velocity (v). From this analogy, de Broglie gave the following relation between wavelength (λ) and momentum (p) of a material particle:

$$\lambda = h/mv = h/p$$

2.2.5 Heisenberg's uncertainty principle

It states that it is impossible to determine simultaneously, the exact position and exact momentum (or velocity) of an electron. Uncertainties in the products of "conjugate pairs" (momentum/position) and (energy/time) were defined by Heisenberg as having a minimum value corresponding to Planck's constant divided by 4π:

$$\Delta x . \Delta p \geq \frac{h}{2\pi}$$

where Δ refers to the uncertainty in that variable and h is Planck's constant. If an experiment is designed to locate the position of a particle with great precision, it is not possible to measure its momentum precisely. Its future action (trajectory) cannot be predicted with certainty. Similarly, if the momentum is measured precisely, the position of the particle is not known with certainty. The effect of Heisenberg's uncertainty principle is significant only for microscopic objects and is negligible for macroscopic objects.

2.2.6 Quantum mechanical model of atom

Quantum mechanics is a theoretical science that deals with the study of the motions of the microscopic objects that have both observable wave like and particle-like properties. An implication, explored by Schrödinger in 1927, is that electrons can be treated as matter waves. Their motion can be likened to wave motion. The waves associated with electrons must correspond to certain allowable patterns. These patterns can be described by mathematical equations. The solutions of these wave equations are known as wave functions (ψ). The Schrödinger wave equation for one particle with coordinates (x, y, z) and energies (E, U) is

$$\frac{\partial^2 \psi}{\partial x^2} + \frac{\partial^2 \psi}{\partial y^2} + \frac{\partial^2 \psi}{\partial z^2} + \frac{8\pi^2 m}{h^2}(E - U).\psi = 0$$

Wave functions contain a set of three quantum numbers, and when specific values of these numbers are assigned, the result is called orbital. The physical interpretation of an orbital is that it can be used to represent a region in space where an electron is likely to be found. The term "orbit" used in the Bohr's theory suggests a definite two-dimensional pathway that electrons follow. The term "orbital" is intended only to outline the general region, in which electrons are likely to be found. Although the wave function has no physical significance, the square of the wave function represents the electron charge density or the probability of finding an electron at any given point in the atom. There are many orbitals in an atom. Electrons occupy an atomic orbital which has definite energy. An orbital cannot have more than two electrons. The orbitals are filled in increasing order of energy. All the information about the electron in an atom is stored in orbital wave function ψ. The probability of finding electron at a point within an atom is proportional to square of orbital wave function ψ^2, known as probability density which is always positive. From ψ^2, it is possible to predict the region around the nucleus where electron most probably is found.

2.3 Characteristics of quantum numbers. Atomic orbitals

To produce acceptable solutions to the Schrödinger wave equation it is necessary to assign integral values to quantum numbers. Atomic orbitals can be specified by giving their corresponding energies and angular momentums which are quantized (with specific values). The quantized values can be expressed in terms of quantum numbers. These are used to get complete information about electron, that is, its location, energy, spin, etc.

2.3.1 Principal quantum number (n)

It gives the principal energy level or shell to which the electron belongs, determining energy and size of orbital. It can have any positive integral value except zero, that is, $n = 1, 2, 3, 4$, etc. The various principal energy shells are also designated by the letters, K, L, M, N, O, P, etc., starting from the nucleus. In wave mechanical model, atoms have a series of energy levels called *principal energy levels*, represented by principal quantum numbers. The energy increases as the value of n increases. The principal quantum number gives us the following information: (i) it gives the average distance of the electron from the nucleus; (ii) it completely determines the energy of the electron in hydrogen atom and hydrogen like particles; and (iii) the maximum number of electrons present in any principal shell is given by $2n^2$, where n is the number of the principal shell. Electrons with the same value of n are in the same energy shell.

2.3.2 Azimuthal or orbital angular quantum number (ℓ)

It determines the shape and type of the orbital (mainly). It is found that the spectra of the elements contain not only the main lines but there are many fine lines also present. This number helps to explain the fine lines of the spectrum. The Azimuthal quantum number gives the following information: (i) the number of subshells present in the main shell; (ii) the angular momentum of the electron present in any subshell; (iii) the relative energies of various subshells; and (iv) the shapes of the various subshells present within the same principal shell. This quantum number is denoted by the letter ℓ. For a given value of n, it can have any value ranging from 0 to $n - 1$. It cannot be negative and it cannot be any large than $n - 1$. In atom each principal energy level contains one or more types of orbitals, called sublevels. A sublevel is defined as a group of electrons in an atom all having the same principal quantum number and also the same orbital quantum number. For example, for the first shell (K), $n = 1$, ℓ can have only one value, that is, $\ell = 0$ (1s); for $n = 2$, the possible value of ℓ can be 0 and 1 (2s, 2p). Subshells corresponding to different values of ℓ are represented by the following symbols:

Value of ℓ	0 1 2 3 4 5 . . . . ., $n-1$
Subshell's notation	s p d f g h . . . .

2.3.3 Magnetic orbital quantum number (m or m_ℓ)

It determines the number of preferred orientations of the electrons present in a subshell. Since each orientation corresponds to an orbital, therefore, the magnetic orbital quantum number determines the number of orbitals present in any subshell. The magnetic quantum number is denoted by letter m or m_ℓ and for a given value of ℓ, it can have all the positive or negative values ranging from $-\ell$ to $+\ell$ including 0. Thus, for energy value of ℓ, m has $2\ell+1$ value. For example, for $\ell=0$ (s-subshell), m_ℓ can have only one value, that is, $m_\ell=0$. This means that s-subshell has only one orientation in space. In other words, s-subshell has only one orbital called s-orbital. If $\ell=1$, $m_\ell=0\pm1$: three p-orbitals for each value of n; if $\ell=2$, $m_\ell=0$, ±1, ±2: five d-orbitals for each value of n; if $\ell=3$, $m_\ell=0$, ±1, ±2, ±3: seven f orbitals for each value of n. The value of m does not ordinarily affect the energy of an electron, but it does determine the orientation in space of the volume that can contain the electron, and it does determine the numbers of orbitals in each sublevel.

If two electrons in an atom have the same principal quantum number, the same angular quantum number and the same magnetic quantum number the electrons are said to be in the same orbital. An orbital can be empty or it can contain one or two electrons but never more than two. If two electrons occupy the same orbital, they must have opposite spins. Available values of quantum numbers are listed in Table 2.1.

Table 2.1: Values of quantum numbers.

n	ℓ	m_ℓ	Atomic orbitals	Number of electrons
1	0	0	1s	2
2	0, 1	0; −1, 0, +1	2s, 2p	8
3	0, 1, 2	0; −1, 0, +1; −2, −1, 0, +1, +2	3s, 3p, 3d	18
4	0, 1, 2, 3	0; −1, 0, +1; − 2, − 1, 0, +1, +2; −3, −2, −1, 0, +1, +2, +3	4s, 4p, 4d, 4f	32

2.3.4 Spin quantum number (s or m_s)

It helps to explain the magnetic properties of substances. A spinning electron behaves like a micromagnet with a definite magnetic moment. If an orbital contains two electrons, the two magnetic moments oppose and cancel each other. Spin quantum number is used to distinguish each electron with the same n, ℓ, and m_ℓ values. The spin quantum number may have a value of $-\frac{1}{2}$ or $+\frac{1}{2}$ only. The value of m_s does not depend on the value of any other quantum number. It indicates that the electron is spinning on its axis in one direction or the opposite. Therefore, each orbital is designated by the three quantum numbers n, ℓ, m_ℓ.

s-Orbital is present in the s-subshell. For this subshell, $l = 0$ and $m_l = 0$. Thus, s-orbital with only one orientation has a spherical shape with uniform electron density along all the three axes. The probability of 1s electron is found to be maximum near the nucleus and decreases with the increase in the distance from the nucleus. In 2s electron, the probability is also maximum near the nucleus and decreases to zero probability. s-Orbital is spherically symmetrical because when $\ell = 0$, m must also be 0, so there can be only one orbital of the s type.

p-Orbitals are present in the p-subshell for which $\ell = 1$ and m_l can have three orientations −1, 0, +1. Thus, there are three orbitals in the p-subshell which are designated as p_x, p_y, and p_z orbitals depending upon the axis along which they are directed. The shape of a p-orbital is dumb-bell consisting of two portions known as lobes. It is obvious that unlike s-, a p-orbital is directional and it influences the shapes of molecules where it participates.

d-Orbitals are present in d-subshell, where $\ell = 2$ and $m_l = -2, -1, 0, +1$, and $+2$. This means that there are five orientations leading to five different orbitals (Figure 2.2).

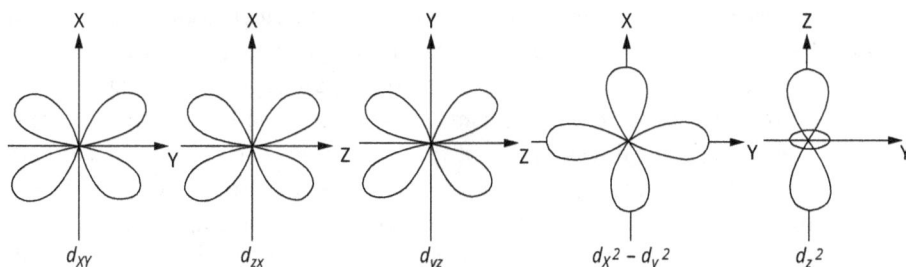

Figure 2.2: Shapes of d-orbitals.

2.4 Electronic structure of atoms

2.4.1 Aufbau principle

The principle states: In the ground state of the atoms, the orbitals are filled in order of their increasing energies. In other words, electrons first occupy the lowest energy orbital available to them and enter into higher energy orbitals only after the lower energy orbitals are filled. The order in which the energies of the orbitals increase and hence the order in which the orbitals are filled is as follows: 1s, 2s, 2p, 3s, 3p, 4s, 3d, 4p, 5s, 4d, 5p, 6s, 4f, 3d, 6p, 7s, 5f, and 6d.

2.4.2 Pauli exclusion principle

According to this principle, no two electrons in an atom can have the same set of four quantum numbers. Pauli exclusion principle can also be stated as: only two electrons may exist in the same orbital and these electrons must have opposite spins.

2.4.3 Hund's rule of maximum multiplicity

It states that pairing of electrons in the orbitals belonging to the same subshell (p, d, or f) does not take place until each orbital belonging to that subshell has got one electron each, that is, it is singly occupied (Figure 2.3).

2.4.4 Electronic configuration of atoms

By considering the existence of quantum numbers and a set of rules the probable assignment of electrons to orbitals can be made. The distribution of electrons into orbitals of an atom is called its electronic configuration. It gives information about the number of electrons in each shell, subshell, and orbital of an atom. The subshells are filled in order of increasing energy. The spin of the electron is represented by its direction. Electrons with similar spin repel each other which (spin-pair repulsion). Electrons will therefore occupy separate orbitals in the same subshell to minimize this repulsion and have their spin in the same direction, e.g., if there are three electrons in a p subshell, one electron will go into each p_x, p_y, and p_z orbital. Electrons are only paired when there are no more empty orbitals available within a subshell in which case the spins are opposite to minimize repulsion. If there are four electrons in a p subshell, one p-orbitals contains two electrons with opposite spin and two orbitals contain per one electron only. Electrons pair up and occupy the lower energy levels first.

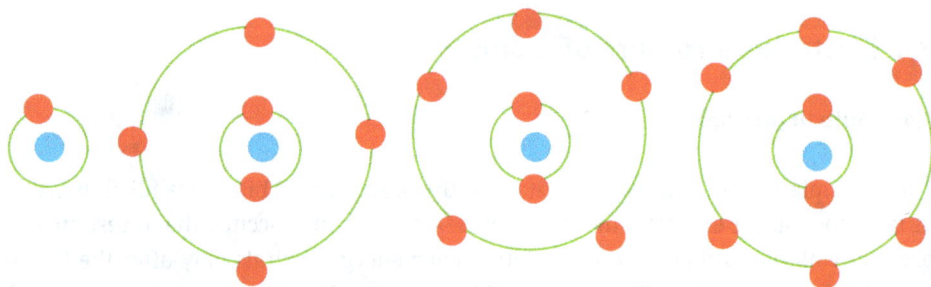

Figure 2.3: Electronic configuration of H, C, N, and O atoms.

The transition metals fill the 4s-subshell before the 3d-subshell but lose electrons from the 4s first and not from the 3d-subshell (the 4s-subshell is lower in energy). The periodic table is split up into four main blocks depending on their electronic configuration: *s-block elements* – have their valence electron(s) in an s-orbital; *p-block elements* – have their valence electron(s) in a p-orbital; *d-block elements* – have their valence electron(s) in a d-orbital; *f-block elements* – have their valence electron(s) in an f orbital.

Table 2.2 presents the "capacity" of the first three energy levels, that is, the maximum number of electrons at a given shell with a principal quantum number n.

Table 2.2: Maximum electronic capacity at the first three energy levels.

n	ℓ	m_ℓ	s	Number of e⁻ at subshell	Number of e⁻ at shell	Electron configuration
1	0 (s)	0	±½	2	2	$1s^2$
2	0 (s)	0	±½	2	8	$2s^2 2p^6$
	1 (p)	−1, 0, +1	±½	6		
3	0 (s)	0	±½	2	18	$3s^2 3p^6 3d^{10}$
	1 (p)	−1, 0, +1	±½	6		
	2 (d)	−2, −1, 0, +1, +2	±½	10		

From the table it can be seen that, depending on the value of n, the number of orbitals at a given energy level is equal to n^2. And since there are two electrons in one orbital, the maximum electron capacity at the energy level will be $2n^2$.

2.4.5 Completely filled and half-filled subshells

The completely filled and half-filled subshells are stable because they have symmetrical distribution of electrons in them. Thus, chromium and copper have the following electron configurations, which are different to what may be expected: Cr is [Ar] $3d^5 4s^1$ not [Ar] $3d^4 4s^2$ and Cu is [Ar] $3d^{10} 4s^1$ not [Ar] $3d^9 4s^2$. This is because the [Ar] $3d^5 4s^1$ and [Ar] $3d^{10} 4s^1$ configurations are energetically more stable.

2.5 Atomic nucleus

Almost the entire mass of the atom (99.95%) is concentrated in the nucleus. The sizes of the nuclei are in the order of 10^{-14}–10^{-15} m, and of the atoms – 10^{-10} m. The atomic nuclei consist of protons and neutrons. The masses of the proton and neutron are close and almost 2,000 times greater than the mass of the electron: $m_e = 9.1 \times 10^{-31}$ kg; $m_p = 1,836.15.m_e = 1.673 \times 10^{-27}$ kg; $m_n = 1,838.68.m_e = 1.675 \times 10^{-27}$ kg. Protons were discovered in 1919 during Rutherford's experiments, which revealed the separation of hydrogen nuclei, later called protons. The electric charge of a proton is equal in magnitude to the charge of an electron ($e = 1.60217733 \times 10^{-19}$ C), but it is positive. Neutrons are electrically neutral particles with a mass slightly greater than the mass of protons. They were discovered only in 1932 by Chadwick, although their existence was predicted by Rutherford as early as 1921.

A *chemical element* is a collection of atoms with the same number of protons in the nucleus. Until the creation of the modern theory of the structure of atom, it was believed that each chemical element consisted of one type of atoms. Later it was found that almost all chemical elements are made up of several varieties of atoms. Individual varieties of atoms of a chemical element possess an equal number of protons in their nucleus, but differ in the number of neutrons. Such varieties of atoms of a chemical element are called *isotopes* ("iso" – the same, "topos" – place). Isotopes have the same chemical and almost identical physical properties, determined by the electron shells.

Almost all chemical elements that occur in nature are a mixture of several isotopes. The ratio between the amounts of isotopes for a given element is constant, with usually one of the isotopes predominating in the mixture. The presence of such mixtures affects the atomic mass. That is why for most chemical elements it is not an integer. At present, more than 1,500 isotopes are known, of which not more than 300 are stable, and the rest are radioactive. Of the 107 known chemical elements, about 92 are found in nature (next to uranium, U). The rest, called transuranic elements, were artificially obtained through nuclear reactions. Artificially obtained chemical elements are radioactive, easily decay and do not occur in nature. The difference between nuclear and chemical reactions is that in the latter the nuclei of the atoms do not change, which is why in these reactions the transformation of one chemical element into another does not take place.

2.5.1 Nuclear forces

The binding of nucleons in a nucleus indicates the presence of a special type of interaction forces, the so-called nuclear forces. They are different from the known types of forces – gravitational, electric, and magnetic. They are the most powerful attractive forces that act at close distance. Due to their positive charge, protons repel each other, but the atomic nucleus is stable and does not decay, indicating that nuclear forces are much greater than electrical forces and are hundreds of times superior to them. As the number of protons in the nucleus increases, so does the electrical repulsion between them, which leads to splitting of the nucleus. A larger number of neutrons are needed to maintain the stability of the nucleus. This explains why stable nuclei with a large atomic number have more neutrons than protons.

Radioactive nuclei (radionuclides) undergo spontaneous transformations due to internal causes of the nucleus. They are unstable and even without external influence decay and turn into nuclei of other elements, which emits energy in the form of electromagnetic radiation or particles (α-particles, β-particles or γ-rays), Figure 2.4. The time interval of radioactive nuclei life is from 10^{-9} s to 10^{22} years. The three types of radioactive rays differ from each other in their ability to ionize atoms and in their penetrating ability.

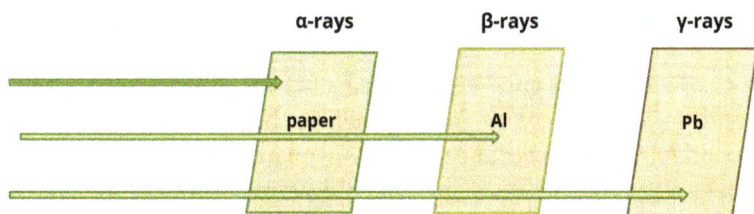

Figure 2.4: α-, β-, and γ-rays.

α-Rays are a type of radioactive radiation with high ionizing and relatively low penetrating power. They have a positive charge and are a stream of helium nuclei. β-Rays are radioactive radiation with low ionizing capacity. Their velocity is in the range from 1×10^8 m/s to 2.7×10^8 m/s. They are a stream of electrons with high kinetic energy. Gamma rays have the highest penetrating and lowest ionizing power. They have a speed of 3×10^8 m/s. They are electromagnetic waves or a stream of photons with a short wavelength and high frequency. They are physically similar to X-rays, but have more energy than them.

Artificial radioactivity was discovered in 1934 by Irene and Frédéric Joliot-Curie, who bombarded aluminum with α-particles and obtained ^{30}P, which decays with the emission of positrons ($T_{1/2} = 2.5$ m). The reaction they observed was as follows:

$$^{27}\text{Al} + {}^{4}\text{He} \rightarrow {}^{30}\text{P} + \text{n}$$

$$^{30}\text{P} \rightarrow {}^{30}\text{Si} + \beta^{+} + \text{n}$$

The entirety of the available experimental data on the mutual transformations of atomic nuclei and elementary particles shows that the following regularities are observed in radioactive decay:

1. Nuclei with a small atomic number Z are most stable when the number of neutrons in them is equal to the number of protons ($N = Z$).
2. Stable nuclei with a large atomic number contain more neutrons than protons ($N > Z$).
3. All nuclei with more than 82 protons are unstable. The most energetically stable are the nuclei of the middle part of the periodic table. Heavy and light nuclei are more unstable.

As Z increases, the electrical repulsion forces increase, and the nuclear forces remain constant. Because of this, nuclei with a very large Z ($Z > 100$) are unstable, as the electrical forces disintegrate them. It is this instability that limits the number of known elements in nature and explains the natural radioactivity of heavy nuclei ($Z > 83$).

This means that the following processes are considered to be possible:

1. decay of heavy nuclei into lighter ones – fission reactions;
2. fusion of light nuclei with each other in heavier – thermonuclear reactions.

In both processes, a huge amount of energy is released.

In all radioactive decays, the laws of conservation of charge, mass, and energy are observed. In ordinary nuclear reactions (reactions without the formation of antiparticles), the mass number remains constant, which is an expression of the law of conservation of the number of nucleons.

Chapter 3
Atomic properties and the periodic table

3.1 Genesis of periodic classification

The discovery of the periodic law has a long history. In the second half of the eighteenth century, intensive study of chemical elements began, as a result of which new elements were discovered. While by the middle of the eighteenth century 15 elements were known, by the end of the century their number had reached 30, and by the end of the 60 s of the nineteenth century, 63 elements had been discovered and described. The growing number of elements raises the question of their arrangement and classification. First, Lavoisier divided the chemical elements into metals and nonmetals.

Dobereiner's triads: In 1829, Dobereiner arranged certain elements with similar properties in groups of three in such a way that the atomic mass of the middle element was nearly the same as the average atomic masses of the first and the third elements (Li, Na, K; Ca, Sr, Ba; Cl, Br, I). The triads given by Dobereiner were helpful in grouping some elements with similar characteristics together, but he could not arrange all the elements known at that time into triads.

Newlands' law of octaves: Newlands' law of octaves states that when elements are arranged in the order of increasing atomic masses, every eighth element has properties similar to the first. Newlands called it the law of octaves because similar relationship exists in the musical notes also. This classification was successful only up to calcium. After that, every eighth element did not possess the same properties as the element lying above it in the same group. When noble gas elements were discovered, their inclusion in the table disturbed the entire arrangement.

Mendeleev's periodic law: *The physical and chemical properties of the elements are a periodic function of their atomic masses.* Mendeleev arranged the elements known at that time in order of increasing atomic masses and this arrangement was called periodic table. In the periodic table, the elements with similar characteristics are arranged in vertical rows called *groups* and horizontal rows known as *periods*. Mendeleev's periodic table made the study of the elements quite systematic in the sense that if the properties of one element in a particular group are known, those of others can be predicted. This helped in the discovery of these elements later. Mendeleev corrected the atomic masses of some elements by their expected positions and properties.

Shortcomings in Mendeleev's periodic table: (i) Hydrogen has been placed in group IA along with alkali metals. But it also resembles halogens of group VII A in many properties. Thus, its position is controversial; (ii) although the elements in the Mendeleev's periodic table have been arranged in order of their atomic masses, in some cases the

https://doi.org/10.1515/9783111712246-003

element with higher atomic mass precedes the element with lower atomic mass; (iii) the isotopes of an element have different atomic masses but same atomic number. Since periodic table has been framed on the basis of increasing atomic masses of the elements different positions must have been allotted to all the isotopes of a particular element; (iv) according to Mendeleev, the elements placed in the same group must resemble in their properties. But there is no similarity among the elements in the two sub-groups of a particular group; (v) in some cases, elements with similar properties have been placed in different groups; (vi) lanthanoids and actinoids were placed in two separate rows at the bottom of the periodic table without assigning a proper reason; and (vii) no proper explanation has been given as to why the elements placed in group show resemblance in their properties.

Modern periodic law: *physical and chemical properties of the elements are the periodic function of their atomic numbers.* Present long form of periodic table, called modern periodic table, is based on this law. In it, the elements have been arranged in order of increasing atomic numbers.

3.2 Structure of the periodic table of chemical elements

Each chemical element in the periodic table occupies one cell, in which the chemical symbol of the element, its name, its relative atomic mass and its atomic number, are given. The basic structural units of the periodic table are the periods and groups (Figure 3.1).

Groups: The long form of periodic table also consists of the vertical rows called groups. There are in all 18 groups in the periodic table, where each group is an independent group. All the elements present in a group have same general electronic configuration of the atoms. The elements in a group are separated by definite gaps of atomic numbers (2, 8, 8, 18, 18, 32). The atomic sizes of elements in group increase down the group due to increase the number of shells. The physical properties of elements such as melting point, boiling point, density, solubility, follow a systematic pattern. The elements in each group have generally similar chemical properties.

Periods: Horizontal rows in a periodic table are known as periods. There are in all seven periods in the long form of periodic table. In all the elements present in a period, the electrons are filled in the same valence shell. The atomic sizes generally decrease from left to right.

Characteristics of s-block elements: General electronic configuration: ns^{1-2}. All the s elements are soft metals. They have low melting and boiling points. They are highly reactive, form ionic compounds and are heat and electricity conductors. Most of them impart colors to the flame.

H													p-elements					He
Li	Be												B	C	N	O	F	Ne
Na	Mg					d-elements							Al	Si	P	S	Cl	Ar
K	Ca	Sc	Ti	V	Cr	Mn	Fe	Co	Ni	Cu	Zn		Ga	Ge	As	Se	Br	Kr
Rb	Sr	Y	Zr	Nb	Mo	Tc	Ru	Rh	Pd	Ag	Cd		In	Sn	Sb	Te	I	Xe
Cs	Ba	Ln*	Hf	Ta	W	Re	Os	Ir	Pt	Au	Hg		Il	Pb	Bi	Po	At	Rn
Fr	Ra	Ac*	Rf	Db	Sg	Bh	Hs	Mt	Ds	Rg	Cn		Nf	Fl	Mc	Lv	Ts	Og

f-elements

Ln*	La	Ce	Pr	Nd	Pm	Sm	Eu	Gd	Tb	Dy	Ho	Er	Tm	Yb	Lu
Ac*	Ac	Th	Pa	U	Np	Pu	Am	Cm	Bk	Cf	Es	Fm	Md	No	Lr

Figure 3.1: Periodic table of chemical elements.

Characteristics of p-block elements: General electronic configuration: ns^2np^{1-6}. The compounds of these elements are mostly covalent in nature. They show variable oxidation states. In moving from left to right in a period, the nonmetallic character of the elements increases. The reactivity of elements in a group decreases downward. Metallic character increases down the group. At the end of each period is a noble gas with a closed valence shell ns^2np^6 configuration.

Characteristics of d-block elements: The d-block elements are known as transition elements. They have incompletely filled d-orbitals in their ground state. General electronic configuration: $(n–1)d^{1-10}ns^{0-2}$. They are all metals with high melting and boiling points. The compounds of d-elements are generally paramagnetic in nature. They mostly form colored ions, exhibit variable valence (oxidation states). They are used as catalysts.

Characteristics of f-block elements: They are known as inner transition elements because in the transition elements of d-block, the electrons are filled in $(n–1)$d-subshell while in the inner transition elements of f-block the filling of electrons takes place in $(n–2)$f subshell, which happens to be one inner subshell. General electronic configuration is $(n–2)f^{1-14}(n–1)d^{0-1}ns^2$. The two rows of elements at the bottom of the periodic table, called lanthanides Ce($Z = 58$)–Lu($Z = 71$) and actinides Th($Z = 90$)–Lr ($Z = 103$). These two series are called inner transition elements (f-block elements). They are all metals. Their properties are quite similar. Most of the elements of the actinoid series are radioactive in nature. The elements after uranium are transuranic elements.

Metals: Metals comprise more than 78% of all known elements and appear on the left side of the periodic table. Metals are solids at room temperature. Metals usually have high melting and boiling points. They are good conductors of heat and electricity. They are malleable and ductile.

Nonmetals: They are located at the top right-hand side of the periodic table. Nonmetals are usually solids or gases at low temperature with low melting and boiling points. They are poor conductors of heat and electricity. The nonmetallic character increases from left to right across the periodic table. Most nonmetallic solids are brittle and are neither malleable nor ductile.

Metalloids: The elements (e.g., silicon, germanium, arsenic, antimony, and tellurium) show the characteristic, of both metals and nonmetals. These elements are also called semimetals.

Noble gases: These are the elements present in group 18. Each period ends with noble gas element. All members are of gaseous nature and because of the presence of all the occupied filled orbitals, they have little tendency to take part in chemical combination. They are called inert gases.

3.3 Periodic trends in properties of elements

Atomic radii: It is defined as the distance from the center of the nucleus to the outermost shell containing the electrons. Depending upon whether an element is a nonmetal or a metal, different types of radii are used. (a) *Covalent radius:* It is equal to half of the distance between the centers of the nuclei of two atoms held together by a purely covalent single bond. (b) *Ionic radius:* It may be defined as the effective distance from the nucleus of an ion up to which it has an influence in the ionic bond. (c) *van der Waals radius:* Atoms of noble gases are held together by weak van der Waals forces of attraction. The van der Waals radius is half of the distance between the centers of nuclei of atoms of gases. (d) *Metallic radius:* It is defined as half of the internuclear distance between the two adjacent metal ions in the metallic lattice.

The trends of atomic radius can be explained by the *electron shell theory*.

Variation in a period: Along a period, the atomic radii of the elements generally decrease from left to right. Atomic radii decrease as the atomic number increases (increased positive nuclear charge) but at the same time extra electrons are added to the same principal quantum shell. The larger the nuclear charge, the greater the pull of the nuclei on the electrons which results in smaller atoms (Figure 3.2).

Variation in a group: The atomic radii of the elements in every group increase downwards. Atomic radii increase as there is an increased number of shells going down the group. The electrons in the inner shells repel the electrons in the outermost shells, shielding them from the positive nuclear charge. This weakens the pull of the nuclei on the electrons resulting in larger atoms.

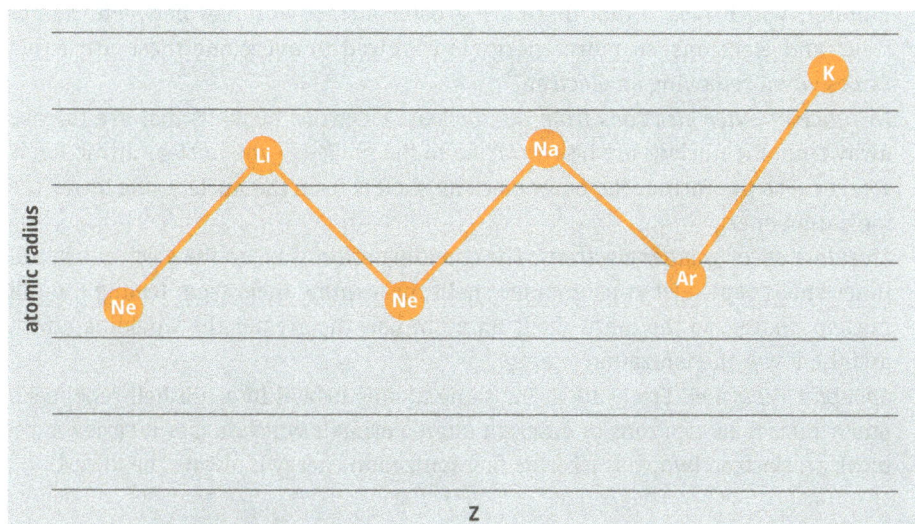

Figure 3.2: Dependence of the atomic radius on the sequence number of the element Z.

The ionic radii show predictable patterns. Ionic radii increase with increasing negative charge and decrease with increasing positive charge. *Negatively charges ions* are formed by atoms accepting electrons while the nuclear charge remains the same. The outermost electrons are further away from the positively charged nucleus and are therefore held only weakly to the nucleus which increases the ionic radius. The greater the negative charge, the larger the ionic radius. *Positively charged ions* are formed by atoms losing electrons. The nuclear charge remains the same but there are fewer electrons which undergo a greater electrostatic force of attraction to the nucleus which decreases the radius. The greater the positive charge, the smaller the ionic radius.

Ionization energies: The ionization energy of an element is the amount of energy required to remove 1 mol of electrons from 1 mol of gaseous atoms of an element to form 1 mol of gaseous ions

$$X° + I = X^+ + e^-$$

Ionization energies are measured under *standard conditions* which are 298 K and 101 kPa. The units of I are kilojoules per mole (kJ/mol). The *first ionization energy* (I_1) is the energy required to remove 1 mol of electrons from 1 mol of atoms of an element to form 1 mol of positive 1+ ions: $Ca(g) \rightarrow Ca^+(g) + e^-$.

Ionization energies show periodicity – a trend across a period of the periodic table (Figure 3.3). As could be expected from their electron configuration, the group 1 metals have a relatively low ionization energy, whereas the noble gases have very high ionization energies. The size of the first ionization energy is affected by four factors:

–   *Size of the nuclear charge*: The nuclear charge increases with increasing atomic number, which means that there are greater attractive forces between the nucleus and electrons, so more energy is required to overcome these attractive forces when removing an electron.

–   *Distance of outer electrons from the nucleus:* Electrons in shells that are further away from the nucleus are less attracted to the nucleus – the nuclear attraction is weaker – so the further the outer electron shell is from the nucleus, the lower the ionization energy.

–   *Shielding effect of inner electrons:* The shielding effect is when the electrons in full inner shells repel electrons in outer shells, preventing them from feeling the full nuclear charge, so the more shells an atom has, the greater the shielding effect, and the lower the ionization energy.

–   *Spin-pair repulsion:* Electrons in the same atomic orbital in a subshell repel each other more than electrons in different atomic orbitals which makes it easier to remove an electron (which is why the first ionization energy is always the lowest).

Ionization energy increases across a period and decreases down a group. It decreases down a group due to the following factors: The number of protons in the atom is increased, so the nuclear charge increases. But, the atomic radius of the atoms increases by adding more shells of electrons, making the atoms bigger. So, the distance between the nucleus and outer electron increases. The shielding by inner shell electrons increases as there are more shells of electrons. These factors outweigh the increased nuclear charge, meaning it becomes easier to remove the outer electron.

The ionization energy across a period generally increases due to the following factors: across a period, the nuclear charge increases. This causes the atomic radius of the atoms to decrease, as the outer shell is pulled closer to the nucleus, so the distance between the nucleus and the outer electrons decreases. The shielding by inner shell electrons remains reasonably constant as electrons are being added to the same shell. It becomes harder to remove an electron as you move across a period, more energy is needed.

There is a slight decrease in I_1 between Be ($1s^2 2s^2$) and B ($1s^2 2s^2 2p_x^1$) as the fifth electron in B is in the 2p subshell, which is further away from the nucleus than the 2s-subshell of Be. There is a slight decrease in I_1 between N ($1s^2 2s^2 2p_x^1 2p_y^1 2p_z^1$) and O ($1s^2 2s^2 2p_x^2 2p_y^1 2p_z^1$) due to spin-pair repulsion in the $2p_x$ orbital of O. In oxygen, there are 2 electrons in the $2p_x$ orbital, so the repulsion between those electrons makes it slightly easier for one of those electrons to be removed.

Electron affinity: Electron affinity is the energy what can be spent to change a neutral atom into a negative charge ion:

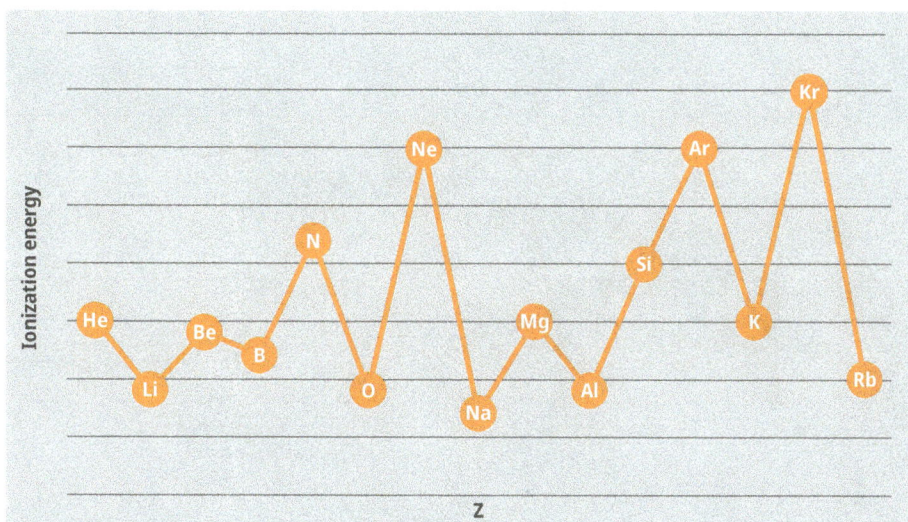

Figure 3.3: Ionization energy of the elements (kJ/mol).

$$X + e^- = X^- + A$$

Electron affinity tends to decrease in going from the top to the bottom of a group (Figure 3.4). Thus, it is not surprising that electron affinity tends to increase from left to right across a given period on the periodic table. Just as low ionization energy is a measure of metallic behavior, a low negative value of electron affinity is a characteristic of active nonmetals. For majority of the elements the electron affinity is negative. For example, the electron affinity for halogens is highly negative because they can acquire the nearest noble gas configuration by accepting an extra electron. In contrast, noble gases have large positive electron affinity because the extra electron has to be placed in the next higher principal quantum energy level thereby producing highly unstable electronic configuration.

Electron affinity depends on the following factors: *(i) atomic size:* as the size of an atom increases, the distance between its nucleus and the incoming electron also increases and electron affinity becomes less negative; *(ii) nuclear charge:* with the increase in nuclear charge, force of attraction between the nucleus and the incoming electron increases and thus electron affinity becomes more negative; and *(iii) symmetry of the electronic configuration:* the atoms with symmetrical configuration (having fully filled or half-filled orbitals in the same subshell do not have any urge to take up extra electrons because their configuration will become unstable. In that case the energy will be needed and electron affinity will be positive. For example, noble gas elements have positive electron affinity. Electron affinity becomes more negative with an increase in the atomic number across a period and becomes less negative as we go down a group.

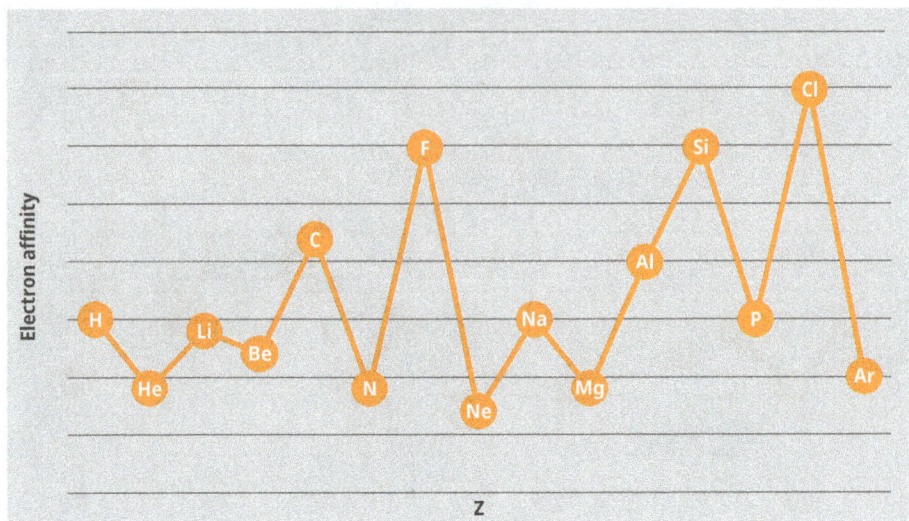

Figure 3.4: Change in the electron affinity of the elements from period I to period III (kJ/mol).

Electronegativity: A qualitative measure of the ability of an atom in a chemical compound to attract shared electrons to itself is called electronegativity. Unlike ionization enthalpy and electron affinity, it is not a measurable quantity. Electronegativity describes the ability of an atom to complete for electrons with another atom to which it is bonded. The electronegativity (χ) is related to ionization energy (I) and electron affinity (A), since these quantities reflect the ability of an atom to lose or gain an electron. The most widely used electronegativity scale is based on an evaluation of bond energies and was devised by Pauling: $\chi = I + A$. The basis of this scale is explored further. Pauling's electronegativities are dimensionless (no units) numbers ranging from 1 for very active metals to 3.98 for F, the most active nonmetal. Another electronegativities scale was proposed by R. Malliken:

$$\chi = \frac{1}{2}(I + A)$$

As a rule, metals have electronegativities less than about 2; metalloids, about equal to 2; and nonmetals, greater than 2. The electronegativity of any given element is not constant. It varies depending on the element to which it is bound. Electronegativity generally increases across a period from left to right and decreases down a group (Figure 3.5).

Factors on which electronegativity depends are: *(i) Nuclear charge:* attraction exists between the positively charged protons in the nucleus and negatively charged electrons. An increase in the number of protons leads to an increase in nuclear attraction for the electrons in the outer shells. Therefore, an increased nuclear charge results in an increased electronegativity. Electrons closer to the nucleus are more strongly attracted toward its positive nucleus. Those electrons away from the nucleus are less strongly attracted toward the nucleus. Therefore, an increased radius gives a decreased electronegativity. *(ii) Atomic radius:* the atomic radius is the distance between the nucleus and electrons in the outermost shell. Electrons closer to nucleus are more strongly attracted toward its positive nucleus. Those electrons further away from the nucleus are less strongly attracted toward nucleus. Therefore, an increased atomic radius results in a decreased electronegativity. *(iii) Shielding:* filled energy levels can shield (mask) the effect of the nuclear charge causing the outer electrons to be less attracted to the nucleus. Therefore, the addition of extra shells and subshells in an atom will cause the outer electrons to experience less of the attractive force of the nucleus. Sodium (period 3) has higher electronegativity than cesium (period 6) as it has fewer shells and therefore the outer electrons experience less shielding than in Cs. So, an increased number of inner shells/subshells will result in a decreased electronegativity.

Relative electronegativity increase

H														p-elements				He
Li	Be											B	C	N	O	F	Ne
Na	Mg				d-elements							Al	Si	P	S	Cl	Ar
K	Ca	Sc	Ti	V	Cr	Mn	Fe	Co	Ni	Cu	Zn	Ga	Ge	As	Se	Br	Kr
Rb	Sr	Y	Zr	Nb	Mo	Tc	Ru	Rh	Pd	Ag	Cd	In	Sn	Sb	Te	I	Xe
Cs	Ba	Ln*	Hf	Ta	W	Re	Os	Ir	Pt	Au	Hg	Il	Pb	Bi	Po	At	Rn
Fr	Ra	Ac*	Rf	Db	Sg	Bh	Hs	Mt	Ds	Rg	Cn	Nf	Fl	Mc	Lv	Ts	Og

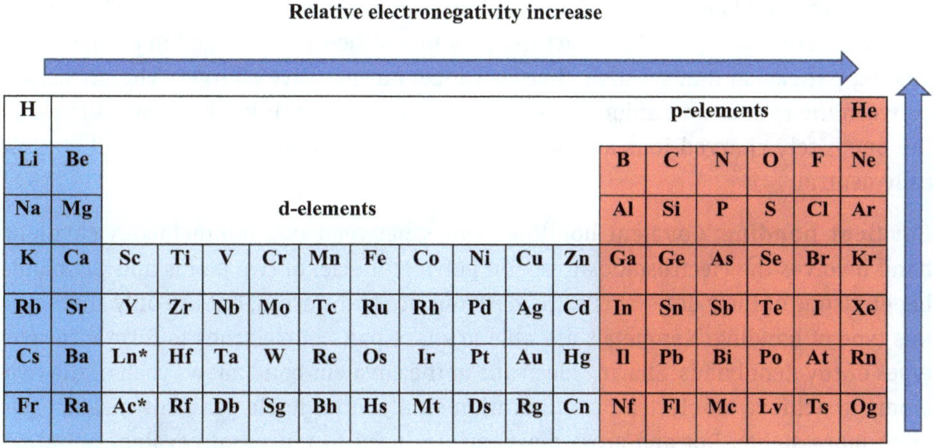

Figure 3.5: Relative electronegativity of the elements.

The differences in Pauling electronegativity values can be used to predict whether a bond is covalent or ionic in character. Single covalent bonds are formed by sharing a pair of electrons between two atoms. In diatomic molecules the electron density is shared equally between the two atoms, H_2, O_2 and Cl_2. Both atoms will have the same electronegativity value and have an equal attraction for the bonding electron pair leading to formation of a covalent bond, which leads to a *nonpolar molecule*. When atoms of different electronegativities form a molecule, the shared electrons are not equally distributed in the bond. The more electronegative atom will draw the bonding electron pair toward itself. A molecule with partial charges forms. The more electronegative atom will have a partial negative charge (δ^-). The less electronegative atom will have a partial positive charge (δ^+). This leads to a *polar covalent molecule*.

Ionic bonding: If there is a large difference in electronegativity of the two atoms in a molecule, the least electronegative atom's electron will transfer to the other atom. This in turn leads to an *ionic bond* – one atom transfers its electron (cation) and the other gains that electron (anion). As a general rule, metals are on the left of the periodic table and nonmetals are on the right-hand side. Ionic bonding involves the transfer of electrons from a metallic element to a nonmetallic element. Transferring electrons usually leaves the metal and the nonmetal with a full outer shell. Once the atoms become ions, their electronic configurations are the same as a stable noble gas. K^+ has the same electronic configuration as Ar: $[2,8,8]^+$. Cl^- also has the same electronic configuration as Ar: $[2,8,8]^-$. Cations and anions are oppositely charged and therefore are attracted to each other. The ionic bond is the electrostatic attraction formed between the oppositely charged ions, which occur in all directions. This form of attraction is very strong and requires a lot of energy to overcome. This causes high melting points in ionic compounds. The ions form a lattice structure which is an evenly distributed crystalline structure. Ions in a lattice are arranged in a regular repeating pattern so that positive charges cancel out negative charges. The attraction between the cations and anions occurs in all directions. Each ion is attracted to all of the oppositely charged ions around it. Therefore, the final lattice is overall electrically neutral.

Covalent bonding: Covalent bonding occurs between two nonmetals. A covalent bond involves the electrostatic attraction between nuclei of two atoms and the bonding electrons of their outer shells. No electrons are transferred, but are only shared in this type of bonding. Nonmetals are able to share pairs of electrons to form different types of covalent bonds. Sharing electrons in the covalent bond allows each of the two atoms to achieve an electron configuration similar to a noble gas. This makes each atom more stable. Dot and cross diagrams are used to represent covalent bonding, showing that just the outer shell of the atoms are involved. To differentiate between

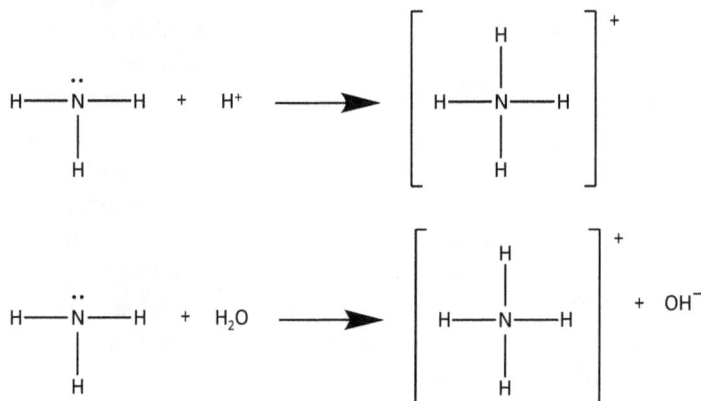

Figure 3.6: Dative bonding.

the two atoms involved, dots for electrons of one atom and crosses for electrons of the other atom are used.

Dative bonding: In simple covalent bonds the two atoms involved share electrons. Some molecules have a lone pair of electrons that can be donated to form a bond with an electron-deficient atom. An electron-deficient atom is an atom that has an unfilled outer orbital. So, both electrons are from the same atom. This type of bonding is a dative covalent bonding or coordinate bond. An example of a dative bond is in an ammonium ion (Figure 3.6). The hydrogen ion H^+ is electron-deficient and has space for two electrons in its shell. The nitrogen atom in ammonia has a lone pair of electrons which it can donate to the hydrogen ion to form a dative covalent bond.

AlCl$_3$ is also formed using dative covalent bonding. At high temperatures, it can exist as a monomer (AlCl$_3$). At lower temperatures, it forms a dimer (Al$_2$Cl$_6$).

Chapter 4
Chemical bonds and the structure of molecules

4.1 Definition and main characteristics of chemical bond

The identical or different atoms cooperate among themselves with formation of steadier systems – molecules, ions, radicals, crystals. The first successful theory of chemical bonding was formulated by Lewis. Lewis began developing his ideas in 1902, when it was widely believed that chemical bonding involved electrostatic attraction between ion-like entities. This seemed satisfactory for compounds such as NaCl that were known to dissociate into ions when dissolved in water, but it failed to explain the bonding in nonelectrolytes such as CH_4. Atomic orbitals had not yet been thought of, but the concept of "valence" electrons was known, and the location of the noble gases in the periodic table suggested that all except helium possess eight valence electrons. It was also realized that elements known to form simple ions such as Ca^{2+} or Cl^- do so by losing or gaining whatever number of electrons is needed to leave eight in the valence shell of each. Lewis sought a way of achieving this octet in a way that did not involve ion formation. The idea that the noble-gas configuration is a particularly favorable one, which can be achieved through formation of electron-pair bonds with other atoms is known as the *octet rule*. The noble gas configuration (Ne, s^2p^6) is achieved when two fluorine atoms (s^2p^5) are able to share an electron pair, which becomes the covalent bond. Notice that only the outer (valence shell) electrons are involved.

4.1.1 Basic parameters of chemical bonding

The basic parameters of the chemical bond are the length of bond, the valence corner and the energy. The *length of bond* is a distance between the centers of atoms in a molecule. The stronger a chemical bond, the shorter is the bond's distance. The *valence corner* is the corner formed by lines connecting centers of bonding atoms. The *energy of bonding* is the energy required to break the bond. The bond energy is the amount of work that must be done to pull two atoms completely apart. This is almost similar to the bond dissociation energy required to break the chemical bond. This energy is a measure of the strength of chemical connections between atoms in a molecule. The energy of bonding is designated E, units of measurements kJ/mol. Numerically, the energy of bonding is equal to the amount of heat which is allocated as a result of formation the bonding from atoms under standard conditions.

https://doi.org/10.1515/9783111712246-004

4.1.2 Types of chemical bonding

There are several types of the chemical bond: covalent bonding, covalent polar bonding, ionic bonding, metallic bonding, hydrogen bonding, and intermolecular bonding. Electronegativity is one of atom's properties. It is the ability of an atom in a molecule to attract shared electrons to it. The type of chemical bonding and its polarity depends on the difference between the electronegativity values of the atom forming the bond. According to the difference in electronegativity values we can determine the type of the bonding as following: If $0 < \Delta\chi < 0.4$ – covalent bond; $0.5 < \Delta\chi < 1.9$ – polar covalent bond; $\Delta\chi > 2.0$ – ionic bond.

4.2 Ionic chemical bond

Ever since the discovery early in the nineteenth century that solutions of salts and other electrolytes conduct electric current, there has been general agreement that the forces that hold atoms together must be electrical in nature. Electrolytic solutions contain ions having opposite electrical charges; opposite charges attract, so perhaps the substances, from which these ions come, consist of positive and negatively charged atoms held together by electrostatic attraction. It turns out that this is not true generally, but a model built on this assumption does a fairly good job of explaining a rather small but important class of compounds, called ionic solids. The most well-known example of such a compound is NaCl, which consists of two interpenetrating lattices of Na^+ and Cl^- ions arranged in such a way that every ion of one type is surrounded (in three-dimensional space) by six ions of the opposite charge. One can envision the formation of a solid NaCl unit by a sequence of events, in which one mole of gaseous Na atoms loses electrons to one mole of Cl atoms, followed by condensation of the resulting ions into a crystal lattice:

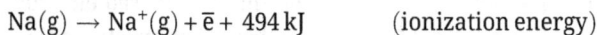

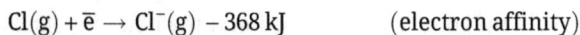

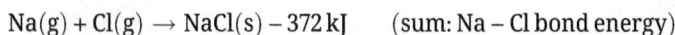

$$Na(g) \rightarrow Na^+(g) + \bar{e} + 494\,kJ \qquad \text{(ionization energy)}$$
$$Cl(g) + \bar{e} \rightarrow Cl^-(g) - 368\,kJ \qquad \text{(electron affinity)}$$
$$Na^+(g) + Cl^-(g) - 498\,kJ \qquad \text{(lattice energy)}$$
$$Na(g) + Cl(g) \rightarrow NaCl(s) - 372\,kJ \qquad \text{(sum: Na – Cl bond energy)}$$

Since the first two energies are known experimentally, as is the energy of the sum of the three processes, the lattice energy can be found by difference. It can also be calculated by averaging the electrostatic forces exerted on each ion over the various directions in the solid, and this calculation is in good agreement with observation, confirming the model. The sum of the three energy terms is clearly negative, and corresponds to the liberation of heat in the net reaction, which defines the Na–Cl bond energy. The ionic solid is more stable than the equivalent number of gaseous atoms because the three-dimensional NaCl structure allows more electrons to be closer to more nuclei. This is the criterion for the stability of any kind of molecule; especially about the ionic.

General characteristics of ionic compounds: (i) *physical state*: they generally exist as crystalline solids, known as crystal lattice. Ionic compounds do not exist as single molecules like other gaseous molecules, for example, H_2, N_2, and Cl_2; (ii) *melting and boiling points*: since ionic compounds contain high interionic force between them, they generally have high melting and boiling points; (iii) *solubility*: they are soluble in polar solvents such as water but do not dissolve in organic solvents like benzene, CCl_4; (iv) *electrical conductivity*: in solid state they are poor conductors of electricity but in molten state or when dissolved, they conduct electricity; and (v) *ionic reactions*: ionic solids produce ions in a solution that gives fast reactions with oppositely charged ions.

4.3 Covalent bonding

The shared-electron pair model, introduced by Lewis, showed how chemical bonds could form in the absence of electrostatic attraction between oppositely charged ions. It has become the most popular and generally useful model of bonding in all substances other than metals. A chemical bond occurs when electrons are simultaneously attracted to two nuclei, thus acting to bind them together in an energetically stable arrangement. The covalent bond is formed when two atoms are able to share a pair of electrons. If the combining atoms are same the molecule is known as homoatomic. If they are different, they form heteroatomic molecules.

4.3.1 Nonpolar covalent bonds

When the atoms joined by covalent bond are the same like; H_2, O_2, Cl_2, the shared pair of electrons is equally attracted by two atoms and thus the shared electron pair is equidistant to both of them. Alternatively, we can say that it lies exactly in the center of the bonding atoms. As a result, no poles are developed and the bond is called as nonpolar covalent bond. The corresponding molecules are known as nonpolar molecules.

4.3.2 Polar bond

When covalent bonds formed between different atoms of different electronegativity, shared electron pair between atoms gets displaced to highly electronegative atoms. In HCl molecule, since electronegativity of Cl is high as compared to H, the electron pair is displaced more toward Cl atom, thus chlorine will acquire a partial negative charge (δ^-) and H atom will have a partial positive charge (δ^+) with the magnitude of charge same as on chlorination. Such covalent bond is called polar covalent bond. Due to polarity, the polar molecules are also known as dipole molecules and they possess dipole moment. Dipole moment is defined as the product of magnitude of the positive or negative charge and the distance between the charges.

HF consists of two atoms. They are not so different that electrons are completely transferred, but they are different enough so that unequal sharing of electrons results, forming a polar covalent bond, $\Delta\chi(H–F) = \chi(F)–\chi(H) = 4.0–2.1 = 1.9$. *Bond polarity* means that the electrons in the bonds are not shared equally. F atom has a stronger attraction than H atom for the shared electrons. This bonding produces partial positive and negative charges on the F and H atoms.

4.3.3 Dipole moment

When nonidentical atoms are joined in a covalent bond, the electron pair will be attracted more strongly to the atom that has the higher electronegativity. As a consequence, the electrons will not be shared equally; the center of the negative charges in the molecule will be displaced from the center of the positive charge. Such bonds are polar in nature and possess partial ionic characters. They may confer a polar nature on the molecule as a whole. A polar molecule acts as an electric dipole, which can interact with electric fields that are created artificially or that arise from nearby ions or polar molecules. Dipoles are represented as arrows pointing in the direction of the negative end. The magnitude of interaction with the electric field is given by the permanent electric dipole moment of the molecule. The dipole moment corresponding to an individual bond (or to a diatomic molecule) is given by the product of the quantity of charge displaced (q) and the bond length (l):

$$\mu = q \times l$$

In SI units, q is expressed in coulombs and l in meters, so μ has the dimensions of C·m. If one entire electron charge is displaced by 100 pm (typical bond length), then $\mu = (1.6022 \times 10^{-19} \text{ C}) \times (10^{-10} \text{ m}) = 1.6 \times 10^{-29} \text{ C·m} = 4.8 \text{ D}$. The quantity denoted by D, the Debye unit, is still commonly used to express dipole moments. It was named after Debye, the physicist who pioneered the study of dipole moments and of electrical interactions; he won a Nobel Prize in 1936.

4.4 Valence bond theory

Valence bond theory was introduced by Heitler and London (1927) and developed by Pauling and others. It is based on the concept of atomic orbitals and the electronic configuration of the atoms. Let us consider the formation of H_2 molecule based on valence-bond theory. As these two atoms come closer, new attractive and repulsive forces begin to operate: (i) the nucleus of one atom is attracted toward its own electron and the electron of the other and vice versa; and (ii) repulsive forces arise between the electrons of two atoms and nuclei of two atoms. Attractive forces tend to bring the two atoms closer whereas repulsive forces tend to push them apart. According to orbital overlap concept, covalent bond formed between atoms results in the overlap of orbitals belonging to the atoms having opposite spins of electrons. Formation of H_2 as a result of overlap of the two atomic orbitals of hydrogen atoms is shown below:

$$\boxed{\uparrow} + \boxed{\downarrow} = \boxed{\uparrow\downarrow}$$

Stability of molecular orbitals depends upon the extent of the overlap of the atomic orbitals.

4.4.1 Types of orbital overlaps

Depending upon the type of overlapping, the covalent bonds are of three types, known as sigma (σ), pi (π) and delta (δ) bonds.

Sigma (σ bond): sigma bond is formed by the end to end (head-on) overlap of bonding orbitals along the internuclear axis. The axial overlap involving these orbitals is of three types: *s-s overlapping:* in this case, there is overlap of two half-filled s-orbitals along the internuclear axis; *s-p overlapping:* this type of overlapping occurs between half-filled s-orbitals of one atom and half-filled p-orbitals of another atoms; *p-p overlapping:* this type of overlapping takes place between half-filled p-orbitals of the two approaching atoms (Figure 4.1).

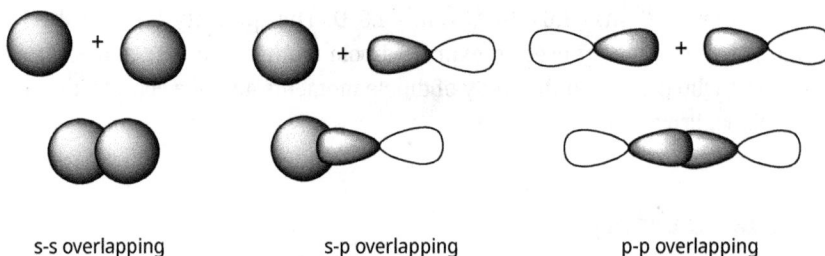

s-s overlapping   s-p overlapping   p-p overlapping

Figure 4.1: Sigma (σ) covalent bonds.

Pi (π bond): π bond is formed by the atomic orbitals when they overlap in such a way that their axes remain parallel to each other and perpendicular to the internuclear axis. The orbital formed is due to lateral overlapping or side wise overlapping.

Delta (δ bond): δ bond is formed with four areas of crossing.

Sigma bond (σ bond) is formed by the axial overlapping of the atomic orbitals while the π-bond is formed by side wise overlapping. Since axial overlapping is greater as compared to side wise. Thus, the sigma bond is said to be stronger bond in comparison to a π-bond.

4.5 Hybridization

Hybridization is the process of intermixing of the orbitals of slightly different energies so as to redistribute their energies resulting in the formation of new set of orbitals of equivalent energies and shape. Orbitals with almost equal energy take part in the hybridization. The number of hybrid orbitals produced is equal to the number of atomic orbitals mixed. The geometry of a covalent molecule can be indicated by the type of hybridization. The hybrid orbitals are more effective in forming stable bonds than the pure atomic orbitals. Only orbitals of valence shell take part in the hybridization. Orbitals, involved in hybridization, should have almost equal energy. Promotion of electron is not a necessary condition prior to hybridization. In some cases, filled orbitals of valence shell also take part in hybridization.

4.5.1 Types of hybridization

(i) *sp hybridization:* When one s- and one p-orbital hybridize to form two equivalent orbitals, the orbital is known as sp hybrid orbital, and the type of hybridization is called sp hybridization. Each of the hybrid orbitals formed has 50% s-character and 50%, p-character. This type of hybridization is also known as diagonal hybridization (Figure 4.2);

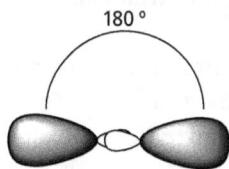

180 °

Figure 4.2: sp hybridization.

(ii) sp^2 *hybridization:* In this type, one s- and two p-orbitals hybridize to form three equivalent sp^2 hybridized orbitals. All the three hybrid orbitals remain in the same plane making an angle of 120°. A few compounds in which sp^2 hybridization takes place are BF_3, BH_3, BCl_3, carbon compounds containing double bonds, etc. (Figure 4.3);

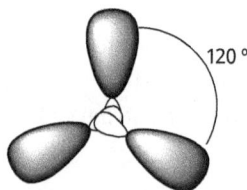

120 °

Figure 4.3: sp^2 hybridization.

(iii) *sp^3 hybridization*: In this type, one s and three p-orbitals in the valence shell of an atom get hybridized to form four equivalent hybrid orbitals. There is 25% s-character and 75% p-character in each sp^3 hybrid orbital. The four sp^3 orbitals are directed toward four corners of the tetrahedron. The angle between sp^3 hybrid orbitals is 109.5°. A compound in which sp^3 hybridization occurs is CH_4. The structures of NH_3 and H_2O molecules can also be explained with the help of sp^3 hybridization (Figure 4.4).

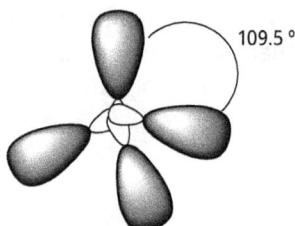

109.5 °

Figure 4.4: sp^3 hybridization.

4.5.2 Valence shell electron pair repulsion theory (VSEPR)

Sidgwick and Powell in 1940, proposed a simple theory based on repulsive character of electron pairs in the valence shell of the atoms. It was further developed by Nyholm and Gillespie (1957). The main postulates are:

(i)   The exact shape of molecule depends upon the number of electron pairs (bonded or nonbonded) around the central atoms.

(ii) The electron pairs have a tendency to repel each other since they exist around the central atom and the electron clouds are negatively charged.

(iii) Electron pairs try to take such position which can minimize the repulsion between them.

(iv) The valence shell is taken as a sphere with the electron pairs placed at maximum distance.

(v)   A multiple bond is treated as if it is a single electron pair and the electron pairs which constitute the bond as single pairs.

The form of NH_3 and H_2O molecules is presented in Figure 4.5.

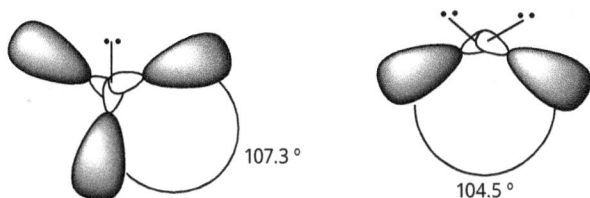

107.3 °

104.5 °

Figure 4.5: Valence shell electron pair repulsion theory.

4.5.3 Resonance

In certain cases, molecules can be represented by more than one reasonable Lewis structures that differ only in the location of π electrons. Electrons in σ bonds have a fixed location and so they are said to be *localized*. In contrast, π electrons that can be drawn in different locations and are said to be *delocalized* (C_6H_6 and CO_3^{2-}). Collectively these Lewis diagrams are then known as *resonance structures* or resonance contributors. The "real" structure has characteristics of each of the contributors, and is often represented as the resonance hybrid. In a way, the resonance hybrid is a mixture of the contributors (Figure 4.6).

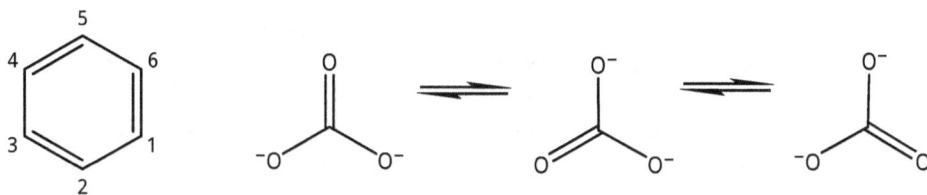

Figure 4.6: Delocalized bonds.

4.6 Molecular orbital theory

In the molecular orbital theory, when atoms join to form a molecule new regions of a high electron probability (*molecular orbitals*) are established for the molecule. The formation of molecular orbitals can be explained by the linear combination of atomic orbitals. Combination takes place either by addition or by subtraction of wave function.

The combining atomic orbitals must have almost equal energy. The atomic orbitals must have the same symmetry about the molecular axis. The combining atomic orbitals must overlap to the maximum extent. The molecular orbital formed by addition of atomic orbitals is called *bonding molecular orbital* while molecular orbital formed by subtraction of atomic orbitals is called *antibonding molecular orbital*. Bonding molecular orbitals correspond to high electron probability or the electron charge density in the internuclear region between atoms. Antibonding molecular orbitals concentrate electron probability or the charge density in regions away from the internuclear region. The numbers and kinds of molecular orbitals in a molecule are related to the corresponding atomic orbitals from which they arise (Figure 4.7).

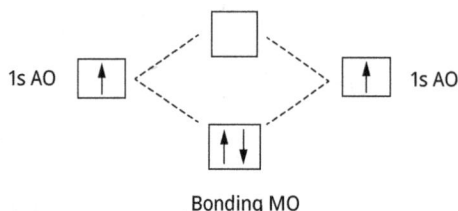

1s AO

Bonding MO

Figure 4.7: Molecular orbitals.

Using the molecular orbital theory to describe chemical bonding requires some rules. These rules pertain to the particular molecular orbitals that arise when atomic orbitals are combined and the manner, in which electrons are assigned to these molecular orbitals: (1) The number of molecular orbitals produced is equal to the number of the atomic orbitals combined; (2) of the two molecular orbitals produced when two atomic orbitals are combined, one is a bonding molecular orbital at a lower energy than the original atomic orbitals. The other is an antibonding orbital at a higher energy; (3) electrons normally seek the lowest energy molecular orbitals available to them in a molecule; (4) the maximum number of electrons that can be assigned to a given molecular orbital is two (the Pauli exclusion principle); (5) electrons enter molecular orbitals of identical energies singly before they pair up (the Hung's rule); and (6) formation of a bond between atoms requires that the number of electrons in bonding molecular orbitals exceed the number of electrons in antibonding orbitals.

4.6.1 Types of molecular orbitals

The molecular orbitals are of two types, known as: *sigma (σ)*, which are symmetrical around the bond axis and *pi (π)*, which are not symmetrical, because of the presence of positive lobes above and negative lobes below the molecular plane.

Bond order is defined as half of the difference between the number of electrons present in bonding and antibonding molecular orbitals: $\mathbf{BO = ½[N_b - N_a]}$. The bond order may be a whole number, a fraction or even zero. It may also be positive or negative. Bond order is inversely proportional to *bond length*. Thus, greater the bond order, smaller will be the bond length. If all the molecular orbitals have paired electrons, the substance is diamagnetic. If one or more molecular orbitals have unpaired electrons, it is paramagnetic, for example, O_2 molecule.

4.6.2 Bonding in some homonuclear molecules

Hydrogen molecule H_2 is formed by the combination of two H atoms. Each H atom has one electron in its atomic orbital, so the energy diagrams of molecular orbitals of He_2^+ and He_2 can be represented (Figure 4.8).

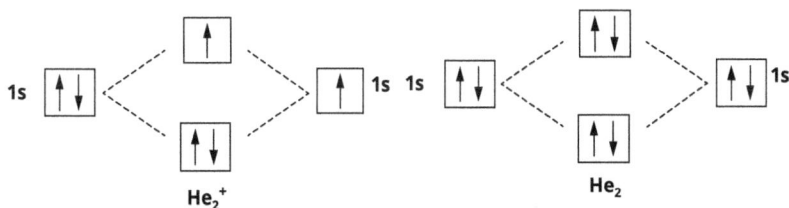

Figure 4.8: Energy diagrams of molecular orbitals of He_2^+ and He_2.

Helium molecule (He_2) is formed by the combination of two He atoms. Each helium atom contains two electrons; thus, in He_2 molecule there would be four electrons. The electrons will be accommodated in σ1s and σ*1s molecular orbitals: BO = ½ [2–2] = 0.

To apply the molecular orbital method to elements of the second period requires that molecular orbitals be formed from atomic orbitals of the second principal electronic shell. We will limit our discussion to diatomic molecules. Also, we note again that two molecular orbitals, one bonding and one antibonding, are produced for every pair of atomic orbitals in the separated atoms. Because there are four orbitals in each atom ($2s$, $2p_x$, $2p_y$, $2p_z$) we need to deal with eight new molecular orbitals – four bonding and four antibonding. We can describe bonding in diatomic molecules of the second period elements (Li_2, Be_2, B_2, C_2, N_2, O_2, F_2, and Ne_2) and we can think of the first level electrons as not involved in the bonding (as nonbonding electrons). This allows us simply to consider the assignment of the valence shell electrons of the two atoms. Energy diagrams for these orbitals of diatomic molecules of the second period are illustrated in Figures 4.9–4.12.

Because the 2p level is at higher energy than 2s in the separated atoms, we expect the molecular orbitals formed by combinations of 2p atomic orbitals to be at higher energy than those formed from 2s-orbitals. Concerning the relative placement of the σ^b2p and π^b2p molecular orbitals, more extensive overlap occurs when p atomic orbitals are combined end to end than side by side, suggesting that σ^b2p lies at a lower energy than π^b2p. This is a situation, confirmed by experiment, when the energy difference between 2s and 2p atomic orbitals is large (for O, F, and Ne). When this difference is smaller

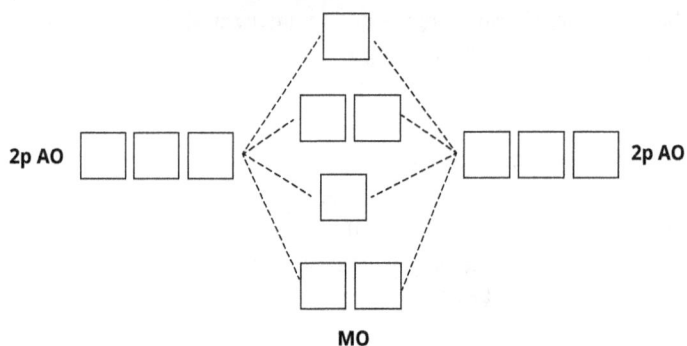

Figure 4.9: Molecular orbital diagram for Li_2, Be_2, B_2, C_2, and N_2.

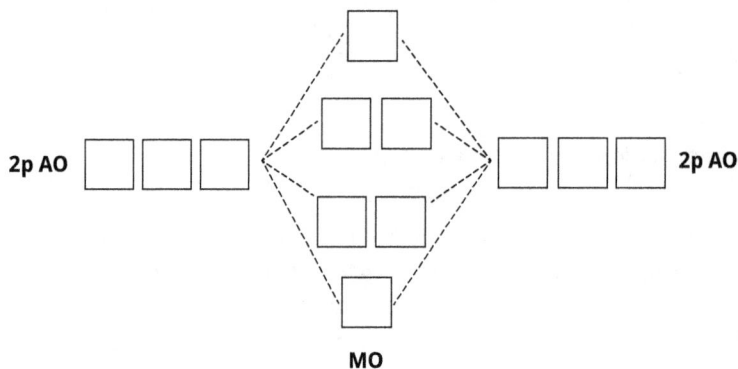

Figure 4.10: Molecular orbital diagram for O_2, F_2, and Ne_2.

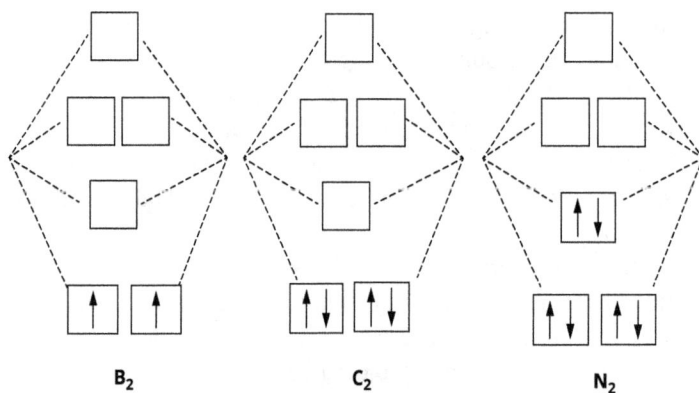

Figure 4.11: Energy diagrams of B_2, C_2, and N_2 diatomic molecules of the second period.

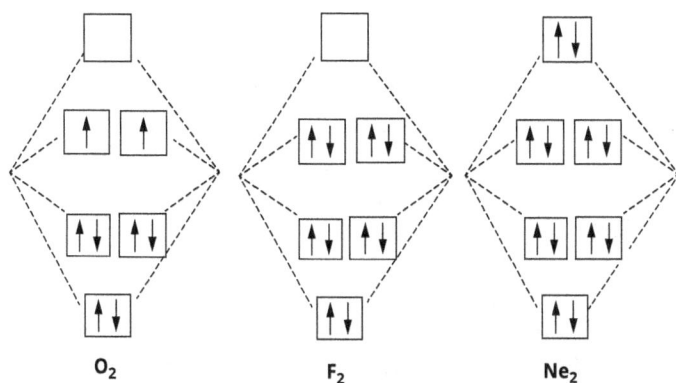

Figure 4.12: Energy diagrams of O_2, F_2, and Ne_2 diatomic molecules of the second period.

(from Li through N), there is a mixing of 2s and 2p atomic orbitals in the formation of molecular orbitals that results in a reversal of the $\sigma^b 2p$ and $\pi^b 2p$ levels.

4.7 Intermolecular interactions

4.7.1 van der Waals interactions

There are several types of intermolecular forces. They are also called *van der Waals forces*. In speaking of the electronic structure of an atom or molecule we refer to the probability of an electron being in a certain region at a given instant of time. An event that may occur is that an instantaneous displacement of electrons toward one region of an atom or molecule. This displacement causes a normally nonpolar species to become polar, thus an instantaneous dipole is formed. Following this, electrons in a neighboring atom or molecule may be displaced also leading to a dipole. This is a process of induction and newly formed dipole is called an induced dipole. Taken together these two events lead to the intermolecular force of attraction. Although it is proper to call this an instantaneous dipole-induced dipole interaction, the terms dispersion force. The ease with which an electron cloud is distorted by the external electric field (and hence the ease with which a dipole is induced) is called *polarizability*. In a polar substance, molecules tend to become oriented with the positive end of one dipole directed toward the negative ends of neighboring dipoles. This additional partial ordering of molecules can cause a substance to persist as a solid or liquid at temperatures higher than otherwise expected. All molecules experience intermolecular attractions, although in some cases those attractions are very weak. Even in a gas like hydrogen H_2, if you slow the molecules down by cooling the gas, the attractions are large enough for the molecules to stick together eventually to form a liquid and then a solid. In the hydrogen's case the attractions are so weak that the molecules have to be cooled to 21 K (–252 °C) before the attractions are enough to condense the hydrogen as a liquid. Helium's intermolecular attractions are even weaker – the molecules will not stick together to form a liquid until the temperature drops to 4 K (–269 °C).

Intermolecular forces are the forces of attraction and repulsion between interacting particles. The attractive forces decrease with the increase of distance between dipoles. The interaction energy is proportional to $1/r^6$ where r is the distance between the polar molecules. Intermolecular forces include: forces between two permanent dipoles *(Keesom forces)*, forces between a permanent dipole and a corresponding induced dipole (*Debye forces*) and forces between two instantaneously induced dipoles (*London dispersion forces*).

4.7.1.1 Dipole-dipole orientation interactions

This is the force of attraction which exists between the polar molecules (CO, HI, HBr, NH_3, and H_2O). The strength of attraction depends upon the charge of the dipole moment and the size of the polar molecule. Molecules with net dipole moments tend to align themselves so that the positive end of one dipole is near the negative end of another and vice versa. HCl, which is partially held together by dipole-dipole interactions, is a gas at room temperature and 1 atm pressure. Since the dipole-dipole interaction is significantly influenced by the temperature, its increase disrupts the orientation of the molecules and leads to a decrease in orientation forces.

Kesom (1912) showed that polar molecules tend to orient each other in such a way that their charges of different signs are at the closest distance and their intermolecular potential energy is minimal. According to Kesom's theory, the average potential energy of mutual attraction of two identical dipole molecules at sufficiently high temperatures and at large (compared to the size of the molecules) distances between them can be calculated by the formula:

$$E_{or} = -\frac{2\mu^4}{3kTr^6}$$

where μ is the dipole moment of the molecule, k is the Boltzmann constant, and T is the absolute temperature. In this case, the process of orientation of the dipole molecules takes place, conditioned by the orientational forces of attraction (Kesom forces), which are more pronounced if the dipole moment is greater. At the same time, the polar molecules, influencing each other with their own electric fields cause each other a corresponding deformation, as a result of which induced dipoles are also produced. The latter interact with each other in the same way as permanent dipoles. They are a source of the so-called induction forces, which have their part in the mutual attraction between polar molecules.

4.7.1.2 Ion-induced dipolar interactions

What would happen if we mix HCl with argon, which has no dipole moment? The electrons on an argon atom are distributed homogeneously around the nucleus of the atom. But these electrons are in constant motion. When an argon atom comes close to a polar HCl molecule, the electrons can shift to one side of the nucleus to produce a very small dipole moment that lasts for only an instant. By distorting the distribution of electrons around Ar atom, the polar HCl molecule induces a small dipole moment on this atom, which creates a weak dipole-induced dipole force of attraction between the HCl molecule and the Ar atom.

This force is very weak, with a bond energy of about 1 kJ/mol. In this type of interaction permanent dipole of the polar molecule induces dipole on the electrically neutral atom by deforming its electronic cloud. Interaction energy is proportional to $1/r^6$ where r is the distance between two molecules.

Ion-induced dipolar interactions were studied by Debye in 1920. When polar and nonpolar molecules are present, polar molecules, the electric fields of the polar molecule induce dipole in the nonpolar molecule. Obviously, this is associated with a certain deformation of the nonpolar molecule. Since polarization in this case occurs under the influence of an electric field, the resulting dipole is designated as an induced dipole. When the field is removed, the induced dipole disappears and the molecule becomes nonpolar again. The greater the dipole moment of the induced dipole, the stronger the electric field. Therefore, the electrons that are located in the outermost electron shell are of great importance, since they are the farthest from the atomic nucleus. The induced dipoles interact with the permanent dipoles of the polar molecules.

The induction effect depends on the constant dipole moment of one molecule and on the tendency of the other molecule to polarize, that is, on polarizability. The energy of intermolecular attraction due to the induction effect does not depend on the temperature and is determined by the formula:

$$E_{ind} = -\frac{2\alpha\mu^2}{r^6}$$

where α is the polarizability of the molecule.

4.7.1.3 London forces and dispersion interactions

Neither dipole-dipole nor dipole-induced forces can explain the fact that helium becomes a liquid at temperatures below 4.2 K. By itself, a helium atom is perfectly symmetrical. But movement of the electrons around the nuclei of a pair of neighboring helium atoms can become synchronized so that each atom simultaneously obtains an induced dipole moment. These fluctuations in electron density occur constantly, creating an induced dipole-induced dipole force of attraction between pairs of atoms. As might be expected, this force is relatively weak in helium – only 0.076 kJ/mol. But atoms or molecules become more polarizable as they become larger because there are more electrons to be polarized. It has been argued that the primary force of attraction between molecules in solid I_2 and in frozen CCl_4 is induced dipole-induced dipole attraction. In 1930, London proposed that temporary fluctuations in the electron distributions within atoms and nonpolar molecules could result in the formation of short-lived instantaneous dipole moments, which produce attractive forces, called London dispersion forces between otherwise nonpolar substances. As we know, in nonpolar molecules there is no dipole moment because their electronic charge cloud is symmetrically distributed. But it is believed that at any instant of time, the electron cloud of the molecule may be distorted so that an instantaneous dipole or momentary dipole is produced in which one part of the molecule is slightly negative than the other part. This momentary dipole induces dipoles in the neighboring molecules. Thus, the forces of attraction exist between them and are exactly the same as between permanent dipoles. These forces of attraction are known as London forces or dispersion forces. These forces are always attractive and the interaction energy is inversely proportional to the sixth power of the distance between two interacting particles, $1/r^6$, where r is the distance between the two particles.

According to the theory, the average potential energy of interaction between two molecules with a spherically symmetric charge distribution is calculated by the formula:

$$E_{disp} = -\frac{3Ia^2}{4r^6}$$

where a is the polarizability of the molecule and I is the ionization energy of the molecule. Dispersion forces are universal. They act not only in molecules with a spherically symmetric distribution of charge, but also in random molecules and atoms.

4.7.2 Hydrogen bonding

When highly electronegative elements like nitrogen, oxygen, and fluorine are attached to hydrogen to form covalent bond, the electrons of the covalent bond are shifted toward the more electronegative atom. Thus, partial positive charge develops on hydrogen atom which forms a bond with the other electronegative atom. This bond is known as hydrogen bond and it is weaker than the covalent bond. For example, in HF molecule, hydrogen bond exists between hydrogen atom of one molecule and fluorine atom of another molecule. The most common substance, in which hydrogen bonding occurs, is ordinary water. Tetrahedral arrangement by hydrogen bonds is the structural arrangement in the crystalline water or ice. As ice melts, only a fraction of the hydrogen bonds is broken.

4.7.2.1 Types of H-bonds
There are two types: *(i) intermolecular hydrogen bond,* typical for HF molecules, water molecules, etc. and *(ii) intramolecular hydrogen bond,* in which hydrogen atom is in

between the two highly electronegative F, N, O atoms present within the same molecule (*o*-nitrophenol, *o*-chlorophenol, and *o*-hydroxybenzoic acid); see Figures 4.13 and 4.14. When the temperature rises, the hydrogen bond breaks, so it manifests itself mainly in a condensed state, which leads to the association of molecules, for example, $(HF)_n$ and $(H_2O)_n$.

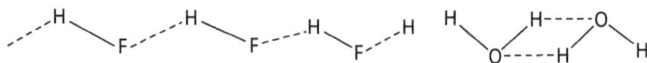

Figure 4.13: Examples of intermolecular hydrogen bonds in the molecules of HF and H_2O.

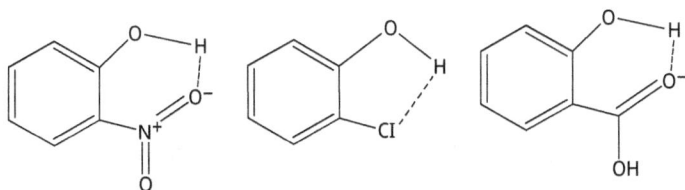

Figure 4.14: Examples of intramolecular hydrogen bonds in the molecules of *o*-nitrophenol, *o*-chlorophenol, and *o*-hydroxybenzoic acid.

4.7.2.2 Importance

The hydrogen bond is a rather strong intermolecular force, with energies in the range of 15–40 kJ/mol. For instance, van der Waals interaction corresponds to energies of about 2–20 kJ/mol. Liquids, in which hydrogen bonding occurs, exhibit stronger than usual intermolecular forces, and these liquids generally have high heats of vaporization. Hydrogen bonding plays an important role in crystallization, dissolution, biochemical and polymerization processes. Hydrogen bonds make life possible. Living organisms are maintained through a series of chemical reactions involving complex structures, such as DNA and proteins. Certain bonds in these structures must be capable of breaking and reforming with ease. Only hydrogen bonds have just the right energies to permit this.

Chapter 5
Complex compounds

5.1 Basic concepts

When solutions of two or more salts are mixed in stoichiometric proportions, new compounds called molecular or addition compounds are formed. For example:

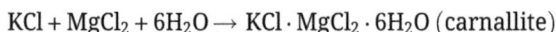

$$KCl + MgCl_2 + 6H_2O \rightarrow KCl \cdot MgCl_2 \cdot 6H_2O \text{ (carnallite)}$$

$$K_2SO_4 + Al_2(SO_4)_3 + 24H_2O \rightarrow K_2SO_4 \cdot Al_2(SO_4)_3 \cdot 24H_2O \text{ (potash alum)}$$

Such compounds are broadly grouped into two categories.

(a) **Double salts:** the addition compounds, which are stable only in crystalline state and lose their identity in solution form, are called double salts. These salts in solution form give the same ions as are given by the compounds from which these were formed:

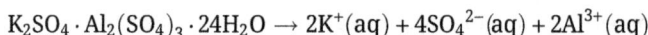

$$K_2SO_4 \cdot Al_2(SO_4)_3 \cdot 24H_2O \rightarrow 2K^+(aq) + 4SO_4{}^{2-}(aq) + 2Al^{3+}(aq)$$

A special place among the double salts is occupied by alums, the composition of which can be expressed by the following general formula: $M^+M^{3+}(SO_4)_2 \cdot 12H_2O$, where $M = Na^+$, K^+, $NH_4{}^+$, Rb^+, Cs^+, and $M^{3+} = Cr^{3+}$, Al^{3+}, Fe^{3+}, etc. They eventually dissociate into simple ions:

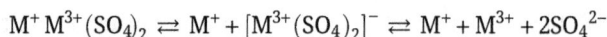

$$M^+M^{3+}(SO_4)_2 \rightleftharpoons M^+ + [M^{3+}(SO_4)_2]^- \rightleftharpoons M^+ + M^{3+} + 2SO_4{}^{2-}$$

In their solutions, the concentration of complex ions of the type $[M^{3+}(SO_4)_2]^-$ is very low. However, the presence of complex ions, although in an insignificant concentration, shows that there is no fundamental difference between double and complex salts, which is why double salts occupy an intermediate position between simple and complex salts.

(b) **Complex compounds or coordination compounds:** The addition compounds, which do not lose identity even in solution form (i.e., they are stable in solid as well as in solution form), are called complex compounds. These compounds do not furnish all the ions which are given by the constituent salts in solution form, for example:

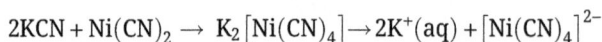

$$2KCN + Ni(CN)_2 \rightarrow K_2[Ni(CN)_4] \rightarrow 2K^+(aq) + [Ni(CN)_4]^{2-}$$

The molar conductivity of such a solution would correspond only to three ions in the solution (molar conductivity of a solution is directly proportional to the number of

https://doi.org/10.1515/9783111712246-005

ions present in the solution). In the above example, the addition compound (of KCN and $Ni(CN)_2$) does not give Ni^{2+} and CN^- ions in aqueous solution.

The coordination compounds consist of central ions and ligands.

Central ion, center of coordination, is a cation or a neutral metal atom to which one or more neutral molecules or anions are attached. Since the central ion act as an acceptor and thus has to accommodate electron pairs donated by the donor atom of the ligand, it must have empty orbitals. This explains why the transition metals having empty d-orbitals, form coordination compounds very readily. Thus, in $[NI(NH_3)_6]^{2+}$ and $[Fe(CN)_6]^{3-}$, Ni^{2+} and Fe^{3+} are the central ions.

Ligands are any atom, ion or molecule which is capable of donating at least a pair of electrons to the central atom. Further in a ligand, the particular atom which actually donates the electron pair is called the donor atom. The common donor atoms are N, O, S, and halogens. The ligand containing one, two or more donor atoms are known as unidentate, bidentate and multidentate ligands, respectively. Examples of *unidentate ligands* are F^-, Cl^-, Br^-, I^-, CN^-, SCN^-, NO_2^-, NH_3, H_2O, pyridine, $(C_2H_5)_3N$, acetone, etc. Generally, a monodentate ligand is capable of forming only one coordinate bond to the metal ion. The common examples of *bidentate ligands* are ethylenediamine ($NH_2CH_2CH_2NH_2$, both N atoms are donor atoms), dimethylglyoxime ($C_4H_8N_2O_2$, both N atoms are donor atoms), oxalate ($C_2O_4^{2-}$, both the donor atoms are O) and glycinate ($NH_2CH_2CO_2^-$, donor atoms are N and O). These have two donor atoms which can simultaneously coordinate with the metal atom. Ethylenediaminetetraacetate EDTA with a formula $(CH_3CO_2^-)_2NCH_2CH_2N$ $(CH_3CO_2^-)_2$ is an example of *hexadentate ligand.*

Coordination number (CN) is the total number of ligands attached to the central ion/ atom. Thus, the coordination number of silver and copper ions in the complexes [Ag $(NH_3)_2]^+$ and $[Cu(H_2O)_4]^{2+}$ are 2 and 4, respectively. Some common coordination numbers are 2, 3, 4, and 6.

Coordination sphere is the central metal atom and the ligands directly attached to it. Coordination sphere is written inside square brackets, $[Co(NH_3)_6]^{3+}$. The central metal atom and the ligands inside the brackets behave as a single entity.

Oxidation number is a number (numerical value) which represents the electric charge on the central metal atom of a complex ion. For example, the oxidation numbers of Fe, Co, and Ni in $[Fe(CN)_6]^{4-}$, $[Co(NH_3)_6]^{3+}$, and $[Ni(CO)_4]$ are +2, +3, and 0, respectively. The charge carried by a complex ion is the algebraic sum of charges carried by the central ion and the coordinated ligands.

Complex ion or complex species is an electrically charged or even a neutral species, which is formed by the combination of simple cations and anions or sometimes neutral molecules, which does not show all the properties of those constituent ions or neutral molecules, for example, $[Ni(CN)_4]^{2-}$.

5.2 Nomenclature and main types of complexes

In general, a complex ion is represented as $[M(L)_n]^{x-/x+}$, where M is the metal ion, L are ligands, n is the coordination number of metal ion, and x is the net charge on the complex. There are basically four types of complexes. These are:
(i)   complex with a cation as complex ion – $[Cr(NH_3)_6]Cl_3$;
(ii) complex with an anion as complex ion – $K_3[Fe(CN)_6]$;
(iii) complex with a cation and an anion both as complex ions – $[Pt(Py)_4]$ $[PtCl_4]$;
(iv) neutral complexes – $[Ni(CO)_4]$.

The complex may be of any type (mentioned above). The method of systematic naming of coordination compounds by IUPAC follows the same set of rules. Cation is named first followed by the name of anion. There has to be a gap between the cation's name and anion's name. There should be no gap anywhere else: the complete name of coordination sphere is written as one word without spacing and punctuation. Only the first letter of the name of complex should be capital and the rest all letters have to be small. The number of cations or anions in a complex are not mentioned. Ligands are named first in the alphabetical order (regardless of their charge) and then the central metal atom or ion followed by the oxidation state of the metal in parenthesis in Roman numeral (except for zero). Zero is written in Arabic numeral (0). If a ligand is repeated more than once, then prefixes di, tri, tetra, penta, hexa, etc., are used, for example, –[–(NH_3)_5] is written as pentaammine. In deciding the alphabetical order for ligands, prefixes di, tri, tetra, etc., are not considered. Alphabetical order is decided by the first letter of the ligand's name. In case of ligands having di, tri, and tetra in their names (inbuilt, which actually means 2, 3, and 4) or organic ligands, if they are repeated more than once, prefixes used are bis (for two), tris (for three), tetrakis (for four), pentakis (for five), etc. After writing bis or tris, the name of the ligand is to be written in parenthesis, for example, $(en)_2$ is written as bis (ethylenediamine). Negative ligands end with "o" and positive with "ium." There is no special ending for neutral ligands. For naming anion complexes, suffix "ate" is put after the name of metal atom, that is, written as metalate, for example, platinate, ferrate, nickelate, cobaltate, cuprate, wolfrate, argentate, aurate, iridate, chromate, osmate, etc. For those metals whose Latin names are available (except for mercury), the suffix "ate" is used after that, while for those whose Latin names are not available, the suffix is used after normal metal names. No such suffix is used for metal in cation complexes and neutral complexes. If any lattice components such as water or solvent of crystallization are present, these follow the name and are preceded by the number of these groups in Arabic numerals, separated from the name of these groups by a hyphen. Bridging complexes are the one which have *bridging ligands* and the bridging ligands are the one which joins two metal ions. For naming bridging ligands, μ is placed before the name of the ligand in order to distinguish it from the normal ligand. If there are two or more bridging ligands of the same kind, this is indicated by di–μ–, tri–μ–, etc. If more than one kind of bridging ligands are present, then they are listed in the alphabetical order. If a bridging group bridges more than two metal atoms it is shown as μ3, μ4, μ5, or μ6 to indicate how many atoms it is bonded to. There are some ligands which have more than one donor atom (*ambidentate ligands*) and such ligands can coordinate with the metal atom by any donor atom. For example, −ONO is called nitrito and −NO$_2$ is called nitro, and −SCN is called thiocyanato and −NCS is called isothiocyanato. These may also be named systematically as thiocyanato −S and thiocyanato −N (to indicate which atom is bonded to the metal). When writing the formula of complexes, the complex ion should be enclosed by square brackets. The metal is named first, then the coordinated groups are listed in the following order: negative, neutral, and positive ligands (alphabetically arranged according to the first symbol within each group).

Naming of complex: For complexes with metal-to-metal bridging, prefix bi- is placed before the name of the metal atom or ion. For example, octaamminedichlorobi-platinum(II) chloride.

5.3 Isomerism

The compounds having the same molecular formula but different physical and chemical properties on account of different structures are called isomers and the phenomenon as isomerism. Isomerism in coordination compounds may be divided into two main types: structural isomerism and stereoisomerism.

5.3.1 Structural isomerism

It is displayed by compounds that have different ligands within their coordination spheres. The different types of structural isomerism of coordination compounds are:
(i)   *Ionization isomerism:* This type of isomerism arises when the coordination compound gives different ions in solution. For example, there are two isomers of the formula Co(NH$_3$)$_5$BrSO$_4$.
(ii) The violet isomer pentaamminebromo-cobalt(III) sulfate [Co(NH$_3$)$_5$Br]SO$_4$ gives a white precipitate of BaSO$_4$ with BaCl$_2$ solution. The isomer pentaamminesulfatocobalt(III) bromide [Co(NH$_3$)$_5$SO$_4$]Br is red. Other examples of ionization isomerism: [Pt(NH$_3$)$_4$Cl$_2$]Br$_2$ and [Pt(NH$_3$)$_4$Br$_2$]Cl$_2$; [Co(NH$_3$)$_4$Cl$_2$]NO$_2$ and [Co(NH$_3$)$_4$ClNO$_2$]Cl.
(iii) *Hydrate isomerism:* This type of isomerism arises when different number of water molecules are present inside and outside the coordination sphere. This isomerism is best illustrated by the three isomers that have the formula CrCl$_3$.6H$_2$O: (a) [Cr(H$_2$O)$_6$]Cl$_3$ with violet color. All the six water molecules are coordinated to Cr. It has three ionizable Cl$^-$ ions; (b) [Cr(H$_2$O)$_5$Cl]Cl$_2$.H$_2$O with green color. Five water molecules are coordinated to Cr. It has two ionizable Cl$^-$ ions. One water molecule outside the coordination sphere can be easily lost; and (c) [Cr(H$_2$O)$_4$Cl$_2$]Cl.2H$_2$O with green color. Four water molecules are coordinated to Cr. It has one ionizable Cl$^-$ ion. Other ex-

amples of hydrate isomerism are: $[Co(NH_3)_4(H_2O)Cl]Br_2$ and $[Co(NH_3)_4(Br)_2]Cl·H_2O$; $[Cr(en)_2(H_2O)Cl]Cl_2$ and $[Cr(en)_2Cl_2]Cl·H_2O$.

(iv) *Coordination isomerism:* This type of isomerism is observed in the coordination compounds having both cationic and anionic complex ions. The ligands are interchanged in both the cationic and anionic ions to form isomers. Some examples are: tetraammine Pt(II) tetrachlorocuprate(II) $[Pt(NH_3)_4][CuCl_4]$ and tetraammine Cu(II) tetrachloroplatinate(II) $[Cu(NH_3)_4][PtCl_4]$; hexaammine Cr(III) trioxalatocobaltate(III) $[Cr(NH_3)_6][Co(C_2O_4)_3]$ and hexaammine Co(III) trioxalatochromate(III) $[Co(NH_3)_6][Cr(C_2O_4)_3]$; hexaammine chromium(III) hexathiocyanatochromate(III) $[Cr(NH_3)_6][Cr(SCN)_6]$ and tetraamminedithiocyanato Cr(III) diamminetetrathiocyanatochromate(III) $[Cr(NH_3)_4(SCN)_2][Cr(NH_3)_2(SCN)_4]$.

(v) *Linkage isomerism:* This type of isomerism occurs in complex compounds which contain ambidentate ligands like, SCN^-, CN^-, and CO. These ligands have two donor atoms but at a time only one atom is directly linked to the central metal atom of the complex. These types of isomers are proven by IR spectroscopy. $[Co(NH_3)_5NO_2]Cl_2$ and $[Co(NH_3)_5ONO]Cl_2$ are linkage isomers as is linked through nitrogen or through oxygen. $[Co(NH_3)_5NO_2]Cl_2$ is yellow and $[Co(NH_3)_5ONO]Cl_2$ is red. Other examples are: dipyridyldithiocyanato Pd(II) $[Pd(dipy)(SCN)_2]$ and dipyridyldiisothiocyanato Pd(II) $[Pd(dipy)(NCS)_2]$.

(vi) *Polymerization isomerism:* This type of isomerism exists in compounds having same stoichiometric composition but different molecular compositions. The molecular compositions are simple multiples of the simplest stoichiometric arrangement. In the following three compounds, the second and third compounds are polymers of the first: (i) $[Pt(NH_3)_2Cl_2]$; (ii) $[Pt(NH_3)_4][PtCl_4]$; (iii) $[Pt(NH_3)_3Cl]_2[PtCl_4]$. (ii) and (iii) compounds are actually not the examples of polymerization, that is, (i) compound is not acting as a monomer in the formation of (ii) and (iii) compounds.

(vii) *Coordination position isomerism:* This type of isomerism is exhibited by polynuclear complex by changing the position of ligands with respect to different metal ions in the complex.

5.3.2 Stereoisomerism

Compounds are stereoisomers when they contain the same ligands in their coordination spheres but differ in the way that these ligands are arranged in space. Stereoisomerism is of two types, viz., geometrical isomerism and optical isomerism:

(I) *Geometrical isomerism:* This isomerism is due to ligands occupying different positions around the central metal atom or ion. The ligands occupy positions either adjacent or opposite to one another. This type of isomerism is also known as *cis-trans* isomerism (Figures 5.1 and 5.2).

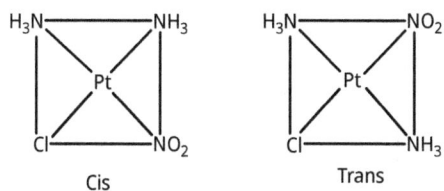

Figure 5.1: Geometrical isomers of Pt(II) planar complex.

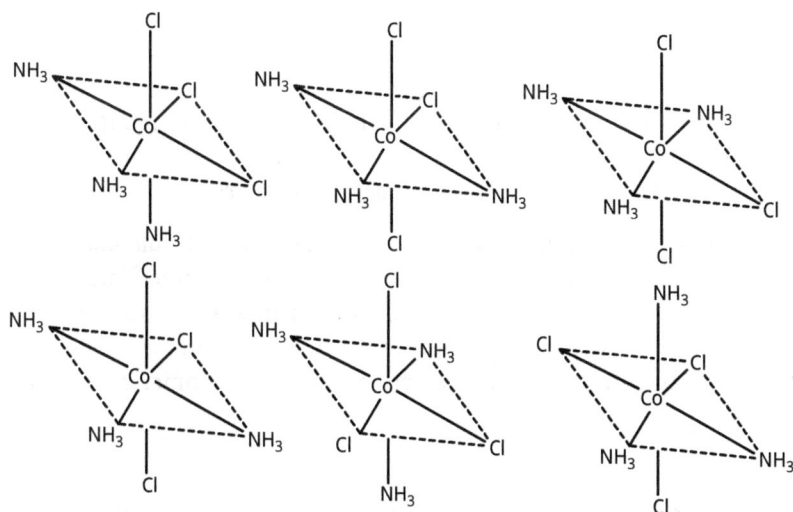

Figure 5.2: Geometrical isomers of Co(III) octahedral complex.

Geometrical isomerism is characteristic for complexes with a composition of $[Pt(NH_3)_3Cl_3]^+$ (Figure 5.3). A configuration with three identical ligands located along one meridian (with three places in the horizontal plane) is called *mer*-isomer. The other configuration with face-centered ligands occupying three places at the corners of the triangular face of the octahedron and on the opposite triangular face is called *fac*-isomer (from English facial). Such isomeric compounds, as a rule, differ in color, crystal shape and in their relation to certain reagents. Mer-$[Pt(NH_3)_3Cl_3]Cl$ is soluble in water and greenish-yellow in color, while fac-$[Pt(NH_3)_3Cl_3]Cl$ is pale yellow and shows poor solubility.

Geometrical isomerism is very much common in coordination number 4 and 6 isomerism. Square planar complexes (coordination number four) exhibit geometrical isomerism. (i) Complexes with general formula, Ma_2b_2 (where both a and b are mono-dentate) can have cis- and trans-isomers; (ii) complexes with general formula Ma_2bc can also have cis- and trans-isomers; (iii) complexes with general formula, Mabcd, can have three isomers; and (iv) square planar complexes having unsymmetrical biden-tate ligands can also show geometrical isomerism, for example, Pt glycinato complex,

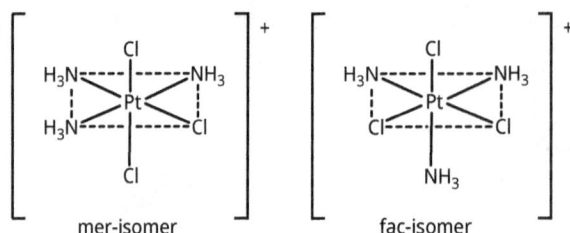

Figure 5.3: *Mer-fac* isomerism.

$[Pt(Gly)_2]$. Geometrical isomerism is not observed in complexes of coordination number 2 and 3. It is not observed in complexes of coordination number 4 of tetrahedral geometry. The complexes of general formulae Ma_3b or Mab_3, or Ma_4 of square planar geometry do not show geometrical isomerism. The complexes of general formula Ma_6 and Ma_5b of octahedral geometry do not show geometrical isomerism.

(II) *Optical isomerism:* Optical isomers have the same molecular formula and structural formula but they have a different arrangement of atoms in three-dimensional space. Optical isomers are two non-superimposable, mirror images of each other. As in organic chemistry, optical isomers of complexes are called enantiomers. They can be distinguished by their effect on plane-polarized light. Each isomer rotates the light equally but in opposite directions.

5.4 Dissociation of complex compounds in solution: stability of complexes

5.4.1 Factors affecting the stability of complexes

The stability of complex compounds depends on the properties and nature of the central metal ion, the nature of the ligands, external factors such as temperature, composition of the medium, and others. The higher the charge density on the central metal ion, the greater is the stability of the complex. In other words, the greater the charge and smaller the size of an ion (i.e., charge/radius of an ion), the greater is the stability of its complexes. The greater the base strength of a ligand, the greater is the ease with which it can donate its lone pair of electrons and therefore, the greater is the stability of the complexes formed by it. Chelating ligands form more stable complexes as compared to monodentate ligands.

The complex compounds are usually soluble in water and are strong electrolytes. In their dissociation, they give simple and complex ions:

$$K_3\left[Fe(CN)_6\right] \rightarrow 3K^+ + \left[Fe(CN)_6\right]^{3-}$$
$$\left[Ag(NH_3)_2\right]OH \rightarrow \left[Ag(NH_3)_2\right]^+ + OH^-$$

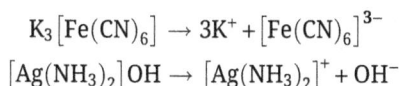

Complex ions, in turn, also dissociate into ions, and their degree of dissociation is small:

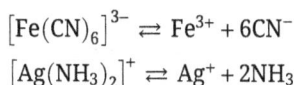

$$\left[Fe(CN)_6\right]^{3-} \rightleftarrows Fe^{3+} + 6CN^-$$
$$\left[Ag(NH_3)_2\right]^+ \rightleftarrows Ag^+ + 2NH_3$$

Depending on the electrolyte dissociation of the complex ion, the complexes are divided into *stable* and *unstable*. For example, the solution of $K_3[Fe(CN)_6]$ does not give the characteristic reactions for Fe^{3+} and CN^- ions. This shows that the dissociation of $[Fe(CN)_6]^{3-}$ is negligible and this complex ion is stable. In the solution of $[Ag(NH_3)_2]^+$, free Ag^+ can be confirmed, indicating that $[Ag(NH_3)_2]^+$ is less stable.

5.4.2 Dissociation and stability constants

In solutions of complex compounds, there is a complex system of dynamic equilibria that depends on both the nature of the complex and the nature of the solvent. In an aqueous solution, complex compounds manifest themselves as electrolytes, but their dissociation should be considered from two sides:

a) *primary dissociation*
The complex compound under the action of the solvent dissociates into a complex ion and ions from the outer sphere. Dissociation proceeds as with strong electrolytes:

$$K_3[AlF_6] \rightarrow 3K^+ + [AlF_6]^{3-}$$

b) *secondary dissociation*
The complex ion dissociates as a weak electrolyte, that is, to a negligible degree:

$$[AlF_6]^{3-} \rightleftarrows Al^{3+} + 6F^-$$

By using the law of mass action, the equilibrium constant is expressed, which for the given system is a dissociation constant and is a quantitative criterion for the stability of the complex ion:

$$K_D = \frac{\left[Al^{3+}\right] \cdot \left[F^-\right]^6}{\left[[AlF_6]^{3-}\right]}$$

The more stable the complex ion, the lower its dissociation constant. Thus, compounds of the same type have different values of the dissociation constant, for example, for $[Ag(NO_2)_2]^-$ ion $K_D = 1.3 \times 10^{-3}$, for $[Ag(NH_3)_2]^+ - K_D = 6.8 \times 10^{-8}$, for the most stable $[Ag(CN)_2]^- - K_D = 1 \times 10^{-21}$.

Recently, the use of the reciprocal value of the dissociation constant, called the *stability constant* (β), has been preferred to characterize complex compounds:

$$\beta = 1/K_D$$

The stability constant β determines the stability of the complex ion. The complex ion is more stable if the value of its stability constant is greater. For $[Ag(NH_3)_2]^+$ ions $\beta = 1.69 \times 10^7$, and for $[Ag(CN)_2]^-$ ions $\beta = 1 \times 10^{21}$. Therefore, of the two complex ions, the second is more stable.

The stability constant β essentially represents the constant of formation of a given complex ion. Obtaining the complexes in an aqueous solution, where the metal ions are hydrated, is a step-by-step process of transforming the aqua-complexes into new complexes. All reactions in aqueous solution involving metal ions can be considered as processes of exchange of water molecules from the internal coordination sphere with other ligands. For example, the interaction of $[Zn(H_2O)_4]^{2+}$ with NH_3 takes place in four stages, with one ligand exchanged in each of them:

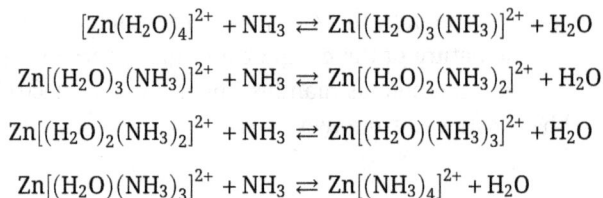

$$[Zn(H_2O)_4]^{2+} + NH_3 \rightleftharpoons Zn[(H_2O)_3(NH_3)]^{2+} + H_2O$$

$$Zn[(H_2O)_3(NH_3)]^{2+} + NH_3 \rightleftharpoons Zn[(H_2O)_2(NH_3)_2]^{2+} + H_2O$$

$$Zn[(H_2O)_2(NH_3)_2]^{2+} + NH_3 \rightleftharpoons Zn[(H_2O)(NH_3)_3]^{2+} + H_2O$$

$$Zn[(H_2O)(NH_3)_3]^{2+} + NH_3 \rightleftharpoons Zn[(NH_3)_4]^{2+} + H_2O$$

Each of these processes is characterized by its own equilibrium or stability constant, equal to K_1, K_2, K_3, and K_4 (the so-called *stepwise stability constants*). The *overall stability constant* is equal to the product of those of the individual steps:

$$\beta_4 = K_1.K_2.K_3.K_4$$

In general, if the complexation goes through n intermediate stages, the stability constant is

$$\beta_n = K_1.K_2.K_3. \ \ldots \ K_n$$

Usually, as the number of ligands increases, the value of the stepwise stability constants decreases, which is also true for the general case, which means that each subsequent ligand binds weaker than the previous one. In the dissociation of the complex ion, it is exactly the opposite – the first ligand is the easiest to separate and the first dissociation constant has the highest value.

The substitution of monodentate ligands with polydentate leads to the formation of chelated complexes, and this is associated with an increase in the stability of the complex. A typical example of this is the formation of the next complex ions $[Ni(NH_3)_6]^{2+}$,

[Ni(en)$_3$]$^{2+}$, and [Ni(EDTA)]$^{2-}$ containing monodentate, bidentate and polydentate ligands. When NH$_3$ is included:

$$[Ni(H_2O)_6]^{2+} + 6NH_3 \rightleftharpoons [Ni(NH_3)_6]^{2+} + 6H_2O$$

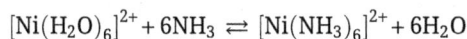

$$\beta = \frac{c[Ni(NH_3)_6]^{2+}}{c[Ni(H_2O)_6]^{2+} \cdot c^6[NH_3]}$$

In this case, $\beta = 4.8 \times 10^7$ L/mol and lg $\beta = 7.7$, typical of a complex with monodentate ligand.

When a bidentate ligand such as ethylenediamine is included (en = H$_2$N–CH$_2$–CH$_2$–NH$_2$):

$$[Ni(H_2O)_6]^{2+} + 3en \rightleftharpoons [Ni(en)_3]^{2+} + 6H_2O$$

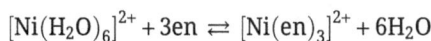

$$\beta = \frac{c[Ni(en)_3]^{2+}}{c[Ni(H_2O)_6]^{2+} \cdot c^3[en]}$$

In this process, $\beta = 2.0 \times 10^{18}$ L/mol, and lg $\beta = 18.3$, that is, a value significantly higher than that for the complex with the monodentate ligand ammonia.

The use of a hexadentate ligand EDTA leads to the formation of a more stable complex:

$$[Ni(H_2O)_6]^{2+} + EDTA^{4-} \rightleftharpoons [Ni(EDTA)]^{2-} + 6H_2O$$

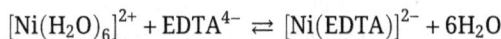

$$\beta = \frac{c[Ni(EDTA)]^{2-}}{[Ni(H_2O)_6]^{2+} \cdot c[EDTA^{4-}]}$$

In this case, $\beta = 1 \times 10^{19}$ L/mol, lg $\beta = 19$. This is the highest value of β compared to the other values for [Ni(NH$_3$)$_6$]$^{2+}$, [Ni(en)$_3$]$^{2+}$. This stabilization is explained by a decrease *in Gibbs energy* ΔG, since the more negative the isobaric potential, the greater the equilibrium (stability) constant:

$$\Delta G° = -RT.\ln K_C \quad \text{or} \quad \Delta G° = -RT.\ln \beta$$

The Gibbs energy depends on entropy and enthalpy: $\Delta G = \Delta H - T\Delta S$. Therefore, the question arises which of the two factors is more significant for the greater decrease in ΔG in the above complexing processes. *Since the heats of formation of* the above three complexes turn out to be close, thus that the entropy factor is significant. In fact, it can be seen that the progression of the second and third reactions is associated with an increase in the number of particles.

5.5 Chemical bonding in complex compounds

Several theoretical methods are used to explain the nature of chemical bonding and the structure of coordination compounds: valence bond method (VBM), molecular orbital method (MOM), crystal field theory (CFT), and ligand field theory (LFT), which combines some principles of MOM and CFT.

5.5.1 Valence bond method

5.5.1.1 Structure of the complexes with coordination numbers 2, 4, and 6

This theory was proposed by Pauling. According to this theory, the structures of the complexes can be explained on the basis of hybridization. Orbitals of central metal atom/ion are hybridized and each coordinating groups (i.e., ligands) give a pair of electrons. These pair of electrons are accommodated in the vacant hybrid orbitals of the metal atom/ion. For example, combination of one s- and three p-orbitals of the valence shell gives four hybrid orbitals of equivalent energy. This is known as sp^3-hybridization. Besides this, dsp^2 (square planar), d^2sp^3 and sp^3d^2 (octahedral) hybridizations are also very common.

The important features of valence bond theory are as follows. In this approach, the basic assumption made is that the metal-ligand bond arises by the donation of pair of electrons by ligands to the metal atom/ion. In order to accommodate these electrons, the metal atom/ion must possess requisite number of vacant orbitals of comparable energy. These orbitals of metal atom/ion undergo hybridization to give a set of hybrid orbitals of equal energy. In complex formation, Hund's rule of maximum multiplicity is strictly followed. However, in complex formation redistribution of electrons and complete pairing of electrons may take place. In this way, some orbitals are vacated and made available for hybridization. A strong ligand donates a pair of electrons easily, whereas a weak ligand with difficulty. Under the influence of strong ligands, the metal electrons are forced to pair up contrary to Hund's rule. Sometimes, the unpaired $(n-1)$ d-electrons pair up as per the need prior to bond formation thus making some $(n-1)$ d-orbitals vacant. The central metal atom then makes available a number of empty orbitals (equal to its coordination number) for the formation of coordinate bonds with suitable ligand orbitals. With the approach of the ligands, metal-ligand bonds are formed by the overlap of these hybrid orbitals with those of the ligands, that is, by donation of electron pair by the ligands to the empty hybridized orbitals. Consequently, these bonds are of equal strength and directional in nature. The complex species may or may not be colored, if the unpaired electrons are present in it while it would be colorless if no unpaired electrons are present. A substance which does not contain any unpaired electron is not attracted by a magnet and is said to be diamagnetic. On the other hand, a substance which contains one or more unpaired electrons in the d-orbitals, is attracted by a magnetic field (exception O_2 and NO). It is said to be paramagnetic. On the basis of value of magnetic moment, μ (in Bohr magneton, B.M.), we can predict the number of unpaired electrons present in the complex. If we know the number of un-paired electrons in the metal complex, then it is possible to predict the geometry of the complex species. Octahedral, square planar and tetrahedral complexes are formed. For predicting the type of structure (geometry) of a complex species, the electronic configuration of the central ion would be helpful (Figure 5.4).

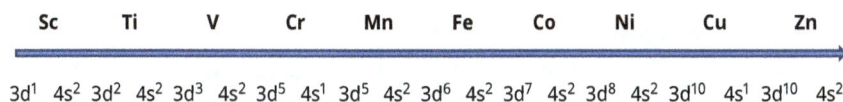

Sc	Ti	V	Cr	Mn	Fe	Co	Ni	Cu	Zn
$3d^1\ 4s^2$	$3d^2\ 4s^2$	$3d^3\ 4s^2$	$3d^5\ 4s^1$	$3d^5\ 4s^2$	$3d^6\ 4s^2$	$3d^7\ 4s^2$	$3d^8\ 4s^2$	$3d^{10}\ 4s^1$	$3d^{10}\ 4s^2$

Figure 5.4: Electronic configuration of the central ions.

For coordination number 4, the hybridizations possible are sp^3 and dsp^2, having tetra-hedral and square planar geometries respectively, while for coordination number 6, the hybridizations possible are d^2sp^3 and sp^3d^2, having octahedral geometry in both the cases. The common types of hybridization are presented in Table 5.1.

Table 5.1: Common types of hybridization.

CN	Hybridization	Geometry	Examples
2	sp	Linear	$[Ag(NH_3)_2]Cl$ and $K[Ag(CN)_2]$
3	sp^2	Trigonal planar	$K[HgI_3]$
4	sp^3	Tetrahedral	$[MnX_4]^{2-}$, $[FeCl_4]^{2-}$, $[NiX_4]^{2-}$, $[ZnCl_4]^{2-}$, and $[Ni(CO)_4]$, where $X = Cl^-$, Br^-, I^-
4	dsp^2	Square planar	$[Cu(NH_3)_4]^{2+}$, $[Pt(NH_3)_4]^{2+}$, and $[Ni(CN)_4]^{2-}$
5	sp^3d	Square pyramidal	$[SbF_5]^{2-}$
5	dsp^3	Trigonal bipyramidal	$Fe(CO)_5$ and $[CuCl_5]^{3-}$
6	sp^3d^2	Octahedral	$[Fe(H_2O)_6]^{2+}$, $[Co(H_2O)_6]^{2+}$, $[FeF_6]^{3-}$, and $[Ni(NH_3)_6]^{2+}$
6	d^2sp^3	Octahedral	$[Cr(NH_3)_6]^{3+}$, $[Fe(CN)_6]^{3-}$, $[Fe(CN)_6]^{4-}$, and $[Co(NH_3)_6]^{3+}$

5.5.1.2 Four coordinated complexes

The complexes in which the coordination number of metal atom or ion is 4 are known as four coordinated complexes. For the formation of four coordinated com-plex, it is essential for the metal ion to have four equal hybrid orbitals so as to accept four electron pairs from the coordinating ligands. This can be achieved in two ways:

(i) If one s- and three p-orbitals hybridize, four sp^3 hybrid orbitals are formed and the complex involves sp^3 hybridization. Due to sp^3 hybridization, the structure will be tetrahedral.

(ii) On the other hand, if one d, one s, and two p-orbitals hybridize, four dsp^2 hybrid orbitals result and the complex involves dsp^2 hybridization. The structure of the complex is square planar.

5.5.1.2.1 Tetrahedral complexes

The metal ions like Co^{2+}, Fe^{2+}, and Mn^{2+} with Cl^- ions as ligands can form tetrahedral complexes.

(i) *Formation of $[CoCl_4]^{2-}$:* Let us consider Co^{2+} which in coordination with Cl^- forms the complex $[CoCl_4]^{2-}$. The orbital diagram of the metal ion Co^{2+}, having configuration $3d^7$, and the complex $[CoCl_4]^{2-}$ is presented in Table 5.2.

Table 5.2: Formation of $[CoCl_4]^{2-}$.

	3d	4s	4p
Co atom in the ground state	↑↓ ↑↓ ↑ ↑ ↑	↑↓	☐ ☐ ☐
Co^{2+} ion	↑↓ ↑↓ ↑ ↑ ↑	☐	☐ ☐ ☐
Formation of $[CoCl_4]^{2-}$	↑↓ ↑↓ ↑ ↑ ↑	↿⇂ Cl^-	↿⇂ ↿⇂ ↿⇂ Cl^- Cl^- Cl^-

Here, the 3d-orbital of Co^{2+} ion is not available for accepting the lone pair of electrons of Cl^- ions. And, since Cl^- ion acts as a weak ligand, it can't force the metal electrons to pair up. Hence, one 4s- and three 4p-orbitals of this cation hybridize to form four sp^3 hybrid orbitals. There are four electron pairs donated by four Cl^- ions. The orbital diagram for $[CoCl_4]^{2-}$ can be shown.

The calculated magnetic moment corresponds to three unpaired electrons and hence the complex is paramagnetic. Therefore, the 3d-orbitals remain undisturbed and sp^3 hybridization occurs resulting in the tetrahedral structure of the complex. The complex may/may not be colored.

(ii) *Formation of $[Ni(CO)_4]$:* Oxidation state of nickel in this complex is zero. Its outer electronic configuration is $3d^8 4s^2$. The hybrid sp^3 orbitals accommodate four pair of electrons from four CO molecules and the resulting tetrahedral complex is diamagnetic due to the absence of unpaired electrons (Table 5.3). The complex is colorless.

Table 5.3: Formation of [Ni(CO)$_4$].

	3d	4s	4p
Atomic orbitals of Ni in (Z = 28) ground state	[↑↓][↑↓][↑][↑][↑]	[↑↓]	[ ][ ][ ]
Hybridized sp^3 orbitals of Ni	[↑↓][↑↓][↑↓][↑↓][↑↓]	[ ]	[ ][ ][ ]
Formation of [Ni(CO)$_4$]	[↑↓][↑↓][↑↓][↑↓][↑↓]	[↑↓]	[↑↓][↑↓][↑↓]

(iii) *Formation of [Zn(NH$_3$)$_4$]$^{2+}$*: Complexes of Zn^{2+} are invariably tetrahedral because they involve sp^3 hybridization. In this case, 3d-orbitals are fully occupied, therefore, they cannot participate in hybridization. Four sp^3 orbitals can be formed by mixing of one 4s and three 4p-orbitals. The four sp^3 hybrid orbitals, accommodate four pairs of electrons from four NH$_3$ molecules. The resulting [Zn(NH$_3$)$_4$]$^{2+}$ complex will be tetrahedral and diamagnetic as there are no unpaired electrons (Table 5.4). The complex would be colorless.

Table 5.4: Formation of [Zn(NH$_3$)$_4$]$^{2+}$.

	3d	4s	4p
Zn atom (Z = 30) in ground state	[↑↓][↑↓][↑↓][↑↓][↑↓]	[↑↓]	[ ][ ][ ]
Zn^{2+} ion with four empty sp^3 hybridized orbitals	[↑↓][↑↓][↑↓][↑↓][↑↓]	[ ]	[ ][ ][ ]
Formation of [Zn(NH$_3$)$_4$]$^{2+}$ with four NH$_3$ molecules	[↑↓][↑↓][↑↓][↑↓][↑↓]	[↑↓]	[↑↓][↑↓][↑↓]

5.5.1.2.2 Square planar complexes

The metal ions like Cu^{2+}, Pt^{2+} can form such complexes.

(i) *Formation of [Ni(CN)$_4$]$^{2-}$*: For understanding the formation of square planar complexes, let us consider the [Ni(CN)$_4$]$^{2-}$ complex ion. The electronic configuration of Ni is 3d^{8}4s^2 and of Ni^{2+} ion is 3d^8. The orbital diagrams for Ni^{2+} ion and for [Ni(CN)$_4$]$^{2-}$ are presented in Table 5.5.

From the experiments, it is known that [Ni(CN)$_4$]$^{2-}$ is diamagnetic and hence the two unpaired d-orbital electrons must become paired. As the CN$^-$ ion acts as a strong ligand, it will force the metal electrons to pair up resulting in the dsp^2 hybridization. Thus, the shape of the complex [Ni(CN)$_4$]$^{2-}$ is square planar. This complex is colorless.

Table 5.5: Formation of $[Ni(CN)_4]^{2-}$.

	3d	4s	4p
Ni atom in ground state	↑↓ ↑↓ ↑ ↑ ↑	↑↓	☐ ☐ ☐
Ni^{2+} ion	↑↓ ↑↓ ↑ ↑ ↑	☐	☐ ☐ ☐
Formation of $[Ni(CN)_4]^{2-}$	↑↓ ↑↓ ↑↓ ↑↓ ↑↓(CN⁻)	↑↓(CN⁻)	↑↓ ↑↓(CN⁻ CN⁻) ☐

(ii) *Formation of $[Cu(NH_3)_4]^{2+}$:* Copper in $[Cu(NH_3)_4]^{2+}$ is in +2 oxidation state, as Cu^{2+} ion. The formation of $[Cu(NH_3)_4]^{2+}$ may be explained through hybridization (Table 5.6).

Table 5.6: Formation of $[Cu(NH_3)_4]^{2+}$.

	3d	4s	4p
Cu atom ($Z = 29$) in ground state	↑↓ ↑↓ ↑↓ ↑↓ ↑↓	↑	☐ ☐ ☐
Cu^{2+} ion	↑↓ ↑↓ ↑↓ ↑↓ ↑	☐	☐ ☐ ☐
dsp^2-hybridized orbitals of Cu^{2+}	↑↓ ↑↓ ↑↓ ↑↓ ☐	☐	☐ ☐ ↑
Formation of $[Cu(NH_3)_4]^{2+}$	↑↓ ↑↓ ↑↓ ↑↓ ↑↓	↑↓	↑↓ ↑↓ ↑

As $[Cu(NH_3)_4]^{2+}$ contains one unpaired electron, it is paramagnetic. The complex may or may not be colored. Other examples of square planar metal complexes include $[Cu(NH_3)_4]^{2-}$, $[PtCl_4]^{2-}$, $[Pt(NH_3)_4]^{2+}$, etc.

5.5.1.3 Six coordinated complexes

In octahedral complexes, two types of complexes occur, that is, inner orbital complex and outer orbital complex. The distinction between inner and outer orbital complex is based purely on magnetic measurements. In inner orbital complex, $(n-1)$d-orbitals are used for hybridization, whereas in outer orbital complex, nd orbitals are used.

All the complexes containing configuration of metal ion d^0, d^1, d^2, and d^3 form inner orbital complexes, that is, d^{2}sp^3 hybridization. These octahedral complexes contain the same number of unpaired electrons as that present in the free metal ion. If the complex has d^4, d^5, and d^6 configuration of metal ion, two types of complexes (i.e., inner and outer orbital complexes) are possible. Most of the complexes of d^7, and d^8 configuration of metal ion form outer orbital complex (i.e., sp^3d^2 hybridization) but in exceptional cases, a few complexes of this system form inner orbital complexes (i.e., d^2sp^3 hybridization), for example, $[Co(NO_2)_6]^{4-}$. The inner and outer orbital complexes may be distinguished by magnetic measurements. Outer orbital complexes are more reactive or labile and consequently ligands may be substituted easily but the inner orbital complexes are sometimes called inert or nonlabile and substitution of ligands is fairly difficult. Usually, we call the ionic complexes as outer orbital complexes and the covalent complexes as the inner orbital complexes. These two classes of complexes are also called spin free and spin paired complexes respectively (by Nyholm) and high spin (HS) and low spin (LS) complexes respectively by Orgel. A low spin complex ion is an ion in which there is maximum pairing of electrons in the orbitals of the metal atom. A high spin complex ion is an ion in which there is minimum pairing of electrons in the metal atom orbitals.

Inner orbital octahedral complexes

Formation of $[Cr(NH_3)_6]^{3+}$: The oxidation state of chromium in ion is + 3. The electronic configuration of Cr is [Ar] $3d^5\,4s^1$ (Table 5.7).

Table 5.7: Formation of $[Cr(NH_3)_6]^{3+}$.

	3d					4s	4p		
Cr atom (Z = 24) in ground state	↑	↑	↑	↑	↑	↑			
Cr^{3+} ion	↑	↑	↑						
d^2sp^3-hybridized orbitals of Cr^{3+} ion	↑	↑	↑						
Formation of $[Cr(NH_3)_6]^{3+}$	↑	↑	↑	↑↓	↑↓	↑↓	↑↓	↑↓	↑↓

Cr^{3+} ion provides six empty orbitals to accommodate six pairs of electrons from six molecules of ammonia. The resulting complex involves d^2sp^3 hybridization and is thus octahedral. The presence of three unpaired electrons in the complex explains its paramagnetic character. The complex may or may not be colored.

Outer orbital complexes

Formation of ferricyanide ion: Oxidation state of Fe in this complex ion is + 3 (Table 5.8).

The resulting complex involves d^2sp^3 hybridization and is thus octahedral. Further, due to one unpaired electron, it should be slightly paramagnetic. The complex may or may not be colored.

Table 5.8: Formation of $[Fe(CN)_6]^{3+}$.

Fe^{3+} ion	↑ ↑ ↑ ↑ ↑		☐	☐ ☐ ☐
d^{2}sp^3 hybridized orbitals of Fe^{3+} ion	↑↓ ↑↓ ↑ ☐ ☐		☐	☐ ☐
Formation of $[Fe(CN)_6]^{3+}$	↑↓ ↑↓ ↑ ↑↓ ↑↓	↑↓	↑↓ ↑↓ ↑↓	

5.5.2 Crystal field theory

CFT is an electrostatic model that assumes that the metal-ligand bond in a metal complex is purely ionic, formed of electrostatic interactions between the metal ion and ligand. CFT was developed by physicist Hans Bethe in 1929 for crystalline solids. It explains bonding in metal complexes, electronic spectra, and magnetism. CFT is based on the splitting of crystal fields and is used for describing the effect of the electric field of neighboring ions on the energy of the valence orbitals of an ion in a crystal. It explains the breaking of degeneracy in transition metal complexes due to the presence of ligands. Crystal field splitting is splitting the five-degenerate d-orbitals of a metal ion into different orbitals with different energies in the presence of the crystal field of ligands.

5.5.2.1 Postulates of crystal field theory

In CFT, we assume that the metal ion is surrounded by an electric field created by the ligands surrounding the metal ion. The forces of attraction between the central metal ion and the ligand are considered purely electrostatic. The metal ion is targeted by the negative end of the dipole of the neutral molecule ligand. The transition metal ion is a positive charge ion equal to the oxidation state. It is surrounded by a specific number of ligands, which may be negative ions or neutral molecules having lone pairs of electrons. Ligands act as point charges that are responsible for generating an electric field. This electric field changes the energy of the orbitals on the metal atom or ions. The repulsive force between the central metal ion and ligand is responsible for the electrons of the metal ion occupying the d-orbitals as far as possible from the direction of approach of the ligand. There is no interaction between the metal orbital and the ligand orbitals.

In an isolated metal atom or ion, all the orbitals have the same energy, which means all the d-orbitals are degenerate. If the central metal atom or ion is surrounded by the spherical symmetrical field of negative charges, the d-orbitals degenerate. However, the energy of orbitals is raised due to the repulsion between the field and the electron on the metal atom or ion (Figure 5.5).

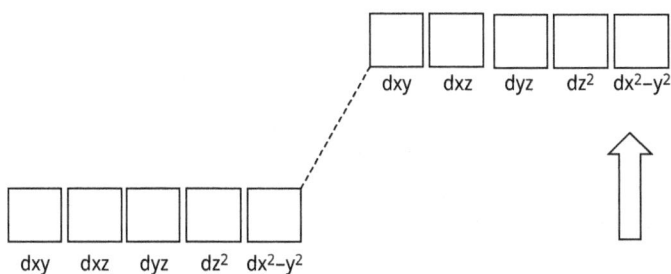

Figure 5.5: The d-orbitals energy of the central ion, surrounded by the spherical symmetrical field.

The d-orbitals are affected differently in most transition metal complexes, and their degeneration is lost due to the field produced by the unsymmetrical ligand.

Octahedral complexes are formed when a central metal ion forms six-coordinate bonds with ligands. In these octahedral complexes, ligands are represented by negative charges and metal ions by a positive charge. In an octahedral complex, there are two positions – axial and equatorial positions. The equatorial position has five neighboring bond pairs/lone pairs, and the axial position has four neighboring bond pairs/lone pairs. In octahedral complexes, we observe repulsions between ligands and the d-orbitals when the ligand approaches metal ions, raising their energy relative to the free ion. The orbitals $d_{x^2-y^2}$ and d_{z^2} are strongly repelled by the ligands compared to the other three orbitals. This results in lower energies of d_{xy}, d_{xz}, d_{yz} orbitals. The orbitals having lower energies are known as t_{2g} and the orbitals with higher energy are known as e_g orbitals (Figure 5.6).

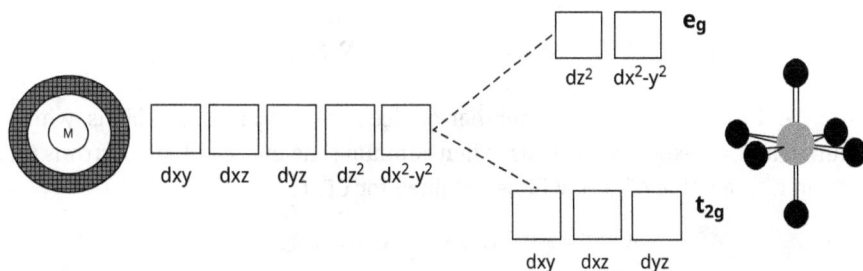

Figure 5.6: The d-orbitals energy of the central ion in octahedral complexes.

Tetrahedral complexes are formed when a central metal ion forms four coordinate bonds with ligands. Tetrahedral complexes have opposite splitting patterns compared to octahedral patterns. The orbitals d_{xy}, d_{xz}, d_{yz} are strongly repelled by the ligands compared to the other two orbitals. This results in lower energies of $d_{x^2-y^2}$ and d_{z^2} orbitals (Figure 5.7).

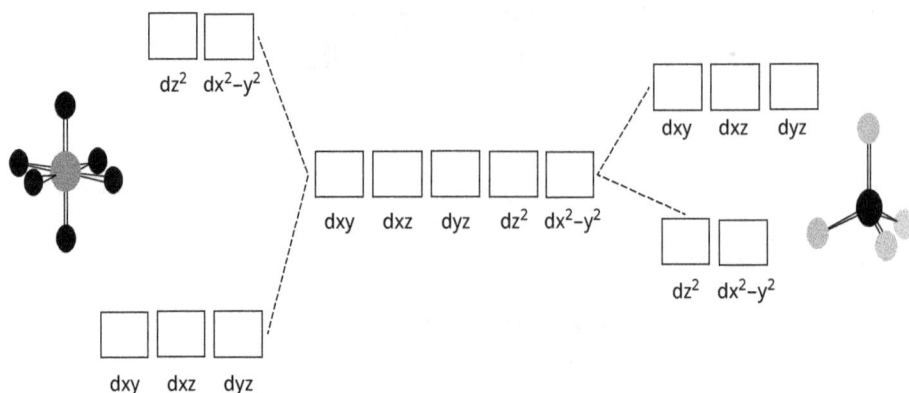

Figure 5.7: Splitting of d-orbitals in octahedral and tetrahedral crystal fields.

5.5.2.2 Crystal field stabilization energy

The difference in energy between two sets of d-orbitals is known as crystal field splitting energy or crystal field stabilization energy (CFSE). The more the CFSE is, the more stable the complex is. CFSE is calculated by using the formula

$$\Delta E = E_{\text{ligand field}} - E_{\text{isotropic field}}$$

The CFSEs for octahedral and tetrahedral complexes are different.

For octahedral complexes, e_g orbitals have an energy level of $+0.6\Delta_0$ or $3/5 \, \Delta_0$ above the average, and t_{2g} orbitals have an energy level of $-0.4\Delta_0$ or $-2/5\Delta_0\Delta$ below the average:

$$\text{CFSE} = -0.4\Delta_0 N(t_{2g}) + 0.6\Delta_0 N(e_g)$$

where $N(t_{2g})$ and $N(e_g)$ represent the number of electrons at t_{2g} and e_g orbitals, which is maximum 6 and 4, respectively. Thus, when summing the energy of all electrons (in configuration d^{10}), a value of zero will be obtained for CFSE:

$$\text{CFSE} = -0.4\Delta_0 \cdot 6 + 0 \cdot 6\Delta_0 \cdot 4 = 0$$

As can be seen from the formula, each electron that enters the e_g level will increase its energy by $0.6\Delta_0$ units compared to the initial one, and when it enters the t_{2g} level, it will decrease it by $0.4\Delta_0$ units. Weak field ligands are high spin complexes with a low Δ_0 value, such as $[Fe(CN)_6]^{4-}$ and $[Co(NH_3)_6]^{3+}$.

The crystal field splitting energy for tetrahedral complexes is $\Delta_t = 49\Delta_0$, which is 49 times the octahedral complexes. Electrons do not pair in tetrahedral complexes because of the narrow energy gap. As a result, tetrahedral complexes have a high spin structure, such as $[Fe(H_2O)_6]^{2+}$ and $[CoF_6]^{3-}$.

The energy difference, in addition to the parameter Δ_o, can also be denoted by D_q, whose numerical value is one order of magnitude higher than Δ_o or $D_q = 10\Delta_o$. Thus, the energy of the electrons becomes $6D_q$ and $-4D_q$, respectively, or:

$$\text{CFSE} = -4D_qN(t_{2g}) + 6D_qN(e_g)$$

The way t_{2g} and e_g levels are filled depends on the total number of electrons in the d-orbitals, on the value of Δ_o and on the value of the energy required to pair the electrons, denoted by P.

The crystal field stabilization energies for octahedral complexes are mentioned below.

With a number of electrons from 1 to 3 and from 8 to 10, the values of Δ_o and P are irrelevant for the order of filling of the levels t_{2g} and e_g.

If the number of electrons is from 1 to 3, according to the principle of minimum energy, the orbitals t_{2g} are always occupied (Table 5.9).

Table 5.9: The values of crystal field splitting energy for one to three electrons in an octahedral field.

M ion	Ti^{3+}	V^{3+}	V^{2+} and Cr^{3+}
d^n	dγ ◯◯	dγ ◯◯	dγ ◯◯
	dε (↑)◯◯	dε (↑)(↑)◯	dε (↑)(↑)(↑)
	d^1	d^2	d^3
CFSE	$-4D_q$	$-8D_q$	$-12D_q$

If the number of electrons is from 8 to 10 after filling the low-energy t_{2g} levels with six electrons, the rest should be located at the e_g levels. In Table 5.10, the CFSE values for the number of electrons from 8 to 10 are calculated.

When the electrons are between four and seven, two ways of filling are possible and they depend on the values of Δ_o and P; see Table 5.11. In a weak field, when $\Delta_o < P$ the electrons are located at both levels individually according to Hund's rule, in which a high-spin complex is obtained. In a strong ligand field (high value of Δ), $\Delta_o > P$, the electrons first occupy the low-energy orbitals T_{2g} and a low-spin complex is obtained.

It has been found that the crystal field stabilization energy represents only 5–10% of the total energy of complex formation. However, the CFSE is an important energy characteristic for complex compounds, reflecting their stability.

Table 5.10: The values of crystal field splitting energy for 8–10 electrons in an octahedral field.

M ion	Ni^{2+}	Cu^{2+}	Zn^{2+}
d^n	$d\gamma$ (↑)( )(↑)	$d\gamma$ (↑↓)( )(↑)	$d\gamma$ (↑↓)( )(↑↓)
	$d\varepsilon$ (↑↓)(↑↓)(↑↓)	$d\varepsilon$ (↑↓)(↑↓)(↑↓)	$d\varepsilon$ (↑↓)(↑↓)(↑↓)
	d^8	d^9	d^{10}
CFSE	$-12D_q + 3P$	$-6D_q + 4P$	$0 + 5P$

Table 5.11: The values of crystal field splitting energy for four to seven electrons in an octahedral field.

M ion	Cr^{2+}, Mn^{3+}	Mn^{2+}, Fe^{3+}	Fe^{2+}, Co^{3+}	Co^{2+}
d^n in a weak field	$d\gamma$ (↑)( )	$d\gamma$ (↑)(↑)	$d\gamma$ (↑)(↑)	$d\gamma$ (↑)(↑)
	$d\varepsilon$ (↑)(↑)(↑)	$d\varepsilon$ (↑)(↑)(↑)	$d\varepsilon$ (↑↓)(↑)(↑)	$d\varepsilon$ (↑↓)(↑↓)(↑)
	d^4	d^5	d^6	d^7
d^n in a strong field	$d\gamma$ ( )( )	$d\gamma$ ( )( )	$d\gamma$ ( )( )	$d\gamma$ (↑)( )
	$d\varepsilon$ (↑↓)(↑)(↑)	$d\varepsilon$ (↑↓)(↑↓)(↑)	$d\varepsilon$ (↑↓)(↑↓)(↑↓)	$d\varepsilon$ (↑↓)(↑↓)(↑↓)
	d^4	d^5	d^6	d^7
CFSE weak/strong	$-6D_q$ / $-16D_q + P$	0 / $-20D_q + 2P$	$-4D_q + P$ / $-24D_q + 3P$	$-8D_q + P$ / $-18D_q + 3P$

As a result of research on numerous complex compounds, it has been found that the most abundant ligands can be arranged in the so-called *spectrochemical row*, according to their increasing ability to induce Δ (Figure 5.8).

At the beginning are those which create the weakest field. With a weak ligand field (low Δ value), the d-orbitals are filled so that the number of electrons with parallel spins is maximum (Hund's rule). In this case, the central ion is in a high-spin state (weak crystal field). In a strong ligand field (high value of Δ), the orbitals with lower energy are occupied by the maximum possible number of d-electrons for a given central ion. In this case, the central ion is in a low-spin state. A strong field ligand is capable of forcing the electrons of the metal atom/ion to pair up (if required). Pairing is

dz^2 dx^2-y^2

$I^- < Br^- < S_2^- < SCN^- < Cl^- <$
$NO_3^- < N_3^- < F^- < OH^- < C_2O_4^{2-} < H_2O <$
$NCS^- < CH_3CN < NH_3 < NO_2^- < CN^- < CO$

dxy dxz dyz

dz^2 dx^2-y^2

dxy dxz dyz

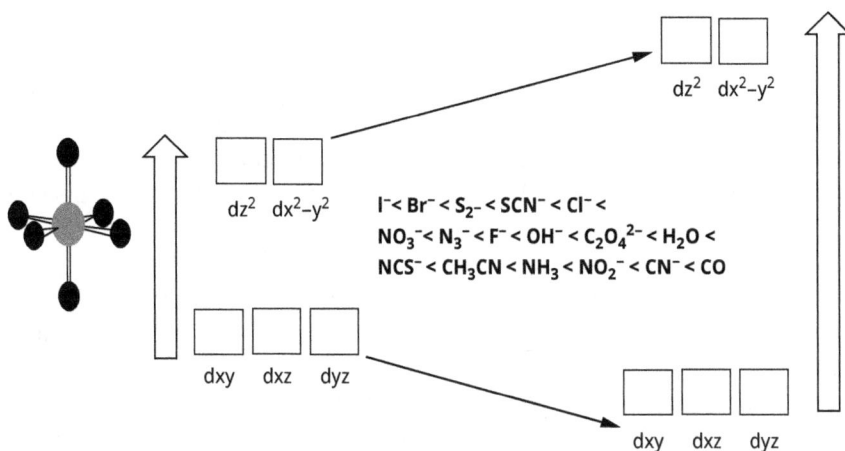

Figure 5.8: Spectrochemical row of the ligands.

done only to the extent which is required to cause the hybridization possible for that coordination number. A weak field ligand is incapable of making the electrons of the metal atom/ion to pair up.

The main factors giving rise to high values of Δ_o are the high effective charge of the central ion and the low electronegativity of ligands containing an easily polarizable lone electron pair capable of forming $d\pi$-$p\pi$ or $d\pi$-$d\pi$ bonds, by overlapping their vacant $p\pi$ or $d\pi$ orbitals with the vacant t_{2g} orbitals of the metal. The values of Δ_o for metal ions increase in order:

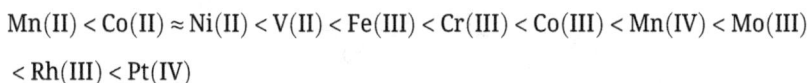

$$Mn(II) < Co(II) \approx Ni(II) < V(II) < Fe(III) < Cr(III) < Co(III) < Mn(IV) < Mo(III)$$
$$< Rh(III) < Pt(IV)$$

The CFT provides explanation to color, arranges ligands according to their strength, explains distortion of complexes and anomalies in their physical properties.

Nevertheless, there are some *limitations of CFT*. CFT could not account for the covalent bonding found in some transition metal complexes. It also fails to explain the order of ligands in the spectrochemical series. This is because ligands are considered point charges, and anionic ligands should have a stronger splitting effect. But anionic ligands are present at the bottom of the spectrochemical series. For example, the theory does not explain why water is a stronger ligand than hydroxide ions. No contribution is considered for s-and p-orbitals, which is required in certain cases. This is a critical drawback because p-bonding is found in numerous compounds. There is no discussion about the orbitals of the ligands in the transition metal. Thus, the theory fails to explain properties related to ligand orbitals and their interaction with metal ones.

5.5.3 Molecular orbital method

This method has the same approach for both complex and other compounds. The interaction of the atomic orbitals of the complexing agent with the atomic or molecular orbitals of the ligands is considered. The chemical bond is realized at the newly obtained multicenter molecular orbitals. Interacting orbitals must have appropriate symmetry and be energetically close.

The formation of an octahedral complex of a d-element is discussed below, given that the bonds are only of σ-type, that is, only such orbitals are involved in ligands, the orientation of which corresponds with the axis of binding of the ligand to the complexing agent.

The complexing agent as a 3d-element has nine valence orbitals: one s-, three p- and five d-orbitals. Of these, six are symmetrically suitable for σ bonding: One s-, three p-orbitals and two of the d-orbitals – $d_{x^2-y^2}$ and d_{z^2}, which are of appropriate orientation. These six orbitals of the metal ion with six more ligand's orbitals form twelve molecular orbitals – six bonding and six anti-bonding orbitals. The three d-orbitals (d_{xy}, d_{xz}, d_{yz}), which are not involved in binding, remain as nonbonding; see Figure 5.9.

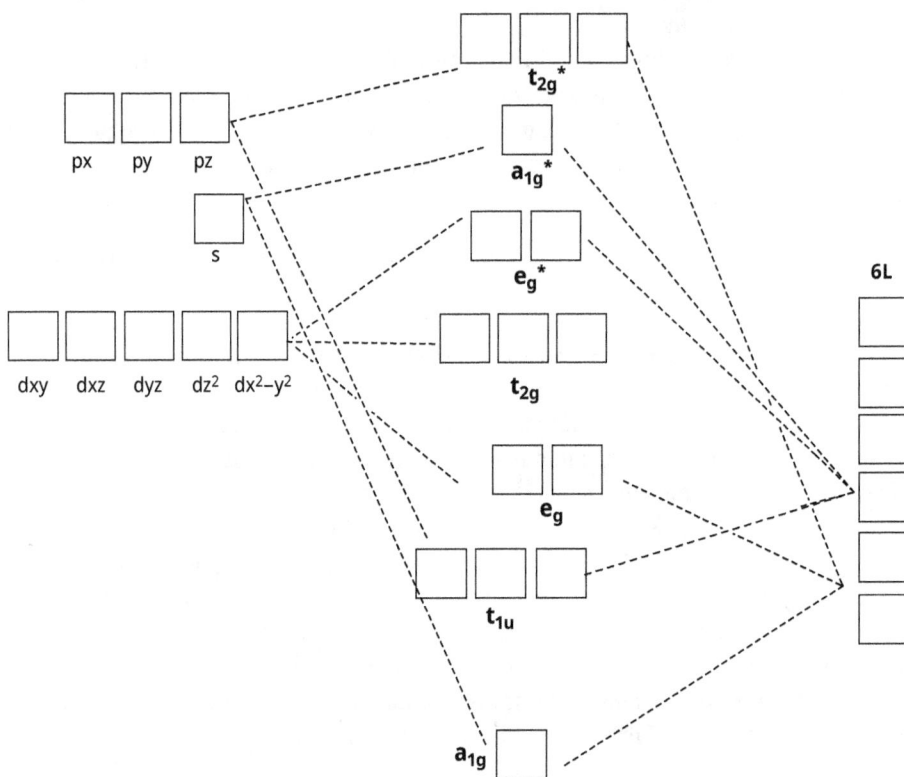

Figure 5.9: Energy diagram of the molecular orbitals of an octahedral complex.

The spherical symmetry of the s-orbital allows it to interact with the six ligands, resulting in two seven-center orbitals σ_{sp} and $\sigma_{sp}{}^*$, which are denoted by the symbol a_{1g}. The three p-orbitals of the metal interact with the p-orbitals of the ligand and form three bonding and three anti-bonding three-center molecular orbitals, which are denoted by the symbol t_{1u}. The two d-orbitals $d_{x^2-y^2}$ and d_{z^2}, which are of appropriate orientation, form two-fold degenerate molecular orbitals, whose symbol is e_g. The three d-orbitals (d_{xy}, d_{xz}, d_{yz}) remain as nonbonding, marked with t_{2g}.

5.5.4 Ligand field theory

With the introduction of the concept of molecular orbitals, CFT gradually developed into LFT. The combination of CFT and MOM is known as ligand field theory. LFT is the most advanced modern theory of the interactions in coordination chemistry.

This theory takes into account the overlap of the atomic orbitals of the ligands with those of the central atom and the delocalization of the electron density. The wave functions of the molecular complex, corresponding to the coordination polyhedron, include a large number of orbitals. For example, in an octahedral complex of fluoride ions with a transition metal, one s-, three p-, and five d-orbitals of the central atom and one s-, and three p-orbitals of the six fluoride ions, or a total of 33 orbitals are considered. Some of these orbitals are incompatible in energy or symmetry, and in fact, only the orbitals of the ligands containing their own (lone electron) pair enter the group orbital, which overlap by σ-type with the unpopulated orbitals of the metal, directed towards the vertices of the octahedron.

The stability of some complexes is also enhanced by the ability of ligands with free antibonding orbitals such as CO, NO, PF_3, CN^-, etc. to form additional π-bonds.

Table 5.12: Differences between crystal field theory and ligand field theory.

Crystal field theory	Ligand field theory
CFT describes the effect of the electric field of on the energy of the valence orbitals of the ion.	LFT describes bonding, orbital arrangement, and other characteristics of coordination complexes.
It only describes the electrostatic interactions happening between metal ions and ligands.	It describes both electrostatic interactions as well as covalent bonding between cations and ligands.
It only helps with the electronic structure of transition metals.	It helps with electronic and optical characteristics of transition metals.
This theory is quite unrealistic.	This theory is realistic in comparison to CFT.

For example, the CO molecule simultaneously participates with a free electron pair in the formation of a σ-bond and accepts an electron from a nonbinding AO of the central atom, similar in symmetry to the free antibonding π^*-orbital. The study of π-acids proves that the LFT is the most advanced modern theory of the covalent interaction.

The main differences between the CFT and the LFT are listed in Table 5.12.

Chapter 6
Chemical thermodynamics

6.1 Main terms

Thermodynamics is the science that studies the mutual transitions of heat and work in equilibrium systems, as well as in the conversion to equilibrium.

The object of study of thermodynamics is *thermodynamic systems*, which are macroscopic objects, interacting with each other and separated from the environment by a real or imaginary surface. Systems are: *open*, in which there is an exchange of energy and matter with the surroundings (presence of reactants in an open); *closed*, where there is an exchange of energy but no exchange of matter (presence of reactants in a closed vessel made of conducting material); *isolated*, in which there is no exchange of either energy or substance with the surroundings (presence of reactants in a thermos-flask, or substance in an insulated closed vessel). According to the type of constituents, thermodynamic systems are divided into two groups – homogeneous and heterogeneous. A system is said to be *homogeneous* when all the constituents present are in the same phase and are uniform throughout the system (mixture of two miscible liquids). A mixture is said to be *heterogeneous* when it consists of two or more phases and the composition is not uniform (mixture of insoluble solid in water). The state of a thermodynamic system means its macroscopic or bulk properties which can be described by state variables: pressure (P), volume (V), temperature (T), amount (n) etc., known as *state functions* or *state parameters*.

If at least one of the parameters of the system changes over time, it means that a *process* is taking place in the system. The processes are: *irreversible* or *reversible*, and in the latter the transition of the system from one state to another and vice versa occurs in the same way, and after returning to the initial state, no macroscopic changes in the environment are observed. An equilibrium process is a continuous sequence of equilibrium states. In such a process, thermodynamic parameters change infinitely slowly. Any rapid change in any of the parameters would bring the system into a non-equilibrium state. The nonequilibrium process is realized when the system passes through nonequilibrium states. All real processes are to some extent nonequilibrium.

When the operation is carried out at constant temperature ($dT = 0$), the process is said to be *isothermal process*. A process in which no transfer of heat between system and surroundings takes place is called *adiabatic process*. When the process is carried out at constant pressure ($dP = 0$), it is said to be *isobaric process*. A process, carried out at constant volume, is known as *isochoric process*. If a system undergoes a series of changes and finally returns to its initial state, it is said to be *cyclic process*. When in a process, a change is brought in such a way that the process could, at any moment, be reversed by an infinitesimal change, the process is called *reversible process*.

https://doi.org/10.1515/9783111712246-006

6.2 Basic thermodynamic principles and quantities

The state of the system is described using macroscopic parameters (T, V, P), as well as using *parameter-dependent thermodynamic functions*. There are: *state functions* (depend only on the state of the system and do not depend on the way this state is obtained) and *transition functions* (depend on how a change occurs in the system). When a property is considered as dependent on other variables, that is, it cannot be measured directly, but only calculated, it is called a function of state. Examples of state functions are: *internal energy U, enthalpy H, Gibbs free energy G*, and *entropy S*. Examples of transition functions are *heat Q* and *work A*.

Internal energy (U) is the sum of all the forms of energies that a system can possess. In thermodynamics, it may change, when heat passes into or out of the system; work is done on or by the system; matter enters or leaves the system. The internal energy of a system is the sum of all kinetic and potential energies of all components of the system. By definition, the change in internal energy is the final energy of the system minus the initial energy of the system:

$$\Delta E = E_{\text{final}} - E_{\text{initial}}$$

If $\Delta E < 0$, $E_{\text{final}} < E_{\text{initial}}$ the system released energy into the surroundings. When energy is exchanged between the system and the surroundings, it is exchanged as either heat (Q) or work (W). The internal energy of a system is independent of the path by which the system achieved that state. Internal energy, E, is a state function. Heat Q and work W are not state functions.

Change in internal energy by doing work: Let us bring the change in the internal energy by doing work. Let the initial state of the system is state A and temperature T_A, internal energy is U_A. On doing some mechanical work the new state is called state B and the temperature T_B. It is found to be $T_B > T_A$. U_B is the internal energy after change:

$$\Delta U = U_B - U_A$$

Change in internal energy by transfer of heat: Internal energy of a system can be changed by the transfer of heat from the surroundings to the system without doing work:

$$\Delta U = Q$$

where Q is the heat absorbed by the system. It can be measured in terms of temperature difference. Q is positive when heat is transferred from the surroundings to the system and Q is negative when heat is transferred from system to surroundings.

When change of state is done both by doing work and transfer of heat:

$$\Delta U = Q + W$$

When heat is absorbed by the system from the surroundings, the process is *endothermic*. When heat is released by the system into the surroundings, the process is *exothermic*.

First law of thermodynamics (law of conservation of energy) states that energy can neither be created nor be destroyed and is therefore conserved. Energy can be transferred from one form to another. The energy of an isolated system is constant:

$$\Delta U = Q + W$$

According to the first principle, when the system transitions from state 1 to state 2, the sum of the work done on it W and the amount of heat received from it Q is determined only by states 1 and 2 and does not depend on the way in which the transition between them takes place:

$$U_2 - U_1 = \Delta U = \Delta W + \Delta Q$$

Enthalpy (H) is defined as total heat content of the system at constant pressure. It is equal to the sum of internal energy and pressure-volume work.

$$H = U + PV$$

Units of enthalpy are kJ. Heat is a form of energy, which is temperature and pressure dependent. Enthalpy is a state function. It is an extensive property. Enthalpy is reversible. The enthalpy change for a reaction is equal in magnitude, but opposite in sign, to ΔH for the reverse reaction. ΔH for a reaction depends on the state of the products and the state of the reactants:

$$\Delta H = H_{final} - H_{initial}$$

$$\Delta H = H_{products} - H_{reactants}$$

A process is endothermic when ΔH is positive (>0). A process is exothermic when ΔH is negative (<0). Change in enthalpy is the heat absorbed or evolved by the system at constant pressure ($\Delta H = Q_p$). For exothermic reaction (system loses energy to surroundings), ΔH and Q_p both are negative. For endothermic reaction (system absorbs energy from the surroundings) ΔH and Q_p both are positive.

An *extensive property* is a property whose value depends on the quantity or size of matter present in the system. For example: mass, volume, enthalpy etc. are known as extensive properties. *Intensive properties* do not depend upon the size of the matter or quantity of the matter present in the system. For example: T, density, pressure etc. are called intensive properties.

The amount of energy required to raise the temperature of a substance by 1 K (1 °C) is its *heat capacity* (C in units of J/K). The increase in temperature is proportional to the heat transferred.

q = coeff. ΔT and $q = C.\Delta T$, where the coefficient C is heat capacity. C is directly proportional to the amount of substance and $C_m = C/n$. It is the heat capacity for 1 mol of the substance.

Molar heat capacity is defined as the quantity of heat required to raise the temperature of a substance by 1 °C (K or °C).

Specific heat capacity is defined as the heat required to raise the temperature of one unit mass of a substance by 1 °C (K or °C):

$$Q = C.m.\Delta T$$

where m is the mass of the substance and ΔT is the rise in temperature.

Enthalpy of fusion is the heat energy or change in enthalpy when one mole of a solid at its melting point is converted into liquid state.

Enthalpy of vaporization is defined as the heat energy or change in enthalpy when one mole of a liquid at its boiling point changes to gaseous state.

Enthalpy of sublimation is defined as the change in heat energy or change in enthalpy when one mole of solid directly changes into gaseous state at a temperature below its melting point.

Enthalpy of formation ΔH_f is defined as the enthalpy change for the reaction in which a compound is made from its constituent elements in their elemental forms.

Standard enthalpy of formation ΔH_f^0 is defined as the change in enthalpy in the formation of 1 mol of a substance from its constituting elements under standard conditions of temperature at 298 K and 1 atm pressure.

Enthalpy of combustion is defined as the heat energy or change in enthalpy that accompanies the combustion of 1 mol of a substance in excess of air or oxygen.

The systems with the lowest energy are known to be the most sustainable. Therefore, chemical processes must proceed in the direction of reducing the internal energy of the system, that is, in the direction of exothermic reactions, since in this case the internal energy of the final products of the reaction will be lower than that of the initial ones. In fact, most chemical reactions occur with the release of heat, but there are also endothermic reactions that occur spontaneously:

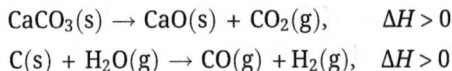

$$CaCO_3(s) \rightarrow CaO(s) + CO_2(g), \quad \Delta H > 0$$
$$C(s) + H_2O(g) \rightarrow CO(g) + H_2(g), \quad \Delta H > 0$$

Therefore, the enthalpy factor (ΔH) cannot always serve as a criterion for determining the possibility of a chemical reaction taking place. In other words, there must be another factor that determines the spontaneous course of the process in the absence of a thermal effect ($\Delta H = 0$), or even if the process is endothermic ($\Delta H > 0$). Many processes take place without the introduction of energy from an external source. Such processes are called spontaneous.

Spontaneous process is a process which can take place by itself or has a tendency to take place. Spontaneous process need not be instantaneous. Its actual speed can vary from very slow to quite fast. A few examples of spontaneous process are: (i) common salt dissolves in water of its own; and (ii) carbon monoxide is oxidized to carbon dioxide of its own. Spontaneous processes occur without any outside input or influence. Spontaneous processes can occur due to physical or chemical properties: metals oxidizing in the presence of oxygen; water melting at temperatures above 0 °C; salt dissolving in water; gas or heat diffusing across a room (from an area of higher concentration to lower concentration); osmosis (passive movement of water across a membrane); skiing downhill. Processes spontaneous in one direction are nonspontaneous in the reverse direction. Gas molecules trapped on one side of valve stay there as they their volume is physically compressed; once the valve is opened, the gas spontaneously diffuses to occupy both bulbs. Gas molecules cannot move in the reverse direction (back into one bulb) without outside intervention such as pressure to move the gas. Spontaneous processes can be both exothermic and endothermic and since heat is dependent on T, at various temperatures. The standard entropy can be calculated.

Entropy (S) is a measure of degree of randomness or disorder of a system. Entropy of a substance is minimum in solid state while it is maximum in gaseous state. The change in entropy in a spontaneous process is expressed as ΔS. Entropy, like enthalpy, is a state function, which means that the entropy value of an element of compound is based on its physical state. Units of entropy are J. The same substance can have different enthalpy and entropy values for its various physical states. These reference values are taken under standard conditions (1 atm, 25 °C) for 1 mol of the substance. Both enthalpy and entropy are determinants of spontaneity – they help determine whether a process is spontaneous or not. Since entropy and enthalpy have different units (J and kJ, respectively), one must be converted to another when used in the same equation so that the units match. Otherwise, values will be off by 1,000 or 10^3.

Second law of thermodynamics states that the system tries to move from states with less probability to states with more probability or entropy in the universe increases in spontaneous processes. The following is often considered the basic formulation of the second law of thermodynamics: Heat cannot pass by itself from a less heated object to a more heated one.

Entropy is a measure of how many microstates are present in a particular substance based on its physical state. Microstate W is a representation of the order of a substance – think of it as parts or units. Entropy can be calculated from the number of microstates using the following equation:

$$S = k.\ln W$$

where k is the Boltzmann constant, $k = R/N_A = 1.38 \times 10^{-23}$ J/K; R is the universal gas constant, N_A is Avogadro's number. Since W can only be a natural number (1, 2, 3, . . .), entropy according to Boltzmann's formula must be a positive quantity. Therefore, entropy is a measure of the disorder or probability of a given state of a system. If the system goes from a more ordered to a less ordered state, then the entropy S increases and the change in entropy ΔS in the process will be a positive value ($\Delta S > 0$). If the reverse process is observed, then the entropy in the system decreases and ΔS will have a negative value ($\Delta S < 0$).

Third law of thermodynamics states that the entropy of a pure crystalline substance at absolute zero is 0. At absolute zero, the substance is a perfect lattice and thus has only one microstate. Since $W = 1$, $S = k.\ln 1$ or $S = k.0$, thus $S = 0$. Other factors that increase the microstates and thus entropy: (i) higher temperature tends to increase entropy; (ii) larger volumes (typically due to a physical state change or increase in temperature) will increase entropy; (iii) the number of mobile molecules or gaseous moles increase entropy; (iv) physical states vary in entropy with solids having lowest entropy, followed by liquids and gases (gases > liquids > solids); and (v) formation of a solution (solute dissolving in solvent) increases entropy. The opposite processes will thus decrease entropy: (i) lowering temperatures; (ii) recrystallization of a solution; (iii) restoring physical states (e.g., solids have the lowest entropy); and (iv) decreasing volumes, particularly of gasses.

In an isothermal process that takes place thermodynamically reversibly (very slowly), the change in entropy ΔS can be expressed as the ratio of the total amount of heat to the absolute temperature T:

$$\Delta S = Q/T$$

where Q is the amount of heat introduced or released and T is the absolute temperature.

The above expression expresses mathematically the second principle of thermodynamics for reversible processes. When combining the above equation with the equation of the first principle ($Q = \Delta U + P\Delta V$), the **basic equation of thermodynamics** is obtained:

$$T\Delta S = \Delta U + P\Delta V$$

Entropy is an extensive (additive) quantity. In fact, if a system is divided into two parts and the amount of heat received from it is the sum of the amounts of heat received by the two parts, it follows from the above definition that the entropy of the system is the sum of the entropies of the two parts.

Gibbs energy (G) and spontaneity: The so-called Gibbs free energy together with temperature, entropy, and enthalpy all together contribute to predicting if a process will be spontaneous. The new thermodynamic function, the Gibbs energy or Gibbs function G, is defined:

$$G = H - TS$$

$$\Delta G = \Delta H - T\Delta S$$

ΔG gives criteria for spontaneity at constant pressure and temperature. If ΔG is negative (<0) the process is spontaneous. If ΔG is positive (>0) the process is nonspontaneous. Table 6.1 summarizes how signs of H, S, and T affect spontaneity:

Table 6.1: Spontaneous and nonspontaneous processes.

ΔH	ΔS	$-T\Delta S$	ΔG	Result
–	+	–	–	Spontaneous at all temperatures in forward direction (nonspontaneous in reverse direction)
+	–	+	+	Nonspontaneous at all temperatures in forward direction (spontaneous in reverse direction)
–	–	+	+ or –*	Spontaneous at low temperatures; nonspontaneous at high T
+	+	–	+ or – *	Spontaneous at high T; nonspontaneous at low T

*Temperature dependent.

Gibbs free energy and equilibrium: Gibbs free energy can help determine progress of a reaction at equilibrium:

$$\Delta G = \Delta G° + RT \ln K$$

At equilibrium, $\Delta G = 0$ means $0 = \Delta G° + RT \ln K$. Equation is rearranged to solve:

$$\Delta G° = -RT \ln K$$

where R is the gas constant (8.314 J/K.mol), T is the temperature (K), K is the equilibrium constant. K and $\Delta G°$ are also related (Table 6.2).

What may be determined: Which is favored more at equilibrium: reactants/ products; Are reactants/products equally favored?

Table 6.2: Spontaneous and nonspontaneous processes.

K	$\ln K$	$\Delta G°$	Favor at equilibrium
=1	0	0	Products and reactants equally favored
>1	+	–	Products are favored
<1	–	+	Reactants are favored

Chapter 7
Chemical kinetics

7.1 Basic concepts in chemical kinetics

Chemical kinetics is the branch of science that deals with the rate of reaction, factors affecting the rate of reaction and the reaction mechanism. Different reactions occur at different rate. Chemical reactions involve redistribution of bonds – breaking of bonds in the reactant molecules and making of bonds in the product molecules. The rate of a chemical reaction depends upon the strength of the bonds and the number of bonds to be broken during the reaction. Reactions involving strong bond breaking occur at relatively slower rate, while those involving weak bond breaking occur at relatively faster rate. On the basis of rate, reactions are classified as:

- instantaneous or extremely fast reactions with half-life of the order of fraction of seconds;
- extremely slow reactions with half-life of the order of years; and
- reactions of moderate or measurable rate.

Ionic reactions are instantaneous. If a drop of $AgNO_3$ solution is added to a solution of NaCl or HCl, a white precipitate of AgCl immediately appears. This is because of the fact that in aqueous solutions an ionic compound exists as its constituent ions and no bond needs to be broken. The half-life period of ionic reactions is of the order of 10^{-10} s:

$$Na^+ + Cl^- + Ag^+ + NO_3{}^- \longrightarrow AgCl \downarrow + Na^+ + NO_3{}^-$$

Free radicals are very unstable and reactive due to the presence of unpaired electrons. Reactions involving free radicals also occur instantaneously:

$$CH_3 + Cl_2 \longrightarrow CH_3Cl + \cdot Cl$$

$$CH_3 + \cdot CH_3 \longrightarrow H_3C - CH_3$$

Some molecular reactions, involving reactants containing odd electrons, complete within a fraction of a second. The speed of such reactions is attributable to the tendency of the odd electron molecule (paramagnetic in nature) to transform into a stable spin-paired diamagnetic molecule by dimerization, such as the dimerization of nitrogen dioxide into nitrogen tetraoxide:

$$NO_2 + NO_2 \longrightarrow N_2O_4$$

There are some molecular reactions which are known to be extremely slow. Their half-lives are of the order of several years, for instance:

https://doi.org/10.1515/9783111712246-007

$$4\text{Fe} + x\text{H}_2\text{O} + 3\text{O}_2 \longrightarrow 2\text{Fe}_2\text{O}_3 \cdot x\text{H}_2\text{O}$$

$$\left[\text{Cr}(\text{H}_2\text{O})_6\right]^{3+} + \text{I}^- \longrightarrow \left[\text{Cr}(\text{H}_2\text{O})_5\text{I}\right]^{2+} + \text{H}_2\text{O}$$

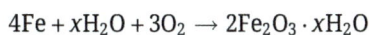

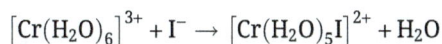

Most molecular reactions especially organic reactions occur at a measurable rate. The half-life of such reactions is of the order of minutes, hours, and days. Chemical kinetics deals with the rates of only those reactions which occur with a measurable rate, that is, which are neither too fast nor too slow.

7.2 Rate of chemical reactions

The *rate of the reaction* is defined as the change in the molar concentration of the reactants or the products per unit of time. For a reaction of A → B, where A is the reactant and B is the product, the graph concentration = f(time) is shown in Figure 7.1.

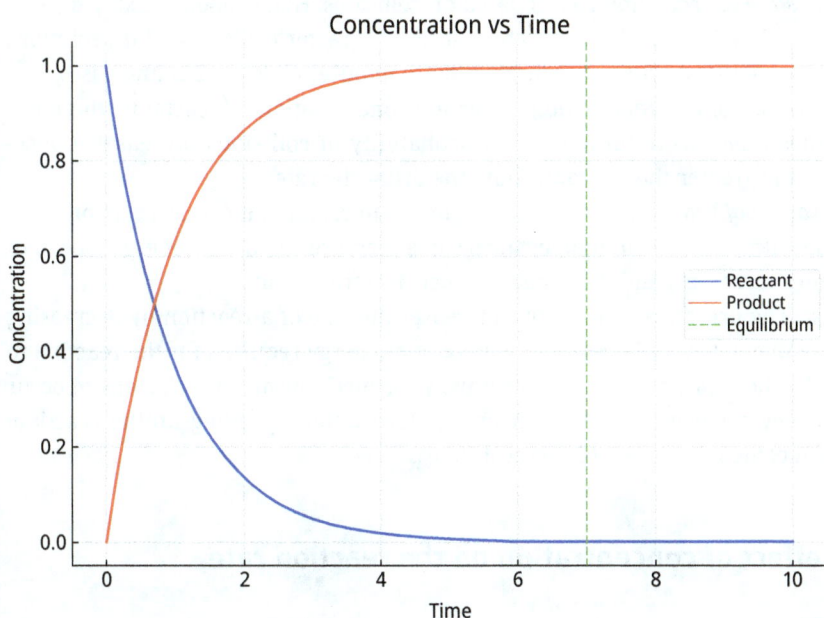

Figure 7.1: Concentration as a function of time.

7.2.1 Types of rates

–   *Instantaneous rate* is the rate calculated for one specific data point of time and the corresponding concentration:

$$v = \lim(\Delta C/\Delta t) = \pm (dC/dt)$$

–   *Average rate* is the rate calculated at a variety of time intervals (and corresponding concentrations) throughout the reaction. It gives an overall impression of the rate in the reaction.

7.2.2 Factors affecting the rate

–   *Number of collisions:* The greater the number of collisions between reactants, the faster they react with one another, and the more rapid the rate;
–   *Physical state of reactants:* This is based on collisions. Homogenous mixtures (solutions) are already in a physical state that favors greater collisions. Thus, solutions or aqueous physical states yield faster rates than pure solids, liquids and gases;
–   *Concentration of reactants:* The greater the concentration of reactants, the more molecules there are interacting and probability of collisions increases. Consequently, the greater the concentration, the faster the rate;
–   *Temperature of reactants:* Generally, higher temperatures increase reaction rates because with increases in temperature come increases in kinetic energy (speed of molecules increases) and thus collisions occur more readily.
–   *Catalyst:* The presence of a catalyst increases the rate of a reaction by decreasing the activation energy (the minimum amount of energy required for the reaction to proceed). The catalysts create new pathways or mechanisms for collisions to occur so that they occur earlier at lower energy levels, thus speeding up the reaction. They can come in a variety of physical states.

7.3 The effect of concentration on the reaction rate

The rate of homogeneous reaction can be expressed in mol/L.s:

$$v = (C_2 - C_1)/(t_2 - t_1) = \pm \Delta C/\Delta t$$

where C_1 and C_2 are the initial and final molar concentrations of the substance at time t_1 and t_2, respectively. The rate is typically a numerical value but can be expressed as the decline of reactant (molecule A) or the increase of product (molecule B). The rate or slope of A is negative and the rate or slope of B is positive or increasing. The rate of decline of the reactant concentration and the increase of the product concentration over time are typically equal.

7.3.1 Rate laws

In contrast to the "rate expression" the rate law only focuses on the reactants (not products as well) and the coefficients of reactants are arbitrary. The rate law expresses the relationship of the rate as proportional to the rate constant k; concentrations of the reactants C; and orders of the reactants raised expressed as exponents. This basic principle was first put forth quantitatively in 1863 by Guldberg and Waage in their statement of the *law of mass action*, which asserts that, for a general reaction of the type:

$$a\,A + bB = cC + d\,D$$

the rate of the reaction at a given temperature is

$$v = k \cdot C_A^a \cdot C_B^b$$

where C_A and C_B are the molar concentrations of the reactants; a and b are the coefficients in the balanced chemical equation; and k is a proportionality constant (rate constant), characteristic of each specific reaction.

7.3.2 Physical meaning of the rate constant

If the concentrations of reactive substances A and B are 1 mol/L, k is equal to the reaction rate $v = k$. Rate constant k does not depend on reagent concentrations and time. It characterizes the nature of reactive substances and depends on the temperature and the presence of catalysts.

7.3.3 Determining orders of reactions

Reaction orders (exponents) can be zero order, first order, and second order. Orders express how the concentration of a reactant can affect the rate. Different orders will change the rate differently.

If the reactant changes in concentration and there is no relation to the change in rate, then it is considered to be zero order. If the concentration of A doubles from 1 M to 2 M, the rate stays the same: $[1]^0 = 1$ and $[2]^0 = 1$. If the reactant concentration changes by the same proportion as the rate, then it is first order. If the concentration of A doubles, the rate doubles: $[1]^1 = 1$ and $[2]^2 = 2$.

If the reactant concentration changes and the rate changes by a factor of a square, it is second order reaction and the exponent is the order. If the concentration of A doubles from 1 M to 2 M, the rate quadruples $[2]^2 = 4$. If the concentration of A triples from 1 M to 3 M, the rate changes $[3]^2 = 9$. The shape of a rate-concentration graph can reveal the reaction order with respect to a particular reactant (Table 7.1).

Table 7.1: Reaction order and rate laws.

	Zero order	First order	Second order
Rate law	Rate = $k[A]^0 = k[1] = k$	Rate = $k[A]^1$	Rate = $k[A]^2$
Units of rate constant k	M/s or M s^{-1}	1/s or s^{-1}	1/M s or M^{-1} s^{-1}

The shape of the resulting rate-concentration graph reveals the reaction order with respect to the reactant (Figure 7.2).

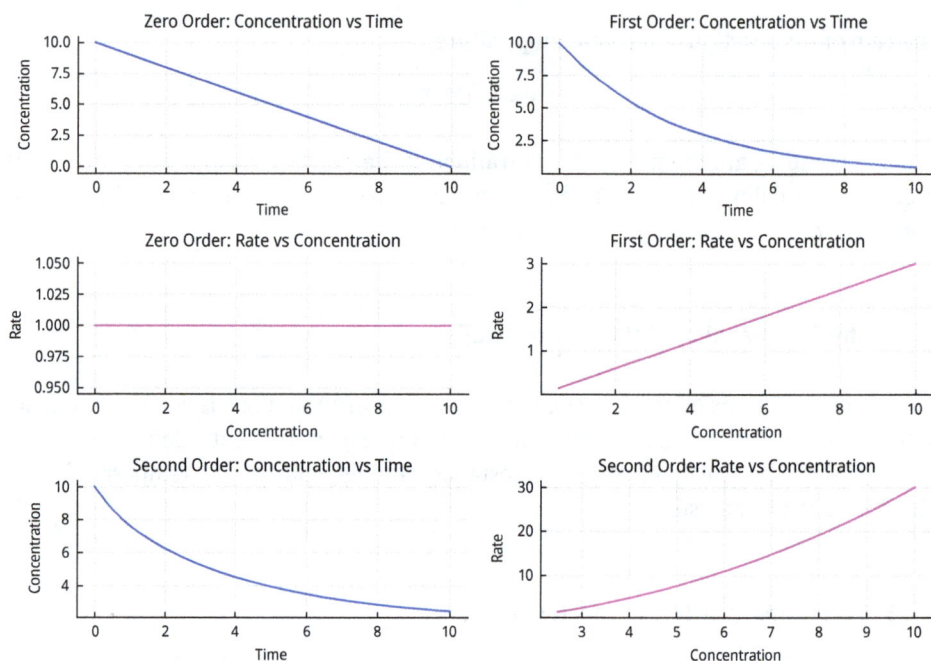

Figure 7.2: Rate-concentration graphs.

The rate-concentration graphs show that in the case of zero order reaction, the horizontal line indicates that the reaction rate is independent of the reactant concentration. In the first order reaction, the straight line passing through the origin signifies that the reaction rate is directly proportional to the reactant concentration. In the second order reaction, the curved plot indicates that the reaction rate is proportional to the square of the reactant concentration.

7.4 The effect of temperature on reaction rate

Molecular collisions are a fundamental requirement for chemical reactions, and concentrations are important because they determine the frequency of these collisions. Nevertheless, most collisions do not result in a chemical reaction, because the energy of collision is insufficient to surmount the barrier to reaction. This barrier, the energy required to rearrange atoms in going from reactants to products, is very substantial for most reactions. It is called the *activation energy*, or the enthalpy of activation (E_{act}). The activation energy of a reaction is usually denoted by E_{act} and is given in units of kilojoules per mole. Only an extremely small fraction of molecules, having energies of collision greater than E_{act} (the minimum amount needed for reaction to occur), are likely to have a series of successive collisions. This fraction increases exponentially with the temperature.

7.4.1 Energy changes: activation energy

During the course of chemical reactions, it is necessary to break the bonds between the atoms in the molecules of the reacting substances and to create new bonds between the atoms in the molecules of the reaction products. The energetic changes that accompany the individual stages of these chemical interactions in the cases of exothermic and endothermic reactions are different (Figure 7.3).

The first stage of any chemical reaction which involves the absorption of energy necessary to break the bonds between the atoms of the reacting substances is called the activation of molecules. As a result of this reaction, a state is reached in which the old bonds are not completely broken, and the new ones are not fully formed. The energy required to obtain it, coming out of the average energy of the initial molecules, is precisely the activating energy of the process. The activated complex has maximum energy in the course of the reaction, so it is unstable. It differs from molecules in the energy of the bond, in the length of the bond and in the angles between the bonds, which in the activated complex are distorted. Further in the course of the reaction, new bonds are created in the molecules of the reaction products. This process is accompanied by a general decrease in energy, which reaches the level of the average energy of the molecules of the reaction products.

7.4.2 Distribution curves

The strong dependence of the rate of the chemical reaction on temperature cannot be explained only by an increase in the number of collisions between molecules with an increase in temperature. The reason is more complex and consists in a sharp increase in the number of "successful impacts," that is, impacts that end with the formation of

Exothermic Reaction

Endothermic Reaction

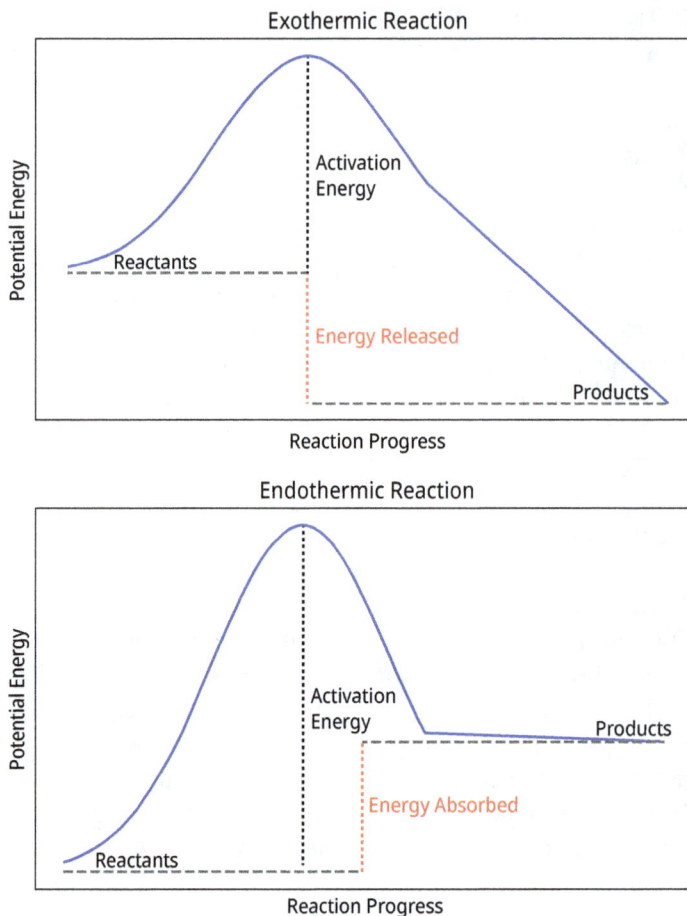

Figure 7.3: The energetic changes of exothermic and endothermic reactions.

reaction products. The influence of temperature on the reaction rate is clearly presented with the help of the so-called *distribution curves*.

By increasing the temperature of the reaction, the kinetic energy of the particles also increases and thus a greater proportion of colliding particles will have the sufficient energy to overcome the activation energy and to react. The Maxwell-Boltzmann distribution graph (Figure 7.4) displays the distribution of velocities of the reactant particles' kinetic energy which gives an understanding of the effect of the factors of reaction rates:

$$dN_E = f(E)$$

where dN_E is the number of molecules at temperatures $T_2 > T_1$.

As shown above, the shaded region for T_2 is a lot larger than that for T_1, which infers that there are a greater proportion of particles that have overcome the activa-

Effect of Temperature on Molecular Energy Distribution

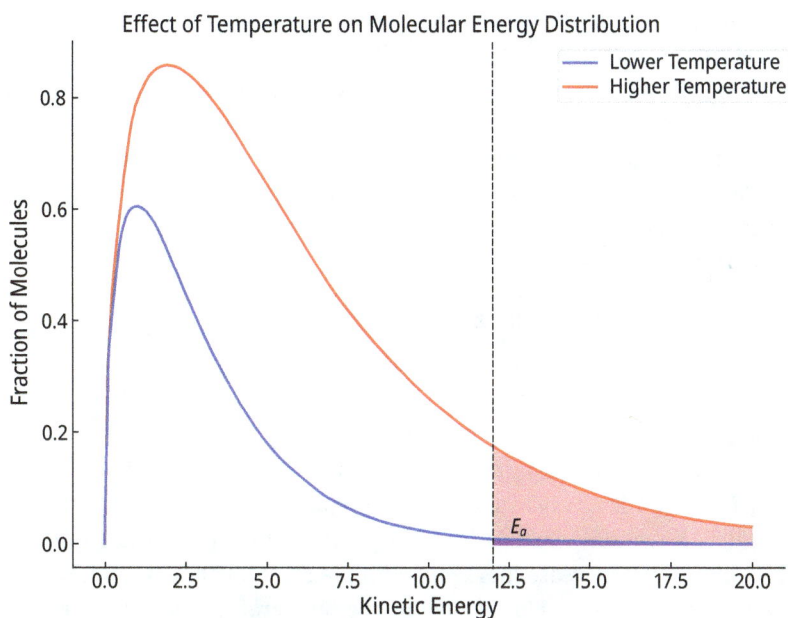

Figure 7.4: The Maxwell-Boltzmann distribution graph.

tion energy at a higher temperature. Only molecules with energy higher than E_{act} are involved in the chemical interaction. These are the active molecules (shaded areas). The activation energy E_{act} is necessary to reduce the strength of the chemical bonds. The increase in T increases the area of active molecules possessing energy higher than energy barrier E_{act}.

The temperature dependence of the rate of a chemical reaction was empirically determined by Dutch scientist J. H. Van't Hoff. He found that the fraction of molecules with enough energy to react and the rate of multiple reactions increased 2–4 times as the temperature rose for every 10 °C.

The first quantitative studies of the temperature dependence of rate constants were published in the last half of the nineteenth century, and various empirical formulas were proposed. A common rule of thumb is that the rate of a reaction doubles if the temperature is raised by 10 °C. The mathematical expression for this rule can be written as follows:

$$v_2 = v_1 \cdot \gamma^{(T_2 - T_1)/10}$$

where v_1 is the reaction rate at temperature T_1; v_2 is the reaction rate at temperature T_2; γ is the temperature coefficient, equal to 2–4. However, this rule does not explain the mechanism by which temperature affects a particular reaction rate value and does not describe the whole set of laws. It is logical to conclude that as the temperature rises, the chaotic motion of the particles increases and this leads to an increase in their collisions. However, this does not particularly affect the collision efficiency of the molecules, as it is mainly dependent on the activation energy. Also, their spatial correspondence to each other plays an important role in the efficiency of particle collisions. The Van't Hoff rule has limited applications. It is observed in gas reactions and is not observed in reactions with solids. Many reactions do not obey the rule, for instance, reactions occurring at high temperatures, and very rapid and very slow reactions.

The above expression can be represented as follows:

$$v_2/v_1 = k_2/k_1 = \gamma^{(T_2-T_1)/10}$$

$$\gamma = (v_2/v_1)^{10/(T_2-T_1)}$$

where v_2, v_1, k_2, k_1 are, respectively, the rates and rate constants of the reaction at the given temperatures, and γ is the temperature coefficient of the reaction rate.

The law of mass action is always stated to be applicable to a given temperature, and it appears not to have temperature involved in its statement. Yet the rates of chemical reaction invariably increase markedly with the increase in temperature. Because concentrations will be negligibly affected by the temperature, the temperature-sensitive factor in the law of mass action must be the rate constant, k. As a good approximation, and given the nature of the reagents, the temperature dependence of the chemical reaction rate obeys the Arrhenius equation:

$$k = A \cdot e^{-E_{act}/RT}$$

or in logarithmic form:

$$\ln k = \ln A - E_a/RT$$

where A is a multiplier; E_{act} is the activation energy; k is proportional to the fraction of molecules (or collisions) that have the required enthalpy of activation; R is the gas constant (8.314 J/mol K); T is the temperature in K; e is the exponential function. The temperature-independent factor A is called the preexponential factor (proportionality constant). Experimental molar activation energy values are usually in the range from 50 to 200 kJ/mol, somewhat smaller than energies required breaking chemical bonds. These magnitudes seem reasonable if we depict the activation process as partially breaking one bond while partially forming another. The multiplier A is constant for a given reaction and practically does not depend on temperature:

$$A = P \cdot z$$

where z is a number proportional to the number of impacts between the particles of the reactants, depending on the mass and size of the particles, and P is a steric multiplier indicating the probability of impacts in a direction that is favorable for the rearrangement of chemical bonds, that is, for the formation of the product.

Obviously, the temperature cannot change the concentrations of the reactants. This means that the only thing that can change is the rate constant, k. The variation of the rate constant k with temperature T is given in Figure 7.5.

The rate constant of the reaction increases exponentially with increasing temperature according to the Arrhenius equation.

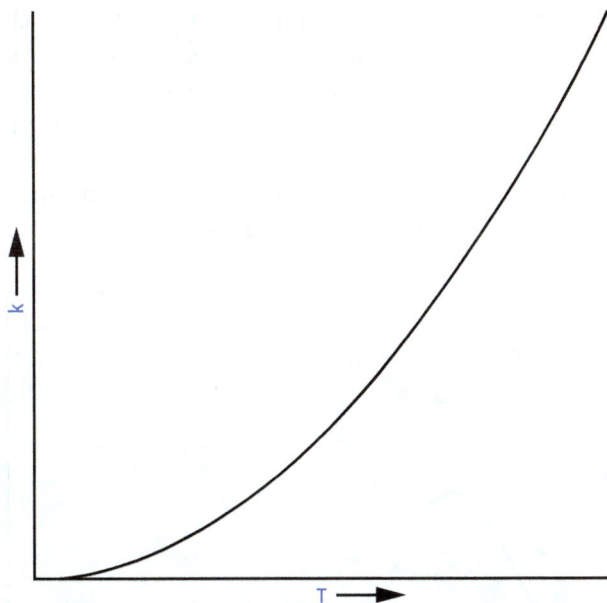

Figure 7.5: The rate constant k as a function of temperature.

Chapter 8
Catalysis

8.1 Influence of catalysts on the rate of chemical processes

The rate of chemical reactions is actively studied by the influence of catalysis, which shows how and why relatively small amounts of some substances increase the rate of interaction of others. Such substances, that can accelerate the reaction but are not consumed in it, are called catalysts. Catalysts change the mechanism of chemical interactions on their own, helping to create new transition states characterized by low energy barrier heights (Figure 8.1). That is, they contribute to a decrease in activation energy and therefore an increase in the number of particle collisions.

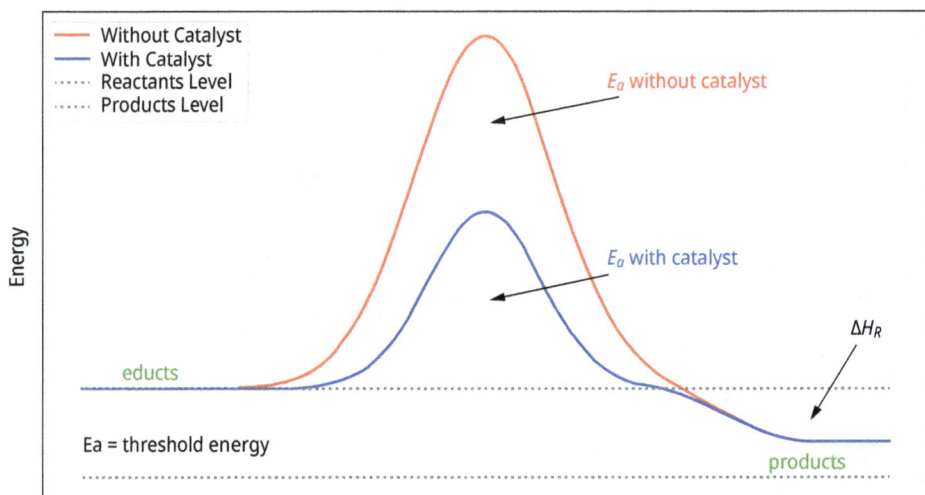

Figure 8.1: Dependence of energy of the process in the presence and absence of a catalyst.

Catalyst is a substance that increases the rate of a chemical reaction without itself being consumed. A catalyst causes a lower activation energy, which in turn makes the rate constant larger and, thus, the reaction rate higher. Each catalyst has its own specific way of functioning. The catalyst is not consumed, but rather used and then regenerated. The catalyst cannot cause a reaction that is energetically impossible. According to chemical thermodynamics, it is possible to catalyze only spontaneous processes in which $\Delta G < 0$. The catalyst cannot change the equilibrium constant of a reaction ($\Delta G = 0$). If a catalyst catalyzes a forward reaction, it must also catalyze the reverse reaction. A catalyst speeds up the forward and reverse reactions. Thus, a reaction has the same yield with or without a catalyst, but the product forms faster.

https://doi.org/10.1515/9783111712246-008

As an example, the following model reaction will be considered:

$$A + B \rightarrow AB + Q$$

It takes place in the presence of catalyst K. It is assumed that the catalyst changes the reaction pathway, resulting in intermediates, which further break down, and the catalyst is restored. Substance A interacts more easily with catalyst K than with substance B, forming [A . . . K] (*activated complex*) that passes into an intermediate compound AK:

$$A + K \rightarrow [A \ldots K] \rightarrow AK + q_1$$

The compound AK readily interacts with the substance B, forming [B . . . AK] (*activated complex*), which is degraded to the substances AB and K:

$$B + AK \rightarrow [B \ldots AK] \rightarrow AB + K + q_2$$

The described process is two-stage and its rate is determined by the rate of the slower stage, the so-called rate-determining stage. The essential thing is that if, for example, the catalyst is positive, then the rate of the slower (rate-determining) stage is significantly higher than the rate of the noncatalyzed reaction. The AK intermediate is different from the activated complex [A . . . K] by the fact that when the AC is obtained, a characteristic energy minimum is observed in the energy diagram of the reaction. The energy minimum usually corresponds to a compound with a defined composition.

The well-studied example of catalysis involves the interaction of ammonia with oxygen, which without catalyst runs to N_2:

$$4NH_3 + 3O_2 = 2N_2 + 6H_2O$$

while in the presence of a catalyst (Pt) NO is obtained:

$$4NH_3 + 5O_2 = 4NO + 6H_2O$$

Hydrogen peroxide decomposes to form oxygen and water:

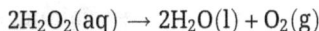

$$2H_2O_2(aq) \rightarrow 2H_2O(l) + O_2(g)$$

However, this reaction is very slow and it does not change for a long time. The addition of only a few grains of magnesium oxide stimulates the reaction.

8.2 Types of catalysis

Catalysis can be divided into three classes: *homogeneous, heterogeneous,* and *enzyme catalysis*. In homogeneous catalysis all substances involved in the reaction, including the catalyst, occur in the same physical state. In heterogeneous catalysis the catalyzed reaction occurs at the boundary between two phases (usually on the surface of a solid catalyst). Enzyme catalysis is a special case of homogeneous catalysis. Biological catalysts work specifically on a given substrate so that the physical shape of the enzyme and its substrate fit together precisely in a "lock and key" mechanism. Enzymes usually end in "ase" such as lactase enzyme, which works on the substrate lactose. Other examples are alcohol dehydrogenase, salivary amylase, etc.

In homogeneous catalysis, the reaction rate depends on concentration, while in heterogeneous catalysis it depends on the surface of the catalyst. A homogeneous catalyst exists in solutions with the reaction mixtures, so it must be a gas, liquid, or soluble solid. Consider the reactions of sulfuric acid manufacture: the key step, the oxidation of sulfur dioxide to sulfur trioxide, occurs so slowly that it is not economical:

$$SO_2(g) + \tfrac{1}{2}O_2(g) \rightarrow SO_3(g)$$

In the presence of nitrogen monoxide, however, the reaction speeds up dramatically:

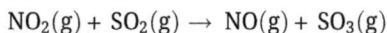

$$NO(g) + \tfrac{1}{2}O_2(g) \rightarrow NO_2(g)$$

$$NO_2(g) + SO_2(g) \rightarrow NO(g) + SO_3(g)$$

Note that NO_2 acts as an intermediate (formed and then consumed) and NO as a catalyst (used and then regenerated).

A *heterogeneous catalyst* speeds up a reaction in a different phase. Most often solid catalysts, interacting with gaseous or liquid reactants, have enormous surface areas. A very important organic example of a heterogeneous catalysis is the hydrogenation, the addition of H_2 to C=C bonds to form C–C bonds. The petroleum, plastics, and food industries employ this process. The simplest hydrogenation converts ethylene (ethene) to ethane:

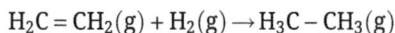

$$H_2C = CH_2(g) + H_2(g) \rightarrow H_3C - CH_3(g)$$

In the absence of a catalyst, the reaction is very slow. But, at high H_2 pressure (high concentration of H_2) and in the presence of finely divided Ni, Pd, Pt metal, it is rapid even at ordinary temperatures. In the rate-determining step, the adsorbed H_2 splits into two H atoms that become weakly bonded to the catalyst's surface (catM–H atoms bound to metal surface):

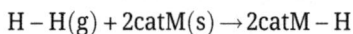

$$H - H(g) + 2catM(s) \rightarrow 2catM - H$$

Then, C_2H_4 adsorbs and reacts with the H atoms to form C_2H_6. Thus, the catalyst acts by lowering the activation energy of the slow step as part of a different mechanism.

Enzyme catalysis is also classified as *micro-heterogeneous catalysis*, because biocatalysts are basically protein substances with huge molecules. With the help of enzymes, biochemical reactions in the body can be accelerated up to 1,000,000 times. Biocatalysts-enzymes affect the rate of strictly defined reactions, i.e. they show very high selectivity. They accelerate reactions at room temperature. They function effectively at optimum temperatures from 25 to 35 °C. At elevated temperatures, they lose their activity, as denaturation of proteins occurs. Optimum pH conditions are also needed for enzyme catalysis. Normally, pH from 7.2 to 7.4 is required for the enzymatic reactions. All biochemical reactions occurring in living organisms depend on catalysts.

Chapter 9
Chemical equilibrium

9.1 Features of chemical equilibrium

Reactions can occur in only one direction (unidirectional), or in two directions (bidirectional). A bidirectional reaction can proceed in both the forward and reverse directions, as recognized by the presence of two arrows instead of one. When the reaction proceeds at the same rate in both the forward and the reverse directions, it is said to be at equilibrium.

If, in the progression of the reaction, the concentrations of the starting substances are constantly decreasing, and those of the products are constantly increasing, then the reaction is defined as *irreversible*. The time for an irreversible reaction, that is, for its "completeness," is different for different reactions and is determined primarily by the nature of the reacting substances, as well as by the conditions – temperature, concentration, pressure, catalysts, etc. Therefore, practically irreversible reactions are defined as those for which no conditions can be found for the interaction of the products in the opposite direction until the starting substances are obtained. For example, $KClO_3$ decomposes when heated to potassium chloride KCl and oxygen O_2:

$$2KClO_3 \longrightarrow 2KCl + 3O_2$$

Practically irreversible under ordinary conditions are reactions with a pronounced exothermic effect, redox reactions, in which the potential difference between the oxidizer and the reducer is large, ionic reactions, in which the products are precipitate, gas or weak electrolyte, etc. For example, when mixing equal volumes and equimolar concentrations of solutions of HCl and NaOH, complete neutralization occurs and NaCl and water (weak electrolyte) are obtained. This reaction is irreversible, since mixing salt with water does not give back hydrochloric acid and sodium base due to the impossible hydrolysis process. Irreversible reactions are widely used in chemical analysis, where the object of determination must react quantitatively, that is, $\geq 99\%$. The introduction of the concept of irreversible process is related to the practical impossibility of realizing one of the directions of chemical interaction.

Most of the chemical reactions under the ordinary or special conditions proceed bidirectionally or *reversibly*. A deeper study of chemical processes and the conditions for their occurrence leads to the conclusion that reversible reactions are not the exception, but rather the rule.

In a chemical reaction the *chemical equilibrium* is defined as the state at which there is no further change in concentration of the reactants and the products. At equilibrium the rate of the forward reaction is equal to the rate of the backward reaction. The mixture of reactants and products in the equilibrium state is called an equilib-

https://doi.org/10.1515/9783111712246-009

rium mixture. Based on the extent to which the reactions proceed to reach the state of equilibrium, these may be classified into three groups: (i) the reactions which proceed almost to completion, and the concentrations of the other reactants are negligible; (ii) the reactions in which most of the reactants remain unchanged, that is, only small amounts of the products are formed; and (iii) the reactions in which the concentrations of both the reactants and products are comparable when the system is in equilibrium.

An example of equilibrium in a chemical process in homogeneous system, where the reactants are chemically transformed into products, is the following:

$$N_2O_4(g) \leftrightharpoons 2NO_2(g)$$

The dynamic nature of chemical equilibrium can be demonstrated in the synthesis of ammonia by Haber's process:

$$N_2 + 3H_2 \leftrightharpoons 2NH_3$$

Haber started his experiment with the known amounts of N_2 and H_2 at high temperature and pressure. At regular intervals of time, he determined the amount of ammonia present. He also found out the concentrations of unreacted N_2 and H_2. After a certain time, he found that the composition of mixture remains the same even though some of the reactants are still present. After some time, the two reactions occurred at the same rate and the system reached a state of equilibrium.

Concentration-time curves: a plot of concentration versus time shows that as the concentration of the reactant decreases and product concentration increases, eventually the slopes (rates) taper off and plateau, becoming constant, as equilibrium is achieved. A plot of rate versus time shows the rates (rate laws) becoming equal when reaching equilibrium (Figure 9.1).

The molar concentrations of the reactants and products are not the same as at the beginning of the reaction. The concentration of the reactant will decrease as the concentration of the product decreases. As the two rates become equal, the molar concentrations of the reactants and products plateaus and become constant (not the same, but constant).

Law of chemical equilibrium: At a constant temperature, the rate of a chemical reaction is directly proportional to the product of the molar concentrations of the reactants each raised to a power equal to the corresponding stoichiometric coefficients as represented by the balanced chemical equation. We refer back to the law of mass action to write k expressions: the ratio of the products to the reactants held to power of their respective coefficients (since k relates to rate). For the reversible reaction:

$$n_1A_1 + n_2A_2 + n_3A_3 + \cdots \leftrightharpoons m_1B_1 + m_2B_2 + m_3B_3 + \cdots$$

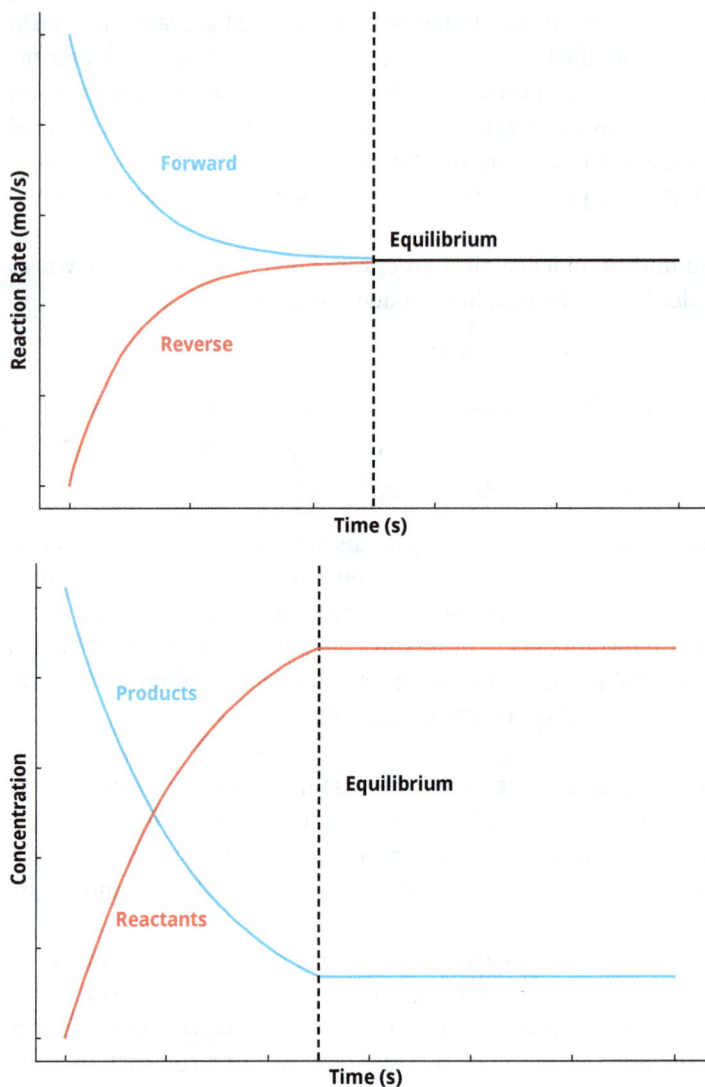

Figure 9.1: Reaction rate-time and concentration-time curves.

the equilibrium constant is

$$K_c = \frac{c^{m_1}{}_{B_1} \cdot c^{m_2}{}_{B_2} \cdot c^{m_3}{}_{B_3} \cdots}{c^{n_1}{}_{A_1} \cdot c^{n_2}{}_{A_2} \cdot c^{n_3}{}_{A_3} \cdots} \quad \text{and} \quad K_p = \frac{P^{m_1}{}_{B_1} \cdot P^{m_2}{}_{B_2} \cdot P^{m_2}{}_{B_3} \cdots}{P^{n_1}{}_{A_1} \cdot P^{n_2}{}_{A_2} \cdot P^{n_3}{}_{A_3} \cdots} = \text{const}$$

Rate constants are not based on experimental data and analysis of how concentration of only reactants affects rate, but rather, depend of the concentrations of both the reactants and products as well as their coefficients, which serve as the exponents in the equation. Recall that the exponents in unidirectional reactions (first, second, zero orders of the reaction) were also based on experimental data.

Only reactants and products in aqueous (aq) and gases (g) physical states are found in the k expression; solids (s) are pure substances and are not found in k expression.

The *equilibrium constant K* has no units in bidirectional equilibrium reactions. Values that are used to calculate K can be in either in molarity (for aqueous solutions) or pressure units (for gases). If a reaction occurs in more than one step, then the equilibrium constant for the overall reaction in the mechanism is the product of the equilibrium constants for each of the individual steps in the reaction (i.e., $K_1 . K_2 . K_3 = K$). K values are based on a given temperature. Therefore, the temperature value is usually provided.

The equilibrium constant depends on the nature of the reagents and on the temperature change. In endothermic processes, the increase in temperature is associated with an increase in the equilibrium constant, and in exothermic processes with its reduction The equilibrium constant does not depend on the concentrations of the reagents, on the pressure, and on the presence of catalysts.

If an external stress is applied to a system at equilibrium, the system will respond in such a way as to compensate for or reverse that disturbance in order to restore equilibrium. Equilibrium will "shift" to right or left, depending on the influencing factor, to restore equilibrium.

The behavior of a system at chemical equilibrium, when subjected to a change in concentration, temperature and pressure, illustrates the *principle of Le Châtelier*. Le Chatelier's principle states that if a stress is brought to bear on a system in equilibrium, a reaction will occur in the direction which tends to relieve the stress. Some predictions based on Le Chatelier's principle are the following:

- Increasing the concentration of one component of an equilibrium leads to a reaction that consumes that component.
- Increasing the volume of a gas equilibrium or adding solvent to a solution equilibrium leads to a reaction that produces more molecules.
- Increasing the temperature of an equilibrium system causes a reaction that absorbs the heat.
- A catalyst is a substance that increases the rate of a reaction. When a reaction is reversible, the catalyst increases both the forward and reverse reaction, leaving the equilibrium constant unaffected.

9.2 Influence of conditions on chemical equilibrium

Effect of concentration change: when the concentration of any of the reactants or products in a reaction at equilibrium is changed, the composition of the equilibrium changes so as to minimize the effect. If a reactant is added, the shift is right (to use up the reactants added). If the reactant is subtracted, the shift is left (to produce more reactants). If a product is added, the shift is left (to use up excess products). If the product is subtracted, the shift is right (to produce more products).

Consider the reversible equilibrium reaction of $FeCl_3$ and NH_4SCN:

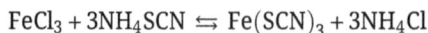

$$FeCl_3 + 3NH_4SCN \leftrightarrows Fe(SCN)_3 + 3NH_4Cl$$

The product $Fe(SCN)_3$ is a compound with intensively red color. The equilibrium constant for the reaction can be expressed:

$$K_c = \frac{[Fe(SCN)_3][NH_4Cl]^3}{[FeCl_3][NH_4SCN]^3}$$

The K_c is constant at a constant temperature. Accordingly, if more $FeCl_3$ or NH_4SCN is added and so their concentrations increased, K_c will decrease. To remain constant at the same temperature the concentrations of $Fe(SCN)_3$ and NH_4Cl should increase. Thus, the equilibrium will shift to the right direction and the color of the solution will become intensively red. If more $Fe(SCN)_3$ or NH_4Cl is added, increasing their concentrations, thereby K_c will increase. To remain constant at the same temperature the concentrations of $FeCl_3$ and NH_4SCN should decrease. Thus, the equilibrium will shift to the left direction and the intensity of the solution color will reduce.

Effect of pressure change: If the number of moles of gaseous reactants and products are equal, there is no effect of pressure. When the total number of moles of gaseous reactants and total number of moles of gaseous products are different, the equilibrium will shift. On increasing pressure, the total number of moles per unit volume increases, thus the equilibrium will shift in a direction in which the number of moles per unit volume will be less. If the total number of moles of the products is more than the total number of moles of the reactants, the low pressure will favor the forward reaction. If the total number of moles of the reactants is more than the total number of moles of the products, the high pressure is favorable for the forward reaction. Recall that the pressure and volume are inversely proportional (as one increases, the other decreases) and thus we can judge two factors at once. If the pressure increases or the volume decreases, the equilibrium will shift to side with fewest moles of gas. If the pressure decreases or the volume increases, the equilibrium will shift to side with most moles of gas. Consider the gas phase equilibrium:

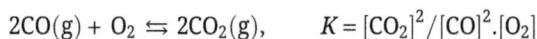

$$2CO(g) + O_2 \leftrightarrows 2CO_2(g), \qquad K = [CO_2]^2/[CO]^2.[O_2]$$

If we add more CO and so increase P_{CO}, the Le Chatelier's principle states that the system will respond in such a way as to decrease P_{CO}. It does this by reacting to consume CO, producing more CO_2 and incidentally consuming O_2.

Another example of gas phase equilibrium is the following:

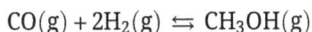

$$CO(g) + 2H_2(g) \leftrightarrows CH_3OH(g)$$

If we add more CO and so increase P_{CO}, the Le Chatelier's principle states that the system will respond in such a way as to decrease P_{CO}. It does this by reacting to consume CO, producing more CH_3OH and incidentally consuming H_2. A similar argument applies to solution-phase equilibrium: increasing the concentration of one species will shift the equilibrium to consume that species.

Effect of inert gas addition: If the volume is kept constant there is no effect on equilibrium after the addition of an inert gas. This is because the addition of an inert gas at constant volume does not change the partial pressure or the molar concentrations. The reaction quotient changes only if the added gas is involved in the reaction.

Effect of temperature change: When the temperature of the system is changed (increased or decreased), the equilibrium shifts in opposite direction in order to neutralize the effect of the change. In exothermic reactions, the low temperature favors the forward reaction but practically the very low temperature slows down the reaction and thus a catalyst has to be used. In case of endothermic reactions, the increase in the temperature will shift the equilibrium in the direction of the endothermic reaction. The change in temperature depends on if the reaction is exothermic or endothermic. Equilibrium is shifted and K is influenced. If K increases, the rate of the reaction increases, so the temperature affects how quickly equilibrium is reached. Exothermic reactions respond best to lower temperatures while endothermic reactions prefer higher temperatures. Note that energy (heat) is given in chemical equations in units of kJ and is written as a reactant or a product.

When the T of a system at equilibrium is increased, a reaction occurs in the direction that absorbs the heat. Consider the following equilibrium:

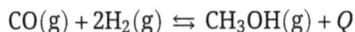

$$CO(g) + 2H_2(g) \leftrightarrows CH_3OH(g) + Q$$

The equilibrium is exothermic when the reaction proceeds to the right. Thus, if the T is increased, we expect the equilibrium to shift such as to absorb heat. Shifting to the right would liberate more heat, thus we expect a shift toward CO and H_2.

Another example of equilibrium is the following:

$$CaCO_3(s) \leftrightarrows CaO(s) + CO_2(g) - Q$$

The straight reaction is endothermic and is favored by the increase in temperature.

Effect of a catalyst: The catalyst has no effect on the equilibrium composition of the reaction mixture. Thus, the catalyst increases the speed of both the forward and backward reactions to the same extent in a reversible reaction. The catalyst does not alter K. It does not shift equilibrium; it only helps to reach equilibrium faster by decreasing the activation energy (E_a) for both forward and reverse reactions.

Heterogeneous equilibrium is an equilibrium system in which reactants and products exist in two or more states of matter. An example of a heterogeneous equilibrium, where the reactants and the products are in different physical states, is

$$H_2O(l) \leftrightarrows H_2O(g)$$

A feature of a heterogeneous chemical equilibrium is that solids of the straight and reverse reactions do not participate in the equilibrium constant equation. Their concentrations or pressures are taken as a unit and do not enter into the expression of K_c, for instance:

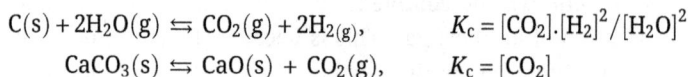

$$C(s) + 2H_2O(g) \leftrightarrows CO_2(g) + 2H_{2(g)}, \qquad K_c = [CO_2].[H_2]^2/[H_2O]^2$$
$$CaCO_3(s) \leftrightarrows CaO(s) + CO_2(g), \qquad K_c = [CO_2]$$

Chapter 10
Disperse systems

10.1 General characteristics of disperse systems: classification

Systems, which consist of two phases, one of which is scattered or dispersed in other are called *disperse systems*. The phase, which is scattered, is a disperse phase (discontinuous phase). The phase in which scattering/dispersion is done is called disperse medium or continuous phase.

There are two types of dispersed systems *by state:* homogeneous (single-phase mixtures of components) and heterogeneous (with different phases).

Dispersion systems are mainly classified *by the degree of dispersion* or by the size of the particles of the dispersed phase (Table 10.1).

Table 10.1: Classification of dispersions by size of individual units in dispersed phase.

Dispersions	Diameter of the particles (m)	Characterization
Coarse	$10^{-7}–10^{-5}$	S/L – suspension (dispersion of solid particles in a liquid)
		L/L – emulsion (dispersion of liquid drops in another liquid)
		Unstable systems
Colloid	$10^{-8}–10^{-7}$	Relatively stable systems
		Making colloid is an energy-dependent process
Molecular	$<10^{-9}$	Molecular-dispersion or ion-dispersion
		Stable systems

The *true solution* is homogeneous. The solute particles in true solutions have diameters lesser than 1 nm being of molecular dimensions. These do not settle down when the solution is left standing. The particles are invisible and cannot be separated through filter paper, e.g., sodium chloride in water, sugar in water. *Colloid systems* (colloids) are dispersion systems, mostly bicomponent, looking as physically uniform, but in fact they are not molecularly dispersed. Colloid consists of two separate phases: dispersed phase (internal phase) and continuous phase (dispersion medium, solvent). The size of dispersed phase is between 1 and 100 nm, which is the range between molecular size (solutions) and mechanical (suspensions). Colloid systems may be solid, liquid, or gaseous. *Suspensions* are dispersion systems with two distinct phases. The particles bigger than 1,000 nm are visible to the naked eye. Suspensions are heterogeneous systems that are not stable and can stay only for a limited period. The particles have a tendency to settle down under the influence of gravity. Their particles cannot pass through ordinary filter paper and membranes, e.g., sand in water, oil in water.

https://doi.org/10.1515/9783111712246-010

Other signs by which dispersed systems are divided into groups are the nature and physical state of the dispersed phase and dispersed medium. With this classification, nine types of disperse systems are possible, which are presented in Table 10.2. The most important in chemistry are dispersed systems in which the medium is a liquid.

Table 10.2: Types of dispersed systems depending on the basic three states/phases.

Dispersed system	Dispersed phase	Dispersion medium	Examples
Gas	Gas	Gas	Air
	Liquid	Gas	Cloud
	Solid	Gas	Smoke
Liquid	Gas	Liquid	Foam
	Liquid	Liquid	Emulsion – milk
	Solid	Liquid	Suspension
Solid	Gas	Solid	Solid foam
Coarse	Liquid	Solid	Solid emulsion
	Solid	Solid	Suspension – minerals, alloys

In emulsions the phases are immiscible, e.g., oil in water (O/W): dispersed phase is nonpolar or water in oil (W/O): dispersed phase is polar.

10.2 Solutions: mechanism of dissolution

Solvation, called dissolution, represents an attraction and association of molecules of a solvent with molecules/ions of a solute. Molecule/ion of the solute surrounded by solvent is called a solvation complex. Solvation depends on hydrogen bonding and van der Waals forces. Solvation of a solute by water is called hydration. Dissolution is distinct from the solubility. Solvation is a kinetic process and is quantified by its rate. The typical unit for dissolution rate is mol/s.

Solubility quantifies the dynamic equilibrium state achieved when the rate of dissolution equals the rate of precipitation. The units for solubility express a concentration: mass per volume (mg/mL), molarity (mol/L). Sugar or salt dissolved in water forms stable mixtures of two components – a solution. The majority component is called a *solvent* while the minority component – a *solute*. There is really no fundamental distinction between them.

Classification of solutions: Solutions are *homogeneous* (single-phase) mixtures of components. Solutions can exist as gases and solids as well. Solid solutions are very common (natural minerals and metallic alloys) (Table 10.3).

Table 10.3: Classification of solutions.

Depending on the state	Depending on the solute quantity
Liquid	*Saturated*
Solvent and solute – liquid	Solution with maximum possible amount of solute
Solvent – liquid; solute–gas	
Solvent – liquid; solute–solid	*Unsaturated*
	Solution contains less than the maximum amount of solute
Gases	
Gaseous mixtures (air)	
	Supersaturated
Solid solutions	Solution contains more dissolved solute than it would contain at lower
Natural minerals and many	temperature – unstable
alloys	

When the solid is added to the solvent, they initially form *unsaturated solution*. When maximum possible amount of the solute has dissolved, solution becomes *saturated*. A *supersaturated solution* can be formed from saturated by filtering off/lowering the *T*. When crystal is added to supersaturated solution, the solute particles form a precipitate.

All substances have at least a slight tendency to dissolve in each other. This raises two important and related questions: why do solutions tend to form? What factors limit their mutual solubility? According to the early stages of chemistry, there is no difference between the chemical processes and dissolution. There are two opposite conceptions about the nature of the solutions: *physical view* (Van't Hoff and Arrhenius) according to which the dissolution is a physical process and *chemical view* (Mendeleev) according to which the dissolution is a chemical interaction between the particles of the substance and the solvent, forming compounds, solvates. Till today these two conceptions exist. A relatively correct position can be formulated like this: dissolution is a complex physical-chemical process. The solutions are mixtures of two or more substances, but they are also the product of their interaction. Solutions occupy intermediate position between mixtures and chemical compounds. The similarities with chemical compounds are uniform composition, thermal effects, and modifying the volume. The similarities with mixtures are composition changeability in some concentration interval, *T*, and *P*, as well as their easy separation of component parts. Chemical compounds have strictly defined composition, but in hydrates the quantity is not strictly defined, for instance $H_2SO_4 \cdot xH_2O$, $x = 1$, 2, or 4 depending on concentration and *T*. Sometimes the bonds in hydrates are so hard that water enters in the composition, for instance, $CaSO_4 \cdot 2H_2O$ – gypsum, $Na_2B_4O_7 \cdot 10H_2O$ – borax. The formation of hydrates changes the properties of the solute. Waterless $CuSO_4$ is colorless; meanwhile the solution of this salt has a blue color.

Solubility is the maximum amount of the solute that can dissolve in a solvent at a given temperature and pressure. It can be expressed by: mass of solute per volume (g/L); mass of solute per mass of solvent (g/g); moles of solute per volume (mol/L). There is a limit to how much solute can dissolve in a given quantity of solvent. The factors affecting solubility are energetic factors, temperature and pressure – for gases. The solubility of most solids increases with increasing the temperature. The energy effect of dissolution is expressed by the deconstruction of the crystal structure and the distribution of particles in the solvent. To overcome the forces of attraction energy is needed, the processes are endothermic. Dissolution is a chemical interaction (donor-acceptor, H bonds), the processes are exothermic. Upon dissolution it shall be carried out endo and exothermic chemical processes. The total energy effect is the sum of energy effects of all processes.

10.3 Methods of expressing the concentration of solutions

Concentration is a quantity of solute used to prepare a given amount of solution. Different ways of expressing concentration are in use. The choice is usually a matter of convenience in a particular application. The concentration can be expressed by: "parts-per" concentration; weight/volume and volume/volume basis; molarity: mole/volume basis; mole fraction: mole/mole basis; molality: mole/weight basis.

Weight-percent concentration (ω_B, %): "Cent" is the Latin-derived prefix relating to the number 100. "Parts per 100" is known as "percent." It can be calculated by

$$\omega_B = m_B/m$$

where m_B is the mass of the solute and m is the mass of the solution. Solution, containing 5 g of NaCl and 95 g of H_2O, is a 5% solution of NaCl. ω_B is expressed by [g_B/g], [kg_B/kg]. It does not depend on T and P. ω_B [%] denotes the mass of the solute m_B in 100 parts of the solution: ω_B, % = $(m_B/m) \times 100$. There are the signs ‰ (permille) and ppm (part per million): ω_B, ppm = $(m_B/m) \times 10^6$.

Volume-percent concentration (φ_B) means the volume of solute (V_B) in a given volume of solution (V). The volume V_i of the component i is the sum of the volumes of all the components:

$$\varphi_B = V_B/V \quad \text{or} \quad \varphi_i = V_i/\Sigma V$$

A solution, 5% by volume means 5 mL of the solute dissolved in 95 mL of the solvent. This concentration is mostly used in commercial and industrial applications.

Mole fraction: mole/mole basis (χ_B): This is the most fundamental measure, since it makes no assumptions about the volumes. The moles of the substance (n_B) are in a mixture with the moles of the solvent (n_A). The mole fraction of the substance (χ_B) and of the solvent (χ_A) are defined as

$$\chi_B = n_B/(n_A + n_B) \quad \text{and} \quad \chi_A = n_A/(n_A + n_B)$$

where n_B is the number of moles of substance B, n_A is the number of moles of the solvent A. The sum of the mole fractions is $\chi_B + \chi_A = 1$. Mole fractions run from zero (substance not present) to unity (the pure substance).

Molarity: mole/volume basis (C_B): Molar concentration (molarity) is the number of moles of the solute (n_B) per 1 L of the solution:

$$C_B = n_B/V = m_B/M.V$$

where m_B is the mass of solute (g); M is the molar mass of solute (g/mol); V is the volume of the solution (L). The measurements are mol/L, mmol/L, etc. For example, 0.1 M = 0.1 [mol/L]. Although molar concentration is widely employed, it suffers from one serious defect: since volumes are T-dependent (substances expand on heating), so are molarities, for instance 0.1 M solution at 0 °C will have another concentration at 50 °C. Molarity is not the preferred measure in applications where physical properties of the solutions and the effect of temperature on these properties is of importance.

Molality: mole/weight basis (C_m): Molality is the number of moles of solute (n_B) per 1 kg of the solvent. One molal solution contains 1 mol of solute per 1 kg of solvent. It can be expressed as follows:

$$C_m = n_B/m = m_B/M.m$$

where m_B is the mass of solute (g), M is the molar mass of solute (g/mol), and m is the mass of solvent (kg). Molality is a hybrid concentration unit, retaining the convenience of mole measure for the solute, but expressing it in relation to a T-independent mass rather than a volume. Molality, like mole fraction, is used in applications dealing with physical properties. The measurements of C_m are mol/kg, mmol/g, kmol/kg, etc.

Conversion between concentration measures. Anyone doing practical chemistry must be able to convert one concentration into another. Any conversion involving molarity requires knowledge of the density of the solution. Molar concentration (molarity) and weight-percent concentration are connected:

$$C_B = 10.\omega_B.\rho/M \quad \text{and} \quad \omega_B = C_B.M/10.\rho$$

where ω_B is weight-percent concentration, %; $\rho = m/1{,}000$; V is the density of the solution, g/mL.

10.4 Colligative properties

Many important physical properties of solutions depend on solvent concentration: lowering of vapor pressure; depression of freezing point; elevation of boiling point; osmotic pressure. Because they are tied together (Latin, coligare), they are referred to as the *colligative*. These properties depend on the concentration of solute molecule or ions in solution but not on the chemical identity of the solute. For example, addition of ethylene glycol to water lowers the freezing point of water below 0 °C. The magnitude of freezing point lowering is directly proportional to the number of solute molecules added to a quantity of solvent. If 0.01 mole of ethylene glycol is added to 1 kg of water, the freezing point is lowered to −0.019 °C while on adding 0.020 mol of ethylene glycol to 1 kg of water, the freezing point is lowered to −0.038 °C.

Vapor pressure of solutions: Raoult's law: Evaporation is the tendency of molecules to escape into the gas phase from a solid or liquid. The vapor pressure of the solution reduces as the concentration of solute increases and that of solvent decreases. The partial pressure of water P_w in the vapor space will initially be zero. Gradually, P_w will rise as molecules escape from the substance and enter the vapor phase. P_w continues to rise. Finally, a balance is reached when P_w stabilizes at a fixed value P_{vap} known as the *equilibrium vapor pressure* of the liquid or solid. The vapor pressure is a direct measure of the escaping tendency of molecules from a condensed state. When solute and solvent are volatile, both substances will contribute to vapor pressure. If the solution is ideal, the relative contributions will be proportional to the mole fractions.

Ideal solutions: An ideal solution of substances A and B is one in which both substances follow Raoult's law for all values of mole fractions. Such solution occurs when the substances (solvent and solute) are chemically similar so that the intermolecular forces between A and B molecules are similar to those between two A molecules or between two B molecules and the interactions are all the same. For example, solution of benzene C_6H_6 and toluene $C_6H_5CH_3$ are ideal. Ethylene bromide and ethylene chloride, *n*-hexane and *n*-heptane, *n*-butyl chloride and *n*-butyl bromide, carbon tetrachloride and silicon tetrachloride are few other examples of ideal solutions. Note the similarity in their structural formula.

If two volatile liquids form an ideal solution, each will make a contribution to the overall vapor pressure and χ is directly proportional to the vapor pressure (Figure 10.1). The total vapor pressure over an ideal solution equals the sum of the partial vapor pressures, each of which is given by Raoult's law. The linear relation between the mole fraction and its vapor pressure is expressed by *Raoult's law* (1886):

$$P_{total} = P_A + P_B = \chi_A P_A^0 + \chi_B P_B^0$$

where P_{total} is the total vapor pressure of a solution containing A and B; P_A and P_B are the partial pressures resulting from molecules A and B in the vapor; χ_A and χ_B are the mole fractions of A and B; P_A^0 and P_B^0 are the vapor pressures of pure A and pure B. If the solute is nonvolatile, it does not contribute to the vapor pressure over the solution.

The vapor pressure of the solvent is

$$P_A = \chi_A.P_A^0 = n_A.P_A^0/(n_A + n_B)$$

where $\chi_A = n_A/(n_A + n_B)$ is the mole fraction of the solvent (χ_A).

The above equation is linear and is valid also for the component B:

$$P_B = \chi_B.P_B^0 = n_B.P_B^0/(n_A + n_B)$$

where $\chi_B = n_B/(n_A + n_B)$ is the mole fraction of the solute.

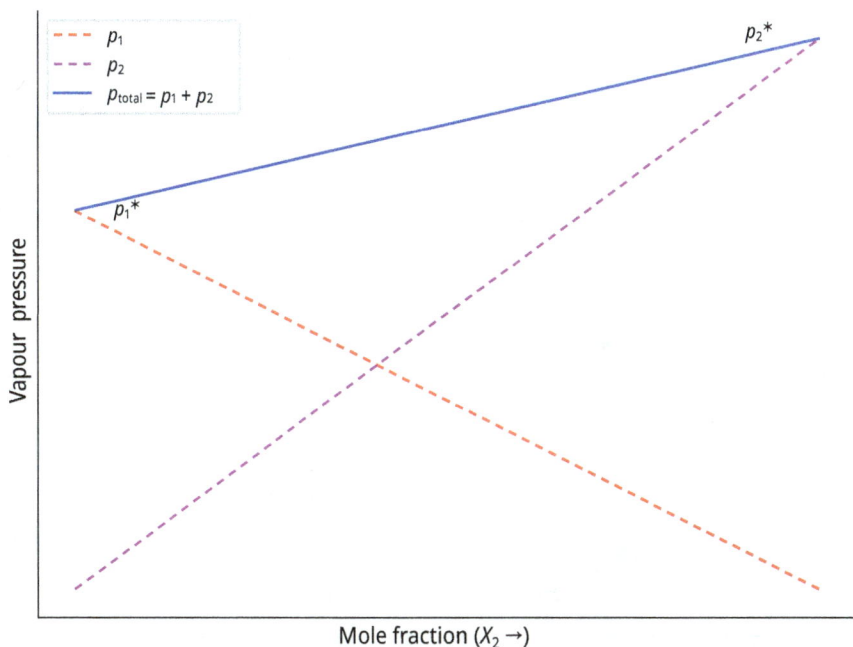

Figure 10.1: Dependence of total and partial vapor pressures on the composition for an ideal binary solution.

The $P = f(\chi)$ diagram (p-composition) at $T =$ const represents straight lines for the ideal solutions joining 0 with P of the pure components. The total P of the solution is equal to the sum of the partial P of the two components (the line joining the points $p_A{}^0$ and $p_B{}^0$). The total P is higher than that of hard volatile component and lower than P of easy volatile. Such solutions are called *zeotropic* and found relatively rare. Usually, the total and partial vapor pressures deviate more or less than the linear dependency both in positive and in negative direction for the real solutions.

   Real or nonideal solutions: Those solutions which do not obey Raoult's law over entire range of composition and deviate from ideal behavior, are real or nonideal solution. They are divided into two types: (a) solutions showing positive deviations from ideal behavior for those type of solutions, $P_A > \chi_A P_A{}^0$ and $P_B > \chi_B P_B{}^0$ and (b) solutions showing large negative deviations from ideal behavior and the vapor pressure of each component is considerably less than that predicted by Raoult's law, for these types of solutions, $P_A < \chi_A P_A{}^0$ and $P_B < \chi_B P_B{}^0$. The nature of the deviation depends on the interaction between the molecules. If prevailing are the forces of repulsion, vapor pressure is higher than the calculated. If prevailing are the forces of attraction, vapor pressure is lower than the calculated. The dependence of total and partial vapor pressures on the composition for a nonideal solution with positive deviations from Raoult's law and for a nonideal solution with negative deviations from Raoult's law is shown in Figures 10.2 and 10.3.

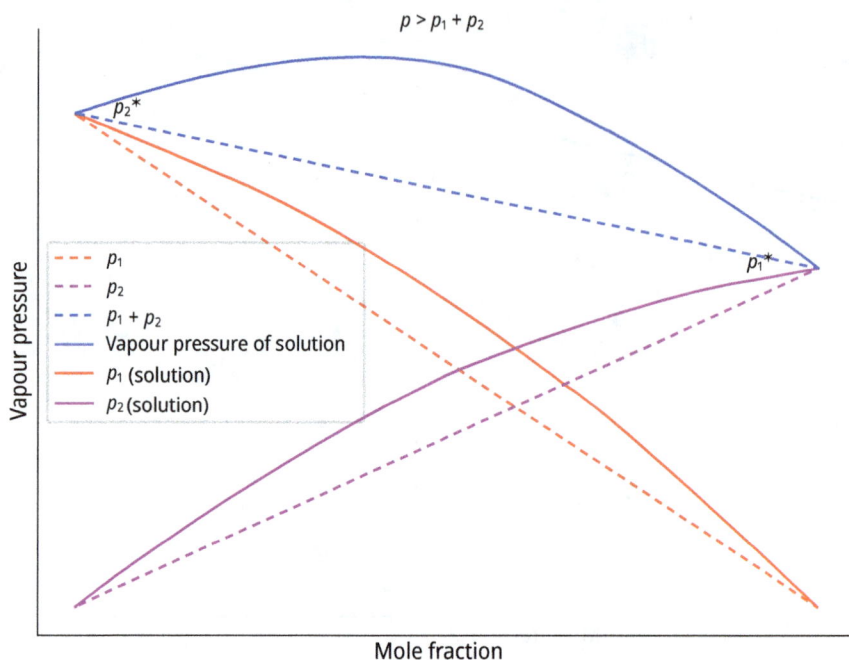

Figure 10.2: Dependence of total and partial vapor pressures on the composition for a nonideal solution with positive deviations from Raoult's law.

Positive deviations occur when the interaction between unlike molecules is weaker than the interaction between like molecules. Solute-solvent interactions are weaker than interactions among pure liquids and more energy is required. Molecules have a higher tendency to escape than expected. Solutions of polar ethanol and nonpolar hexane, water-benzene, water-methanol, acetaldehyde-carbon disulfide, water-propyl alcohol, ethyl alcohol-chloroform, and cyclohexane-carbon tetrachloride show positive deviation with attractive forces between A–A or B–B > A–B. Such solutions are formed hardly. The solutions of this type are characterized by positive enthalpy of mixing ΔH_{mix} and positive volume of mixing ΔV_{mix}.

Negative deviations occur when the interaction between unlike molecules is stronger than the interaction between like molecules. The solutions of this type are characterized by negative enthalpy of mixing and negative volume of mixing. The solute and solvent are very much similar (solvent has affinity for the solute forming H-bonding) thus leading to strong interactions between the solute and solvent. The observed vapor pressure is lower than the value predicted by Raoult's law. The solutions of HNO_3 in water, pyridine–acetic acid, acetone–chloroform show negative deviation where the attractive forces between A–A or B–B < A–B.

Azeotropic mixture: Some liquids on mixed from azeotropes which are binary mixtures having the same composition liquid and vapor phase and boil at a constant

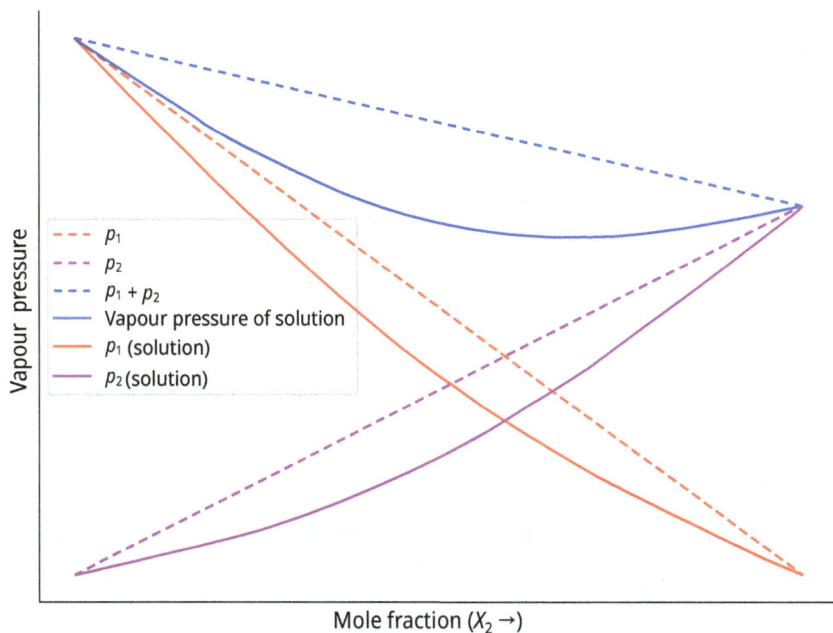

Figure 10.3: Dependence of total and partial vapor pressures on the composition for a nonideal solution with negative deviations from Raoult's law.

temperature. In such cases, it is not possible to separate the components by fractional distillation. There are two types of azeotropes called as minimum boiling azeotrope and maximum boiling azeotrope, respectively. Solutions of ethanol and water show such a large deviation from Raoult's law that there is a maximum in the vapor pressure curve and hence a minimum in the boiling point. Ethanol-water mixtures (obtained by fermentation of sugars) are rich in water. Fractional distillation is able to concentrate the alcohol to at best, the azeotropic composition of approximately 95% by volume of ethanol. Once this composition has been achieved, the liquid and vapor have the composition, and no additional fractional occurs. Other methods of separation have to be used for preparing 100% C_2H_5OH. There are also solutions that show a large negative deviation from ideality and, therefore, have a minimum in their vapor pressure curves. This leads to a maximum on the boiling point diagram. HNO_3 and H_2O form examples of this class of the azeotrope. This azeotrope has the approximate composition, 68% nitric acid and 32% water by mass, with a boiling point of 393 K.

Boiling diagrams (P = const): A typical boiling diagram of a binary volatile mixture whose components A and B are *infinitely soluble* in each other and do not form an azeotropic mixture is presented in Figure 10.4. The lower curve gives the boiling point of the liquid mixture as a function of composition. Below the lowest curve none of the sample has evaporated and the system is entirely a liquid. Above the upper curve the

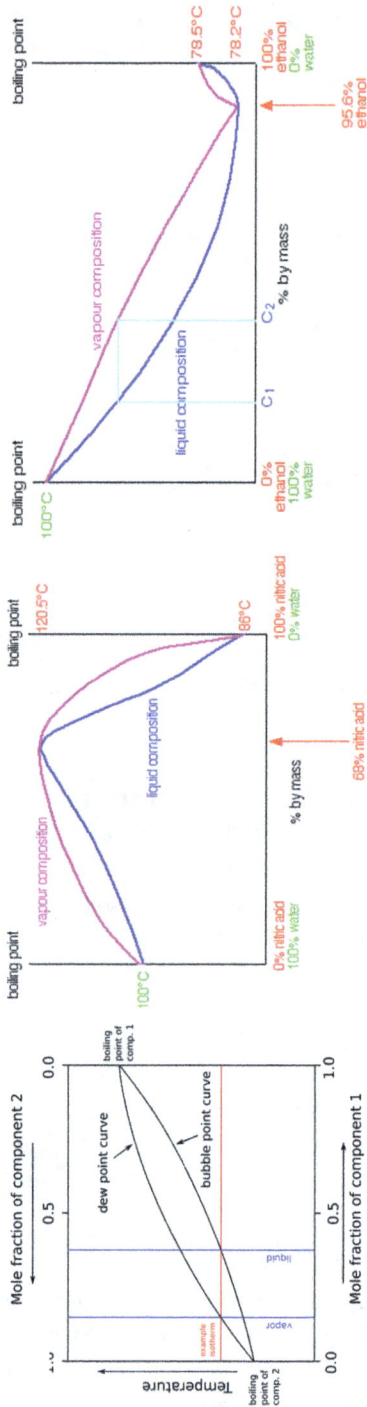

Figure 10.4: Boiling diagrams of an ideal zeotropic solution and of azeotropic solutions with unlimited solubility of A and B with negative and positive deviation from Raoult's law.

sample has fully evaporated and the system is entirely a gas. Between the two curves the system is composed of two phases in equilibrium. Two curves separate the diagram in three areas: vapor area; area of liquid; area of vapor and liquid.

This phase diagram for ideal zeotropic solution shows that fractional distillation can be applied. Example of zeotropic mixture is the mixture of benzene and methylbenzene.

Boiling diagrams for nonideal solutions: There are two types: (a) with negative deviations – boiling points are higher than expected (high boiling azeotrope – HCl–H_2O); (b) with positive deviations – the solution would boil at a lower temperature than expected (low boiling azeotrope – EtOH–H_2O) (Figure 10.4). The extreme points are called azeotrope points (T and composition). Azeotrope composition lies above or below the boiling points of the pure substances depending upon the deviations from the Raoult's law (negative and positive deviation).

In case of a negative deviation from Raoult's law, the components A and B form azeotrope composition with high boiling temperature. Azeotrope mixture has the highest boiling temperature – harder volatile than the components. The first distillation separates the pure components and finally azeotrope composition.

In case of a positive deviation from Raoult's law, the components A and B form azeotrope composition with low boiling temperature – minimum. The first distillation separates the azeotrope composition, then pure components. In these mixtures it is impossible to reach a complete separation of the pure components.

The above diagrams of boiling of binary mixtures at constant external pressure provide useful information about the systems. They provide an answer to the questions: are the components of the mixture limited or indefinitely soluble in each other; do the components of the system form an azeotropic mixture and, if so, what is its composition at the given external pressure.

Nonvolatile solute: A nonvolatile solute is a substance that does not easily vaporize. When dissolved in a solvent, it does not contribute to the vapor pressure of the solution.

Relative lowering of vapor pressure: The addition of a nonvolatile solute to a solvent at a given temperature and pressure results in the lowering of the vapor pressure of the solvent (Figure 10.5). Partial pressure of the liquid solvent (A) determines the total pressure:

$$p_A = \chi_A \cdot P_A{}^0$$

where p_A is the pressure of the solution, which is always lower than that of the pure solvent $P_A{}^0$ ($\chi_A \leq 1$). From this equation, it is clear that the relative vapor pressure lowering is proportional to the mole fraction of the nonvolatile solute in the solution. It is independent of the nature of the solute. The relative vapor pressure lowering is, therefore a colligative property.

The normal boiling point is the T at which the liquid is in equilibrium with vapor at 1 atm. By constructing a horizontal line, interrupts of each curve indicate T_{bp} of

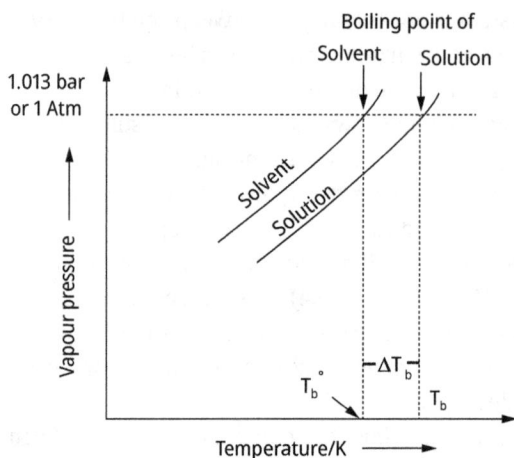

Figure 10.5: Lowering of the vapor pressure.

each solution. The solution T_{bp} at which the vapor pressure is 1 atm will be higher. The boiling point will be raised.

The *elevation of the boiling point* is directly proportional to the molality:

$$\Delta T_{bp} = E_{bp} \cdot C_m$$

The constant E_{bp} (*boiling point elevation constant*) is a property of the solvent and independent of the nature of the solutes. $E_{bpH2O} = 0.52\ °C$. Dilute water solutions of nonionic solutes ($C_m = 1$) boil at 100.52 °C. This law is accurate for dilute solutions. For more concentrated solutions, the boiling point is increased but not precisely in proportion to the solute molality.

The boiling point of the solution is higher than that of pure water (Figure 10.5). On the contrary, the freezing point of the solution is lower than that of the pure solvent (Figure 10.6). The freezing point is the T at which solid and liquid can all together coexist. The triple point (where solid, liquid, and vapor coexist at the same T) is shifted to lower temperature for the solution.

The *decrease in freezing point* is proportional to the molal concentration of solute. There is a linear relation between the freezing point depression and the molal concentration of the solute:

$$\Delta T_{fp} = E_{fp} \cdot C_m$$

where E_{fp} is the freezing-point constant of the solvent. K_{fp} is a property of the solvent, independent of the nature of the solutes. $E_{fpH2O} = -1.86\ °C$. Dilute water solutions of nonionic solutes ($c_m = 1$) crystalize at (–1.86 °C), as $T_{fpH2O} = 0\ °C$.

The method for determining the decrease in crystallization temperatures of solutions is called *cryoscopy*, and the method for determining the increase in boiling

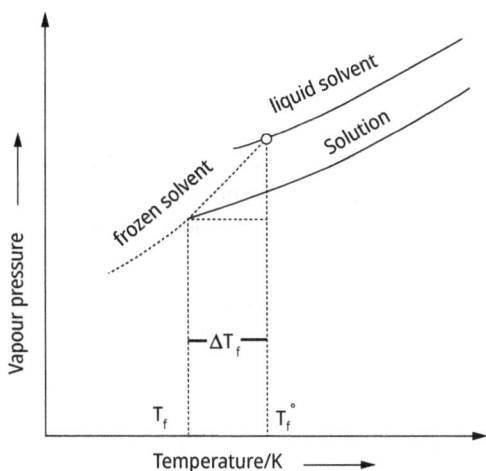

Figure 10.6: Lowering of the freezing point.

points of solutions is *called ebullioscopy*. Cryoscopy and ebullioscopy find a variety of applications and explain a number of natural phenomena. In chemical laboratories, for example, the decrease in T_{fr} and the increase in T_b in solutions make it possible to quickly and accurately determine the relative molecular weight of the solute.

Osmosis and osmotic pressure: There is a natural tendency of solutes in a solution to diffuse from a higher concentration to a lower concentration so as to bring about a uniform distribution throughout. Certain membranes allow solvent molecules to pass through them but not solute molecules, particularly not those of large molecular weight. Such a membrane is called semipermeable and might be an animal bladder, a vegetable tissues or a piece of cellophane.

Osmosis is the phenomenon of solvent flow thorough a semipermeable membrane to equalize the solute concentrations on both sides of the membrane.

Osmotic pressure is a colligative property of a solution equal to the pressure that, when applied to the solution just stops osmosis. It is denoted by π. The osmotic pressure π of a solution is related to the molar concentration of the solute:

$$\pi = c.R.T$$

where R is the gas constant, $R = 8.314$ J/mol.K and T is the absolute temperature, K. Van't Hoff equation for dilute solution is similar to ideal gas equation, $\pi V = nRT$, where π is the osmotic pressure.

The Van't Hoff equation is valid for dilute solutions of low molecular weight non-electrolytes. In electrolyte solutions, the experimentally measured osmotic pressure is greater than theoretically calculated. To avoid this difference, Van't Hof introduced the correction, namely: $\pi = i.C.R.T$, where i is called the isotonic (isoosmotic) coefficient, determined experimentally. In solutions of strong electrolytes, the experimentally measured osmotic pressure can be 2, 3, and 4 times greater than theoretically calculated, that is: $i = 2$ (NaCl → Na$^+$ + Cl$^-$ and HCl → H$^+$ + Cl$^-$); $i = 3$ (BaCl$_2$ → Ba^{2+} + 2Cl$^-$ and H$_2$SO$_4$ → 2H$^+$ + SO$_4^{2-}$); $i = 4$ (AlCl$_3$ → Al^{3+} + 3Cl$^-$).

Isotonic solutions: A pair of solutions having same osmotic pressure is called isotonic solution. When two solutions having the same osmotic pressure ($\pi_1 = \pi_2$) are put into communication with each other through a semi permeable membrane, there will be net transference of solvent from one solution to the other, but there is a dynamic equilibrium between two solutions. According to Van't Hoff equation, it is evident that isotonic solutions must have the same molar concentration, if both are at same temperature. Human blood is most often with an osmotic pressure of 7.7 atm and is isotonic of a 0.9% solution of sodium chloride in water (the so-called saline).

Chapter 11
Electrolyte solutions

11.1 Electrolytes and nonelectrolytes

Some liquids and solutions conduct electric current. Pure liquid water is a very poor conductor of electric current. The addition of certain solutes to water results in an aqueous solution that is a good electrical conductor. Yet there are some solutes that do not enhance the electrical conductivity of water and still others that render it only weakly conducting. If ions are present in an aqueous solution, the solution will conduct electricity. A substance that forms an aqueous solution that conducts electricity is called an *electrolyte*. A solution that conducts is assumed to contain ions: cations (positive) and anions (negative). A substance that forms an aqueous solution that does not conduct electricity is a *nonelectrolyte*. The most soluble molecular substances (e.g., sugar and alcohol) are nonelectrolytes. Characterizing a substance as a nonelectrolyte means that it does not form ions when it dissolves.

Substances which on dissolution in water are ionized almost completely to form no products other than positive and negative ions are *strong electrolytes*:

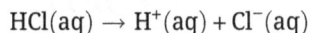

$$HCl(aq) \rightarrow H^+(aq) + Cl^-(aq)$$

Weak electrolytes are substances that dissolve in water, partially dissociate, and react with water to only a slight extent, producing low concentrations of ions:

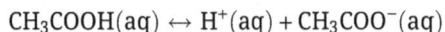

$$CH_3COOH(aq) \leftrightarrow H^+(aq) + CH_3COO^-(aq)$$

11.2 pH, acids, bases, and salts

11.2.1 Acids, bases, and salts

According to *Arrhenius theory*, acids are substances that dissociate in water to give hydrogen ions $H^+(aq)$. Bases are substances that produce $OH^-(aq)$ after dissociation in water. Strong acids completely dissociate in water, whereas weak acids dissociate partially. Strong bases dissociate to metal cations and hydroxide anions in water, whereas weak bases react only partially to produce hydroxide anions. Examples of Arrhenius acids (in water) are HCl, H_2SO_4, etc. Examples of Arrhenius bases (in water) are $NaOH$, NH_3, etc. The degree to which solute molecules are dissociated into ions is denoted by α, the *degree of dissociation*. For nonelectrolytes $\alpha = 0$, for weak electrolytes $\alpha = 0.01$–0.03, and for strong electrolyte solutions, especially at low concentrations, $\alpha = 1$. There are some limitations of the Arrhenius concept: (i) According to the Arrhenius concept, an acid gives H^+ ions in water, but the H^+ ion does not exist independently because of its

https://doi.org/10.1515/9783111712246-011

very small size ($\sim H^{-18}$ m radius) and intense electric field; (ii) It does not account for the basicity of substances like ammonia, which does not possess a hydroxide ion.

According to *Brønsted-Lowry theory*, an acid is a substance that is capable of donating a hydrogen ion H^+ and bases are substances capable of accepting a hydrogen ion H^+. In other words, acids are proton donors and bases are proton acceptors. The acid-base pair that differs only by one proton is called a conjugate acid-base pair. Let us consider the example of the ionization of HCl in water. Here, water acts as a base because it accepts the proton. Cl^- is a conjugate base of HCl, and HCl is the conjugate acid of base Cl^-. Similarly, H_2O is a conjugate base of an acid H_3O^+ and H_3O^+ is a conjugate acid of a base H_2O. Brønsted's acids and bases are generally the same acids and bases as in the Arrhenius model, but the model of Brønsted and Lowry is not restricted to aqueous solutions. Brønsted's model introduces the notion of conjugate acid-base pairs. It is logical that if something (acid) exists and may lose a proton, then the product of such a proton loss is by definition a base since it has the capability to add a proton. Likewise, any compound with a pair of electrons may behave as a Brønsted base. It is possible for the same compound to be able to behave as a Brønsted base and as a Brønsted acid. Usually, a compound is called an acid or base depending on the circumstances. Theoretically, any compound that has a hydrogen atom in it may behave as a Brønsted acid. Under the Brønsted-Lowry model, an acid-base reaction is always a reaction between an acid and a base, giving their conjugate base and acid:

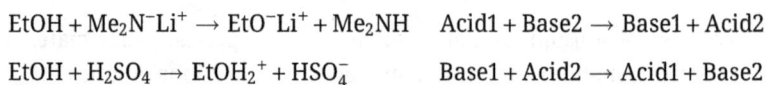

$$EtOH + Me_2N^-Li^+ \rightarrow EtO^-Li^+ + Me_2NH \quad Acid1 + Base2 \rightarrow Base1 + Acid2$$

$$EtOH + H_2SO_4 \rightarrow EtOH_2^+ + HSO_4^- \quad Base1 + Acid2 \rightarrow Acid1 + Base2$$

Generally, the reactions proceed to form weaker acids and bases.

According to *Lewis theory*, an acid is a substance that accepts an electron pair, and a base is a substance that donates an electron pair. Electron-deficient species like $AlCl_3$, BH_3, H^+ etc., can act as Lewis acids, while species like H_2O, NH_3 etc., can donate a pair of electrons and act as Lewis bases. For a species to function as a Lewis acid, it needs to have an accessible empty orbital. For a species to function as a Lewis base, it needs to have an accessible electron pair. Examples of Lewis acids are BF_3, $AlCl_3$, SbF_5, Na^+, H^+, S^{6+}, etc. Examples of Lewis bases are F^-, H_2O, Me_3N, C_2H_4, Xe, etc. Since H^+ and any cation from a solvent auto-dissociation is a Lewis acid, and anything that can add H^+ or a solvent-derived cation is a Lewis base, the Lewis acid concept effectively includes the ones discussed previously. Acid-base reactions under the Lewis model are the reactions of forming adducts between Lewis acids and bases:

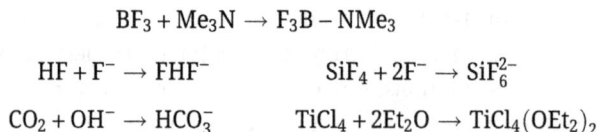

$$BF_3 + Me_3N \rightarrow F_3B - NMe_3$$

$$HF + F^- \rightarrow FHF^- \qquad SiF_4 + 2F^- \rightarrow SiF_6^{2-}$$

$$CO_2 + OH^- \rightarrow HCO_3^- \qquad TiCl_4 + 2Et_2O \rightarrow TiCl_4(OEt_2)_2$$

Acids and bases combine in neutralization reactions to form salts and water. Neutralization is typically an exothermic, heat-releasing reaction. While acid and base "neutralize" one another, that does not guarantee a neutral pH value. The extent of neutralization and changes in pH are dependent on the strength and concentration of acids and bases. A solution of hydrochloric acid has the typical properties of an acid. A solution of sodium hydroxide has the typical properties of a base. When these two solutions are mixed, the acid and base neutralize each other's properties in a chemical reaction:

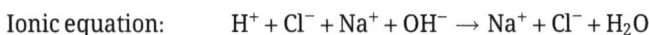

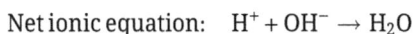

$$HCl + NaOH \rightarrow NaCl + H_2O$$

Ionic equation: $\qquad H^+ + Cl^- + Na^+ + OH^- \rightarrow Na^+ + Cl^- + H_2O$

Net ionic equation: $\quad H^+ + OH^- \rightarrow H_2O$

11.2.2 Water as an electrolyte

A general Arrhenius acid-base reaction is the reaction between H^+ and OH^- to produce water. Water is a poor electrical conductor. According to the Arrhenius theory, these ions, which arise from the ionization of water molecules themselves, are H^+ and OH^-. One water molecule acts as an acid; it loses a proton. Another water molecule acts as a base:

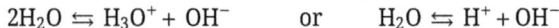

$$2H_2O \rightleftharpoons H_3O^+ + OH^- \qquad \text{or} \qquad H_2O \rightleftharpoons H^+ + OH^-$$

We can describe the equilibrium of the ionization of water through a thermodynamic equilibrium constant expression:

$$K_D = \frac{[H^+] \cdot [OH^-]}{[H_2O]} \qquad K_D \cdot [H_2O] = [H^+] \cdot [OH^-]$$

The product of K and $[H_2O]$ is a new constant called K_w – the *ion product of water*:

$$K_w = K_D[H_2O] = = 1.0 \times 10^{-14} = [H^+]\,[OH^-]$$

The product of $[H^+]\,[OH^-]$ is always equal to 1.0×10^{-14}. So, $[H^+] = K_w/[OH^-]$. At 25 °C in pure water or a neutral aqueous solution, the concentrations of hydronium and hydroxide ions are equal. These facts can be expressed as $[H^+] = [OH^-] = 1.0 \times 10^{-7}$ M. Such a solution is called a neutral solution. At $[H^+] > [OH^-]$, the solution is acidic, and at $[H^+] < [OH^-]$, the solution has a basic character.

In most cases, $[H^+]$ is in the range of 1–10^{-14} mol/L. Because numbers of this magnitude are inconvenient to work with, an alternate system for expressing the acidity of dilute solutions has been devised. This system is based on a quantity called *pH*. pH determines how acidic or basic a solution is. pH is calculated on a logarithmic scale of 10 such that each 1 increment increase in pH results in a 10-fold difference in $[H^+]$. The pH scale is actually based on negative logarithms for the values of H_3O^+ (the hydronium ion) or H^+ (protons) molar concentrations in a given substance. The formula is thus:

$$pH = -\log[H_3O^+] = -\lg[H^+]$$

and the presence of hydronium ions or protons is measured according to their molar concentrations (mol/L) in the solution. The pH scale is logarithmic, or exponential, meaning that the numbers represent exponents, and thus an increased value of 1 represents not a simple arithmetic addition of 1, but an increase of 1 power. The negative logarithm is specified because the logarithm of any number less than 1 is negative; thus, multiplication by −1 causes the values of pH to be positive over the range. pOH also determines the level of acidity and basicity:

$$pOH = -\lg[OH^-]$$

Their sum is equal to the full scale of 14:

$$pK_w = pH + pOH = 14$$

A low pH corresponds to a high pOH and vice versa since the sum of the two must be equal to 14. If pH < 7 the solution is acidic, if pH > 7, it is basic, and if pH = 7, the solution is neutral (see Figure 11.1).

pH Scale

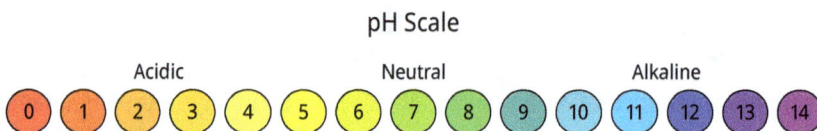

Figure 11.1: pH scale.

11.2.3 pH measurements and indicators

The pH scale is a means of determining the acidity or alkalinity of a substance. There are theoretically no limits to the range of the pH scale, but figures for acidity and alkalinity are usually given with numerical values between 0 and 14. A rating of 0 on the pH scale indicates a substance that is virtually pure acid, while a 14 rating represents a nearly pure base. A rating of 7 indicates a neutral substance. For 1 M HCl, the concentration $[H^+] = 1$ mol/L, which means pH = $-\lg[H^+]$ = $-\lg 1 = 0$. For 1 M NaOH, the molar concentration $[H^+] = 10^{-14}$ M or $[OH^-] = 1$ mol/L, which means pH = $-\lg 10^{-14} = 14$. At 25 °C (298 K), in pure water and in all aqueous neutral solutions, $[H^+] = [OH^-] = 10^{-7}$ mol/L, which means pH = $-\lg 10^{-7} = 7$.

The most precise pH measurements are made with *electronic pH meters*, which can provide figures accurate to 0.001 pH. However, simpler materials are also used. The best known among these is *litmus paper*, which turns blue in the presence of bases and red in the presence of acids. Litmus is just one of many materials used for making pH paper, but in each case, the change of color is the result of the neutralization of the substance on the paper. For instance, paper coated with *phenolphthalein* changes from colorless to pink in a pH range from 8.2 to 10, so it is useful for testing materials believed to be moderately alkaline.

Acid-base indicators are themselves weak acids (or weak bases) that possess an intense color in the acidic form (HIn) and a different intense color in the salt form (In⁻). The equilibrium between the two-color forms is:

$$\text{HIn} \leftrightarrows \text{H}^+ + \text{In}^- \qquad \text{or} \qquad \text{InOH} \leftrightarrows \text{In}^+ + \text{OH}^-$$

and the equilibrium constant expression is

$$K_D = \frac{c(\text{H}^+) \cdot c(\text{In}^-)}{c(\text{HIn})}$$

If the equilibrium lies far to either side, the solution will have the color corresponding to the predominant form, as the eye will be unable to perceive the small amount of the other form.

Changes in the colors of acid-base indicators – litmus, phenolphthalein, and methyl orange – depending on the pH of the solution, are given in Table 11.1.

Table 11.1: Acid-base indicators.

Indicator	Colors of acid-base indicators		
	Acidic	Neutral	Basic
Litmus	Red	Violet	Blue
	pH = 1.0–5.0	pH = 5.0–8.0	pH = 8.0–14.0
Phenolphthalein	Colorless	Colorless	Pink
			pH = 9.0–14.0
Methyl orange	Red	Orange	Yellow
	pH = 1.0–3.0	pH = 3.0–4.4	pH = 5.0–14.0

A solution that contains a weak acid plus a salt of that acid (CH_3COOH and CH_3COONa), or a weak base plus a salt of that base (NH_4OH and NH_4Cl), is known as *a buffer*. Such a solution has the capability to buffer against (to resist) changes in pH when small amounts of a strong acid or base are added. Only very small changes occur. Some of the most commonly used buffers are: acetate buffer, containing CH_3COOH and CH_3COONa; ammonia buffer, containing NH_3 and NH_4Cl; phosphate buffer, containing KH_2PO_4 and Na_2HPO_4. Many of the processes in living organisms take place in a buffered environment.

11.3 Hydrolysis

11.3.1 Hydrolysis of salts and the pH of their solutions

The reaction of ion exchange of salt with water molecules, as a result of which a weak electrolyte is formed and the pH of the salt solution is changed, is called *hydrolysis*. The reason for hydrolysis is the polarization of a molecule of water by an ion of a salt. Based on the polarization of water molecules, all salts are divided into four groups:

11.3.1.1 Group I: salts of strong acids and strong bases (NaCl, K₂SO₄, Ca(NO₃)₂, and CsClO₄)

Ions of such salts do not form weak electrolytes with ions of water, so there is no condition of hydrolysis. In pure water at 25 °C, pH = 7. When a salt such as NaCl is added to water, complete dissociation into Na^+ and Cl^- ions occurs, but these ions do not influence the ionization of water. The pH of the solution remains 7.

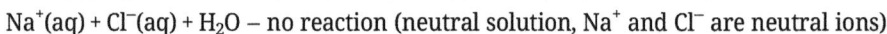

$$Na^+(aq) + Cl^-(aq) + H_2O - \text{no reaction (neutral solution, } Na^+ \text{ and } Cl^- \text{ are neutral ions)}$$

These salts are not subject to hydrolysis because the salt ions, when interacting with water, do not form weak electrolytes. The auto-protolyzed equilibrium of water is not disturbed, and the solutions are neutral. Such behavior is also typical for slightly soluble salts ($CaCO_3$, $Mg_3(PO_4)_2$, etc.), which are not hydrolyzed due to the very low concentration of ions in their aqueous solutions.

11.3.1.2 Group II: salts of strong bases and weak acids (CH₃COONa, KCN, Na₂CO₃, and K₂S)

When sodium acetate is added to water, the pH rises above 7. This means that $[OH^-]$ in the solution increases and $[H^+]$ decreases. Hydrolysis of such salts involves an anion of a weak acid, and the reaction of an aqueous solution of the salt is alkaline.

$$CH_3COONa + H_2O \leftrightharpoons Na^+ + OH^- + CH_3COOH \qquad pH > 7, \text{ basic solution}$$

$$CH_3COO^- + H_2O \leftrightharpoons OH^- + CH_3COOH$$

11.3.1.3 Group III: salts of weak bases and strong acids (NH₄Cl, CuCl₂, ZnSO₄, and Al(NO₃)₃)

When NH_4Cl is added to water, the pH falls below 7. This means that $[H^+]$ in the solution increases and $[OH^-]$ decreases. Cl^- cannot act as an acid. However, as we have already seen, a reaction does occur between NH_4^+ and H_2O. Hydrolysis of such salts involves a cation of the weak base, and the reaction of an aqueous solution of the salt is acidic:

$NH_4Cl(aq) + H_2O \leftrightarrows NH_3 \times H_2O + HCl(aq)$ pH < 7, acidic solution

$NH_4^+ + H_2O \leftrightarrows H_3O^+ + NH_3$

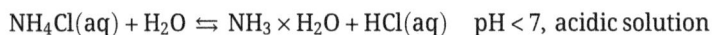

11.3.1.4 Group IV: salts of weak bases and weak acids (CH₃COONH₄, ZnS, Cr₂S₃, and (NH₄)₂CO₃)

Such salts are exposed to hydrolysis both on the cation and on the anion, and the reaction of an aqueous solution of the salt CH_3COONH_4 is nearly neutral because $K_a(CH_3COOH) = K_b(NH_3 \cdot H_2O)$:

$$CH_3COONH_4(aq) + H_2O \leftrightarrows NH_3.H_2O + CH_3COOH$$

$$CH_3COO^- + NH_4^+ + H_2O \leftrightarrows CH_3COOH + NH_4OH$$

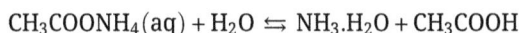

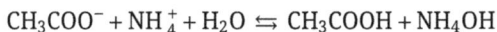

Hydrolysis, as a rule, is a reversible process. It proceeds up to an establishment of balance under the given equilibrium conditions (temperature, molar concentration of the salt). Hydrolysis is characterized by a degree of hydrolysis (h) and the constant of hydrolysis (K_h).

The degree of hydrolysis is defined as the fraction (or percentage) of the total salt that is hydrolyzed at equilibrium. For example, if 90% of a salt in the solution is hydrolyzed, its degree of hydrolysis is 0.90% or 90%. The degree of hydrolysis, h, depends on the nature of the salt, the salt concentration, and the temperature. If $C_M = 0.1$ mol/L and $T = 25$ °C, for electrolytes of group I, $h = 0$; for electrolytes of groups II and III, h is from 0.1% up to 2%; for electrolytes of group IV, $h \approx 100\%$.

The constant of hydrolysis K_h depends on the nature of the salt and on the temperature. The higher the value of K_h, the more the salt is exposed to the hydrolysis reaction (at the given temperature):

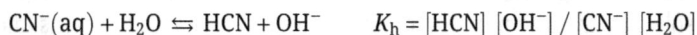

$CN^-(aq) + H_2O \leftrightarrows HCN + OH^-$ $K_h = [HCN] \, [OH^-] \, / \, [CN^-] \, [H_2O]$

The relationship between the degree of hydrolysis, the hydrolysis constant, and the molar concentration of the salt is expressed by an equation similar to Ostwald's dilution law:

$$K_h = \frac{h^2 \cdot c}{(1-h)}$$

In most cases, $h \ll 1$ and $K_h = h^2 \cdot c$.

The hydrolysis of salts depends on different factors (the strength of the electrolytes, dilution, temperature, the introduction into the solution of similarly charged ions, etc.). Hydrolysis is a reversible reaction and shifts according to the principle of Le Châtelier. When diluting the solution, the degree of hydrolysis increases. This dependence does not apply to the salt solutions obtained by the neutralization of a weak acid and a weak base – CH_3COONH_4, $(NH_4)_2CO_3$, and $(NH_4)_2S$. In them, hydrolysis proceeds significantly in both diluted and more concentrated solutions.

Salts of strong acids and weak bases, which contain ions of the transition metals ($ZnCl_2$, $CuSO_4$, and $FeSO_4$), undergo hydrolysis to produce acidic solutions. The hydrolysis occurs only through the first step:

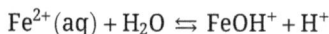

$$Fe^{2+}(aq) + H_2O \leftrightharpoons FeOH^+ + H^+$$

If salt, hydrolyzed by a cationic type, is added to a salt, hydrolyzed by an anionic type, the hydrolysis of the two salts shall be mutually enhanced:

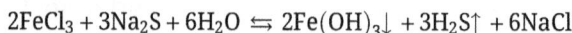

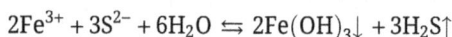

$$2FeCl_3 + 3Na_2S + 6H_2O \leftrightharpoons 2Fe(OH)_3\downarrow + 3H_2S\uparrow + 6NaCl$$

$$2Fe^{3+} + 3S^{2-} + 6H_2O \leftrightharpoons 2Fe(OH)_3\downarrow + 3H_2S\uparrow$$

11.3.2 Hydrolysis of acidic salts

To estimate the pH of the solutions of acidic salts (in which the hydro-anion of an acid is simultaneously exposed to hydrolysis and dissociation) it is necessary to compare the K_h of the salt and the K_a of the weak acid at the appropriate step.

Hydrolysis of Na_2CO_3 proceeds mainly to the first step. Strong dilution, anti-ions (H^+) and heating are factors contributing to the hydrolysis of the salt.

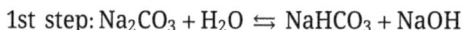

$$1st\ step: Na_2CO_3 + H_2O \leftrightharpoons NaHCO_3 + NaOH$$

$$CO_3^{2-} + HOH \leftrightharpoons HCO_3^- + OH^-$$

Sodium bicarbonate ($NaHCO_3$) is a salt of a weak acid, H_2CO_3, and a strong base, NaOH. In aqueous solution, the anion undergoes hydrolysis to give free OH^- ions and the acid:

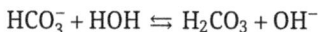

$$2nd\ step: NaHCO_3 + H_2O \leftrightharpoons H_2CO_3 + NaOH$$

$$HCO_3^- + HOH \leftrightharpoons H_2CO_3 + OH^-$$

HCO_3^- hydrolyzes to H_2CO_3 with the formation of a basic solution. The basic properties of the solution are characterized by the constant of hydrolysis, $K_h = 2.22 \times 10^{-8}$. On the other hand, HCO_3^- also dissociates as an acid:

$$HCO_3^- \leftrightharpoons H^+ + CO_3^{2-}$$

The acidic properties of the solution are characterized by the second constant of dissociation: $K_{a2} = 4.7 \times 10^{-11}$. In this case, $K_h > K_{a2}$, and the solution is basic in nature, pH > 7. Finally, $pH(NaHCO_3) < pH(Na_2CO_3)$.

In the aqueous solution of NaH_2PO_4, two reversible processes occur: hydrolysis (pH > 7) and dissociation, resulting in pH < 7:

$$H_2PO_4^- + H_2O \leftrightarrows H_3PO_4 + OH^-$$

$$H_2PO_4^- \leftrightarrows HPO_4^{2-} + H^+$$

In this case, $K_D \gg K_h$, and the solution is acidic, $C(H^+) > C(OH^-)$ with pH < 7.

In the hydrolysis of monohydrogen phosphate, Na_2HPO_4, the solution remains basic, but the pH is lower than in a solution of normal Na_3PO_4. HPO_4^{2-} is a very weak acid, and the hydrolysis here proceeds to a greater extent than the dissociation. In this case, $K_h \gg K_D$, and the solution is basic in nature with pH > 7:

$$HPO_4^{2-} + H_2O \leftrightarrows H_2PO_4^- + OH^-$$

$$HPO_4^{2-} \leftrightarrows PO_4^{3-} + H^+$$

11.3.3 Hydrolysis of normal salts obtained by the neutralization of strong acids and multivalent bases

In the solutions of these salts at ordinary temperature and medium concentrations, the process proceeds only to the first degree, as a result of which hydroxy-cations are obtained. Such salts include $Al_2(SO_4)_3$, $FeCl_3$, $Al(NO_3)_3$, $Cu(NO_3)_2$, $FeCl_2$, $SbCl_3$, $Bi(NO_3)_3$, $Ni(NO_3)_2$, $ZnCl_2$, etc.:

$$Al(NO_3)_3 + H_2O \leftrightarrows Al(OH)(NO_3)_2 + HNO_3$$

$$Al^{3+} + 3NO_3^- + H_2O \leftrightarrows Al(OH)^{2+} + 2NO_3^- + H^+ + NO_3^-$$

$$Al^{3+} + H_2O \leftrightarrows Al(OH)^{2+} + H^+$$

In the solutions of such salts, parts of the OH^- ions are part of the hydroxy-cations, which are weak electrolytes. Therefore, $c(OH^-)$ decreases, $c(OH^-) < c(H^+)$, the character of the solution becomes acidic, and pH < 7. When diluted and when temperature rises, the process of hydrolysis proceeds to the end.

Hydrolysis of $Al(OH)(NO_3)_2$ is also possible:

$$Al(OH)(NO_3)_2 + H_2O \leftrightarrows Al(OH)_2(NO_3) + HNO_3$$

$$Al(OH)^{2+} + 2NO_3^- + H_2O \leftrightarrows Al(OH)_2^+ + NO_3^- + H^+ + NO_3^-$$

$$Al(OH)^{2+} + H_2O \leftrightarrows Al(OH)_2^+ + H^+$$

And the hydrolysis of $Al(OH)_2(NO_3)$, which leads to $Al(OH)_3$:

$$Al(OH)_2(NO_3) + H_2O \leftrightarrows Al(OH)_3 + HNO_3$$

$$Al(OH)_2^+ + NO_3^- + H_2O \leftrightarrows Al(OH)_3 + H^+ + NO_3^-$$

$$Al(OH)_2^+ + H_2O \leftrightarrows Al(OH)_3 + H^+$$

The hydrolysis of aluminum nitrate and its precipitation as aluminum hydroxide are noticeable above a pH of 3.5.

11.4 Solubility product

One common type of a reaction that occurs in aqueous solutions is the precipitation reaction, which results in the formation of an insoluble product, or *precipitate*. The precipitate is an insoluble product that separates from the solution. The equilibrium of solids in water is one of the specific equilibria possible in aqueous solutions. It is a *heterogeneous equilibrium* between the precipitate and the saturated solution. The saturated solution contains only ions formed when the precipitate dissolves. The latter is the proof that these equilibria are characteristic of poorly soluble strong electrolytes. For example, when an aqueous solution of silver nitrate $AgNO_3$ is added to an aqueous solution of sodium chloride (NaCl), a white precipitate of silver chloride (AgCl) is formed:

$$AgNO_3 + NaCl \rightarrow AgCl\downarrow + NaNO_3$$

How can you predict whether a precipitate will form when a compound is added to a solution or when two solutions are mixed? It depends on the *solubility* of the solute, which is defined as the maximum amount of the solute that will dissolve in a given quantity of a solvent at a specific temperature.

To develop a quantitative approach, we consider a saturated solution of silver chloride that is in contact with solid silver chloride, AgCl. The solubility equilibrium can be represented as:

$$AgCl(s) \rightleftharpoons Ag^+(aq) + Cl^-(aq)$$

Because salts like AgCl are considered strong electrolytes, all the AgCl that dissolves in water is assumed to dissociate completely into Ag^+ and Cl^- ions. For heterogeneous reactions, the concentration of the solid is constant (it is 1 mol/L); thus, the equilibrium constant for the dissolution of AgCl is:

$$K_s = [Ag^+] . [Cl^-]$$

where K_s is called a solubility product constant or simply *solubility product*. It equals the product of the equilibrium concentrations of the ions in the compound, where each concentration is raised to a power equal to the number of such ions in the formula of the compound. Like any equilibrium constant, K_s depends on the temperature, but at a given temperature, it has a constant value for various concentrations of the ions. For example:

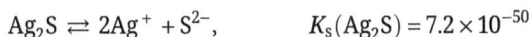

$$BaSO_4 \rightleftharpoons Ba^{2+} + SO_4^{2-}, \qquad K_s(BaSO_4) = 1.8 \times 10^{-10}$$

$$Cu(OH)_2 \rightleftharpoons Cu^{2+} + 2OH^-, \qquad K_s(Cu(OH)_2) = 5.6 \times 10^{-20}$$

$$Ag_2S \rightleftharpoons 2Ag^+ + S^{2-}, \qquad K_s(Ag_2S) = 7.2 \times 10^{-50}$$

The value of K_s is a measure of the solubility of the compound. A smaller value of K_s means a more insoluble ionic compound. In these examples, the most insoluble ionic compound is Ag_2S.

To predict whether a precipitate will form, we use the solubility quotient, often called the *ion product* Q_c. It has the same form as the solubility constant expression, but the concentrations of the ions are initial concentrations instead of equilibrium concentrations. To predict the direction of the reaction, we compare Q_c with K_s.

If $Q_c > K_s$, *precipitation occurs,* and it will continue until the concentration satisfies K_s:

$$c(Ag^+) . c(Cl^-) > K_s(AgCl)$$

The solution is *supersaturated.* A supersaturated solution is one that contains more of the dissolved substance than could be dissolved by the solvent at that temperature, at equilibrium. The supersaturated solution becomes a saturated solution by forming a solid in order to reduce the dissolved substance. The crystals formed are called a precipitate. Ionic solids will precipitate until the system reaches equilibrium.

If $Q_c = K_s$, *the reaction mixture is at equilibrium:*

$$c(Ag^+) . c(Cl^-) = K_s(AgCl)$$

The solution is *saturated.* Saturated solutions contain the maximum concentration of a combination of ions as specified by the K_s value for that combination of ions at a specific temperature. The solution is at equilibrium since opposing processes (dissolving and precipitation) are occurring at equal rates at a constant temperature. No net change in the amount of dissolved solid will occur. This is a state of the chemical system in which as many ions pass from the surface of the crystal to the solution, so many pass back and precipitate, that is, both processes – dissolution and precipitation – take place simultaneously.

If $Q_c < K_s$, *no precipitation occurs,* and more solute can dissolve:

$$c(Ag^+) . c(Cl^-) < K_s(AgCl)$$

The solution is *unsaturated.* An unsaturated solution is a solution that has not dissolved all of the solute possible at that temperature. In this case, more of the ionic solid, if available, will dissolve until the system reaches equilibrium (see Table 11.2).

Table 11.2: Direction of the reaction depending on Q_c and K_s.

Condition	Nature of solution	Possibility of precipitation
Ion product $< K_s$	Unsaturated solution	No precipitation occurs
Ion product $= K_s$	Saturated solution at equilibrium	Solution at equilibrium, no precipitation
Ion product $> K_s$	Supersaturated solution	Precipitation occurs

If a solution of HCl is added to each of the three precipitates of ZnS, FeS, and CuS, the CuS precipitate does not dissolve, but the ZnS and FeS precipitates do dissolve:

$$ZnS + 2HCl = ZnCl_2 + H_2S$$

$$FeS + 2HCl = FeCl_2 + H_2S$$

Sulfide anions, S^{2-} are removed because the very stable weak acid, H_2S is formed:

$$2H^+ + S^{2-} \rightleftharpoons H_2S$$

As a result, the molar concentration of S^{2-} decreases. The equilibrium between precipitates and their ions shifts to the right direction (according to Le Chatelier's principle), and ZnS and FeS precipitates dissolve. For both the solutions, the ion product Q_c is less than the Strontium K_s, that is, $Q_c < K_s$, and the solutions are unsaturated. Dissolving continues until the solutions become saturated.

The CuS precipitate is insoluble in HCl because its solubility product, K_s, has a very low value. Even the low concentration of S^{2-} is sufficient, and the ion product, Q_c, is still greater than the solubility product, K_s, that is, $Q_c > K_s$. The solution is supersaturated, and CuS cannot dissolve.

CuS precipitate can be dissolved in a concentrated solution of HNO_3. Concentrated HNO_3 is a powerful oxidizing agent, and it oxidizes the sulfide ion into elemental sulfur. Consequently, the molar concentration of S^{2-} decreases, $Q_c < K_s$, and CuS dissolves:

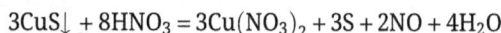

$$3CuS\downarrow + 8HNO_3 = 3Cu(NO_3)_2 + 3S + 2NO + 4H_2O$$

The solubility is influenced by the concentration of common ions. The introduction of a slightly soluble electrolyte with the same ions shifts the equilibrium in the direction of precipitate formation. On the contrary, the continuous removal of any of the common ions will cause the precipitate to dissolve. This can be done in several ways: binding the common ion into a compound that maintains a lower concentration of the common ion than the precipitate, or taking the common ion out of equilibrium through redox processes. For example, adding dilute acid to slightly soluble carbonates causes them to dissolve:

$$CaCO_3\!\downarrow + 2HCl = CaCl_2 + H_2CO_3$$

$$CO_3^{2-} + 2H^+ \rightarrow H_2CO_3$$

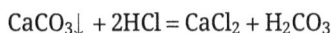

Carbonate ions bind in the weak electrolyte carbonic acid. At the same time, their concentration is constantly decreasing as they are taken out of the equilibrium $CaCO_3 \rightleftarrows Ca^{2+} + CO_3^{2-}$.

As a result of the disturbed equilibrium, a state is reached in which the solubility product cannot be reached, and the precipitate dissolves because $C(Ca^{2+}) \cdot C(CO_3^{2-}) < K_S$ $(CaCO_3)$.

Another typical example of the binding of a common ion into a compound that maintains a lower concentration of the common ion is the complexation process, in which at least one of the common ions binds in a complex compound. For example:

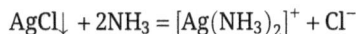

$$AgCl\!\downarrow + 2NH_3 = [Ag(NH_3)_2]^+ + Cl^-$$

Silver ions bind to the complex ion $[Ag(NH_3)_2]^+$, which has a very low degree of dissociation. Their concentration decreases, making it impossible to reach and exceed the solubility product of AgCl, which is a prerequisite for the precipitate to begin to dissolve: $C(Ag^+) \cdot C(Cl^-) < K_S(AgCl)$.

Chapter 12
Colloids

12.1 Methods for preparation of colloids

Colloids are dispersed systems. Dispersed systems are those systems in which the particles of one substance, called the dispersed phase, are distributed throughout the particles of another substance, called the dispersed medium. According to the size of the dispersed phase particles, the dispersed systems are divided into three types: rough-dispersed systems, in which the particle size is greater than 100 nm; colloidal systems, in which the particle size of the dispersed phase is between 1 and 100 nm; and molecular or ionic dispersed systems, in which the particle size is lower than 1 nm. The last systems are also called true solutions or simply solutions.

Colloids consist of two separate phases: a dispersed phase (internal phase) and a continuous phase (dispersion medium, solvent). The size of their particles is between the molecular size (in solutions) and the mechanical dispersion size (in suspensions).

The basic states in which colloid systems may occur are sol and gel. Sols tend to have a lower viscosity and are fluids. If the solid particles form bridged structures possessing some mechanical strength, the systems are called gels. They have a jelly-like consistency.

According to the power of the interaction between the particles of the dispersed phase and the dispersed medium, colloidal solutions are divided into two broad categories: *lyophilic and lyophobic.*

The system is lyophilic (or solvent-loving) when there is a considerable attraction between the dispersed phase and the dispersed medium. The system is said to be hydrophilic if the dispersed medium is water. Lyophilic (hydrophilic) colloids are strongly solvated (hydrated) and therefore are stable and less sensitive to coagulating agents.

The system is lyophobic (solvent-hating) if there is little attraction between the dispersed phase and the dispersed medium. Lyophilic colloidal solutions are obtained from macromolecular compounds.

There are two basic methods for the preparation of lyophobic colloids:
- *Dispersion method* consists of mechanically subdividing larger particles into those with colloidal size.
- *Condensation method* consists of condensing smaller particles into those with colloidal size.

Condensation of smaller particles to form a colloid usually involves chemical reactions such as precipitation reactions, hydrolysis, oxidation-reduction reactions, etc.
- Exchange reactions (colloid can be obtained under suitably chosen conditions):

https://doi.org/10.1515/9783111712246-012

$$AgNO_3 + KI \rightleftharpoons AgI\downarrow + KNO_3$$

$$2HCl + Na_2SiO_3 \rightleftharpoons H_2SiO_3 + 2NaCl$$

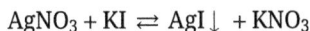

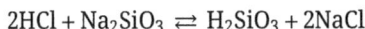

- Hydrolysis of salts, for example, sols of hydroxides and silicic acid, can be obtained at high temperatures:

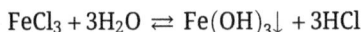

$$FeCl_3 + 3H_2O \rightleftharpoons Fe(OH)_3\downarrow + 3HCl$$

$$Na_2SiO_3 + 2H_2O \rightleftharpoons H_2SiO_3 + 2NaOH$$

- Oxidation-reduction reactions:

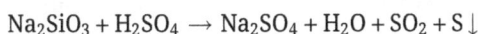

$$2H_2S + O_2 = 2S + 2H_2O$$

$$Na_2SiO_3 + H_2SO_4 \rightarrow Na_2SO_4 + H_2O + SO_2 + S\downarrow$$

- Polymerization reactions, for example, polymerization of divinyl to rubber or polymerization of formaldehyde to polyoxymethylene:

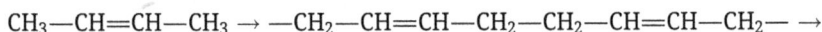

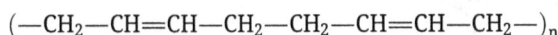

$$CH_3-CH=CH-CH_3 \rightarrow -CH_2-CH=CH-CH_2-CH_2-CH=CH-CH_2- \rightarrow$$

$$(-CH_2-CH=CH-CH_2-CH_2-CH=CH-CH_2-)_n$$

$$H_2C=O \rightarrow -CH_2-O-CH_2-O- \rightarrow (-CH_2O-)_n$$

- Reduction of solubility can be achieved by adding a solvent in which the colloidal component is less soluble, for instance, by adding an ethanol solution of S to water.

The method for the preparation of a sol of $Fe(OH)_3$ is based on the hydrolysis of $FeCl_3$ to $Fe(OH)_3$ at high temperature:

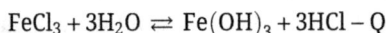

$$FeCl_3 + 3H_2O \rightleftharpoons Fe(OH)_3 + 3HCl - Q$$

Hydrolysis is the opposite of neutralization and is therefore endothermic. At high temperatures, the equilibrium shifts in the direction of the product $Fe(OH)_3$.

As a result of hydrolysis, polynuclear complexes are obtained, which, due to their low solubility, condense into particles of colloidal size. The structure of the colloidal particle is represented in Figure 12.1.

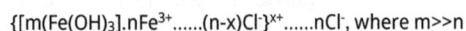

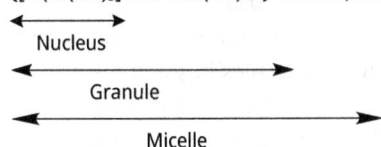

{$[m(Fe(OH)_3).nFe^{3+}......(n-x)Cl^-]^{x+}......nCl^-$, where $m \gg n$

←——————→
    Nucleus

←————————————→
       Granule

←——————————————————→
          Micelle

Figure 12.1: Structure of the colloidal particle of the sol of $Fe(OH)_3$.

The colloidal particle of the sol, obtained by hydrolysis of iron(III) chloride, has a positive charge. Here we have an example of a lyophobic colloid in which the lyophobic colloidal particle is composed of a nucleus and an ionic layer. The nucleus may consist of ions, molecules, or atoms. The ionic layer represents an electric double layer. The reason for the electric charge in lyophobic colloidal solutions – is the selective adsorption of ions from the solution. This is an example with a positive charge on the surface of the particle. The nucleus (electroneutral molecules of $Fe(OH)_3$) adsorbs positive ions from the solution according to the rule of Paneth-Fajans. These ions are Fe^{3+}, because they are in excess and they are common with the ions from the solution. They are called potential ions. These ions attract negative ions and repel positive ions according to electrostatic forces. The potential Fe^{3+} ions, along with a part of the counterions Cl^- form the Stern layer. The nucleus, potential ions, and part of the counterions form the granule. With distance, the number of negative ions decreases and the number of positive ions increases. As a result, the diffusion layer is obtained. The granule and ions from the diffusion layer form the micelle. In the solution, the number of positive and negative ions is the same, which is why the colloidal solution is electroneutral in nature.

12.2 Properties of colloid systems

Colloidal solutions have specific *kinetic, optical,* and *electrical properties.*

12.2.1 Kinetic properties

Brownian movement (motion) is a random (zig-zag) movement of the particles, as shown in Figure 12.2. It arises due to the impact of the molecules of the dispersion medium with the colloid particles. It decreases as the size of the particles increases. When the size of the dispersed particles increases beyond the colloid range, Brownian motion stops. Brownian movement does not depend on the chemical nature of the colloidal solution, its concentration, or time, and it is everlasting. Brownian movement is a demonstration of the endless motion of molecules as postulated by kinetic theory, and it helps in providing stability to the sols by not allowing them to settle down. Suspensions and true solutions do not exhibit Brownian movement.

The colloid solutions show the same kinetic properties as the true solutions. These properties are connected with the disorderly motion of the particles upon heating. Diffusion is the spontaneous process of motion of colloidal particles from a region of higher concentration to a region of lower concentration until the concentration of the system becomes uniform. Sedimentation describes the motion of molecules in solutions or particles in suspension in response to an external force such as gravity, centrifugal forces, or electric forces. In view of their larger size, the colloidal particles do not sediment or precipitate. In coagulation processes, the particles of the dispersed phase adhere to each

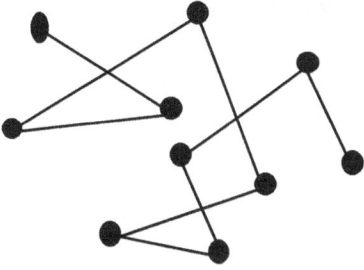

Figure 12.2: Brownian movement.

other, increasing in size. The colligative properties of colloidal solutions (lowering of vapor pressure ΔP; depression of freezing point ΔT_{fr}; elevation of boiling point ΔT_b; and osmotic pressure π) show values less than theoretically predicted.

12.2.2 Optical properties

In terms of their optical properties, colloids differ significantly from true solutions. Their *coloration* is due to selective absorption and diffraction scattering of light.

Most colloidal solutions in reflected light appear cloudy. This phenomenon is called *opalescence*. For example, white silver chloride opalescence is bluish in color. These properties are explained by the fact that the size of colloidal particles is equal to the wavelength of visible light, especially in the blue region. That is why they diffract the light that falls on them, which explains the blue color of diluted milk, lake and sea water, the sky, etc.

Another phenomenon characteristic of all colloidal solutions is associated with the scattering of light by the colloidal particles, the so-called *Tyndall effect*. If a beam of light passes through a colloid, the light is reflected (scattered) by the colloidal particles, and the path of the light can therefore be observed (see Figure 12.3). The scattering of light by colloids was first explained by John Tyndall. The blue color of the sky and seawater is due to the scattering of light by the Tyndall effect.

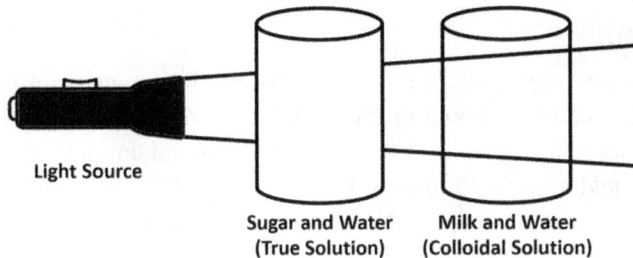

Light Source

Sugar and Water
(True Solution)

Milk and Water
(Colloidal Solution)

Figure 12.3: Tyndall effect.

12.2.3 Electrical properties and the structure of lyophobic colloidal particles

If electricity is passed through a colloidal solution, the particles are oriented toward one of the electrodes – positive or negative one, depending on their charge.

The lyophobic colloid of AgI can be produced by the mixing of $AgNO_3$ and KI solutions:

$$Ag^+ + NO_3^- + K^+ + I^- \leftrightarrows AgI + K^+ + NO_3^-$$

It is important to know which compound is in excess. There are three possible options for this reaction:

i)   The AgI sol bears a positive charge if Ag^+ ions ($AgNO_3$) are in excess

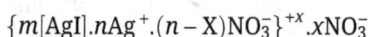

$$\{m[AgI].nAg^+.(n-X)NO_3^-\}^{+X}.xNO_3^-$$

ii) If I^- ions (KI) are in excess, the AgI sol particles acquire a negative charge

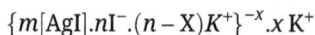

$$\{m[AgI].nI^-.(n-X)K^+\}^{-X}.x\,K^+$$

iii) In case both $AgNO_3$ and KI are in equivalent amounts, the obtained sol particles become bigger and coagulate or sedimentate. The granule does not have an electric charge, and two equally probable formulas can be represented:

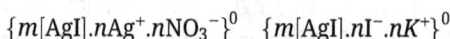

$$\{m[AgI].nAg^+.nNO_3^-\}^0 \quad \{m[AgI].nI^-.nK^+\}^0$$

When Ag^+ ions are in excess, the AgI sol particle adsorbs them and becomes positive. K^+, Ag^+ and NO_3^- ions are in the solution, whereas all I^- ions are connected in AgI. The granule contains the core (AgI), potential ions (Ag^+), and part of the counterions (NO_3^-), and has a positive electric charge. When I^- ions are in excess, the AgI particle adsorbs these ions and acquires a negative charge. I^-, NO_3^-, and K^+ ions are in the solution, while all Ag^+ ions are connected in AgI. The granule contains the core (AgI), potential ions (I^-), and part of the counterions (K^+) and thus has a negative charge. When $AgNO_3$ and KI are equivalent, the granule does not have an electric charge. This is an unstable colloid system and can be easily destroyed.

12.2.3.1 Electrical double layer

According to the micellar theory, the structure of the colloidal particles consists of *a nucleus* (solid phase, a slightly soluble compound) and *an electrical double layer* (with an adsorption and diffusion part), as shown in Figure 12.4. The electrical double layer is made up of *potential-determining ions* (cations or anions) adsorbed on the surface of the nucleus, and oppositely charged ions – *counterions*, that is, from *an adsorption part* (includes all potential-determining ions and part of the counterions) and *a diffusion part* in which the other counter-ions are located. Because of the presence of the electrical double layer, there is an electric potential difference between the charged surface and the bulk solution.

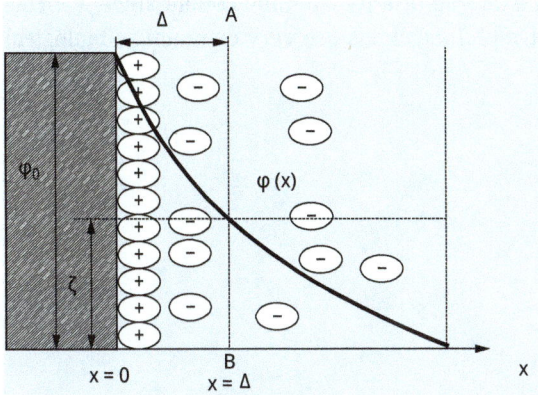

Figure 12.4: Electrical double layer.

The potential on the line aa' is called the *thermodynamic* or *Nernst potential* – φ. The ions from the Stern layer are so strongly bound with the nucleus, that if the colloidal particle is moved through the solution, the surface of shear is obtained. At this place, the layer is torn, and a new electrokinetic or *zeta potential* appears – bb'. This is the reason for the appearance of the electric charge of the particle. The value of the thermodynamic potential does not change, but the zeta potential can change both its value and its charge. It defines the stability of the colloidal solutions.

12.3 Stability of colloidal solutions

The lyophobic colloids are not as stable as lyophilic colloids. The stability of colloids is due to the electrical charge of the particles, which does not permit additional adhesion and precipitation of the colloid solution. The value of the zeta potential determines the stability of the colloidal particles and the possibility of the colloid to exist. If a strong electrolyte is added to a colloid solution, the value of the zeta potential decreases. As a result, the colloidal particles are collected in a larger aggregate, and a sol is transformed into a gel. Many lyophilic colloids can be transformed from gels to sols. This can be realized by washing the ions, which cause a decrease in the zeta potential, or by adding an electrolyte, which recharges the colloidal particles.

Hydrophobic colloids can easily coagulate after the addition of an electrolyte to a solution. Hence, they are sensitive to electrolytes, which causes destabilization of the system. If a hydrophilic colloid is added to a hydrophobic one, the latter is protected and stabilized. The stabilization may be explained by the adsorption of the hydrophilic colloid over the particles of the hydrophobic one. AgCl sol, protected by a gelatinous solution, forms a hydrophilic colloid, which makes a protective cover around the hydrophobic colloidal particles, and they become insensitive to the electrolyte solution of NaCl. Gelatin is a common example of a hydrophilic colloid since it forms strong hydrogen bonds with water. Lyophilic colloids are very common in biological systems and in foods.

Chapter 13
Physicochemical analysis

13.1 Phase rule and equilibrium

Solid-liquid phase diagrams: The graphical representation of a solid/liquid binary system can be simplified by representing it on rectangular coordinates: temperature vs. concentration/composition. It can be constructed by analyzing experiments in which a mixture of known composition is heated above its melting point and then allowed to cool slowly.

Construction of the diagrams is based on the solubility limits determined by *thermal analysis* using *cooling curves*. The cooling curves represent changes in volume, electrical conductivity, crystal structure, etc. Phase diagrams can also be constructed from the data obtained by *differential scanning calorimetry.*

Gibbs' phase rule gives the relationship between the number of phases and components under equilibrium conditions at constant pressure:

$$C + 2 = P + F$$

where P is the number of phases in the system, C is the minimum number of chemical components, and F is the number of degrees of freedom in the system (variance of the system). A phase is any physically separable material in the system: solid, liquid, or vapor. The digit 2 is related to the number of intensive parameters, i.e., those that are independent of mass (such as pressure and temperature) that are being considered. For one-component systems, the maximum number of variables to be considered is two – pressure and temperature. For two-component (binary) systems, the maximum number of variables is three – P, T, C. Three common types of equilibria are possible: (i) *invariant* equilibria, in which neither P nor T can be changed; (ii) *univariant* equilibria, in which either P or T can be changed independently, maintaining the state of the system; and (iii) *bivariant* equilibria, in which both P and T are free to change independently without changing the state of the system.

Cooling curves usually represent the temperature of the mixture as a function of time (Figure 13.1). Cooling curves for pure substances (a) exhibit horizontal line segments representing substance freezing, and when the system is entirely frozen, the T can drop further. The *cooling curve (a)* shows that the cooling proceeds along curve AB until B is reached when the first crystals begin to form. As freezing continues, the T remains constant from B to C until the whole mass has entirely solidified. Further cooling from C causes the T to drop along CD.

Solid/liquid systems are usually investigated at constant pressure, and thus only two variables need to be considered – the vapor pressure can be neglected. This is called a condensed system, represented by the modified phase rule equation:

https://doi.org/10.1515/9783111712246-013

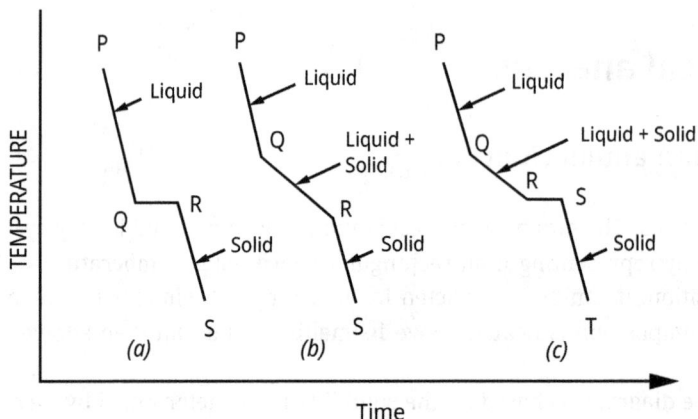

Figure 13.1: Cooling curves.

$$F = C - P + 1$$

where all symbols are the same, the digit 2 is replaced by the digit 1, which stands for *T*. The cooling curve (a) is that of a pure compound. In regions *AB* and *CD*, there is one phase – either a liquid in *AB* or a solid in *CD*:

$$F = C - P + 1 = 1 - 1 + 1 = 1$$

The one-component system has one degree of freedom – *univariant T* can be varied independently without phase disappearance/appearance.

In the region *BC*, there are two phases: liquid and solid:

$$F = C - P + 1 = 1 - 2 + 1 = 0$$

Therefore, the system is *invariant*. T must be constant as long as two phases are present.

The cooling curve (b) is for a binary system consisting of two components. Section *AB* of the curve is similar to that of curve (a) – liquid state:

$$F = C - P + 1 = 2 - 1 + 1 = 2$$

The system is *bivariant*. Section *BC* shows that during the freezing period, *T* does not remain constant but drops along *BC* until the whole mass is solidified, which means:

$$F = C - P + 1 = 2 - 2 + 1 = 1$$

The system is *univariant*, so *T* will vary independently.

Further, the temperature falls along line *CD* – one solid phase:

$$F = C - P + 1 = 2 - 1 + 1 = 2$$

The system is *bivariant*.

The cooling curve (c) is for a binary system whose two components are completely soluble in the liquid state but entirely insoluble in the solid state. The liquid cools along line *AB* until *B* is reached. At point *B*, the component in excess will crystallize, and the temperature will fall along line *BC*. The line *BC* has a slope different from that of *AB*. At point *C*, the two components crystallize at the same time, and *T* remains constant until all the liquid solidifies. This is known as a *eutectic reaction*. Therefore, curve (a) corresponds to a pure component, curve (b) to binary mixtures, and curve (c) to binary eutectic systems (Figure 13.2).

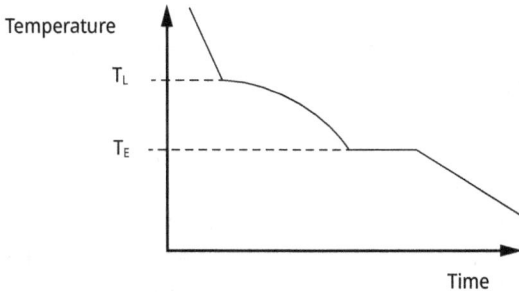

Figure 13.2: Cooling curve of a binary eutectic system.

A pure material has a unique freezing temperature. The liquid cools steadily until the freezing temperature is reached. The heat should maintain a constant temperature until all the liquid has solidified. For a simple binary eutectic melt at temperature T_C – a small amount of the solid is formed. At *eutectic* T_E the remaining liquid (eutectic composition) transforms directly to solid eutectic. The cooling curve remains horizontal until all the liquid has solidified.

Eutectic phase diagram for a two-component eutectic system (temperature vs. composition) is presented in Figure 13.3. The temperature is plotted on the vertical axis, and the composition (% *A*, % *B*) is plotted on the horizontal axis. Mixtures, pure *A*, and pure *B* represent different compositions. The pure *A* plot at 100% *A* corresponds to 0% *B*. The pure *B* plot at 100% *B* corresponds to 0% A. Composition can be expressed as either % of *A* or % of *B*; the total % must add up to 100. For the case shown, we consider pressure to be constant; thus, the temperature is plotted on the vertical axis.

In the diagram shown above, the components are *A* and *B*. The possible phases are pure crystals of *A*, pure crystals of *B*, and a liquid with compositions ranging between pure *A* and pure *B*. If the system contains only pure *A*, the system is one-component, and phase *A* melts at only one *T* – melting *T* of *A*, which is *p. A* –100% *A* and 0% *B*. If the system contains only pure *B*, *B* melts only at the melting *T* of pure *B*, which is *p. B* – 100% *B* and 0% *A*.

Curve X corresponds to 20% *B* in 80% *A*. It is known that the solution crystallizes at a lower *T* than the solvent, which is expressed by "breaking" the curve. This corre-

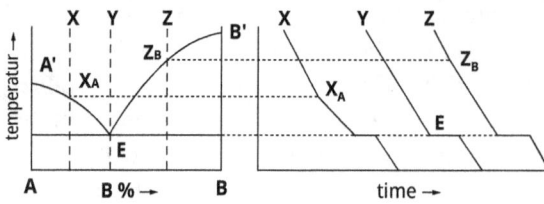

Correlation between state diagram and cooling curves

Figure 13.3: Eutectic phase diagram for a two-component system.

sponds to the start of the crystallization of A (80%), and a new phase appears (crystals of A), represented by the respective point of the diagram curve.

The curves separating the fields of (A + liquid) from liquid and (B + liquid) from liquid are termed *liquidus curves*. The horizontal line separating (A + liquid) from (A + B all solid) and (B + liquid) from (A + B all solid) is termed the *solidus line*. The point E, where liquidus curves and solidus intersect, is termed the *eutectic point*. At point E, the three phases (liquid, crystals of A, and crystals of B) all exist in equilibrium. The term "eutectic" means "easily melting." The eutectic composition freezes at a lower temperature than the melting points of the constituent components. Eutectic means a mixture of solid crystals of the pure components. They are crystals with mixed composition without any chemical interaction between them.

As the solid/liquid systems are usually investigated at constant pressure, and thus only two variables need to be considered, the vapor pressure can be neglected. This is called a condensed system, represented by the modified phase rule equation:

$$F = C - P + 1$$

where the digit 1 stands for T, and the eutectic point is *invariant*. For all compositions between A and B, the melting occurs over various T between the solidus and liquidus. This is true for all compositions except the eutectic. The eutectic melts at only one temperature – T_E.

For a binary system ($C = 2$), $F = 3 - P$. At single-phase equilibrium (solution of A and B):

$$\Gamma - 3 - 1 - 2$$

The field is *bivariant*; $F = 2$ (% and T). This is the region above the liquidus curve.

At 2-phase equilibrium (liquid A + B and solid A or B)

$$F = 3 - 2 = 1$$

The field is *univariant*; F = 1 (% or T). These are the regions below the liquidus curves.

At 3-phase equilibrium (liquid A + B, solid A, and solid B)

$$F = 3 - 3 = 0$$

The system is *invariant* (point E).

Under the solidus line, the system has two phases: crystals of A and crystals of B, which are insoluble in the solid state. The system is *univariant*:

$$C = 3 - 2 = 1$$

Thus, at temperature $T > T_1$, all is liquid; between T_1 and T_E, there are liquid and crystals of A; at T_E, there are liquid, crystals of A, and crystals of B; and at $T < T_E$, all is solid (crystals of A and crystals of B), as shown in Figure 13.4.

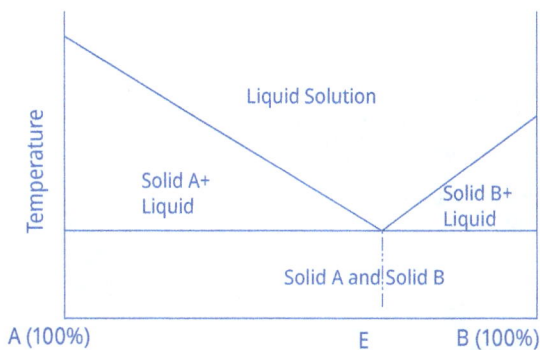

Figure 13.4: Phases in the eutectic phase diagram.

Such a diagram corresponds to an equilibrium system in which there is no chemical interaction between the substances in the system. The components have full solubility in liquid form, complete insolubility in the solid state, and no mixed crystals are formed.

Solid–liquid phase diagram with compounds: If two substances, A and B, form a solid-state compound AB, the crystal lattice of AB is different from that of A or B (see Figure 13.5). Solids A, B, and AB are completely insoluble in each other. The diagram resembles two diagrams set side by side. The right half is the diagram for AB and A, and the left half is the diagram for B and AB. The eutectic areas represent the coexistence of pure A with pure AB and the coexistence of pure B with pure AB.

Thus, the systems $A-AB$ and $AB-B$ may be considered as two separate systems. The maximum is the melting point of the compound AB, having the respective composition. There are two eutectic points: E and E'. E is the eutectic of solid A and solid AB. E' is the eutectic of solid AB and solid B.

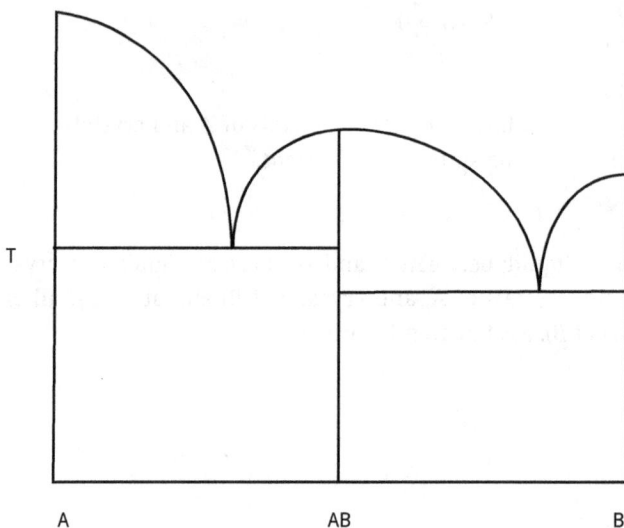

Figure 13.5: Solid-liquid phase diagram with chemical interaction.

13.2 Freezing mixtures

Inorganic salts form eutectic mixtures with water. These mixtures are called freezing mixtures.

Water can coexist with ice in an equilibrium state under normal pressure at a temperature of 0 °C. However, when we mix water, ice, and salt (NaCl), this temperature decreases. In the phase diagram, the vertical axis shows the temperature, and the horizontal axis shows the salt concentration. For each pair of temperature and concentration, the state of the mixture can be found. The lowest temperature at which a mixture of water, ice, and NaCl can exist is ca. − 21 °C, and the salt concentration is approximately 23%. Therefore, if we add salt to a mixture of ice and water, the melting/freezing point of the mixture decreases, and the ice begins to melt. In order for a phase change to occur, the ice draws the heat of fusion from its surroundings, which allows the temperature to decrease.

The left liquidus curve of the phase diagram is the curve of formation of ice. It is a freezing curve of solutions (salt in water) with increasing concentration. The right liquidus line is a curve of the solubility of salt in water. Here, the salt is in equilibrium with the solution. This curve does not start from the pure salt; it starts from the solution containing a portion of salt.

The typical diagram of the water-salt-system is incomplete with respect to the salt because at $T > 160$ °C, this system is not possible (water evaporates). Moreover, the salts melt at high T because of the ionic bonds.

The cross point of the two liquidus lines is the temperature of the double crystallization. The eutectic point is called *cryohydrate,* and the mixture is called a *cryohydrate mixture.* The cryohydrate T is always negative, as water crystallizes at 0 °C. At this temperature, the system can be used for cooling. At the cryohydrate point ($C + 2 = P + F$), there is a four-phase system consisting of crystalline salt, saturated solution, ice, and vapor:

$$F = 4 - 4 = 0$$

The system is *invariant.* This system is non-equilibrium at room temperature. That is why the spreading of salt in cold weather prevents the formation of ice. The mixture (salt-water) freezes at $T \leq 0$ °C. Cryohydrate temperatures of some salts are given in Table 13.1.

Table 13.1: Cryohydrate temperatures of inorganic salts.

Salts	T_{cr} (°C)	Salt in the solution (%)	Composition of cryohydrate
Na_2SO_4	−1.2	3.86	$Na_2SO_4 \cdot 10H_2O$
KCl	−11.1	19.80	$KCl \cdot H_2O$
NaCl	−21.1	22.42	$NaCl \cdot 2H_2O$
$CaCl_2$	−55.5	29.90	$CaCl_2 \cdot 6H_2O$

The salt is a freezing point depressant. This means that when salt is dissolved in water, it lowers the temperature at which the resultant solution will freeze. As the concentration of the salt in the solution increases, the freezing point decreases.

As the solution is cooled, the water component of the solution begins to freeze. Since ice can hold very little salt, the salt that is present is confined to the remaining liquid phase, which becomes more concentrated. At the eutectic point, there is a mixture of ice/snow and concentrated brine. As the concentration of the remaining liquid phase increases, its freezing point is lowered. The salt solution remains in equilibrium until the temperature is lowered to the point at which the solubility limit is reached and the salt precipitates out of the solution. The result is a mixture of recrystallized salt, water, and brine.

Chapter 14
Oxidation-reduction processes

14.1 Basic concepts

The defining characteristic of an oxidation-reduction reaction is that electron(s) have completely moved from one atom/molecule to another. The molecule receiving the electrons has been reduced. The molecule giving the electrons has been oxidized.

Oxidation is defined as the addition of oxygen or an electronegative element to a substance or the removal of hydrogen or an electropositive element from a substance. *Reduction* is defined as the removal of oxygen or an electronegative element from a substance or the addition of hydrogen or an electropositive element to a substance. Oxidation and reduction must always occur together. The total number of electrons associated with the oxidation process must be equal to the total number of electrons associated with the reduction process.

According to the electronic concept, every redox reaction consists of two steps – half-reactions: (i) *Oxidation reaction*: Half-reactions that involve the loss of electrons are called oxidation reactions:

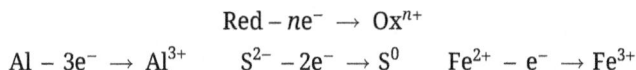

$$\text{Red} - ne^- \rightarrow \text{Ox}^{n+}$$

$$\text{Al} - 3e^- \rightarrow \text{Al}^{3+} \qquad S^{2-} - 2e^- \rightarrow S^0 \qquad \text{Fe}^{2+} - e^- \rightarrow \text{Fe}^{3+}$$

(ii) *Reduction reaction*: Half-reactions that involve the gain of electrons are called reduction reactions:

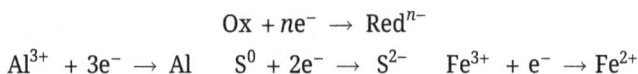

$$\text{Ox} + ne^- \rightarrow \text{Red}^{n-}$$

$$\text{Al}^{3+} + 3e^- \rightarrow \text{Al} \qquad S^0 + 2e^- \rightarrow S^{2-} \qquad \text{Fe}^{3+} + e^- \rightarrow \text{Fe}^{2+}$$

An oxidizing agent is an acceptor of electrons; *a reducing agent* is a donor of electrons.

14.2 Oxidation state

It is the oxidation state of an element in a compound that is the charge assigned to an atom of a compound. It is equal to the number of electrons in the valence shell of an atom that are gained or lost completely or to a large extent by that atom while forming a bond in a compound. Oxidation states (oxidation numbers) are numerical values assigned to atoms that reflect, in a general way, how their electrons are involved in compound formation. Let us consider the case of NaCl. In this compound, the Na atom loses one electron to a Cl atom. The compound consists of the ions Na^+ and Cl^-. The Na is said to be in the oxidation state +1, and the Cl in the oxidation state −1. The oxidation states of atoms in their ionic forms are equal to the ionic charges. In $MgCl_2$, the

https://doi.org/10.1515/9783111712246-014

Mg atom loses two electrons to become Mg^{2+}, and each Cl atom gains an electron to become Cl^-. The oxidation state of Mg is +2 and of Cl is −1.

Rules for assigning oxidation numbers: (i) The oxidation number of an element in its elementary form is zero. For example, H_2, O_2, N_2, etc., have an oxidation number equal to zero. (ii) In a single monoatomic ion, the oxidation number is equal to the charge on the ion. For example, Na^+ ion has an oxidation number of +1, and Mg^{2+} ion has +2. In their compounds, the alkali metals (group IA of the periodic table: Li, Na, K, Rb, Cs, Fr) have an oxidation state of +1, and the alkaline earth metals (IIA) have +2. (iii) Oxygen has an oxidation number of −2 in its compounds. However, there are some exceptions. In compounds such as peroxides Na_2O_2, H_2O_2 the oxidation number of oxygen is −1. In OF_2 the oxidation number of oxygen is +2. In O_2F_2 the oxidation number of oxygen is +1. (iv) In non-metallic compounds of hydrogen like HCl, H_2S, H_2O the oxidation number of hydrogen is +1, but in metal hydrides, the oxidation number of hydrogen is −1 (LiH, NaH, CaH_2, etc.). (v) In compounds of metals and non-metals, metals have a positive oxidation number while nonmetals have a negative oxidation number. For example, in NaCl, Na has a +1 oxidation number while chlorine has −1. (vi) If in a compound there are two non-metallic atoms, the atom with higher electronegativity is assigned a negative oxidation number while the other atom has a positive oxidation number. (vii) The algebraic sum of the oxidation numbers of all atoms in a compound is equal to zero. (viii) In a polyatomic ion, the sum of the oxidation numbers of all the atoms in the ion is equal to the net charge on the ion. For example, in CO_3^{2-} the sum of carbon and three oxygen atoms is equal to −2. Fluorine (F_2) is such a highly reactive non-metal that it displaces oxygen from water. Fluorine has only a −1 oxidation state in its compounds. Many elements in the periodic table can exist in more than one oxidation state. What are the oxidation states in H_2SO_4 The oxidation state of H is +1. The oxidation state of O is −2. $2(+1) + x + 4(-2) = 0$, consequently $x = +6$. The oxidation state of S is +6.

14.3 Types of redox reactions

Redox reactions are divided into three types.

1. *Intermolecular*. This type includes reactions in which the oxidizing element is part of the molecule of one substance, and the reducing element is part of the molecule of another substance. The transfer of electrons takes place between the molecules of different substances:

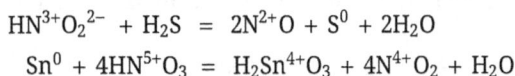

$$HN^{3+}O_2{}^{2-} + H_2S = 2N^{2+}O + S^0 + 2H_2O$$
$$Sn^0 + 4HN^{5+}O_3 = H_2Sn^{4+}O_3 + 4N^{4+}O_2 + H_2O$$

In many redox reactions, the oxidizer or reducer is also involved in the formation of salts. This additional function is more common in acids than in bases:

$$3Cu + 8HNO_3 = 3Cu(NO_3)_2 + 2NO + 4H_2O$$

$$2KMnO_4 + 16HCl = 5Cl_2 + 2MnCl_2 + 2KCl + 8H_2O$$

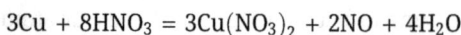

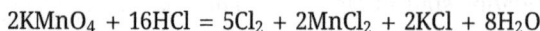

2. *Intramolecular.* In these reactions, the oxidizer and the reducer are part of the molecule of the same substance, and they can be different elements or the same element in different oxidation states; for example:

$$(NH_4)_2Cr_2O_7 = N_2 + Cr_2O_3 + 4H_2O$$

$$2KCl^{5+}O_3^{2-} = 2KCl^{1-} + 3O_2^{0}$$

$$2KNO_3 = 2KNO_2 + O_2$$

3. *Self-oxidation – self-reduction reactions (disproportionation).* Oxidizers and reducers in these reactions are the atoms of the same substance, exhibiting the same oxidation state. Atoms of the same species in a given molecule exchange electrons with each other, with one part of them increasing and another part decreasing their oxidation state; for example:

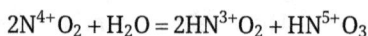

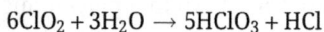

$$Cl_2 + H_2O = HCl + HClO$$

$$2N^{4+}O_2 + H_2O = 2HN^{3+}O_2 + HN^{5+}O_3$$

$$6ClO_2 + 3H_2O \rightarrow 5HClO_3 + HCl$$

14.4 Redox reactions and electrode processes: electrochemical cells

A device in which the redox reaction is carried out indirectly and the decrease in energy appears as electrical energy is called an *electrochemical cell*. An *electrolytic cell* is a cell in which electrical energy is converted into chemical energy. For example, when a lead storage battery is recharged, it acts as an electrolytic cell. There is a metal strip immersed in an aqueous solution containing the metal ions Me^{n+}. A metal strip Me is called an *electrode*. The entire assembly is called a *half-cell*. Consider what happens when a clean piece of copper metal is placed in a solution of silver nitrate. As soon as the copper metal is added, silver metal begins to form and copper ions pass into the solution. The blue color of the solution indicates the presence of copper ions. The reaction may be split into its two half-reactions. Half-reactions separate the oxidation from the reduction, so each can be considered individually:

Oxidation: $Cu(s) \rightarrow Cu^{2+}(aq) + 2e^-$

Reduction: $2Ag^+(aq) + 2e^- \rightarrow 2Ag(s)$

Overall: $\quad 2Ag^+(aq) + Cu(s) \rightarrow 2Ag(s) + Cu^{2+}(aq)$

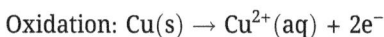

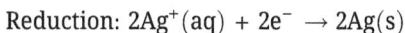

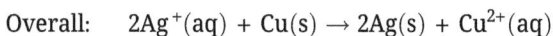

The redox reaction between Cu(s) and $Ag^+(aq)$ involves the movement of electrons. The two half-reactions can be physically separated to make an electrochemical cell. The beaker on one side is called a half-cell and contains a 1 M solution of Cu $(NO_3)_2$ with a piece of Cu metal partially submerged in the solution. The Cu metal is an electrode. The Cu is undergoing oxidation; therefore, the Cu electrode is the anode. The anode is connected to a voltmeter with a wire, and the other terminal of the voltmeter is connected to a silver electrode by a wire. Electrons flow from the anode to the cathode. The Ag is undergoing reduction; therefore, the Ag electrode is the cathode. The half-cell on the other side consists of the silver electrode in a 1 M solution of $AgNO_3$. The circuit is closed using a salt bridge, which transmits the current with moving ions. The salt bridge consists of a concentrated, nonreactive electrolyte solution such as $NaNO_3$. As electrons flow from left to right through the electrode and wire, nitrate ions (anions) pass through the porous plug on the left into the $Cu(NO_3)_2$ solution. This keeps the beaker on the left electrically neutral by neutralizing the charge on the Cu^{2+} ions that are produced in the solution as the Cu metal is oxidized. At the same time, as the nitrate ions are moving to the left, Na^+ cations move to the right, through the porous plug, and into $AgNO_3$ solution on the right. These added cations "replace" the silver ions that are removed from the solution as they are reduced to Ag metal, keeping the beaker on the right electrically neutral. Without the salt bridge, the compartments would not remain electrically neutral, and no significant current would flow. However, if the two compartments are in direct contact, a salt bridge is not necessary. The *cell potential* is created when the two dissimilar metals are connected and is a measure of the energy per unit charge available from the oxidation-reduction reaction. *Galvanic cells*, also known as voltaic cells, are electrochemical cells in which the oxidation-reduction reactions occur spontaneously and produce electrical energy. Galvanic cells have a positive cell potential. These cells are important because they are the basis for the batteries.

It is possible to construct a cell that does work on a chemical system by driving an electric current through the system. These cells are called *electrolytic cells*. Electrolytic cells, like galvanic cells, are composed of two half-cells – one is a reduction half-cell, the other is an oxidation half-cell. The direction of electron flow in electrolytic cells, however, may be reversed from the direction of spontaneous electron flow in galvanic cells, but the definition of both cathode and anode remains the same, where reduction takes place at the cathode and oxidation occurs at the anode. Because the directions of both half-reactions have been reversed, the sign, but not the magnitude, of the cell potential has been reversed. Electrolytic cells are very similar to voltaic (galvanic) cells in the sense that both require a salt bridge, both have a cathode and anode side, and both have a consistent flow of electrons from the anode to the cathode. However, there are also striking differences between the two cells. A *galvanic cell* transforms the energy released by a spontaneous redox reaction into electrical energy that can be used to perform work. The oxidative and reductive half-reactions usually occur in separate compartments that are connected by an external electrical circuit; in addition, a second connection that allows ions to flow between the compartments (porous barrier) is necessary to maintain electrical neutrality. The potential difference between the electrodes (voltage) causes electrons to flow from the reductant to the oxidant through the external circuit, generating an electric current. In an *electrolytic cell*, an external source of electrical energy is used to generate a potential difference between the electrodes that forces electrons to flow, driving a nonspontaneous redox reaction; only a single compartment is employed in most applications. In both kinds of electrochemical cells, the anode is the electrode at which the oxidation half-reaction occurs, and the cathode is the electrode at which the reduction half-reaction occurs.

Three kinds of interactions can take place between the metal atoms on the electrode and the metal ions in the solution: (1) a metal ion Me^{n+} may collide with the electrode and undergo no change; (2) a metal ion may collide with the electrode, gain n electrons, and be converted to a metal atom Me. The ion is reduced; (3) a metal atom Me on the electrode may lose n electrons and enter the solution as the ion Me^{n+}. The metal atom is oxidized. The equilibrium between the metal and its ions, which quickly sets, can be represented as

$$Me(s) \rightarrow Me^{n+}(aq) + ne^-$$

As a result, the electrode acquires a very slight electric charge, and the solution near the electrode acquires the opposite charge, as shown in Figure 14.1. It is impossible to fix the electrode potential between the metal surface and the solution near the electrode.

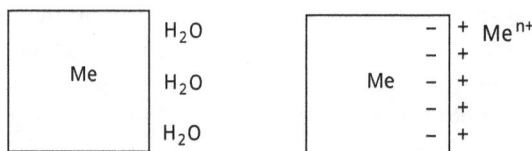

Figure 14.1: Mechanism of the formation of the electric double layer.

If we have a way to assign numerical values to the metal potential, we must compare it to the reference electrode. The reference electrode for potential measurement is the *standard hydrogen electrode*. The standard hydrogen electrode involves H^+ ions in a 1 M solution of H_2SO_4. Hydrogen molecules in the gaseous state are at a pressure of 1 atm. The value of the potential of the standard hydrogen electrode is set at zero:

$$2H^+ + 2e^- \leftrightarrow H_2 \qquad E^0 = 0.0000 \text{ V}$$

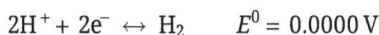

If we connect the metal electrode with the standard hydrogen electrode using the potentiometer and salt bridge, we might be able to calculate the cell potentials of this metal. The symbol for standard electrode potential is E. When a standard hydrogen electrode is combined with a standard zinc electrode, electrons flow from the zinc to the hydrogen electrode. The standard hydrogen electrode acts as the cathode, and the standard zinc electrode acts as the anode. A combination of two half-cells is called an *electrochemical cell*. The reactions that occur in the cell are:

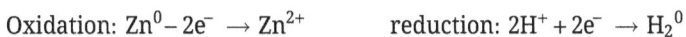

$$\text{Oxidation: } Zn^0 - 2e^- \rightarrow Zn^{2+} \qquad \text{reduction: } 2H^+ + 2e^- \rightarrow H_2^0$$

The measured value of the standard electrode potential for Zn, $E°$, is -0.763 V.

When a standard hydrogen electrode is combined with a standard copper (Cu^0) electrode, the reaction that occurs in the cell is

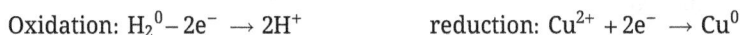

$$\text{Oxidation: } H_2^0 - 2e^- \rightarrow 2H^+ \qquad \text{reduction: } Cu^{2+} + 2e^- \rightarrow Cu^0$$

The measured value of the standard electrode potential for Cu, $E°$, is $+0.337$ V.

Therefore, the potential of the standard hydrogen electrode is equal to zero. Any electrode at which a reduction half-reaction shows a greater tendency to occur than $2H^+ + 2e^- \rightarrow H_2^0$ (1 atm) does, has a positive electrode potential, $E°$. If the tendency for a reduction process is given by $+E°$, the oxidation tendency is simply the negative of this value, that is, $-E°$. By means of numerical values of the half-reaction electrode potential from the table of standard electrode potentials, one can predict whether a spontaneous reaction will occur in the forward direction in each case. If $E°_{cell}$ is positive, the reaction will occur spontaneously in the forward direction. If $E°_{cell}$ is negative, the reaction will proceed spontaneously in the reverse direction. If all the metals are arranged in order of increasing standard electrode potentials, the so-called electrochemical potential line of metals is obtained:

Li, K, Ba, Ca, Na, Mg, Al, Mn, Zn, Cr, Fe, Cd, Co, Ni, Sn, Pb, H, Sb, Bi, Cu, Hg, Ag, Pd, Pt, Au

The metals that come before hydrogen in this order have a negative potential, and those after it have a positive potential. The lower the value of the standard electrode potential of the metal (further to the left in the row), the more chemically active it is, the easier it is to oxidize, and the more difficult it is to reduce its ions. Therefore, the first metals in the row with the most negative potential have the highest chemical activity. They are the strongest reducers and occur in nature mainly in a combined state. The metals which are positioned before hydrogen in the *row of standard electrode potentials* dissolve in HCl(aq) and H_2SO_4(aq), producing H_2(g):

$$Zn(s) + 2HCl(aq) \rightarrow ZnCl_2(aq) + H_2(g)$$

$$Mg(s) + H_2SO_4(aq) \rightarrow MgSO_4(aq) + H_2(g)$$

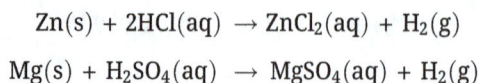

Metals with a positive potential have the least chemical activity and cannot displace hydrogen from acids, except for oxidative acids (HNO_3). In nature, they are more often found in a free state. Copper, mercury, and silver do not dissolve in HCl(aq) and H_2SO_4(aq).

The metals, which are positioned on the left side after Mg, displace the metals that are positioned on the right side of them, such as:

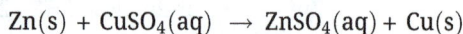

$$Zn(s) + CuSO_4(aq) \rightarrow ZnSO_4(aq) + Cu(s)$$

When a zinc rod is dipped in a copper sulfate solution, a redox reaction begins; hence, zinc is oxidized to Zn^{2+} ions, and Cu^{2+} ions are reduced to metal.

Electrode potential is the potential difference between the electrode and its ions in solution see (Figure 14.2).

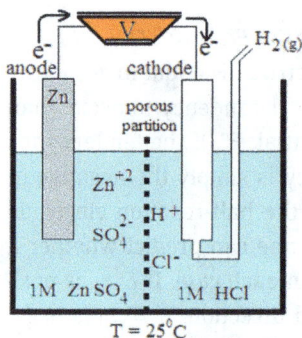

Figure 14.2: Electrode potential measurements.

14.5 Standard electrode potential

Standard electrode potential is the potential of an electrode with respect to the standard hydrogen electrode, as shown in Figure 14.3. The standard potential ($E°$) is a convenient metric for determining the relative strength of redox reactions at 1 M concentrations. It shows how easily a reaction will happen. A very positive value means the reaction is very easy, while a very negative value means it is very difficult. The sign of $E°$ is changed by reversing the reaction. The standard state is defined as the one corresponding to 25 °C (298.15 K), with unit activity for all the substances in an electrochemical zero-current cell under 1 bar of pressure (105 Pa). For a reaction in which H^+ ions participate, the standard state is pH = 0 (approximately 1 mol acid). In the hydrogen electrode used as the standard of electrode potential, 1 atm of hydrogen gas ($aH^+ = 1$)

is slowly contacted with a platinum-black electrode immersed in a strong acid solution of activity $aH_2 = 1$.

The potential is expressed as:

$$E = E^0 + (RT/F) \cdot \ln(aH^+/aH_2)$$

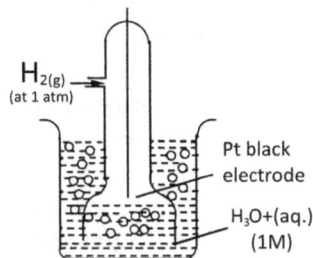

Figure 14.3: Standard hydrogen electrode.

Although reduction potential is usually expressed with reference to the standard, the standard hydrogen electrode is inconvenient to handle. Therefore, a saturated calomel or Ag/AgCl electrode is used as a reference electrode for everyday electrochemical measurements, and experimental potentials are measured against these electrodes or converted into standard values. When the standard value is set to 0, the saturated calomel electrode value is 0.242 V.

The equilibrium electrode potentials depend on the nature of the metal, the concentration of ions in the electrolyte, and the temperature. The first quantitative theory of electrode potentials was created by Nernst. Depending on the ratio between the concentrations of the oxidizer and the reducer under conditions other than standard, the equilibrium redox potential E is calculated according to the *Nernst equation,* where by definition $E^0 = 0$ in the standard state:

$$E = E^0 + \frac{RT}{nF} \ln a_{Me}{}^{n+}$$

In this equation, E^0 is the standard electrode potential; R is the universal gas constant (8.314 J/mol·K); T is the absolute temperature (in K); F is Faraday's constant ($F = 96{,}494$ C/mol); n is the oxidation state of the ions; a is the activity of the ions, in mol/L. With the Nernst equation, it is possible to calculate the cell potential under nonstandard conditions.

A redox reaction takes place only when redox partners exist, and a reactant can be either an oxidant or a reductant depending on its reaction partner. The relative redox capability can be expressed numerically by introducing the reduction potentials E^0 of imaginary half-reactions. The Gibbs free energy change ΔG^0 of a reaction is related to E^0

$$\Delta G^0 = -n.F.E^0$$

where n is the number of transferred electrons. If E^0 of a redox reaction is positive, ΔG^0 is negative, and the reaction occurs spontaneously. Consequently, instead of the free energy change, the difference in reduction potentials can be used to judge the thermodynamic spontaneity of a reaction. The higher the reduction potential of a reagent, the stronger its oxidation ability. The positive or negative signs are based on the expedient of setting the reduction potential of a proton to 0, and it should be understood that a positive sign does not necessarily mean oxidizing, and a negative sign reducing. The series arranged in the order of redox power is called the *electrochemical series*. The electrochemical series is the activity series. It has been formed by arranging the metals in order of increasing standard reduction potential value.

The redox properties of the elements very roughly follow the following general trends: Elements on the left of the periodic table tend to act as reductants; those on the right as oxidants. The noble gases are inert and, as elements, tend not to act as good oxidants or reductants. As one moves toward the left of the periodic table, elements tend to act as good reductants, while those toward the right tend to act as increasingly good oxidants. As one moves down a group of the periodic table, elements tend to act as either weaker oxidants or better reductants.

14.6 Electrolysis

Electrochemistry studies two main types of processes: processes carried out in galvanic cells and accumulators, in which electrical energy is obtained through chemical redox processes, and electrolysis processes, which are carried out under the action of electric current passing through an electrolytic medium. During electrolysis, the energy of the electric current is converted into chemical energy. In other words, the process is the opposite of the process taking place in the galvanic cell. The flow of direct electric current through a solution or melt of an electrolyte is accompanied by a directional movement of its ions. The cations are directed to the cathode and reduced, and the anions to the anode and oxidized. Therefore, during electrolysis, the anode is positively charged (A^+) and the cathode is negatively charged (K^-). *Electrolysis* is defined as a set of redox processes that occur when an electric current is passed through a solution or melt of an electrolyte in which electrodes are immersed (see the electrolysis cell illustration, Figure 14.4).

The nature of the electrode processes in electrolysis depends on the following factors: electrolyte composition, concentrations, electrode material, and electrolysis conditions (temperature, current strength). Depending on the activity of the material, anodes are divided into: *active* – the anode material is easily oxidized (e.g., Fe, Zn, and Cu) and *inert (passive)* – the anode material does not oxidize (graphite, Pt). On the inert anode, the anions of the electrolyte are oxidized.

There are two types of electrolysis: – electrolysis in electrolyte melts and electrolysis in an electrolyte solution. Electrolysis in melts is used for the industrial produc-

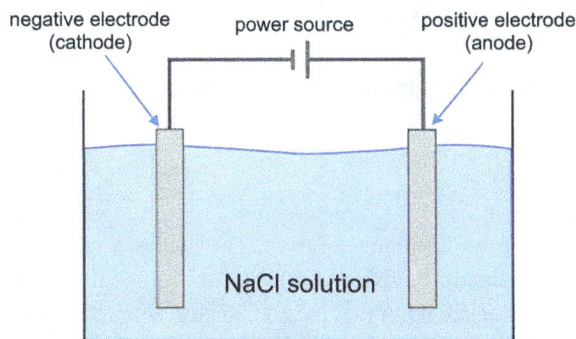

Figure 14.4: Electrolysis in an aqueous solution of NaCl.

tion of active metals (Na, Mg, Al, Li, Ca, Be, and Tl), the separation potential of which from solutions is more negative than the separation potential of hydrogen. Through electrolysis in solutions, metals such as Cu, Zn, Mn, Ni, etc., are obtained and purified (electrometallurgy). Electrolysis is widely used in the application of metal coatings to products, in the production of metal copies of embossed objects, and in the production of various products (chlorine, fluorine, and sodium hydroxide) in the chemical industry.

14.6.1 Electrolysis in melts

If electrodes are immersed in a melt of potassium chloride and a direct electric current is passed through the melt of KCl, the salt ions will begin to move toward the electrodes (cations K^+ toward the cathode, anions Cl^- toward the anode). At the cathode, a process of reduction of cations K^+ takes place, and at the anode – an oxidation of anions Cl^-:

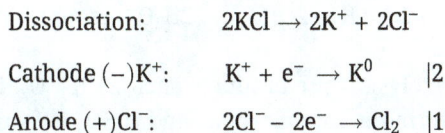

$$\text{Dissociation:} \qquad 2KCl \rightarrow 2K^+ + 2Cl^-$$

$$\text{Cathode } (-)K^+: \qquad K^+ + e^- \rightarrow K^0 \qquad |2$$

$$\text{Anode } (+)Cl^-: \qquad 2Cl^- - 2e^- \rightarrow Cl_2 \quad |1$$

The general equation of the electrolysis process can be represented as follows:

$$2KCl \rightarrow 2K^+ + 2Cl^- \rightarrow 2K + Cl_2$$

Since there are no water molecules in the melts, the ions that make up the electrolyte are subjected to cathode reduction and anode oxidation. By electrolysis of melts, practically all metals can be obtained, regardless of their location in the activity series, including alkali metals.

14.6.2 Electrolysis in aqueous solutions of electrolytes

Electrolysis in aqueous solutions of electrolytes is complicated by the involvement of water molecules.

14.6.2.1 Electrolysis in NaCl solution

Sodium chloride dissociates into ions in an aqueous solution. The solution contains Na^+ and Cl^- ions and, in smaller concentrations, H^+ and OH^- ions (water is a weak electrolyte). Under the action of direct electric current, Na^+ and H^+ are directed to the cathode, and Cl^- and OH^- ions to the anode. Since Na^+ ions are more difficult to reduce than H^+, hydrogen gas H_2 will be released at the cathode. At the anode, the process of oxidation of chloride ions Cl^- takes place, and chlorine gas ($2Cl^- - 2e^- = Cl_2$) will be released; see Figure 14.4. At the same time, due to the discharge of H^+ ions, the equilibrium of electrolytic dissociation of water near the cathode is disturbed, and the solution in the cathode space acquires a basic character. The Na^+ ions remaining in the solution, together with the released OH^- ions, form a solution of NaOH in the electrolysis cell. Thus, during the electrolysis of oxygen-free salts, hydroxides (bases) can be obtained.

14.6.2.2 Electrolysis in Na₂SO₄ aqueous solution

Sodium sulfate dissociates in aqueous solutions: $Na_2SO_4 = 2Na^+ + SO_4^{2-}$. Analogous to the previous example, Na^+ ions are concentrated around the cathode, and the oxygen-containing anion SO_4^{2-} is concentrated around the anode. In this case, not only the cation but also the SO_4^{2-} anion does not undergo a change. H_2 is released at the cathode, and O_2 is released at the anode during the electrolysis of water:

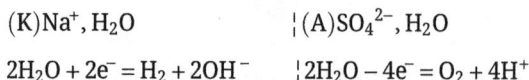

$$(K)Na^+, H_2O \qquad\qquad | (A)SO_4^{2-}, H_2O$$
$$2H_2O + 2e^- = H_2 + 2OH^- \qquad | 2H_2O - 4e^- = O_2 + 4H^+$$

As a result of the electrolysis, four products such as H_2, O_2, NaOH, and H_2SO_4 are formed. NaOH is formed in the cathode zone, and H_2SO_4 is formed in the anode zone of the electrolysis cell.

14.6.2.3 Electrolysis in CuSO₄ aqueous solution

Copper sulfate dissociates in aqueous solutions: $CuSO_4 = Cu^{2+} + SO_4^{2-}$. The following processes take place at the electrodes:

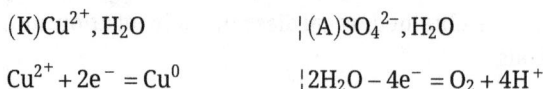

$$(K)Cu^{2+}, H_2O \qquad\qquad | (A)SO_4^{2-}, H_2O$$
$$Cu^{2+} + 2e^- = Cu^0 \qquad\qquad | 2H_2O - 4e^- = O_2 + 4H^+$$

The reduced copper is separated at the cathode. Oxygen is released from water at the anode; that is, the water decomposes, and a solution of sulfuric acid accumulates in the anode zone.

14.6.2.4 Electrolysis in KOH aqueous solution
KOH is dissociated: $KOH = K^+ + OH^-$ and:

$$(K)\ K^+, H_2O \qquad\qquad |\ (A)\ OH^-, H_2O$$
$$2H_2O + 2e^- = H_2 + 2OH^- \quad |\ 2OH^- - 4e^- = O_2 + 2H^+$$

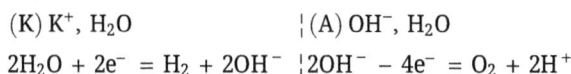

Unlike the previous examples, in this case, oxygen is released from hydroxide ions at the anode. At the cathode, water decomposes, and the concentration of the base in the cathode zone of the electrolysis cell increases.

14.6.2.5 Electrolysis in H₂SO₄ aqueous solution
In the electrolysis of solutions of oxygen-containing acids, H_2 is reduced from the acid at the cathode, and water is decomposed (oxidized) at the anode:

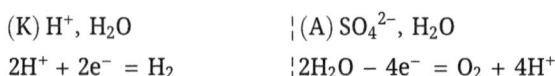

$$(K)\ H^+, H_2O \qquad\qquad |\ (A)\ SO_4^{2-}, H_2O$$
$$2H^+ + 2e^- = H_2 \qquad\qquad |\ 2H_2O - 4e^- = O_2 + 4H^+$$

Electrolysis is widely used in industry. There are various elements and chemical compounds, including organic compounds, that are only produced by electrolysis, such as non-ferrous metals (aluminum and copper), chlorine, NaOH (chlor-alkali process), and the production of green hydrogen through water electrolysis, etc.

Electrolysis is used extensively in metallurgical processes, such as in the extraction (electrowinning) or purification (electrorefining) of metals from ores or compounds and in the deposition of metals from solution (electroplating). Generally, noble metals (least reactive) cannot be extracted from their ores using electrolysis, such as Au, Ag, and Pt. Additionally, some of the less reactive metals, such as Fe, Sn, and Pb, also cannot be extracted by electrolysis.

Section B: **Chemical elements**

——

Section B.I: **Main groups of the periodic table**

Chapter 15
Hydrogen

Hydrogen is the first element in the periodic table, and it does not belong to a specific group. This element has unique chemistry. With an atomic mass of 1.00794 u, it is the lightest chemical element, and its monatomic form (H) is the most abundant chemical substance, forming about 75% of the baryonic mass of the universe. Noncompact stars are mainly composed of hydrogen in a plasma state. At standard temperature and pressure, hydrogen is a gas (H_2) with nonmetallic properties, nontoxic, highly flammable, colorless, tasteless, and odorless. The main part of hydrogen on Earth is in a bound state (water and organic compounds) since it easily forms covalent bonds with most nonmetals.

Hydrogen plays a particularly important role in neutralization processes, in which protons are exchanged between soluble particles in multiple reactions. In its ionic compounds, it can form both negatively charged anions (H^-) and positively charged cations (H^+). At first glance, hydrogen cations look like independent protons, but in reality, they are always more complex structures.

15.1 Isotopes of hydrogen

Isotopes of hydrogen are especially important for the development of chemistry. Although hydrogen was described about 200 years ago, the presence of different isotopes of hydrogen has only been a discovery since the 1930s. Hydrogen has three natural isotopes: protium (^1H), deuterium (^2H), and tritium (^3H), as shown in Figure 15.1. Another four highly unstable isotopes (^4H, ^5H, ^6H, and ^7H) were synthesized under laboratory conditions and do not occur in nature. Its three natural isotopes are quite different in terms of their molar masses, which is why their physical and chemical properties are measurably different. The most abundant in nature is protium (99.985%), followed by deuterium (0.015%) and radioactive tritium (10^{-15}%). In fact, it is the only set of isotopes for which special symbols are used: H for protium, D for deuterium, and T for tritium.

The most common isotope of hydrogen, protium (^1H), consists of a single proton and contains no neutrons. This simplest hydrogen atom is of great importance for theoretical physics. For example, it is the only neutral atom for which the Schrödinger equation has been solved analytically, and the study of energies and bonds in it has played an important role in the development of quantum mechanics. Years after the discovery of nuclear magnetic resonance (NMR), hydrogen was the most extensively studied element using this technique. NMR is of utmost importance for chemists, both to identify compounds and to study their electronic structures. It is also widely used in the field of healthcare for magnetic resonance imaging (MRI).

https://doi.org/10.1515/9783111712246-015

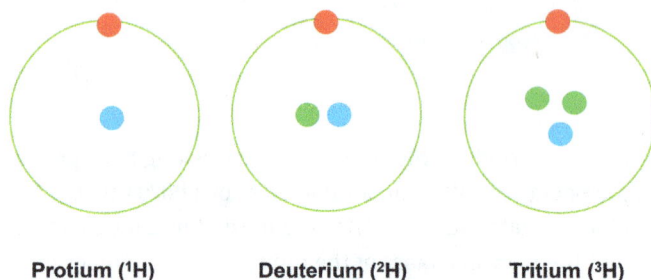

Protium (^{1}H)   Deuterium (^{2}H)   Tritium (^{3}H)

Figure 15.1: Isotopes of hydrogen.

Deuterium is a stable isotope of hydrogen, which is not radioactive but is toxic. Tritium is a radioactive isotope with a half-life of about 12 years. With such a short half-life, it can be expected that this isotope is difficult to survive in nature. In fact, tritium is constantly formed under the influence of cosmic rays in the upper atmosphere. One of the ways to synthesize tritium involves the interaction of neutrons with nitrogen atoms or lithium-6 in nuclear reactors. Tritium has also been used for medicinal purposes. In its radioactive decay, the isotope emits β rays, but not harmful γ rays. It has been used in the production of the hydrogen (tritium) bomb.

As the molar mass of the three isotopes increases from 2 to 6 g/mol, there is a significant increase in boiling points and in the energy of the chemical bond in the row H_2 – D_2 – T_2. Covalent bonds of deuterium and tritium with other elements are stronger. Normal H_2O and "heavy water" D_2O differ in physical properties: unlike water, deuterium oxide melts at 3.8 °C and boils at 101.4 °C. The density of D_2O is about 10% higher than that of H_2O over the entire temperature range, as a result of which heavy water ice cubes will sink into H_2O at 0 °C.

15.2 Synthesis

Hydrogen gas was artificially produced at the beginning of the sixteenth century by mixing metals with acids. Between 1766 and 1781, Henry Cavendish first identified it as an individual substance and found that when it was burned, water was formed, hence the name of the element.

Industrial production of hydrogen is most often carried out in the processing of natural gas, and less often in more energy-intensive technologies (electrolysis of water). In the endothermic process of interaction of natural gas (methane) with steam at high temperature, carbon monoxide and hydrogen are obtained. The two products are difficult to separate, as this requires cooling the mixture to −205 °C, whereby the carbon monoxide condenses. To overcome this problem and to increase hydrogen production, additional steam is injected and catalysts (Ni) are used. Under these conditions, CO is oxidized by an exothermic reaction to CO_2, and the added H_2O is reduced to H_2:

$$CH_4 + H_2O \rightarrow CO + 3H_2$$

$$CO + H_2O \rightarrow CO_2 + H_2$$

Carbon dioxide can be separated from hydrogen in several ways. One of them is cooling the products to the condensation temperature of CO_2 (−78 °C), which is much higher than that of H_2 (−253 °C). However, this process still requires large-scale refrigeration systems. Another method involves passing the gas mixture through a solution of potassium carbonate. Carbon dioxide is an acidic oxide, as opposed to neutral CO, and interacts with K_2CO_3 to produce $KHCO_3$:

$$K_2CO_3 + CO_2 + H_2O \rightleftharpoons 2KHCO_3$$

For industrial purposes, the purity of hydrogen obtained by chemothermal processes is satisfactory. Hydrogen of high purity (99.9%) is obtained electrochemically, namely by electrolysis of KOH or NaOH solutions, as a result of which O_2 and H_2 are released:

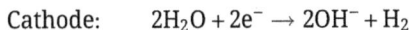

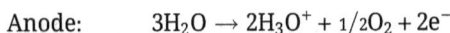

Cathode: $\quad 2H_2O + 2e^- \rightarrow 2OH^- + H_2$

Anode: $\quad 3H_2O \rightarrow 2H_3O^+ + 1/2O_2 + 2e^-$

Other methods of hydrogen production are also used in industry; for example, the interaction of water vapor with molten coal or iron chips, as well as the electrolysis of water:

$$C + H_2O = CO + H_2$$

$$Fe + H_2O = FeO + H_2$$

$$2H_2O = 2H_2 + O_2$$

In laboratory conditions, hydrogen can be obtained by reacting dilute acids (HCl, H_2SO_4) with multiple metals, as well as metals with bases:

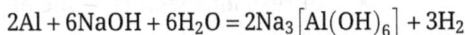

$$Zn + H_2SO_4 = ZnSO_4 + H_2$$

$$2Al + 6NaOH + 6H_2O = 2Na_3\left[Al(OH)_6\right] + 3H_2$$

15.3 Properties

In some variants of the periodic table, hydrogen is located in the group of alkali metals (the H atom, like the atoms of alkali metals, contains one valence electron and forms positive ions). On the other hand, hydrogen can naturally be located in the group of halogen elements (hydrogen, like halogen elements, forms diatomic molecules and negatively charged ions). Some authors put it in both places, and according to others, hydrogen has an independent place in the periodic table, emphasizing its uniqueness. Since the electronegativity of hydrogen is higher than those of alkali metals and lower than those of halogen elements, it makes sense to place hydrogen halfway between the two groups. The ionization potential of atomic hydrogen (13.6 eV) and its electronegativity ($\chi = 2.2$) define it as a nonmetal. Diatomic hydrogen molecules H_2 are very light, mobile, and nonpolar. The intermolecular interaction between them is weak – dispersion forces. Thus, hydrogen is slightly soluble in liquids and very volatile. Under standard conditions, H_2 is a colorless, odorless gas that liquefies at −253 °C and solidifies at −259 °C.

Hydrogen does not exhibit high reactivity, partly due to the high energy of the covalent bond H–H (436 kJ/mol). Therefore, its decay into atoms begins only at 2,000 K. Reactions with H_2, as a rule, take place when heated, where H_2 exhibits both reducing and oxidizing properties:

$$CuO + H_2 = Cu + H_2O$$

$$2Na + H_2 = 2NaH$$

The reaction of hydrogen with oxygen, when heated, occurs with an explosion. When they are mixed in a gaseous state, an explosive reaction is induced with a chain mechanism:

$$2H_2 + O_2 \rightarrow 2H_2O$$

Hydrogen is highly flammable and burns on contact with air. The enthalpy of hydrogen combustion is – 286 kJ/mol. A similar mechanism occurs with the interaction of hydrogen and chlorine when irradiated with light. These explosive mixtures detonate spontaneously by a spark, heating, or sunlight. The self-ignition temperature, at which hydrogen begins to burn spontaneously in the air, is 500 °C. Pure hydrogen-oxygen combustion produces ultraviolet radiation and, at high oxygen concentrations, is almost imperceptible to the naked eye, which makes the detection of hydrogen leakage into the air difficult and dangerous.

Molecular hydrogen reacts with all elements with oxidizing action. At room temperature, spontaneous and powerful reactions with chlorine and fluorine take place, forming the corresponding hydrogen halides – HCl and HF. The rate of reaction of H_2 with halogen elements decreases from top to bottom in the group, with the reaction with F_2 being particularly intense:

$$H_2 + F_2 \rightarrow 2HF$$

The reaction of hydrogen with nitrogen in the absence of a catalyst is a very slow process:

$$3H_2 + N_2 \rightarrow 2NH_3$$

At high temperatures, hydrogen reduces many of the metal oxides to a metallic state:

$$H_2 + CuO \rightarrow Cu + H_2O$$

Usually, precious metals like nickel and iron catalyze reactions involving H_2. In the presence of catalysts, the synthesis of ammonia, methanol, and all reactions involving H_2 in organic chemistry take place. In the case of a catalyst (usually Pd or Pt powder), H_2 reduces the double and triple carbon-carbon bonds to single ones. Thus, ethene (C_2H_4) is reduced to ethane (C_2H_6):

$$H_2C = CH_2 + H_2 \rightarrow H_3C - CH_3$$

Hydrogen reduction is used in converting unsaturated liquid fats with multiple double carbon-carbon bonds into more highly meltable, partially saturated solid fats (margarines) containing fewer double carbon-carbon bonds.

At room temperature, reactions involving atomic hydrogen are possible. At the same time, no special devices are needed for this purpose, and atomic hydrogen can be used at the time of its production. For example, when the amphoteric metal zinc reacts with a solution of a base, hydrogen is formed, which at the time of its separation is in a partial atomic state:

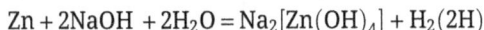

$$Zn + 2NaOH + 2H_2O = Na_2[Zn(OH)_4] + H_2(2H)$$

If nitrates or nitrites are present in the solution, atomic hydrogen reduces them to ammonia:

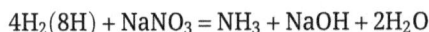

$$4H_2(8H) + NaNO_3 = NH_3 + NaOH + 2H_2O$$

The summary equation of the reaction can be represented as follows:

$$4Zn + NaNO_3 + 7NaOH + 6H_2O = 4Na_2\left[Zn(OH)_4\right] + NH_3$$

This reaction is used in analytical chemistry to determine the content of nitrates and nitrites.

15.4 Use and biological role

Hydrogen is used to produce ammonia, hydrochloric acid, in organic synthesis, and in the production of pure metals. Research is being carried out on the use of hydrogen as a fuel (a substitute for coal, oil, and natural gas).

Hydrogen is the most abundant element in the universe, consisting of 92% atomic hydrogen and 7% helium. It is the third most abundant element on Earth and exists only in a bound state. Free hydrogen is found high in the atmosphere. Hydrogen on

Earth is mainly involved in the composition of water, in gases of volcanic origin (hydrogen sulfide H_2S), and in numerous organic compounds. Water is the main environment for the vital activity of organisms. Most substances involved in metabolic processes dissolve in it. The water content in the organs and tissues of the body is quite high. The physiological solution (saline) is a 0.9% solution of NaCl. Water has a high relative heat capacity and, due to the slow heat exchange with the environment, ensures the maintenance of a constant body temperature. When overheated, water evaporates from the surface of the body. Due to the high heat of evaporation of water, this process is accompanied by energy losses, and the body temperature decreases. The buffer systems (carbonate, phosphate, and hemoglobin) maintain the acid-base balance of the body. The average pH value in the body corresponds to the pH of saline and ranges from 6.8 to 7.4. Individual organs and tissues may have pH values that differ from the physiological pH. For example, the acidity in the stomach is high, and the pH is in the range of 0.9–1.1. There is a weakly basic reaction in the bile (pH 7.5–8.5).

In fact, the existence of life on the Earth is a result of two unique specific properties of hydrogen: its close electronegativity to that of carbon, and its ability to form hydrogen bonds when covalently bonded with nitrogen or oxygen. The low polarity of the carbon-hydrogen bond contributes to the stability of organic compounds. Hydrogen bonds are the basis of all biomolecules: proteins, DNA, RNA, etc. To function properly, most proteins form compact shapes, which is achieved through hydrogen bonds.

15.5 Chemical compounds

Although the diatomic hydrogen molecule is not sufficiently active under standard conditions, H_2 forms compounds with most of the chemical elements. Hydrogen has oxidation states of −1 and +1. In its compounds with halogen elements and other nonmetals, it has a partial positive charge. In its compounds with fluorine, oxygen, or nitrogen, hydrogen is involved in the formation of hydrogen bonds, essential for the stability of many biomolecules.

15.5.1 Hydrides

Hydrogen forms compounds with elements of lower electronegativity (metals and nonmetals), acquiring a partial negative charge in them. These compounds are commonly called *hydrides*. Hydrogen, which forms binary compounds with most elements, has an electronegativity only slightly above the average for the rest of the elements of the periodic table, as a result of which it behaves as a weakly electronegative nonmetal and forms ionic compounds with many metals and covalent compounds with nonmetals. In addition, it also forms metal hydrides with some of the transition metals. Some lanthanides and actinides also form metallic hydrides. The name "hydride" is reserved only for compounds of hydrogen with metals, and in other cases, it is determined by the name of the other element. Hydrides are divided into three main types: *ionic, covalent,* and *metallic.*

15.5.1.1 Ionic hydrides

All ionic hydrides are chemically active white solids and are formed with the most electropositive (alkaline and alkaline earth) metals: NaH, KH, CaH_2, BaH_2, etc. The hydride ion is a strong reducer that reduces water to hydrogen. The reaction of calcium hydride with water is used in organic chemistry as a chemical desiccant for organic solvents:

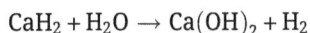

$$CaH_2 + H_2O \rightarrow Ca(OH)_2 + H_2$$

Ionic (salt-like) hydrides are crystalline substances that melt at temperatures of 400–600 °C without decomposition. Their melts conduct electric current.

15.5.1.2 Covalent hydrides

Hydrogen forms compounds containing covalent bonds with all nonmetals (except noble gases) and with very weakly electropositive metals, such as gallium and tin. Almost all simple covalent hydrides are gases at room temperature. There are three subcategories of covalent hydrides: those in which the hydrogen atom is nearly neutral and others in which the hydrogen has a weak positive or weak negative charge (including electron-deficient boron compounds).

The first category of covalent hydrides are gases with low boiling points, such as selenide (H_2Se, −60 °C) and phosphine (PH_3, −90 °C). The largest group of neutral covalent hydrides are those containing carbon. These are *hydrocarbons* that consist of alkanes, alkenes, alkynes, and aromatic hydrocarbons. Many hydrocarbons are large molecules with strong intermolecular forces that allow the existence of liquid or solid states of matter at room temperature. All hydrocarbons are thermodynamically unstable in terms of oxidizing reagents; they react with O_2 to CO_2 and H_2O:

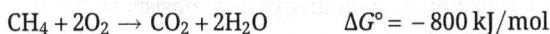

$$CH_4 + 2O_2 \rightarrow CO_2 + 2H_2O \qquad \Delta G° = -800 \text{ kJ/mol}$$

Hydrogen forms multiple compounds with carbon and with heteroatoms, which, due to their relationship with living organisms, are called organic compounds. The study of their properties is the subject of organic chemistry and biochemistry. Millions of hydrocarbons are known, and most of them are formed by complex synthetic mechanisms that rarely involve reactions with pure hydrogen. Most organic compounds contain mainly hydrogen-carbon bonds, determining the basic properties of these compounds.

Ammonia, water, hydrogen fluoride, etc., belong to the second category of covalent hydrides – hydrogen compounds in which hydrogen is in the +1 oxidation state. These compounds differ from other covalent hydrides by their extremely high melting and boiling points, especially characteristic of compounds of the seventh A group. The hydrogen cations in these compounds are attracted by the electron pairs of the halide anions and form a weak bond known as a hydrogen bond or, more precisely, a "proton bridge." Although the hydrogen bond is an intermolecular bond, it is strong enough, although weaker than the covalent bond. For example, the $H_2O \ldots HOH$ proton bridge has an energy of 22 kJ/mol compared to 464 kJ/mol for the covalent O-H bond. In chemistry, the proton bridge is considered a very strong dipole-dipole interaction resulting from the strong polar covalent bonds in the respective molecules.

The third category of covalent hydrides, in which hydrogen has a weak negative charge, includes diborane (B_2H_6), silane (SiH_4), germane (GeH_4), and stanane (SnH_4) (systematically called tin(IV) hydride). These monomeric hydrides containing negatively charged hydrogen are active reducers and react intensively with oxygen. For example, SnH_4 burns to tin(IV) oxide and water:

$$SnH_4 + 2O_2 \rightarrow SnO_2 + 2H_2O$$

They are gaseous substances that interact with water to form acids and hydrogen:

$$SiH_4 + 3H_2O = H_2SiO_3 + 4H_2$$

When heated, covalent hydrides decompose. Some hydrides are polymeric structures in which hydrogen atoms form a bridge between metal atoms, characteristic of the "weak" metals – beryllium, magnesium, aluminum, copper, and zinc, which form these structures.

15.5.1.3 Metal hydrides

The hydrides of some transition metals (d-elements) form a third class of hydrides called metallic. The chemical bond in them is partly ionic, partly covalent, and partly metallic. They resemble both metals in their electrical conductivity and nonmetals (brittle, nonplastic). These compounds most often have a variable nonstoichiometric composition (the so-called berthollides), for example, TiH, TiH_2, $TiH_{1.5}$, $TiH_{1.9}$, etc. The nature of these compounds is complex. For example, titanium hydride consists of $(Ti^{4+})(H^-)_{1.9}(e^-)_{2.1}$. The presence of free electrons determines the metallic character and high electrical conductivity of these compounds.

When heated in an inert atmosphere, these hydrides decompose into metals and hydrogen, burn in air, and interact with acids to form salts and hydrogen:

$$TiH_2 = Ti + H_2$$
$$2TiH_2 + 3O_2 = 2TiO_2 + 2H_2O$$
$$2TiH_2 + 6HCl = 2TiCl_3 + 5H_2$$

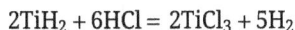

The density of metal hydrides is often less than that of pure metals due to structural changes in the metal crystal lattice, so they are usually quite brittle. The electrical conductivity of metal hydrides is usually lower than that of the corresponding metals. Most metal hydrides are produced by heating the metal with hydrogen at high pressure.

The main use of metal hydrides is in nickel-metal-hydride batteries, which are used in computers, mobile phones, and many wireless electronic devices. These are metal alloys (TiFe, $ZnMn_2$, $LaNi_5$, and Mg_2Ni) bonded to hydrides that reversibly absorb and release H_2 at room temperature. Although hydrides are formed with almost all elements of the main groups, the number and combination of possible compounds are very diverse. In inorganic chemistry, hydrides also serve as bridge ligands, connecting two metal centers in complex compounds. This function is especially common in the elements of the third A group, especially in boron and aluminum.

The properties of hydrides formed by the elements of one period change from basic (NaH and MgH_2) through amphoteric (AlH_3) to acidic (SiH_4, PH_3, H_2S, and HCl). AlH_3 is amphoteric, so it interacts with both the basic hydride NaH and the acidic hydride SiH_4. The reaction equations are similar to the equations of interaction of aluminum oxide with sodium and silicon oxides:

$$NaH + AlH_3 = NaAlH_4$$
$$2AlH_3 + 3SiH_4 = Al_2[SiH_6]_3$$

15.5.2 Water

Water is one of the most abundant substances on this planet. Without water as a solvent for chemical and biochemical reactions, life would be impossible.

Water is one of the few substances that are freely found in nature in its three aggregate states. It can occur in many and varied forms – water vapor, clouds, glaciers, icebergs, small streams, and others. Basically, water is divided into sweet and salty. Humans and animals instinctively recognize the sweet from the salt water, which is not suitable for use.

15.5.2.1 Synthesis
Water is produced by the direct interaction of H_2 and O_2 in the gaseous state:

$$2H_2 + O_2 \rightarrow 2H_2O$$

Water is one of the products in the combustion of organic compounds:

$$CH_4 + 2O_2 \rightarrow CO_2 + 2H_2O$$

15.5.2.2 Properties of water

The hydrogen bonds characteristic of water molecules lead to a very rare feature of water, namely that in the liquid phase, it is denser than in the solid phase. For most substances, the opposite is typical – the solid phase has a higher density than the liquid one. If this were the case with water, lakes, rivers, and seas in some parts of the world where temperatures drop below freezing would freeze from the bottom up, in which marine organisms would not survive in such an environment. The reason for this unusual behavior of water is the widely deployed structure of the ice, which is due to the network of hydrogen bonds. When ice melts, a large part of the hydrogen bonds are not broken but are preserved in the liquid phase. The density of water reaches a maximum at 4 °C, which is explained by the predominant water dimers $(H_2O)_2$ at this temperature. It has been calculated that without the presence of hydrogen bonds in its structure, ice would melt at 173 K (−80 °C), making life on Earth impossible. Hydrogen bonding is also the cause of the high boiling point of water (100 °C at 101.3 kPa). There is no other known compound in which such supercooling and such overheating of its liquid phase is possible (water can remain liquid from −30 to +200 °C). The high solubility of some substances in water is also due to the formation of hydrogen bonds between the molecules of water and the corresponding substances.

Another unique property of water can be traced in its phase diagram. In almost all substances, the solid-liquid line has a positive slope, which means that applying a sufficiently high pressure to the liquid phase will cause solidification of the substance. However, water has an unusual phase pattern because the density of ice is lower than that of liquid water. The Le Chatelier's principle shows that the denser phase is preferred when the pressure increases. Therefore, in the case of water, the increased pressure causes the less dense solid phase to melt to a liquid phase.

On the other hand, water is important in view of its acid-base chemistry, resulting from its self-dissociation, as discussed in a previous section:

$$2H_2O \rightleftharpoons H_3O^+ + OH^-$$

Water also controls the limits of redox processes in an aqueous solution. It is not possible for redox reactions to take place in an aqueous solution beyond the limits of oxidation of water to oxygen or reduction to hydrogen, respectively:

$$H_2O \rightarrow 1/2 O_2 + 2H^+ + 2e^-$$

$$2H_2O + 2e^- \rightarrow H_2 + 2OH^-$$

The oxygen atom in water is a donor of electron pairs; that is, water can act as a Lewis base, for example, in the cases of hydrated metal ions or complex ions, such as $[Ni(:OH_2)_6]^{2+}$, in which the bonding is much stronger than in a simple ion-dipole interaction.

Consequently, the chemistry of this planet is largely determined by the properties of water.

15.5.3 Hydrogen peroxide

The other binary compound of hydrogen and oxygen is H_2O_2.

15.5.3.1 Synthesis

Hydrogen peroxide can be produced only indirectly. A solution of H_2O_2 can be obtained in the laboratory by reacting sodium peroxide with water:

$$Na_2O_2 + 2H_2O \rightarrow 2NaOH + H_2O_2$$

or when barium peroxide reacts with sulfuric acid:

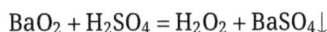

$$BaO_2 + H_2SO_4 = H_2O_2 + BaSO_4\downarrow$$

In the past, hydrogen peroxide was industrially produced by the hydrolysis of ammonium peroxydisulfate or peroxydisulfuric acid:

$$(NH_4)_2S_2O_8 + 2H_2O \rightarrow H_2O_2 + 2(NH_4)HSO_4$$
$$H_2S_2O_8 + 2H_2O \rightarrow H_2O_2 + 2H_2SO_4$$

Another industrial method for obtaining hydrogen peroxide with the participation of organic substances is the catalytic oxidation of isopropyl alcohol to acetone:

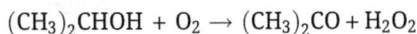

$$(CH_3)_2CHOH + O_2 \rightarrow (CH_3)_2CO + H_2O_2$$

Currently, hydrogen peroxide is produced almost exclusively by the anthraquinone method (Figure 15.2). The process involves the reduction of anthraquinone (e.g., 2-ethylanthraquinone) to the corresponding anthrahydroquinone, usually by hydrogenation using a palladium catalyst. The anthrahydroquinone is then regenerated to the initial anthraquinone by oxidation with oxygen. The cycle is repeated, and hydrogen peroxide is continuously released as a byproduct.

The process of producing hydrogen peroxide directly from the elements has always been of interest. Direct synthesis is difficult to achieve because, from a thermodynamic point of view, the interaction of hydrogen with oxygen results in the production of water. Direct synthesis systems have been developed, most of which are based on the use of finely dispersed metal catalysts, but none of these methods have yet been proposed and used for industrial synthesis.

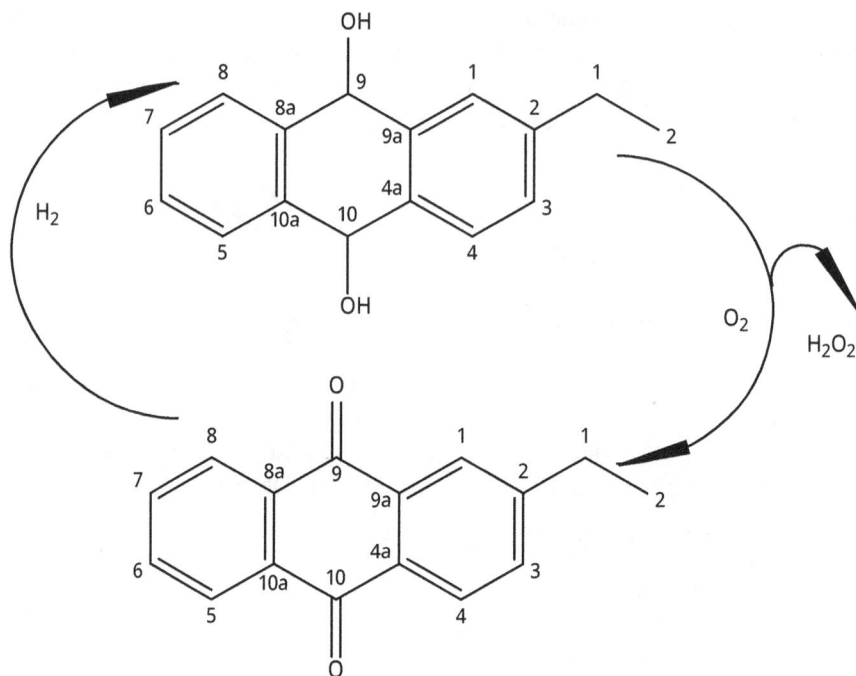

Figure 15.2: Anthraquinone method.

15.5.3.2 Properties

Pure hydrogen peroxide is an almost colorless (slightly bluish) viscous liquid. Its melting point is −11 °C (262.15 K) and its boiling point is 150.2 °C (423.35 K). It is a chemically extremely active substance that should always be handled very carefully. It is advisable to use protective agents when working even with dilute solutions of H_2O_2, since it attacks the skin.

In the H_2O_2 molecule, the formal oxidation state of oxygen is equal to one, but each oxygen atom has two bonds because oxygen atoms form bonds not only with hydrogen but also with each other. The angle of the H–O–O coupling in the gas phase is only 94½° (about 10° smaller than the H–O–H angle in water), and the two H–O units form a dihedral angle of 111° to each other.

Hydrogen peroxide is very unstable thermodynamically, resulting in disproportionation and decomposition to water and oxygen (in the presence of catalysts such as MnO_2, Cr_2O_3, and Fe_2O_3):

$$H_2O_2 \rightarrow H_2O + 1/2 O_2, \qquad \Delta G° = -119.2\,\text{kJ/mol}$$

In its pure state, it decomposes more slowly due to kinetic factors (the reaction has high activation energy). With weak impurities (transition metal ions, metals, blood, dust, glass, etc.), the decomposition is catalyzed. Usually, H_2O_2 is not stored in glass but in plastic containers. In order to prevent the decomposition of H_2O_2, negative catalysts are added, which have a stabilizing effect, such as salicylic acid, barbituric acid, phenacetin, etc. The same role is played by various inorganic acids (H_2SO_4, H_3PO_4, H_3BO_3, HCl, and HF) and salts.

H_2O_2 dissolves in water by forming hydrogen bonds with water molecules and partially dissociates as a very weak acid ($K_D = 2 \times 10^{-12}$) in two degrees:

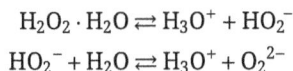

$$H_2O_2 \cdot H_2O \rightleftharpoons H_3O^+ + HO_2^-$$
$$HO_2^- + H_2O \rightleftharpoons H_3O^+ + O_2^{2-}$$

When a concentrated solution of H_2O_2 reacts with hydroxides, metal peroxides are obtained, which are considered as salts of hydrogen peroxide (Li_2O_2 and MgO_2):

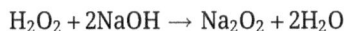

$$H_2O_2 + 2NaOH \rightarrow Na_2O_2 + 2H_2O$$

The oxygen in the composition of H_2O_2 is in an intermediate oxidation state (−1), between that of molecular oxygen (0) and water (−2). That is why H_2O_2 exhibits both oxidative and reducing properties. Hydrogen peroxide can act as an oxidizing agent or reducing agent depending on pH. Oxidation usually takes place in acidic solutions, while reduction occurs in an alkaline medium:

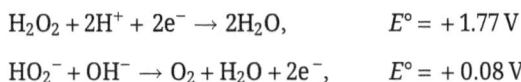

$$H_2O_2 + 2H^+ + 2e^- \rightarrow 2H_2O, \qquad E° = +1.77\,\text{V}$$
$$HO_2^- + OH^- \rightarrow O_2 + H_2O + 2e^-, \qquad E° = +0.08\,\text{V}$$

Oxidizing properties are more characteristic for H_2O_2:

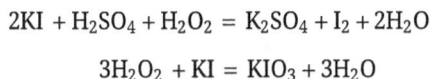

$$2KI + H_2SO_4 + H_2O_2 = K_2SO_4 + I_2 + 2H_2O$$
$$3H_2O_2 + KI = KIO_3 + 3H_2O$$

With oxidizing agents stronger than it, H_2O_2 acts as a reducer and is oxidized to oxygen. Hydrogen peroxide reduces permanganate ions in acidic solutions to Mn(II):

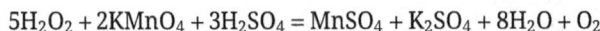

$$5H_2O_2 + 2KMnO_4 + 3H_2SO_4 = MnSO_4 + K_2SO_4 + 8H_2O + O_2$$

Hydrogen peroxide is used in the restoration of antique paintings. Treatment with hydrogen peroxide oxidizes Pb(II) sulfide to white Pb(II) sulfate, thereby restoring the color of the painting:

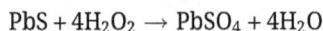

$$PbS + 4H_2O_2 \rightarrow PbSO_4 + 4H_2O$$

Hydrogen peroxide has antiseptic action and is used in dermatology for external use. The 3% solution of hydrogen peroxide is used in medicine as a hemostatic agent. Its antiseptic effect is related to the release of O_2 and its interaction with the blood in the capillary vessels. It heals wounds and inflammations of the skin and mucous membranes. It is also used for cosmetic purposes – for cleaning and whitening the face and teeth, discoloration, and hair coloring. At the same time, its harmlessness has been reported. It improves the general condition of the patient and successfully prevents infectious diseases.

Chapter 16
First main group of the periodic table

16.1 General characteristics of s-elements

The elements of the first and second main groups of the periodic table belong to the s-elements. The first group includes the elements lithium (Li), sodium (Na), potassium (K), rubidium (Rb), cesium (Cs), and francium (Fr), which form the group of *alkali metals.* Their name comes from the Arabic word "alkali," which means base. In the second group are the elements beryllium (Be), magnesium (Mg), calcium (Ca), strontium (Sr), barium (Ba), and radium (Ra). The electronic configuration of the elements of the I and II main groups of the periodic table is: [inert gas] ns^1 and [inert gas] ns^2. Alkali metals are very similar in properties, with only lithium showing some deviations from the general patterns. In the second group, beryllium and magnesium differ significantly from other elements, so they are usually considered separately.

The outer valence electrons of these elements are poorly retained and are easily donated in chemical reactions, as a result of which the s-elements easily form cations. When the respective cations interact with anions, the polarizing effect should be taken into account. One ion plays the role of an external electric field in relation to the other. The cation has a stable configuration of an inert gas. In ionic compounds, the centers of positive and negative charges coincide with the charges of the nuclei, and the internuclear distance is equal to the distance between the poles of the dipole: $l = d$. The cation attracts electrons to itself, and the center of gravity of the negative charge shifts ($l < d$). Thus, the polarity decreases, hence the ionic character decreases as well, and the covalence of the bond increases. s-Elements of group I A exhibit a weak polarizing effect, and their compounds contain mainly ionic-type bonds. In the s-elements of group II A, the polarizing effect of the cations increases.

Metals of main groups I and II dissolve in liquid ammonia to form light blue solutions with high electrical conductivity (at low concentrations). The main carriers of electric current are solvated electrons. As the metal concentration increases, the paramagnetic properties of the solutions decrease. It is not clear whether the electron is bound to NH_3 or with the solvated cation. It is assumed that in addition to solvated electrons, anions (stabilized by solvation) also exist in the solution. There is evidence of the following process taking place in a sodium solution in liquid NH_3:

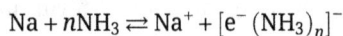

$$Na + nNH_3 \rightleftharpoons Na^+ + [e^- (NH_3)_n]^-$$

https://doi.org/10.1515/9783111712246-016

16.2 General characteristics of s-elements of group IA

The s-elements of the first main group include the elements from hydrogen to francium. The history of alkaline elements began with the discovery of sodium and potassium. Since ancient times, the world has used their compounds, but the elements were discovered only in 1807 by H. Davy. The element lithium was discovered later in 1817 in the mineral petalite. In 1818, Davy obtained the first quantities of the new metal. Rubidium and cesium are almost twins, but with different vapor colors. They have been proven by spectral analyses. Cesium was obtained in 1860, and during the study of its spectrum, two blue bands were found, giving it its name (cesium) – sky blue. In 1861, a new element with red spectral bands was discovered – rubidium – from Latin dark red. Francium is one of the rarest elements. When creating the periodic table, Mendeleev left free space for undiscovered elements. The name of the supposed sixth alkaline metal was exacessium. In 1939, M. Pereille proved that the decay of actinium-227 produces an isotope of the 87th element. This radioactive isotope (half-life of 21.8 min) had the properties of alkali metals and was called francium.

In the electron configurations of these elements, the outer s-subshell is filled. In the hydrogen atom, the 1s-subshell is filled, as a result of which hydrogen differs from its counterparts in the group, both in terms of physical and electrochemical relations. Therefore, the properties of hydrogen are considered in a separate chapter. Alkaline elements are the base for the construction of each of the periods in the periodic table. Their atoms have only one valence electron. They have the lowest ionization potentials and the most negative electrode potentials (see Table 16.1).

Table 16.1: Characteristics of atoms of s-elements of the IA group.

Element	Li	Na	K	Rb	Cs	Fr
Valence electrons	$2s^1$	$3s^1$	$4s^1$	$5s^1$	$6s^1$	$7s^1$
Radius of the atom (nm)	0.155	0.189	0.236	0.248	0.268	0.280
Ionization potential (eV)	5.39	5.14	4.34	4.18	3.89	–
Electronegativity	1.0	0.9	0.8	0.8	0.7	–
Electrode potential $E°$ (V)	−3.05	−2.71	−2.93	−2.92	−2.92	–

Alkali metals are typical metals without any signs of amphoteric properties. They are the strongest reducers of all metals. By giving up their only valence electron, their atoms easily turn into single-charged cations and form numerous ionic compounds, manifesting themselves in their single oxidation state of +1. The properties of alkali metals from lithium to francium are similar. The distribution of electrons and some comparative characteristics of the atoms of the elements in the lithium-francium order are given in Table 16.2.

The low values of the first ionization potentials and the high values of the second ionization potentials are the reason for the constant +1 oxidation state. Despite the

Table 16.2: Distribution of electrons and comparative characteristics of atoms in the IA group.

Distribution of electrons	I_1 (eV)	I_2 (eV)	R_{ion}^+ (Å)
Li 2.1	5.4	75.6	0.60
Na 2.8.1	5.1	47.3	0.95
K 2.8.8.1	4.3	31.8	1.33
Rb 2.8.18.18.8.1	4.2	27.4	1.48
Cs 2.8.8.18.18.8.1	3.9	23.4	1.69
Fr 2.8.8.18.32.18.8.1	3.8	22.5	1.80

fact that in alkali metals the electron affinity has positive values, that is, the addition of an electron is accompanied by the release of energy, they practically never exhibit negative oxidation states. The first ionization potential in the group decreases from the top to the bottom. The ionization potential is highest in lithium, and slightly lower in sodium. The stronger decrease in the first ionization potential after sodium is explained by the appearance of the 3d-subshell. The closest in properties in the group are the metals K, Rb, and Cs. They are complete analogues. Lithium stands out sharply in the group. Sodium occupies an intermediate position. A general rule valid for all groups is that the elements of the major periods are considered to be complete analogues.

The radii of the atoms and ions in the group increase, and their monotonic change is disturbed by potassium. From the ratio of the charge and the radius of the ion, the ion potential is calculated, which characterizes the polarizing action of the cation. The higher the ion potential, the more pronounced the polarizing action of the cation, that is, the greater the tendency to form covalent bonds. This affects the values of the hydration energy of the ions and the enthalpy of dissociation of E_2 molecules (see Table 16.3).

Table 16.3: Characteristics of E^+ ions and E_2 molecules of the first group.

Ion	z/r	$-\Delta H_{hydr}$ (kJ/mol)	Molecule	ΔH_{dissoc} (kJ/mol)
Li^+	1.66	520.5	Li_2	109.45
Na^+	1.05	405.9	Na_2	79.2
K^+	0.75	322.2	K_2	49.79
Rb^+	0.67	300.8	Rb_2	47.28
Cs^+	0.59	255.6	Cs_2	43.52

It can be seen that in alkali metals, the tendency to form covalent bonds decreases from the top to the bottom. Li^+ ion exists in the crystal lattice, but in solution, such an ion is not present. As a result of strong hydration, the ion $[Li(H_2O)_4]^+$ is formed. Water molecules in the complex ion $[Li(H_2O)_4]^+$ are held firmly and are difficult to separate even when heated. The Na^+ and K^+ ions exist in solution because their hydration is weak; see the hydration energy values. In other alkali metals, in some cases, there is a tendency to form covalent bonds. For example, in a gaseous state, there is a covalent bond in the diatomic molecules Na_2, Cs_2, etc. In addition, the bond of alkali metals with carbon, nitrogen, and oxygen in some complex compounds is also of weak covalence.

Compared to the other groups of the periodic table, in the s-elements of group I, the influence of the size and the mass of the atom on chemical and physical properties is most clearly manifested, with some insignificant deviations. During the transition from lithium to francium, the following parameters are successively reduced: the melting point and the heat of sublimation of metals; the energy of formation of the crystal lattice of salts (except for salts with small anions, where the z/r ratio does not change logically); and the ease of thermal decomposition of some salts, as shown in Table 16.4.

Table 16.4: Thermodynamic characteristics of nitrates.

Compound	T_m (°C)	T_{decomp} (°C)
$LiNO_3$		>600
$NaNO_3$		334
KNO_3	313	380
$RbNO_3$	414	>400
$CsNO_3$		

During the transition from lithium to francium, the effective radii of hydrated ions are consistently reduced, and hence the hydration energies, as well as the strength of the covalent bonds in the corresponding E_2 molecules, Table 16.5.

Table 16.5: Effective radii of hydrated ions.

Ion	Li^+	Na^+	K^+	Rb^+	Cs^+
R_{hydr} (Å)	3.40	2.76	2.32	2.28	2.28

From the top to the bottom in the group, the heat of formation of fluorides, hydrides, oxides, and carbides (due to the high energies of the crystal lattices formed by the small cations) also decreases from the top to the bottom, as shown in Table 16.6.

Since the configuration of the E^+ ion is very stable (inert gas configuration), most of the alkali metal compounds are colorless. All alkali metals are strong reducers, and their standard electrode potentials are negative. The largest (in absolute terms) is the potential of lithium. This is due to the high hydration energy of the lithium ion. In melts, the potential of Li^+/Li is the smallest compared to other alkali metals.

Table 16.6: Thermodynamic characteristics of some compounds.

Compound	ΔH_{form} (kJ/mol)	Compound	ΔH_{form} (kJ/mol)
Li_2O	595.8	LiF	612.1
Na_2O	416.0	NaF	573.6
K_2O	363.2	KF	567.4
Rb_2O	330.1	RbF	549.3
Cs_2O	317.6	CsF	530.9

Occurrence: In terms of their content on the Earth, sodium (2.6%) and potassium (2.4%) rank among the top ten elements of the periodic table. They are found in rocks, most often in common compounds with aluminum, silicon, and oxygen. The main minerals of sodium and potassium of industrial importance are halite NaCl (rock salt, table salt), sylvine KCl, sylvinite (KCl with impurities of NaCl), carnallite $KCl \cdot MgCl_2 \cdot 6H_2O$, mirabilite $Na_2SO_4 \cdot 10H_2O$ (Glauber's salt), and cryolite Na_3AlF_6. Lithium, rubidium, and cesium are rare elements; they do not form independent deposits but are often found as impurities with potassium minerals. Francium is an unstable radioactive element, and small amounts of it are obtained artificially.

Preparation: Lithium metal is obtained by the electrolysis of a melt of a eutectic mixture of lithium chloride and other halides (NaCl, NaBr, or $CaCl_2$). A eutectic mixture is defined as a composition of a mixture of two or more substances that has a minimum melting point.

The main method for obtaining sodium is electrolysis in a melt of sodium chloride in a eutectic mixture with KCl, NaF, or $CaCl_2$. If the melting point of NaCl is 801 °C, these mixtures melt at a temperature lower than 600 °C, which facilitates the electrolysis process. In this process, sodium is released at the iron cathode, and chlorine is released at the graphite anode. The second most important method of sodium production is electrolysis in a NaOH melt. The advantages of this method are the low temperature of the process and the possibility of obtaining sodium of high purity, but the disadvantage is the expensive raw material.

Potassium is obtained in several ways:

(1) Electrolysis of a melt of KCl and NaCl to obtain an alloy of potassium with sodium and its separation by vacuum distillation.

(2) Metal-thermal reduction, where Na displaces K in melts of its compounds (KCl and KOH):

$$KCl + Na = NaCl + K$$

This method is suitable for the production of the other elements of the group due to their low melting points and high volatility. Sodium and calcium are most often used as reducers.

(3) Chemical reduction of potassium from potassium chloride using Al or Si in vacuum heating and in the presence of CaO, which contributes to the reaction products:

$$6KCl + 2Al + 4CaO = 3CaCl_2 + CaO \cdot Al_2O_3 + 6K$$

$$4KCl + Si + 4CaO = 2CaCl_2 + 2CaO \cdot SiO_2 + 4K$$

Alkali metals are purified from impurities by vacuum distillation. In air, they quickly oxidize, with rubidium and cesium igniting themselves. This is why alkali metals are usually stored in airtight containers or in kerosene.

16.3 Properties

16.3.1 Physical properties

Alkali metals are silvery-white substances, with the exception of cesium, which has a golden color. They are relatively lightweight, with lithium and sodium being lighter than water. These metals are soft, with low melting points and low density (see Table 16.7).

Table 16.7: Some physical properties of alkali metals.

Metal	T_m (°C)	T_b (°C)	d (g/cm³)	Mohs hardness	$\rho.10^6$ (Ohm cm)
Li	179.0	1327	0.539	0.6	8.55
Na	97.8	883	0.973	0.4	4.34
K	63.5	760	0.893	0.5	6.10
Rb	38.7	703	1.534	0.3	11.6
Cs	28.6	686	1.904	0.2	19.0
Fr	20.0	630	2.440		

Down in the group, melting and boiling temperatures decrease, and the density of metals increases. All these metals crystallize in volume-centered cubic lattices. The parameters of the lattice increase, and therefore the binding forces decrease from the top to the bottom in the group, as a result of which the melting temperatures also decrease. On the other hand, the mass of the nuclei grows, regardless of the increase in volume. In potassium, in comparison with sodium, there is an increase in the atomic radius, in which the influence of the volume is predominant over that of the mass, which leads to a strong decrease in density.

The chemical analysis of these metals is carried out using the flame photometry method because the vapors of alkali metals are intensively colored: lithium – red, sodium – yellow, potassium – violet, rubidium – pink-violet, cesium – light blue. In a hydrogen atmosphere, the vapors of Na have a violet color, that of K – blue-green, and those of rubidium and cesium – green-blue.

The atoms of sodium, potassium, rubidium, and cesium donate their valence electrons under the action of light at ultraviolet irradiation, and it depends on the intensity of light absorption. This property is especially typical for rubidium and cesium, which allows their use in the preparation of devices for the direct conversion of light energy into electrical energy, so-called photocells.

The positive cations of the alkali metals, produced by the donation of valence electrons, are not deformed. They have the configuration of an inert gas and are difficult to polarize. Therefore, their ionic compounds are colorless in solution or white in the solid state. The difficult polarization and the lack of deformation in a given ion are the reasons for the formation of colorless compounds.

16.3.2 Chemical properties

Alkali metals are some of the most chemically active elements, and this activity increases with an increase in their relative atomic mass. They are stored in conditions that exclude interaction with air. The reactions of alkali metals with water are very intense. In the group from the top to the bottom, their activity increases. Thus, lithium interacts slowly with water, sodium – more energetically, and in cesium, the reaction occurs with an explosion.

$$2E + 2H_2O \rightarrow 2EOH + H_2$$

Alkali metals participate in reactions with most elements. Due to their strong reducing properties, alkali metals act as reducers even with respect to hydrogen, which in this interaction manifests itself as an oxidizing agent and acquires an oxidation state of −1:

$$2E + H_2 \rightarrow 2EH$$

Alkali metals interact very actively with oxygen, and their activity is enhanced from Li to Cs. Only Li, in direct interaction with an excess of oxygen, forms the oxide Li_2O. Sodium and potassium give oxides only under special conditions (in cold, in case of oxygen deficiency), while in excess of oxygen, they form peroxides Na_2O_2 and superoxides EO_2:

$$4Li + O_2 \rightarrow 2Li_2O$$

$$2Na + O_2 \rightarrow Na_2O_2$$

$$K + O_2 \rightarrow KO_2$$

Oxides of sodium, potassium, rubidium, and cesium are formed when there is a shortage of oxygen or indirectly when metals react with peroxides and superoxides:

$$Na_2O_2 + 2Na = 2Na_2O$$

$$KO_2 + 3K = 2K_2O$$

They react very actively with the halogen elements, reducing their activity from fluorine to iodine and from cesium to lithium:

$$2E + X_2 \rightarrow 2EX \qquad (X = F, Cl, Br, I)$$

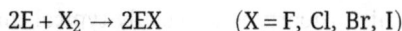

When alkali metals are crushed with sulfur, water-soluble sulfides with an ion-crystal lattice are obtained:

$$2E + S \rightarrow E_2S$$

With nitrogen, the activity of alkaline elements is significantly reduced, and under normal conditions, only Li reacts. The rest of the elements react at very high temperatures to form nitrides:

$$6E + N_2 \rightarrow 2E_3N$$

Lithium is the only alkali metal and one of the few elements in the entire periodic table that reacts directly with nitrogen under normal conditions. Breaking the triple bond in the nitrogen molecule requires an energy of the order of 945 kJ/mol. In order to balance this energy, the energy of the crystal lattice of the product must be very high. Lithium, which has the highest charge density in the group, forms a nitride with a sufficiently high energy of the crystal lattice:

$$6Li + N_2 \rightarrow 2Li_3N$$

Lithium nitride is chemically active and forms ammonia when it reacts with water:

$$Li_3N + 3H_2O \rightarrow 3LiOH + NH_3$$

Alkali metals also interact with other nonmetals to produce carbides, phosphides, etc.:

$$2E + 2C \rightarrow E_2C_2$$
$$3E + P \rightarrow E_3P$$

With metals, they form alloys of different compositions. They energetically dissolve in mercury with the release of heat, forming amalgams. Sodium amalgams are used as strong reducers.

Their activity against acids is even more pronounced than against water, since in an acidic environment, the concentration of hydrogen cations is significantly higher compared to pure water. In nonoxidizing acids, alkali metals displace hydrogen:

$$2E + 2HCl \rightarrow 2ECl + H_2$$

Oxidative acids are reduced by alkali metals to their lower oxidation states:

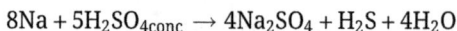

$$8Na + 10HNO_3 \rightarrow 8NaNO_3 + NH_4NO_3 + 3H_2O$$

$$8Na + 5H_2SO_{4conc} \rightarrow 4Na_2SO_4 + H_2S + 4H_2O$$

Alkali metals are strong reducers, which explain many reactions involving their participation:

$$4Na + SiO_2 \rightarrow 2Na_2O + Si$$

16.4 Chemical compounds

16.4.1 Hydrides

When heated, the s-elements react with hydrogen to form hydrides EH, for example:

$$2Na + H_2 \rightarrow 2NaH$$

The hydrides of the alkaline elements (as well as those of the alkaline earth elements) are ionic compounds and are colorless crystalline substances with an ionic crystal lattice containing positive metal ions and negative hydride anions H^-. The thermal resistance of hydrides decreases from Li to Cs, as shown in Table 16.8.

Table 16.8: Melting temperatures of alkali hydrides.

Hydride	LiH	NaH	KH	RbH	CsH
T_m (°C)	691	800	400	400	400

Hydrides are very strong reducers due to the instability of the hydride ion, which easily donates an electron and becomes a neutral hydrogen atom. They react energetically with water, as well as with acidic oxides, acids, and nonmetals:

$$NaH + H_2O \rightarrow NaOH + H_2$$
$$NaH + CO_2 \rightarrow NaCOOH$$

When heated, they decompose by releasing H_2. Their reactivity increases from LiH to CsH.

16.4.2 Oxides

Oxides are produced by direct interaction with oxygen (most often in the case of deficiency). The composition of oxides is E_2O. Li_2O and Na_2O are colorless, K_2O and Rb_2O are yellow, and Cs_2O is orange. The coloration becomes more intense because, as the size of the E^+ ion increases, their polarization also increases. In excess of oxygen, peroxides and superoxides are obtained. Peroxides E_2O_2 contain the peroxide ion O_2^{2-} ($d_{O-O} = 1.5$ Å) and superoxides (EO_2) contain the superoxide ion O^{2-} ($d_{O-O} = 1.3$ Å). Peroxides and superoxides are strong oxidizing agents. Contact with many organic substances causes them to ignite. In their environment, aluminum dust and sawdust burn (when heated). The stability of peroxides and superoxides increases with an increase in the size of the E^+ cation, which stabilizes the crystal lattice.

The oxides are solid, powdery substances with an ionic crystal lattice. They have typical basic properties – they interact with water, acids, and acidic oxides. They are anhydrides of the corresponding hydroxides (bases):

$$Li_2O + H_2O = 2LiOH$$

Peroxides interact with water to release hydrogen peroxide:

$$Na_2O_2 + 2H_2O \rightarrow 2NaOH + H_2O_2$$

They are considered salts of hydrogen peroxide. The above process is a hydrolysis, in which a strong base, NaOH, and a weak acid, H_2O_2, are obtained.

When peroxides react with dilute acids, H_2O_2 is produced:

$$Na_2O_2 + H_2SO_4 \rightarrow H_2O_2 + Na_2SO_4$$

Of interest is the interaction of Na_2O_2 (or K_2O_2) with CO_2:

$$2Na_2O_2 + 2CO_2 \rightarrow 2Na_2CO_3 + O_2$$

$$O_2^{2-} - 2e^- \rightarrow O_2$$

$$O_2^{2-} + 2e^- \rightarrow 2O^{2-}$$

The reaction is redox and is a typical example of a disproportionation process. In this case, the peroxide ion O_2^{2-} is both an oxidizer and a reducer. The reaction of Na_2O_2 with CO_2 is used to regenerate oxygen in closed spaces – submarines, spaceships, etc.

Superoxides react with water to release H_2O_2 and O_2:

$$2KO_2 + 2H_2O \rightarrow 2KOH + H_2O_2 + O_2$$

16.4.3 Hydroxides

Typical basic hydroxides (bases) correspond to the oxides of alkaline metals:

$$E_2O + H_2O \rightarrow 2EOH$$

$$2E + 2H_2O \rightarrow 2EOH + H_2$$

Alkali metal oxides interact with water to form bases; peroxides – to form bases and H_2O_2, and when superoxides react with water, bases, H_2O_2 and oxygen are formed:

$$Li_2O + H_2O = 2LiOH$$

$$Na_2O_2 + 2H_2O = 2NaOH + H_2O_2$$

$$4KO_2 + 2H_2O = 4KOH + 2H_2O_2 + O_2$$

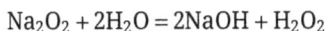

Alkali metal hydroxides are low-melting, colorless crystalline substances. Their melting points decrease from LiOH (473 °C) to CsOH (346 °C). They are water-soluble substances. Their dissolution in H_2O (with the exception of LiOH) is a highly exothermic process. Their aqueous solutions (bases) are highly alkaline and change the color of the indicators to their alkaline color form. The bases are highly hygroscopic; they intensively absorb moisture and CO_2 from the air, forming crystal hydrates (NaOH $\cdot$ H_2O). Solid bases and their concentrated solutions destroy living tissues, so it is not recommended to touch them with hands. Concentrated solutions of bases and their melts interact with glass and porcelain, so they are stored in polyethylene containers:

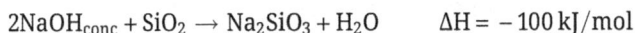

$$2NaOH_{conc} + SiO_2 \longrightarrow Na_2SiO_3 + H_2O \qquad \Delta H = -100\,kJ/mol$$

The bases dissolve platinum in the presence of oxygen, so they are melted in nickel or iron crucibles. They are resistant to heating up to 1,000 °C (except for LiOH, which decomposes at 500 °C):

$$2LiOH \longrightarrow Li_2O + H_2O$$

As typical bases, the hydroxides of alkaline elements interact with acids (neutralization process) and with acidic oxides:

$$NaOH + HCl \longrightarrow NaCl + H_2O$$

$$2NaOH + CO_2 \longrightarrow Na_2CO_3 + H_2O$$

Alkaline bases also participate in oxidation-reduction processes, e.g., in the dissolution of amphoteric metals such as Be, Zn, Al, etc., in which the corresponding salts are obtained and hydrogen is released. In a melt with a solid base (granules), the corresponding beryllates, zincates, meta-aluminates, etc., are obtained, and when using concentrated aqueous solutions of bases – the products are corresponding hydroxo-complexes.

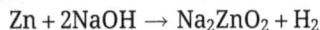

$$Zn + 2NaOH \longrightarrow Na_2ZnO_2 + H_2$$

Of the bases, the one of most practical importance is sodium hydroxide NaOH (sodium base), the so-called caustic soda. It is obtained by electrolysis of a solution of NaCl. When using an iron cathode and a graphite anode, the following processes take place, as shown in Table 16.9.

The sodium base obtained by this method contains impurities of undecomposed sodium chloride, which is undesirable. Pure sodium hydroxide, without impurities of

Table 16.9: Electrolysis of a solution of NaCl.

Cathode	$2H_2O + 2e^- = H_2 + 2OH^-$
Solution	$Na^+ + OH^- = NaOH$
Anode	$2Cl^- - 2e^- = Cl_2$
Total reaction	$2NaCl + 2H_2O = 2NaOH + Cl_2 + H_2$

NaCl, is obtained by electrolysis with a mercury cathode. Likewise, a potassium base, KOH, can also be obtained by electrolysis of KCl in an aqueous solution.

Sodium hydroxide (caustic soda) is produced through two main chemical methods: the causticization method using soda ash and slaked lime, and the ferrite method. In the causticization method, NaOH is formed when a soda solution reacts with calcium hydroxide:

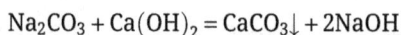

$$Na_2CO_3 + Ca(OH)_2 = CaCO_3\downarrow + 2NaOH$$

The ferrite method involves melting sodium carbonate with Fe_2O_3 at 900 °C:

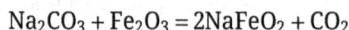

$$Na_2CO_3 + Fe_2O_3 = 2NaFeO_2 + CO_2$$

The resulting sodium ferrite is completely hydrolyzed when treated with hot water to form sodium hydroxide:

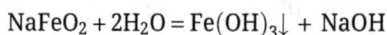

$$NaFeO_2 + 2H_2O = Fe(OH)_3\downarrow + NaOH$$

16.4.4 Salts

Alkali metal salts are numerous and, as a rule, are well soluble in water. Alkali metals form salts with all acids and oxoacids. Salts of weak acids are hydrolyzed in aqueous solutions:

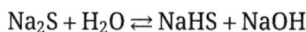

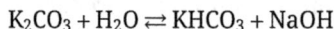

$$Na_2S + H_2O \rightleftharpoons NaHS + NaOH$$

$$K_2CO_3 + H_2O \rightleftharpoons KHCO_3 + NaOH$$

$$K_3PO_4 + H_2O \rightleftharpoons K_2HPO_4 + NaOH$$

The stability of carbonates and nitrates increases from the top to the bottom of the group. For example, Na_2CO_3 melts without decomposition, while Li_2CO_3 decomposes:

$$Li_2CO_3 \rightarrow Li_2O + CO_2$$

Lithium nitrate decomposes similarly to magnesium nitrate:

$$4LiNO_3 \rightarrow 2Li_2O + 4NO_2 + O_2$$

The nitrates of the remaining alkali metals decompose with the release of oxygen:

$$2NaNO_3 \rightarrow 2NaNO_2 + O_2$$

Of the salts of alkali metals, sodium carbonate Na_2CO_3 (soda ash) has the greatest technical application. It is obtained from natural substances $NaCl$ and $CaCO_3$ according to a method developed by Solvay – interaction of an aqueous solution of $NaCl$ and NH_3 and a gaseous CO_2:

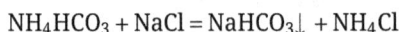

$$NH_3 + CO_2 + H_2O = NH_4HCO_3$$

$$NH_4HCO_3 + NaCl = NaHCO_3\downarrow + NH_4Cl$$

The second reaction occurs with the products due to the low solubility of $NaHCO_3$ in the presence of ammonium chloride. Next comes baking, that is, hydrogen carbonate is separated from the solution, and when heated, it turns into Na_2CO_3:

$$2NaHCO_3 = Na_2CO_3 + CO_2 + H_2O$$

At the same time, the regeneration of ammonia is carried out. Carbon dioxide is returned to the technological scheme, and an additional amount of CO_2 is obtained during the decomposition of limestone:

$$CaCO_3 = CaO + CO_2$$

The calcium oxide (quicklime) obtained in the last reaction is treated with water:

$$CaO + H_2O = Ca(OH)_2$$

The product is $Ca(OH)_2$ (slaked lime), which further interacts with NH_4Cl:

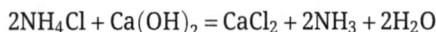

$$2NH_4Cl + Ca(OH)_2 = CaCl_2 + 2NH_3 + 2H_2O$$

The resulting ammonia is returned to the technological scheme. This method of obtaining soda ash is a typical example of a cyclic technological process in which the only waste product is $CaCl_2$. Since 1900, 90% of the world's soda production has been done by this method.

Sodium hydrogen carbonate (bicarbonate) is less soluble in water than sodium carbonate. It is obtained by passing carbon dioxide into a saturated solution of sodium carbonate:

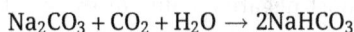

$$Na_2CO_3 + CO_2 + H_2O \rightarrow 2NaHCO_3$$

When heated, sodium hydrogen carbonate turns into sodium carbonate. This reaction is related to the application of sodium hydrogen carbonate as the main component in powder fire extinguishers. The main use of sodium hydrogen carbonate is in the food industry. It is commonly used as a mixture of sodium hydrogen carbonate and calcium dihydrogen phosphate, $Ca(H_2PO_4)_2$, with a little starch as an additive. Calcium dihydrogen phosphate has an acidic reaction and, when moistened, reacts with sodium hydrogen carbonate to produce carbon dioxide:

$$2NaHCO_3 + Ca(H_2PO_4)_2 \rightarrow NaH_2PO_4 + CaHPO_4 + 2CO_2 + 2H_2O$$

It has already been mentioned that alkali metal salts are soluble, although there are also some slightly soluble lithium salts, as well as some complex salts. As a rule, the least soluble salts are those with the greatest similarity in the ionic sizes of the cation and the anion. Thus, a large anion, such as $[Co(NO_2)_6]^{3-}$ will form the least soluble salts with the larger cations of the group. Its salts with lithium and sodium are soluble, while those with potassium, rubidium, and cesium are insoluble. This property can be used as a qualitative reaction to differentiate Na^+ from K^+ ions in solution. Obtaining a bright yellow precipitate indicates the presence of K^+ ions in the solution:

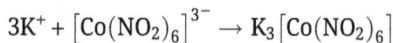

$$3K^+ + \left[Co(NO_2)_6\right]^{3-} \rightarrow K_3\left[Co(NO_2)_6\right]$$

Another large anion that can be used for a qualitative reaction in the precipitation of large cations of the group is the tetraphenylborate ion, $[B(C_6H_5)_4]^-$:

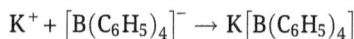

$$K^+ + \left[B(C_6H_5)_4\right]^- \rightarrow K\left[B(C_6H_5)_4\right]$$

In conclusion, the peculiarities of lithium and its specific properties will be considered. The penultimate electron shell of lithium contains two electrons, corresponding to the configuration of helium, which is different from that of the other inert gases. Therefore, lithium has a significantly smaller atomic and ionic radius, and hence a greater ionization energy and electronegativity compared to the other elements of the alkaline group. The strongest reducer is Fr, and the weakest is Li, while standard electrode potentials define lithium as the strongest reducer. This anomaly is due to the very strong hydration of the lithium ion in solution, which is related to its small size. In this regard, lithium is characterized by some differences compared to the other elements of the group. In some chemical properties, Li resembles Mg (diagonal similarity of elements in the periodic table). Like the corresponding magnesium compounds, the salts LiF, Li_2CO_3, Li_3PO_4 are slightly soluble in water. Lithium carbonate and nitrate decompose relatively easily when heated. LiOH decomposes into Li_2O and H_2O at 600–700 °C, while other bases are stable at temperatures higher than 1,000 °C. The peculiarities of lithium are determined by the strong polarizing effect of the small Li^+ cation. This explains the most negative value of the electrode potential of lithium, which is at the beginning of the row of relative activity of metals.

16.5 Usage

Alkali metals have great practical applications. Metallic sodium is the most widely used metal. It is mainly used as a reducer in the metallothermic production of rare metals (Ti, Zr, Ta, Nb). For example, pure titanium is obtained by reducing titanium(IV) chloride with sodium:

$$TiCl_4 + 4Na \rightarrow Ti + 4NaCl$$

Metallic sodium is used in the production of the gasoline additive tetraethyl lead. The synthesis of tetraethyl lead occurs through the interaction of a lead-sodium alloy and ethyl chloride:

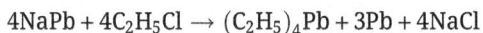

$$4NaPb + 4C_2H_5Cl \rightarrow (C_2H_5)_4Pb + 3Pb + 4NaCl$$

Sodium is used in organic synthesis and in nuclear reactors. In an alloy with potassium, it is used to produce Na_2O_2 for air purification and regeneration in artificial respiration apparatus:

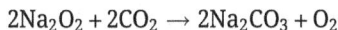

$$2Na_2O_2 + 2CO_2 \rightarrow 2Na_2CO_3 + O_2$$

Potassium exhibits weak radioactivity, as it contains about 0.012% of the radioactive isotope potassium-40. In fact, a significant portion of the radiation produced in organisms comes from this isotope, which has a half-life of 1.3×10^9 years. About 89% of potassium atoms decay by emitting electrons, while another 11% absorb electrons:

$$^{40}_{19}K \rightarrow {}^{40}_{20}Ca + {}^{0}_{-1}e$$

$$^{40}_{19}K + {}^{0}_{-1}e \rightarrow {}^{40}_{18}Ar$$

Metallic potassium is used less frequently than sodium. It is utilized in metallothermic processes, in organic synthesis, to obtain alloys with sodium, mercury, lead, calcium, and other metals, as well as to measure the absorption of X-rays. It produces KO_2, which is used in submarines to regenerate air:

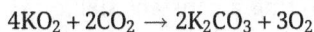

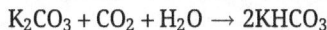

$$4KO_2 + 2CO_2 \rightarrow 2K_2CO_3 + 3O_2$$

$$K_2CO_3 + CO_2 + H_2O \rightarrow 2KHCO_3$$

Lithium is used in alloys for aerospace, in chemical sources of electricity, in thermonuclear energy, in ion engines for rockets, etc. LiOH is used in alkaline batteries. Lithium compounds are used as effective drugs in medicine (psychiatry) for the treatment of bipolar disorders.

Rubidium and cesium are used for the preparation of photocells. Some intermetallic compounds of rubidium and cesium (Rb_3Sb and Cs_3Sb) are used as semiconductor materials for the preparation of photocathodes. Many complex compounds containing rubidium and cesium are used for analytical purposes. Cesium is an element with a large number of isotopes from ^{114}Cs to ^{145}Cs (31 in total). Of these, only ^{133}Cs is stable; the rest are radioactive, with a half-life of 0.57 s (^{114}Cs) to 3×10^6 years (^{135}Cs). The isotope ^{137}Cs is a source of gamma quanta and is used in medicine, but at the same time, it is one of the most dangerous elements in nuclear accidents.

Sodium compounds are widely used in medicine and in many industries. Sodium hydroxide is used in the production of artificial fibers, cellulose, soap, artificial silk, in the purification of petroleum products and vegetable fats, in the production of paints, in inorganic synthesis, etc. Sodium fluoride is used in wood processing. Sodium carbonate is used in the production of aluminum and other metals, in the manufacture of glass, soaps, etc. A large percentage of sodium carbonate production is used in the glass industry. Na_2CO_3 reacts with silica (sand) at about 1,500 °C. The actual formula of the product depends on the stoichiometric ratio of the reactants:

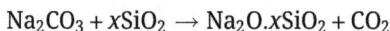

$$Na_2CO_3 + xSiO_2 \rightarrow Na_2O.xSiO_2 + CO_2$$

Potassium compounds are widely used in various industries. Potassium chloride and nitrate are used in agriculture as fertilizers. Potassium compounds are also used in the glass industry, in the production of liquid soap, etc. Sodium and potassium hydroxide, in solid form or in solutions, absorb carbon dioxide from the atmosphere:

$$2NaOH + CO_2 \rightarrow Na_2CO_3 + H_2O$$

Potassium hydroxide is preferred in organic chemistry due to its higher solubility in organic solvents than sodium hydroxide.

16.6 Biological role

Sodium and potassium ions are involved in the biochemical processes of living organisms. A strictly defined concentration of potassium cations is necessary for the normal functioning of the heart. NaCl maintains the osmotic pressure in the blood plasma within the required limits.

As a rule, elements exhibiting a constant stable oxidation state (Na, K, Ca, and Mg) form the fundamental systems in vivo – electrolyte medium, solid supporting structures, etc. Thus, alkali metal ions, which have the least ability to form coordination bonds, participate in the formation of an electrolyte environment in the body and determine the processes of mineral metabolism and transportation of substances due to the difference in osmotic pressure of organs and tissues.

It is known that in inorganic chemistry, the similarities between the elements sodium and potassium are required, but in the biological world, the difference between them is crucial. The difference in the concentrations of alkaline ions inside and outside the cells is the cause of the occurrence of an electrical potential on the cell membrane (see Table 16.10). The potential difference is the basis of many biological processes related to brain activity, the heart, kidneys, eyes, etc.

Table 16.10: Concentrations of Na^+ and K^+ (mmol/l).

Ion	[Na^+] (mmol/L)	[K^+] (mmol/L)
Red blood cells	11	92
Blood plasma	160	10

If in the geosphere the Na^+ and K^+ ions always occur together, and their separation is a difficult task, in the biosphere these ions are distributed on both sides of the cell membranes and relate, respectively, to the extracellular (Na^+) and intracellular (K^+) cations. These ions are constantly moving along the ion channels in both directions against the concentration gradient, that is, from areas with low to areas with high concentrations, in the so-called potassium-sodium pump. The potassium-sodium pump is a biological structure located on the surface of cell membranes. The main function of the potassium-sodium pump is the active transport of ions (Na^+ and K^+) across the membranes. Such a spontaneous process is impossible without the presence of additional energy. The energy is provided by the hydrolysis or reduction of ATP, which leads to the formation of ADP (adenosine diphosphoric acid) and the release of PO_4^{3-}, and hence to the activation of the potassium-sodium pump. The main task of the potassium-sodium pump is to maintain the potassium-sodium balance in all body systems. This balance ensures: firstly, maintaining the necessary osmotic pressure of biological fluids; secondly, preservation of the pH value inherent in each organ and tissue; thirdly, Na^+ and K^+ play an important role in the transmission of nerve impulses. The osmotic pressure in the cell is determined by potassium ions K^+ and their associated anions, and outside the cell by sodium ions Na^+ and their associated anions. The normal osmotic pressure in the cells is equal to the osmotic pressure outside the cells. Solutions that have the same osmotic pressure are called *isotonic*. Isotonic solutions contain the same amount of osmotically active particles. Isotonic solutions are 0.85–0.90% solution of NaCl (0.15 mol/L) or 4.5–5.0% glucose solution (0.30 mol/L).

There is a similarity between NH_4^+ and alkali metal cations. The ammonium ion behaves like a pseudo-alkaline cation, although it is a polyatomic cation containing two nonmetals. The similarity is due to the fact that NH_4^+ is characterized by a low charge and large size, analogous to the cations of alkali metals. In fact, the radius of NH_4^+ (151 pm) is very close to that of the K^+ cation (152 pm). The chemistry of ammonium salts is more similar to rubidium or cesium, and this resemblance to heavier alkali metals is especially characteristic of their crystalline structures. Ammonium chloride at high temperatures has a structure similar to rubidium and cesium chloride, while at low temperatures its structure resembles that of NaCl. The ammonium cation is similar to alkaline cations in terms of precipitation reactions as well. It gives a yellow precipitate with $[Co(NO_2)_6]^{3-}$, analogous to the ions of potassium, rubidium, and cesium:

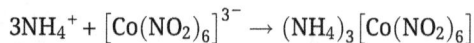

$$3NH_4^+ + \left[Co(NO_2)_6\right]^{3-} \rightarrow (NH_4)_3\left[Co(NO_2)_6\right]$$

However, the similarities do not apply to all chemical reactions in which NH_4^+ is involved. When alkaline nitrates are slightly heated, the corresponding nitrites and oxygen are produced, while heating NH_4NO_3 leads to decomposition, resulting in the release of nitrous oxide and water:

$$2NaNO_3 \rightarrow 2NaNO_2 + O_2$$

$$NH_4NO_3 \rightarrow N_2O + 2H_2O$$

Chapter 17
Second main group of the periodic table

17.1 General characteristics of the s-elements of IIA group

The s-elements of the second main group of the periodic table include the following elements: beryllium (Be), magnesium (Mg), calcium (Ca), strontium (Sr), barium (Ba), and radium (Ra).

Beryllium was discovered in 1798, and its content in the Earth's crust is only $3.8 \times 10^{-4}\%$. In nature, it is mainly found in the composition of the mineral beryl, $3BeO \cdot Al_2O_3 \cdot 6SiO_2$. Its varieties, colored by impurities in green and light blue colors, respectively, are the gemstones, emerald and aquamarine. The multistage technology of beryl processing leads to the formation of $BeCl_2$, from which Be is obtained by electrolysis of a melt in a mixture with NaCl at 350–400 °C.

Magnesium was discovered by Davy in 1808, and its content in the Earth's crust is 1.87%. Magnesium is one of the 10 most common elements in nature. It is a component of many silicates and aluminosilicates (olivine $2MgO \cdot SiO_2$, talc $3MgO \cdot 4SiO_2 \cdot H_2O$), magnesite $MgCO_3$, and carnallite $MgCl_2 \cdot KCl \cdot 6H_2O$. It is most often obtained by electrolysis of molten carnallite.

Calcium, strontium, and barium are jointly known as alkaline earth metals. The origin of this name is related to the fact that the hydroxides of these elements are bases, and their oxides are similar to the oxides of other metals, formerly called "earths." Calcium salts have been known to mankind for a long time, but in the free state, this metal was obtained by Davy only in 1808. Strontium was first isolated in the form of oxide by Crawford in 1790, and in its pure form, it was obtained by Davy in 1808. Barium was discovered by Scheele in 1774 and Davy in 1808. Radium was discovered by M. and P. Curie in 1898. Its content in the earth's crust is only $1.10^{-20}\%$.

Among the alkaline earth metals, calcium is the most common. Like magnesium, it is part of natural silicates and aluminosilicates, as well as in the very common compound $CaCO_3$ (calcite), which has several varieties: limestone, chalk, marble, etc. Calcium carbonate is also part of the mineral dolomite $CaCO_3 \cdot MgCO_3$. Deposits of calcium sulfate in the form of the mineral gypsum $CaSO_4 \cdot 2H_2O$, calcium phosphate in the form of the minerals phosphorite $Ca_3(PO_4)_2$ and apatite $3Ca_3(PO_4)_2 \cdot CaF_2$ (or $Ca_5(PO_4)_3F$), calcium fluoride in the form of the mineral fluorite CaF_2 are often found, as well as calcium nitrate or Norwegian saltpeter $Ca(NO_3)_2$.

The main characteristics of the atoms of the elements of the second main group are presented in Table 17.1. The structure of the valence shell of the atoms of these elements is ns^2, and the charges of the atomic nuclei are one more than in the alkali elements of the corresponding periods. This leads to a decrease in the atomic radius and an increase in ionization potentials and therefore to a decrease in their chemical activity.

https://doi.org/10.1515/9783111712246-017

Table 17.1: Characteristics of atoms of s-elements of the IIA group.

Element	Configuration	I_1 (eV)	I_2 (eV)	EO	$r_{йон}$ E^{2+} (Å)	z/r
Be	$[He]2s^2$	9.32	18.21	1.5	0.31	6.45
Mg	$[Ne]3s^2$	7.64	15.03	1.2	0.65	3.07
Ca	$[Ar]4s^2$	6.11	11.87	1.0	0.94	2.13
Sr	$[Kr]5s^2$	5.69	10.98	0.9	1.10	1.81
Ba	$[Xe]6s^2$	5.21	9.95	0.9	1.29	1.55
Ra	$[Rn]7s^2$	5.28	10.10	–	1.50	1.33

The electronic structure of the outer shell ns^2 determines a characteristic oxidation state of +2 for these elements. The first ionization potential I_1 is higher than in the s-elements of the first group, which is a consequence of the increase in nuclear charge. In addition, in the atoms of the s-elements of the second group, the electron configuration is more stable because it consists of a two-electron filled s-subshell and a free p-subshell. It is the symmetrical 1s-orbital completely filled with two electrons (in He) that is the reason why helium, which in its electron configuration is an analog of the elements of the IIA group, actually falls into the group of inert gases.

Of the elements in the group, beryllium differs strongly in all indicators. Full analogs are Ca, Sr, Ba, and Ra. Magnesium occupies an intermediate position. From the top to the bottom in the group, the tendency to form covalent bonds decreases. Beryllium compounds, both in solution and in solid form, contain covalent bonds. Magnesium also tends to form covalent bonds, while Ca, Sr, and Ba form mainly ionic bonds. In solution, they are mainly found in the form of M^{2+} ions. All elements of the group have high electropositivity, which is confirmed by their high reactivity in the free state and by their ionization potentials. Although the ionization potentials of alkaline earth metals are higher than those of alkali metals, their normal electrode potentials remain close due to the high hydration energy of the M^{2+} ions (Table 17.2).

Table 17.2: Ionization potentials of alkaline earth and alkali metals.

Element	$E°$ (V)	Element	$E°$ (V)
Mg	−2.31	Na	−2.71
Ca	−2.57	K	−2.92
Sr	−2.89	Rb	−2.99
Ba	−2.90	Cs	−2.92
Ra	−2.92		

All M^{2+} ions have significantly smaller radii and are less polarizing than M^+, so they have almost no deviations in the ionic character of salts caused by the polarization of cations. However, Mg^{2+} and particularly Be^{2+} tend to form covalent bonds due to polarization.

Synthesis: Due to their significant chemical activity, these elements do not occur in a free state and are obtained from their compounds by electrochemical reduction. As in the case of alkali metals, the main method for their synthesis is electrolysis in melts, mainly of their chlorides. The technology for obtaining beryllium is more complex since the feedstock (the mineral beryl) is subjected to a number of operations until the formation of relatively pure $BeCl_2$, from which beryllium is obtained by electrolysis in a melt. Metallic calcium is also obtained by electrolysis in melts of its salts. Usually, a melt of $CaCl_2$ and CaF_2 is used in a mass ratio of 3:1. Calcium fluoride is added in order to lower the melting point of the mixture.

In some cases, the elements of the second main group are also obtained by chemical reduction of their oxides and salts at high temperatures using Mg, C, or Al as reducers:

$$BeF_2 + Mg \rightarrow Be + MgF_2$$
$$MgO + C \rightarrow Mg + CO$$
$$4MO + 2Al \rightarrow 3M + Al_2O_3.MO \ (M = Ca, Sr, Ba)$$

Metals and their volatile salts color the flame with a characteristic color: Mg in white, Sr in red, and Ba in green.

17.2 Properties

17.2.1 Physical properties

The metals of the second main group are silvery-white substances, with the exception of beryllium, which is light gray. The general pattern of change in physical properties is analogous to that of alkali metals (Table 17.3).

Table 17.3: Some physical properties of second main group metals.

Element	T_m (°C)	T_b (°C)	d (g/cm^3)	Hardness by Mohs	$\rho.10^6$ (Ohm cm) at 0 °C	Magnetic susceptibility ($\chi.10^{-6}$)
Be	1,284	1,327	1.86	4	6.6	−1.0
Mg	651	1,107	1.74	2.5	4.6	0.55
Ca	851	1,440	1.55	1.5	4.3	1.1
Sr	770	1,380	2.63	1.8	30.7	−0.2
Ba	704	1,540	3.74	3	60	0.9
Ra	700	1,140	5.0	–	–	–

Beryllium is similar in many properties to aluminum (diagonal similarity). It is a light, silvery-white, highly malleable metal that is protected from interaction with oxygen and air moisture by a thin oxide shell on its surface.

The alkaline earth metals have a stronger metallic bond than the alkali metals. This obviously affects their higher melting points and their greater hardness. The ionic radii in the group increase but are smaller than those of the alkali metals. The melting and boiling temperatures in the group as a whole decrease, and this monotony is broken at Mg. From Be to Mg, T_m greatly decreases. The other elements of the group change the type of the crystal lattice, and the T_m decreases again, but slightly. With the exception of radium, the metals are light. The sequence of density change should be monotonous, but deviations in Mg and Ca are observed. In the group, as the volumes and masses of atoms increase, the density should also increase. However, from Be to Mg and from Mg to Ca, the radius changes very strongly, the mass slightly, so such a deviation occurs. Metals are good heat and electrical conductors, the weakest of which is beryllium.

17.2.2 Chemical properties

The metals of the second main group are chemically active. Down in the group, their activity increases because their first and second ionization potentials decrease. Be and Mg are resistant to air because they are covered with an oxide layer. The alkaline earth metals of the Ca subgroup are oxidized in air. Layers of different composition are formed on their surface: EC, EO_2, and E_3N_2. In relation to halogen elements, the least active is Be, which reacts only with fluorine at room temperature. The other metals react with all the halogen elements under normal conditions. Alkaline earth metals interact with sulfur and oxygen, with Be and Mg reacting only when heated. All metals interact with nitrogen and carbon at high temperatures. Be and Mg react weakly to moderately with water when heated, while alkaline earth metals react energetically under ordinary conditions.

They dissolve in liquid ammonia to form amides, but unlike alkali metals, when boiling and separating the solvent from the amides, quite stable ammoniates with a composition of $E(NH_3)_6$ are obtained. Many complex compounds of magnesium and especially of beryllium with different compositions are known. The elements of the calcium subgroup have low complexing ability. However, compared to alkali metals, the elements of the calcium subgroup form much more complex compounds because, with an increase in charge, the polarizing ability of M^{2+} ions increases, as well as their tendency to form complex compounds. An example of this is ammoniates with a defined composition, the formation of which is impossible in alkali metals.

The metals of the group interact with acids. They are powerful reducers, and when interacting with H_2SO_4 and HNO_3, products with minimal oxidation states of sulfur and nitrogen are obtained:

$$4Mg + 10HNO_{3dilut} \rightarrow 4Mg(NO_3)_2 + NH_4NO_3 + 3H_2O$$
$$4Ca + 5H_2SO_{4conc} \rightarrow 4CaSO_4 + H_2S + 4H_2O$$

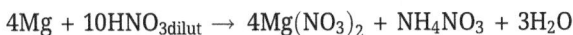

Beryllium: The chemistry of beryllium is significantly different from that of the other elements in the group, which is why covalent bonding predominates in its compounds. The very small beryllium cation has such a high charge density (1,100 C/mm^3) that it polarizes any approaching anion, whereby an overlap of the electron density occurs. Therefore, simple ionic compounds of beryllium, such as tetrahydrates $BeCl_2 \cdot 4H_2O$, actually show a tendency to complexation $[Be(H_2O)_4]^{2+}.2Cl^-$ and they contain ions in their crystal lattice. These tetraaquaberyllate ions, in which the four oxygen atoms of water molecules are covalently bonded to the beryllium ion, also predominate in their aqueous solution. The coordination number for beryllium is four, due to the small size of the beryllium ion.

With acids with nonoxidative action, beryllium interacts with the release of hydrogen, and in concentrated HNO_3 and H_2SO_4 it is passivated like aluminum. Beryllium is amphoteric, so it, as well as its oxide BeO and hydroxide $Be(OH)_2$, also interacts with bases to form oxoberyllates Na_2BeO_2 and hydroxoberyllates $Na_2[Be(OH)_4]$:

$$BeO + 2H_3O^+ + H_2O \rightarrow [Be(H_2O)_4]^{2+}$$
$$BeO + 2OH^- + H_2O \rightarrow [Be(OH)_4]^{2-}$$

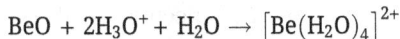

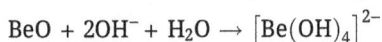

Beryllium, as well as other metals exhibiting amphoteric properties (including aluminum and zinc), is sometimes referred to as a "weak" metal because it is located at the boundary of metals and nonmetals in the periodic table. In this case, in terms of its properties, beryllium shows similarity with boron due to their closeness.

Beryllium hydroxide does not dissolve in water and is a weak base. Therefore, when dissolved, beryllium salts, containing anions of strong acids, are hydrolyzed reversibly. Beryllium salts are characterized by poor resistance to heating. For example, $BaCO_3$ decomposes into BeO and CO_2 at 100 °C, while $BaCO_3$ decomposes at 1,200 °C. Beryllium is the only one of the s-elements that forms stable complex compounds. All the above features in the properties of beryllium are determined by the strong polarizing effect of its small cation.

Magnesium. Magnesium is resistant to air at room temperature, as it is covered with a thin oxide layer, which protects it from additional oxidation. However, when heated, magnesium ignites in the air and burns to form MgO and a small amount of magnesium nitride, Mg_3N_2. When heated, magnesium energetically interacts with halogen elements, sulfur, nitrogen, phosphorus, carbon, silicon, and other elements to form halides, sulfide, nitride, phosphide, carbide, and silicide:

$$2Mg + O_2 = 2MgO$$
$$Mg + Cl_2 = MgCl_2$$
$$3Mg + N_2 = Mg_3N_2$$
$$3Mg + 2P = Mg_3P_2$$
$$2Mg + Si = Mg_2Si$$

Magnesium burns in an atmosphere of carbon dioxide. It also interacts with many other oxides because of its high reducing capacity:

$$2Mg + CO_2 = 2MgO + C$$
$$3Mg + B_2O_3 = 3MgO + 2B$$

Magnesium interacts with many metal oxides and halides; consequently, it is used in the production of many rare metals:

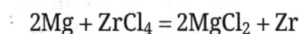

$$3Mg + MoO_3 = 3MgO + Mo$$
$$2Mg + ZrCl_4 = 2MgCl_2 + Zr$$

Magnesium interacts with acids, and in concentrated HNO_3 and H_2SO_4 it is not passivated:

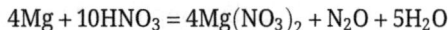

$$Mg + H_2SO_4 = MgSO_4 + H_2$$
$$4Mg + 10HNO_3 = 4Mg(NO_3)_2 + N_2O + 5H_2O$$

It does not react with water at room temperature due to the formation of a slightly soluble hydroxide, $Mg(OH)_2$, on its surface, but it interacts with hot water because, when the temperature increases, the solubility of $Mg(OH)_2$ increases:

$$Mg + H_2O = MgO + H_2$$

In the presence of ammonium chloride, magnesium reacts more easily with water at room temperature because NH_4Cl dissolves the protective layer of $Mg(OH)_2$:

$$Mg(OH)_2 + 2NH_4Cl = MgCl_2 + 2NH_4OH$$

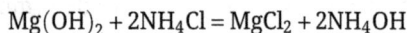

This process indicates that NH_4OH in the presence of ammonium chloride, which inhibits its dissociation, is a weaker base than $Mg(OH)_2$:

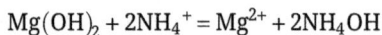

$$Mg(OH)_2 + 2NH_4^+ = Mg^{2+} + 2NH_4OH$$

In this way, magnesium can be stored and dissolved in ammonia. This solution is used for the qualitative and quantitative determination of phosphoric acid ions:

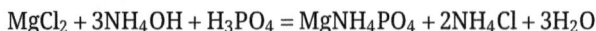

$$MgCl_2 + 3NH_4OH + H_3PO_4 = MgNH_4PO_4 + 2NH_4Cl + 3H_2O$$

The chemistry of magnesium differs from that of the other elements in the group. Of particular importance is that magnesium easily forms compounds containing covalent bonds. This behavior can be explained by its relatively high charge density (120 C/mm³), which for calcium is 52 C/mm³. For example, magnesium metal reacts with organic compounds, such as alkyl halides (C_2H_5Br) in ether (($C_2H_5)_2O$). The magnesium atom itself is inserted between the carbon and the halogen atoms, forming covalent bonds with them:

$$C_2H_5Br + Mg \rightarrow C_2H_5MgBr$$

These organomagnesium compounds are called Grignard reagents (R-MgX) and are widely used as intermediates in synthetic organic chemistry.

Alkaline earth metals: Alkaline earth metals are chemically very active. At room temperature, they are slowly oxidized by the oxygen in the air, and when heated, they burn to form oxides. They interact with water, displacing hydrogen, and form bases such as $Ca(OH)_2$, $Sr(OH)_2$, and $Ba(OH)_2$. With oxygen, nitrogen, halogen elements, hydrogen, and other nonmetals, they interact to form the corresponding ionic oxides, nitrides, halides, hydrides, etc.:

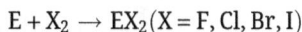

$$2E + O_2 \rightarrow 2EO$$
$$6E + 2N_2 \rightarrow 2E_3N_2$$
$$E + X_2 \rightarrow EX_2 (X = F, Cl, Br, I)$$
$$E + S \rightarrow ES$$
$$2E + H_2 \rightarrow 2EH$$

Calcium, strontium, and barium react energetically with water under ordinary conditions, resulting in the formation of hydroxides and the release of hydrogen:

$$E + 2H_2O \rightarrow E(OH)_2 + H_2$$

Alkaline earth metals form multiple salts but not complex compounds.

Calcium can take oxygen or halogen elements from oxides and halides of less active metals; that is, it exhibits pronounced reducing properties:

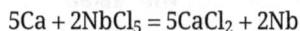

$$5Ca + Nb_2O_5 = 5CaO + 2Nb$$
$$5Ca + 2NbCl_5 = 5CaCl_2 + 2Nb$$

Metallic calcium, obtained from the electrolysis of molten calcium chloride, is used for the reduction of many metals (uranium, zirconium, lanthanum, yttrium, rare earths), for the separation of oxygen and sulfur from alloys, for the absorption of gas residues in vacuum devices, etc.

17.3 Chemical compounds

17.3.1 Hydrides

Hydrides are substances with a composition of EH_2, obtained by direct interaction with hydrogen, as their activity from Mg to Ba decreases, and Be does not react with H_2 under ordinary conditions. For example, when calcium is heated in a hydrogen medium, CaH_2 is formed:

$$Ca + H_2 \rightarrow CaH_2$$

The resulting calcium hydride interacts energetically with water, resulting in the release of hydrogen:

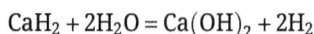

$$CaH_2 + 2H_2O = Ca(OH)_2 + 2H_2$$

The hydrides of alkaline earth elements, analogous to the hydrides of alkali metals, have strong reducing properties. Because of this, they easily react with oxygen and can also take oxygen from oxides of other elements:

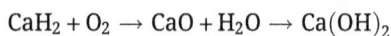

$$CaH_2 + O_2 \rightarrow CaO + H_2O \rightarrow Ca(OH)_2$$

CaO can be obtained from CaH_2 by the reduction of hydride using molybdenum oxide:

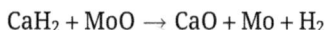

$$CaH_2 + MoO \rightarrow CaO + Mo + H_2$$

17.3.2 Compounds with oxygen

The elements of the second main group react actively with oxygen, but not as much as the alkali elements. Their activity increases up to Ba. The resulting oxides are basic, with the basic character being enhanced from MgO to BaO, while BeO is amphoteric. They are obtained by direct combustion in an oxygen medium or by thermal decomposition of carbonates and nitrates.

At higher temperatures, magnesium and elements from the calcium subgroup form *peroxides* EO_2 and *superoxides* EO_4 with alkaline earth elements:

$$Ba + O_2 \rightarrow BaO_2$$

Calcium peroxide is the product of the interaction of CaO, $Ca(OH)_2$, or calcium salts with a hydrogen peroxide solution:

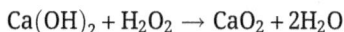

$$Ca(OH)_2 + H_2O_2 \rightarrow CaO_2 + 2H_2O$$

Peroxides have a white color. They can be considered as salts of hydrogen peroxide:

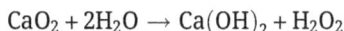

$$CaO_2 + 2H_2O \rightarrow Ca(OH)_2 + H_2O_2$$

The stability of the peroxides in the group increases. Magnesium peroxide is stable in the form of a hydrate. Alkaline earth peroxides are more stable. For example, BaO is oxidized to BaO_2 at 500 °C, and at a higher temperature (>600 °C), the peroxide is reduced to oxide:

$$2BaO + O_2 \rightleftarrows 2BaO_2 \rightarrow 2BaO + O_2$$

H_2O_2 is obtained from barium peroxide, when it reacts with water or acids:

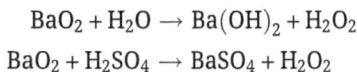

$$BaO_2 + H_2O \rightarrow Ba(OH)_2 + H_2O_2$$
$$BaO_2 + H_2SO_4 \rightarrow BaSO_4 + H_2O_2$$

Superoxides are yellow-colored substances that are less stable than peroxides. They are formed as by-products for the production of peroxides.

Beryllium oxide (BeO) is an amphoteric oxide that reacts with both acidic and basic oxides:

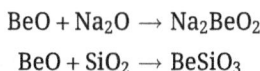

$$BeO + Na_2O \rightarrow Na_2BeO_2$$
$$BeO + SiO_2 \rightarrow BeSiO_3$$

It also reacts with acids and bases:

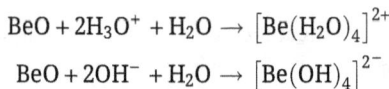

$$BeO + 2H_3O^+ + H_2O \rightarrow \left[Be(H_2O)_4\right]^{2+}$$
$$BeO + 2OH^- + H_2O \rightarrow \left[Be(OH)_4\right]^{2-}$$

Magnesium oxide (MgO) is obtained by burning metallic magnesium (Mg) or by decomposing magnesium hydroxide (Mg(OH)$_2$):

$$Mg(OH)_2 = MgO + H_2O$$

and in the industry at the thermal decomposition of magnesite (MgCO$_3$):

$$MgCO_3 = MgO + CO_2$$

Unlike beryllium oxide, it is not amphoteric; it is heat-resistant and interacts only with acids. Magnesium hydroxide is slightly soluble in water and bases. The solubility of MgO in water is very small, around 10^{-4} g/L at 25 °C. It dissolves easily in acids:

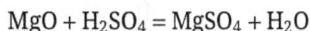

$$MgO + H_2SO_4 = MgSO_4 + H_2O$$

When heated, it reacts with acidic oxides:

$$MgO + SO_3 = MgSO_4$$

17.3.3 Oxides of alkaline earth metals

These oxides, as well as MgO, are very stable compounds. They exhibit a great affinity for oxygen and melt at high temperatures (Table 17.4).

Table 17.4: Melting temperatures of oxides.

Oxide	T_m (°C)
MgO	2,800
CaO	2,585
SrO	2,430
BaO	1,923

Oxides are most often obtained from carbonates, but the decomposition of $BaCO_3$ occurs at 1,625 °C, so BaO is obtained from nitrate:

$$2Ba(NO_3)_2 \rightarrow 2BaO + 4NO_2 + O_2$$

In industry, quicklime (calcium oxide, CaO) is obtained in large quantities by the thermal decomposition of natural limestone:

$$CaCO_3 = CaO + CO_2$$

Calcium oxide (CaO) is a white substance that melts at a temperature of about 3,000 °C, with pronounced basic properties. It interacts with water (converting quicklime to slaked lime), acids, and acidic oxides:

$$CaO + H_2O = Ca(OH)_2$$
$$CaO + 2HCl = CaCl_2 + H_2O$$
$$CaO + CO_2 = CaCO_3$$

All oxides of alkaline earth metals are soluble in water. Their dissolution is accompanied by an interaction and a release of heat. When reacting with water, hydroxides $E(OH)_2$ are formed, and their solubility increases from CaO to BaO:

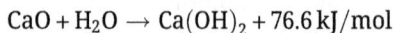

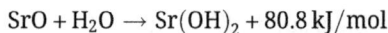

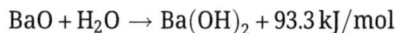

$$CaO + H_2O \rightarrow Ca(OH)_2 + 76.6\,kJ/mol$$
$$SrO + H_2O \rightarrow Sr(OH)_2 + 80.8\,kJ/mol$$
$$BaO + H_2O \rightarrow Ba(OH)_2 + 93.3\,kJ/mol$$

17.3.4 Hydroxides

The hydroxides of the elements of the IIA group have the general formula $E(OH)_2$. These are white crystalline substances. Their solubility in water increases from the top to the bottom in the group: $Be(OH)_2$ is slightly soluble in water with $K_s = 6.3 \times 10^{-22}$; $Mg(OH)_2$ has a solubility of 1×10^{-5} g per 100 g of water and $K_s = 2 \times 10^{-11}$; $Ca(OH)_2$ has a solubility of 0.2 g per 100 g of water and $K_s = 5.5 \times 10^{-6}$; $Sr(OH)_2$ shows a solubility of 0.8 g per 100 g of water and $K_s = 3.2 \times 10^{-4}$; $Ba(OH)_2$ has a solubility of 3–6 g per 100 g of water. From the top to the bottom in the group, the basic character of hydroxides increases: $Be(OH)_2$ is amphoteric; $Mg(OH)_2$ is basic; the rest $E(OH)_2$ are strong bases. All hydroxides, except $Be(OH)_2$, participate in reactions characteristic of strong bases. $Ba(OH)_2$ is the strongest of the listed bases.

They are obtained by dissolving the corresponding oxides in water, but BeO does not dissolve in water, and MgO is difficult to dissolve.

$Be(OH)_2$ is a white, practically water-insoluble substance that is a product of the hydrolysis of beryllium salts.

$Mg(OH)_2$ is a white crystalline substance. It is a medium-strong base. $Mg(OH)_2$ is slightly soluble in water but well soluble in acids:

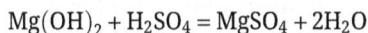

$$Mg(OH)_2 + H_2SO_4 = MgSO_4 + 2H_2O$$

When CO_2 is passed through a suspension of magnesium hydroxide, $Mg(OH)_2$ dissolves to form magnesium hydrogen carbonate:

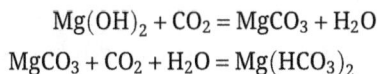

$$Mg(OH)_2 + CO_2 = MgCO_3 + H_2O$$
$$MgCO_3 + CO_2 + H_2O = Mg(HCO_3)_2$$

Magnesium hydroxide is obtained by the interaction of bases or ammonia with solutions of magnesium salts:

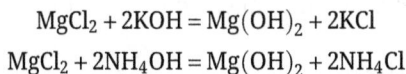

$$MgCl_2 + 2KOH = Mg(OH)_2 + 2KCl$$
$$MgCl_2 + 2NH_4OH = Mg(OH)_2 + 2NH_4Cl$$

Aqueous solutions of alkaline earth metal hydroxides are strong bases. Their salts with strong acids do not hydrolyze, and their aqueous solutions have a neutral reaction. They interact with acids and acidic oxides:

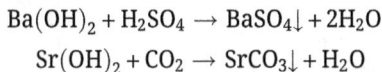

$$Ba(OH)_2 + H_2SO_4 \rightarrow BaSO_4\downarrow + 2H_2O$$
$$Sr(OH)_2 + CO_2 \rightarrow SrCO_3\downarrow + H_2O$$

Calcium hydroxide is a white solid with low solubility in water (1.56 g of $Ca(OH)_2$ is dissolved in 1 L of water at 20 °C). This solution is called *lime water*. It has an alkaline reaction and becomes cloudy in the air due to the absorption of CO_2 and the formation of insoluble calcium carbonate. $Ca(OH)_2$ is a strong base. It easily interacts with acids, acidic oxides, and salts:

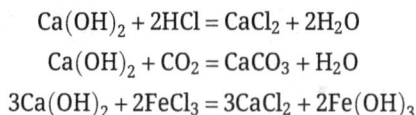

$$Ca(OH)_2 + 2HCl = CaCl_2 + 2H_2O$$
$$Ca(OH)_2 + CO_2 = CaCO_3 + H_2O$$
$$3Ca(OH)_2 + 2FeCl_3 = 3CaCl_2 + 2Fe(OH)_3$$

17.3.5 Salts

The salts of the elements of the IIA group are white crystalline substances. The color is determined either by the coloration of the anion or by defects in the crystal lattice. It should be noted that while all alkali metal salts are soluble in water, many of those in group A elements are insoluble, with the exception of compounds with single-charged anions, such as chlorides and nitrates, which are soluble. Down in the group, the solubility of salts (sulfates, carbonates) decreases. This is explained by the decreasing energy of hydration of E^{2+} ions and by the increasing strength of the crystal lattice. From the top to the bottom, the thermal stability of the salts increases due to the decrease in the polarizing effect of the cation on the anion. Most of the barium salts are anhydrous, while all magnesium salts are crystal hydrates.

The halides of the elements of the IIA group have a common formula EX_2. They are white crystalline substances that are produced by direct interaction between the elements. Fluorides strongly differ from other halides in their solubility in water. They are slightly soluble; for example, CaF_2 has $K_s = 10^{-11}$. F^- ion has a very small size compared to the large size of the metal cations E^{2+}. The lattice energy decreases remarkably quickly because the large cations come into contact with each other without contacts with the fluoride ions. The remaining halides are well soluble in water. Thus, the solubility of $CaCl_2$ is 130 g per 100 g of water. They crystallize from the solutions in the form of crystal hydrates of different compositions (Table 17.5).

Table 17.5: Melting points of chlorides.

Chloride	T_m (°C)
$MgCl_2 \cdot 6H_2O$	715
$CaCl_2 \cdot 6H_2O$	780
$SrCl_2 \cdot 6H_2O$	872
$BaCl_2 \cdot 2H_2O$	960

When heated, the behavior of chlorides depends on the metal cation:

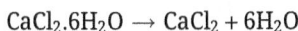

$$BaCl_2.2H_2O \rightarrow BaCl_2 + 2H_2O$$

$$CaCl_2.6H_2O \rightarrow CaCl_2 + 6H_2O$$

$$MgCl_2.6H_2O \rightarrow MgO + 2HCl + 5H_2O$$

The elements of the IIA group form nitrates with the following formulas: $Mg(NO_3)_2.6H_2O$, $Ca(NO_3)_2 \cdot nH_2O$ ($n = 0$–4), $Sr(NO_3)_2 \cdot nH_2O$ ($n = 2$, 4), and $Ba(NO_3)_2 \cdot nH_2O$ ($n = 2$, 4). Barium nitrate is most often obtained as anhydrous, but its crystal hydrates can also be separated. Nitrates are colorless crystalline substances with good solubility in water. When heated, they decompose to form oxides:

$$2Mg(NO_3)_2 \rightarrow 2MgO + 4NO_2 + O_2$$

The sulfates of the elements of the IIA group are white solids. Their solubility in 100 g of water is as follows: $MgSO_4$ – 33.7 g, $CaSO_4$ – 0.202 g, $SrSO_4$ – 0.014 g, and $BaSO_4$ – 0.0002 g. The solubility of sulfates down in the group decreases. The reason for this is that the hydration energy of the ions decreases much faster than the energy of the crystal lattice. Sulfates of strontium and barium are anhydrous. Calcium sulfate crystallizes with two molecules of water ($CaSO_4.2H_2O$) and magnesium sulfate – with 7 molecules of water ($MgSO_4.7H_2O$). The latter can be represented as a complex compound containing one molecule of water in its external coordination sphere $[Mg(H_2O)_6] SO_4.H_2O$.

The carbonates of the elements of the IIA group are white crystalline substances with low solubility in water. $MgCO_3$ is slightly soluble, while $CaCO_3$, $SrCO_3$, and $BaCO_3$ are practically insoluble, and their solubility decreases in the group. The reasons for the decrease in solubility are the same as for sulfates. Carbonates dissolve when CO_2 is passed through their suspensions or when reacting with ammonium chloride:

$$MgCO_3 + CO_2 + H_2O \rightarrow Mg(HCO_3)_2$$

$$CaCO_3 + 2NH_4Cl \rightarrow CaCl_2 + 2NH_3 + H_2O + CO_2$$

When heated, they decompose to the oxide of the corresponding element, EO and CO_2:

$$CaCO_3 \rightarrow CaO + CO_2$$

$$SrCO_3 \rightarrow SrO + CO_2$$

The most important representatives of carbonates are $CaCO_3$ (limestone, chalk, marble) and $MgCO_3$. Carbonates dissolve very easily in acids, including weak organic acids, which are stronger than carbonic acid (such as acetic and formic):

$$CaCO_3\downarrow + 2HCl \rightarrow CaCl_2 + H_2O + CO_2\uparrow$$

$$CaCO_3\downarrow + 2CH_3COOH \rightarrow Ca(CH_3COO)_2 + H_2O + CO_2\uparrow$$

Although slightly, carbonates dissolve in water that contains CO_2. This process is the basis for the dissolution of carbonates in nature and their subsequent deposition in caves as rock formations – stalactites and stalagmites:

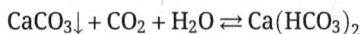

$$CaCO_3\downarrow + CO_2 + H_2O \rightleftharpoons Ca(HCO_3)_2$$

Many other salts of the elements of the IIA group are known: salts of anhydrous acids (nitrides, carbides, etc.); perchlorates, such as $Mg(ClO_4)_2$ (anhydrone), are used as drying agents.

Be and Mg easily react with nitrogen directly at high temperatures, forming nitrides:

$$3Mg + N_2 \rightarrow Mg_3N_2$$

With carbon, beryllium (Be) and magnesium (Mg) form carbides at high temperatures directly:

$$2Be + C \rightarrow Be_2C$$

Beryllium carbide (Be_2C) decomposes in water and releases methane, with sp^3-hybridization of the carbon atom. The stability of magnesium carbide has not been sufficiently studied. Carbides of alkaline earth metals, unlike Be and Mg carbides, decompose in water with the release of acetylene (sp-hybridization). Calcium carbide is obtained indirectly from quicklime CaO and C:

$$CaO + 3C \rightarrow CaC_2 + CO$$

17.4 Application

17.4.1 Beryllium

The element beryllium, being lighter than aluminum, as well as due to its hardness and corrosion resistance, is widely used in aerospace technology. Beryllium is a component of special alloys with high mechanical and chemical resistance. Beryllium oxide ($T_m = 2{,}580$ °C) is one of the best refractory materials. The widespread use of BeO is hindered by its high cost and toxicity (Be and its compounds are poisonous).

17.4.2 Magnesium

The main area of application for metallic Mg is related to the production of various alloys. Usually, these very light and strong alloys contain Al and Zn. They are used in mechanical engineering, rocket, and aircraft construction (duralumin). Pure Mg is used in metallurgy for the production of rare metals: titanium, zirconium, etc. because it reduces the oxides and sulfides of these metals by forming compounds that are difficult to dissolve in molten metals. Mixing powdered MgO with an aqueous solution of $MgCl_2$ results in magnesium cement with good bonding properties. Natural magnesium silicates, talc, and asbestos are widely used. Asbestos has a fibrous structure and is used to make refractory and heat-insulating fabrics.

Magnesium is an extremely important element for photosynthesis in plants, as it is part of the structure of the chlorophyll molecule. Mg^{2+} ions enter the coordination centers of two key enzymes that govern such a global process as photosynthesis, which consists of the conversion of H_2O and CO_2 into hydrocarbons and O_2 under the action of light energy. At the stage of photosynthesis, which takes place in the dark (the so-called "dark stage"), Mg^{2+} is the center of an enzyme containing ribulose-1,5-diphosphate carboxylase. This enzyme, very abundant in the biosphere, governs the binding of atmospheric CO_2 and the creation of biomass. In the "light stage" of photosynthesis, another enzyme – chlorophyll, in the center of which is Mg^{2+}, is photochemically excited and, with the participation of iron-sulfur proteins, reduces CO_2. The next stages of photosynthesis include a whole series of redox reactions involving the molecules ADP, ATP, quinone derivatives, and complexes of Mn^{2+} and Mn^{4+}, as a result of which H_2O, entering the coordination sphere, is oxidized to O_2.

In humans, magnesium is needed for energy production in cells by speeding up the metabolism. Magnesium is an important element for many of the functions in the body. Its main role is to regulate the balance of the nervous and muscular systems. In total, the body contains about 25 g of this element, 60% of which is in the bones, and a quarter in the muscles. The rest is distributed equally in the brain and other key organs such as the heart, liver, and kidneys. Magnesium helps in the distribution of signals to nerve cells, in the production of proteins, and in the regulation of heart rhythm. In addition, magnesium is involved in the gastrointestinal tract and is recommended for babies who suffer from colic. Enough magnesium should also be taken during physical activity. Since the body does not have magnesium reserves, it is necessary to take foods rich in magnesium daily. Most plants, including nuts, contain magnesium. Cereals, vegetables, and dried fruits also provide a sufficient amount of this element. Due to its valuable properties, magnesium is a widespread product in pharmacies and specialized stores for dietary supplements.

17.4.3 Calcium

Metallic calcium is used as a reductant in the metallothermic production of metals, as well as in the production of various alloys with beryllium, magnesium, aluminum, copper, lead, bismuth, etc. Calcium is an additive in iron alloys, which separate carbon and sulfur. Metallic Ca is used to produce calcium hydride, calcium carbide, and other valuable synthetic raw materials.

Due to its great chemical activity, the use of calcium as a metal is limited. However, the application of its compounds in various branches of industry and technology is much wider and more diverse.

Quicklime (CaO) is obtained industrially by the thermal decomposition of $CaCO_3$. Large amounts of CaO are used in the sugar industry to purify sugar solutions. It is one of the raw materials for glass production. Calcium carbide, which is necessary for the synthesis of acetylene, is obtained from CaO and C. Quicklime is also used in the leather and shoemaking industry.

When CaO is dissolved in water, slaked lime is obtained. Slaked lime mixed with sand and water (mortar) is a type of soldering agent that is used to connect bricks, stones, concrete blocks, etc. When dry, the mortar turns into a hard and durable solder. The solidification of this mixture occurs as a result of the absorption of CO_2 by calcium hydroxide over time and the formation of calcium carbonate (carbonization of $Ca(OH)_2$). Unlike gypsum and cement, mortar contains microscopic cavities that allow the wall to "breathe," thus avoiding the formation of mold. Therefore, in rooms where mortar is used, there is always moisture (from the emitted H_2O):

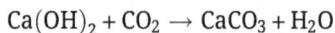

$$Ca(OH)_2 + CO_2 \rightarrow CaCO_3 + H_2O$$

One of the most important uses of calcium carbonate is as a feedstock for the production of soda ash and cement. Cement is obtained by mixing $CaCO_3$ with clay (Al_2O_3 and SiO_2) at 1,500 °C and then grinding it into a fine powder. Cement is a mixture of calcium silicates and aluminates. When mixed with water, it hardens as a result of the following processes:

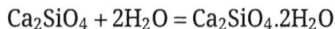

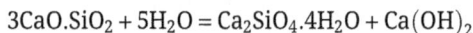

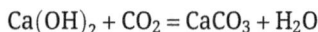

$$Ca_2SiO_4 + 2H_2O = Ca_2SiO_4.2H_2O$$

$$3CaO.SiO_2 + 5H_2O = Ca_2SiO_4.4H_2O + Ca(OH)_2$$

$$Ca(OH)_2 + CO_2 = CaCO_3 + H_2O$$

Calcium sulfate, the so-called gypsum ($CaSO_4.2H_2O$), is also used as a building material. During its heat treatment at a temperature of 150 °C, gypsum loses 75% of its crystallization water and turns into the so-called bonding gypsum ($CaSO_4.\frac{1}{2}H_2O$).

Calcium carbide is used to produce ethyne (acetylene) for acetylene burners:

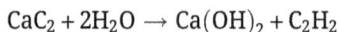

$$CaC_2 + 2H_2O \rightarrow Ca(OH)_2 + C_2H_2$$

Natural waters constantly contain calcium (Ca) and magnesium (Mg) salts. With a high content of these salts, water is called hard, and with an absence or low content, it is called soft. Hard water is not suitable for technical purposes, leading to the premature failure of steam boilers. It is impossible to wash in it:

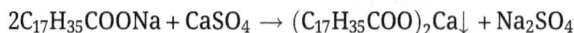

$$2C_{17}H_{35}COONa + CaSO_4 \rightarrow (C_{17}H_{35}COO)_2Ca\downarrow + Na_2SO_4$$

The resulting calcium stearate is slightly soluble, in contrast to sodium stearate in soap.

The issue of reducing the hardness of water, or softening of water, is of significant practical importance and in many cases represents a serious problem. The easiest way is to remove temporary or carbonate hardness since hydrogen carbonates very easily decompose when heated:

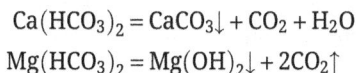

$$Ca(HCO_3)_2 = CaCO_3\downarrow + CO_2 + H_2O$$
$$Mg(HCO_3)_2 = Mg(OH)_2\downarrow + 2CO_2\uparrow$$

Sometimes water purification is carried out using chemical reagents (Na_2CO_3):

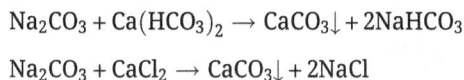

$$Na_2CO_3 + Ca(HCO_3)_2 \rightarrow CaCO_3\downarrow + 2NaHCO_3$$
$$Na_2CO_3 + CaCl_2 \rightarrow CaCO_3\downarrow + 2NaCl$$

Recently, special ion-exchange resins have been used, which exchange cations (cationites) or anions (anionites) when solutions containing salts pass through them. In this way, when hard water passes through ion exchangers, calcium and magnesium cations are absorbed.

17.5 Biological role

Calcium has an important biological role. If alkali metal ions, with the least ability to form coordination bonds, participate in the creation of an electrolyte environment in the body, then Ca^{2+} ions, forming low-soluble compounds, serve as the basis of the "carrier" systems in the body: the skeleton and cartilage. The most important functions of calcium in the human body are related to blood clotting and blood pressure, the conduction of nerve impulses, muscle contractions, etc. The content of Ca^{2+} in the body is approximately 1%. Calcium is the fifth most abundant in vivo element after C, H, O, and N. In mammalian organisms, 95% of calcium is concentrated in hard tissues: bones and teeth, where it is found in the form of fluorapatite $Ca_5(PO_4)_3F$ and hydroxyapatite $Ca_5(PO_4)_3OH$, while in the organisms of birds and mollusks $CaCO_3$ predominates. In the walls of blood vessels and arteries, calcium is present in the form of $CaCO_3$ or in complex with cholesterol, and in the kidneys, it is in the form of oxalates or urates (uric acid salts).

The main function of calcium is structural. The skeleton of a mature man contains about 1.2 kg of calcium. There is a continuous movement of calcium between the skeleton, blood, and other parts of the body, precisely controlled by hormones. The concentration of Ca^{2+} in the body is regulated by the parathyroid hormone, calcitonin, and its absorption is determined by the content of vitamin D in the body. Deficiency of this vitamin leads to reduced absorption of Ca, which manifests itself in the form of rickets disease. In adults, calcium deficiency can lead to impaired mineralization and softening of bones. Osteoporosis causes bones to be fragile and prone to breakage.

Bone loss over the years occurs in all individuals. This usually occurs after 35–40 years and involves a reduction in skeletal mass. Bone loss is greatest in women during menopause as a result of reduced hormone levels. Some studies show that vegetarians are less at risk of osteoporosis, although the factors can be different.

Calcium also plays a role in cell biology. It can bind to a wide range of proteins, altering their biological activity. This is important for the transmission of nerve impulses and muscle contraction. Calcium is also necessary for blood clotting, activating clotting factors.

A low level of calcium in the blood and tissues can lead to muscle spasms. This is more likely to be due to hormonal imbalances in calcium regulation than nutritional deficiencies. Excess calcium in the blood can cause nausea, vomiting, and calcium precipitation in the heart and kidneys. This is usually the result of an excessive overdose of vitamin D. In disease conditions, calcium is deposited in the kidneys, forming so-called "kidney stones," such as oxalates, phosphates, and urates.

It is known that the elements in the biosphere are characterized by distribution in the intra- or extracellular space. Extracellular biogenic elements are Na and Ca and intracellular elements are K and Mg (Table 17.6). The concentration of calcium in cells is very small – about 10^{-7} mol/L, and outside cells, it is ~10^{-3} mol/L. This concentration gradient is preserved, thanks to the Ca pump.

Table 17.6: Biodistribution of cations in intra- and extracellular space.

Bioenvironment	Cations (mmol/L)			
	Na^+	K^+	Ca^{2+}	Mg^{2+}
Blood plasma	152	5	2.5	1.5
Spinal cord	143	4	2.5	1.5
Cell	14	157	–	13

The magnesium ion is a stronger complexing agent than Ca^{2+} ion. It serves as the center of some metalloenzymes. For example, the magnesium-ATP complex enters the substrate of the enzyme kinase, which is responsible for the transfer of phosphate groups. Ca^{2+} ions form low-stability coordination compounds, characterized by low values of formation constants, with a variable coordination number (6 and 8) and a mobile coordination sphere, as well as high ligand exchange rates. Thus, calcium complexes are suitable for participation in signaling systems. That is why they regulate muscle contractions, activate numerous enzymes, and participate in the blood clotting process.

Of the inorganic compounds of magnesium, the so-called bitter salt, $MgSO_4 \cdot 7H_2O$, is used in medicine as a laxative, known as "Epsom salt." Some other Ca^{2+} salts are also used in medicine, such as $CaCl_2$, calcium gluconate (a complex with gluconic acid), and gypsum, $CaSO_4 \cdot 2H_2O$. Among the medicinal preparations of magnesium and calcium, including the complex ones, the most common are antacids. The complexes of magnesium and calcium with aspartic and glutamic acids are drugs that improve the tone of blood vessels.

Strontium and its compounds are of more limited use than calcium. Metallic strontium is used as an additive to alloys of magnesium, aluminum, copper, lead, and nickel. Strontium amalgam is used to produce divalent compounds of rare earth elements.

The physiological role of strontium is not yet fully understood. It has been found that when it enters the bloodstream, Sr is a strong poison. During nuclear explosions, strontium-90 is formed, the radiation of which is very dangerous because it causes leukemia and bone sarcoma. The isotope $_{38}^{89}$Sr, which is released during nuclear reactions, poses the greatest danger to life on Earth.

Strontium is processed in the organism in a similar way to its analog, Ca. In the bodies of animals and humans, Sr accumulates in large amounts in bones and affects the bone formation processes. Sr^{2+} ions are able to replace Ca^{2+} ions in many bioprocesses.

Barium: Metallic Ba is used in lead alloys that replace lead-antimony alloys in polygraphy.

Of the barium compounds, its sulfate, which is used in radiography as a contrast agent in the study of the gastrointestinal tract, is the most widely used in medicine. Barium compounds are poisonous, but $BaSO_4$ has the lowest solubility of the barium compounds. Used in radiography, it is not toxic as Ba ions that enter the body have a low concentration and do not cause complications.

Radium has natural radioactivity. During its radioactive decay, α-particles and electrons are released, and radon is formed. Radium salts are used for research purposes, as well as for the production of radon, which has valuable properties. The α-radionuclide radium-223 meets the requirements for nuclear medical applications.

Chapter 18
Third main group of the periodic table

18.1 General characteristics of the elements

Group IIIA includes the following elements: boron, aluminum, gallium, indium, and thallium. In the outer electronic shell of these elements there are three electrons (s^2p^1). They easily donate these electrons or form three unpaired electrons by the transition of one electron to the p-subshell. If in their outer electron shell their configuration is ns^2np^1, in relation to their penultimate electron shell, they show significant differences: boron has two electrons, aluminum has eight, and the rest of the elements have 18 each, that is, only the last three elements are complete electron analogs. These differences have an impact on the properties of the elements and their compounds. Boron and aluminum have a characteristic oxidation state of +3 in their compounds. In the elements of the Ga subgroup (gallium, indium, thallium), three electrons are also found in the outer electron shell, forming a configuration ns^2np^1, but they are located after a filled 18-electron shell and with an increase in atomic mass, lower oxidation states appear. For example, for Tl, the most stable compounds are those in which its oxidation state is +1. Unlike Al, gallium exhibits rather nonmetallic properties. These properties in the order Ga, In, and Tl weaken, and metallic ones are enhanced. In accordance with the regularities of the periodic table, the metallic properties of these elements are more pronounced than in the elements of the IVA group and they increase with increasing atomic number (Table 18.1).

Table 18.1: Characteristics of atoms of elements of IIIA group.

Element	B	Al	Ga	In	Tl
Valence electrons	$2s^22p^1$	$3s^23p^1$	$4s^24p^1$	$5s^25p^1$	$6s^26p^1$
Radius of the atom (nm)	0.091	0.143	0.139	0.166	0.171
Ionization potential (eV)	8.3	6.0	6.0	5.8	6.1
Electronegativity (χ)	2.0	1.5	1.6	1.7	1.8

Secondary periodicity is one of the important regularities in the properties of the p-elements in the main groups, which is very clearly manifested in group IIIA. Secondary periodicity is a special, non-monotonic character of changes in the various physical and chemical properties of the elements and their compounds depending on their position in the groups, that is, on their belonging to this or that period. Secondary periodicity occurs at the first ionization potentials I_1 of atoms and electronegativity values. It is manifested, firstly, in the anomalously high potentials I_1 in the atoms of the elements of the II period, and secondly, in the increased ionization potentials I_1 of the p-elements of the IV and VI periods. There is a non-monotonic character of the change

https://doi.org/10.1515/9783111712246-018

in these parameters. The factor determining the non-monotonic, secondary-periodic nature of the change in the properties of p-elements is the effect of interaction of valence electrons in multi-electron atoms.

In terms of chemical properties, p-elements of group IIIA are notable for their high reactivity. Under normal conditions, the chemical properties of aluminum, gallium, and indium are similar. They are resistant to corrosion due to the formation of an oxide layer E_2O_3 on their surface. All elements of group IIIA show a very strong affinity for oxygen, and the formation of their oxides is accompanied by the release of a large amount of heat.

18.1.1 Occurrence

The elements of group IIIA are often called "terrestrial" because their compounds are the main constituent of the soil. Boron was discovered in 1808 and is one of the few abundant elements in the Earth's crust, its main minerals being borax $Na_2B_4O_7 \cdot 10H_2O$ and kernite $Na_2B_4O_7 \cdot 4H_2O$. In addition, boron is found in a dissolved state in the form of borax and boric acid in the waters of some lakes.

Aluminum was first chemically obtained in 1825 and later isolated by electrochemical method. It is the most abundant metal in nature. Its content in the Earth's crust is 8.05%. Together with silicon, it makes up the main part of the lithosphere. The most important natural compounds of aluminum are aluminosilicates, bauxite, and corundum. Aluminosilicates make up the bulk of the Earth's crust. The basis of the clay is kaolin $Al_2O_3 \cdot 2SiO_2 \cdot 2H_2O$. Bauxite is a rock ore from which aluminum is obtained. It consists mainly of aluminum oxide hydrates $Al_2O_3 \cdot nH_2O$. Aluminum is also found in nature in the form of nepheline $Na_2O \cdot Al_2O_3 \cdot 2SiO_2$.

Gallium, indium. and thallium are scattered elements. Their content in ores, as a rule, does not exceed thousandths of a percent. The gallium content in the Earth's crust is $1.9 \times 10^{-3}\%$. It was predicted by Mendeleev (ekaaluminum) and was discovered in 1875. Indium was discovered in 1863 and its content in the Earth's crust is $2.5 \times 10^{-5}\%$. Thallium was discovered in 1861 and its content in the Earth's crust is about $10^{-4}\%$.

18.1.2 Physical properties

In accordance with the above, the physical properties of the elements show some peculiarities (Table 18.2). These p-elements, except boron, are metals in their free state.

With the exception of boron, p-elements of group IIIA do not exhibit high melting temperatures. The melting temperatures of the elements in the gallium subgroup are particularly low. The unusually low melting point (29.8 °C) of gallium is explained by the fact that diatomic molecules Ga_2 are located in the nodes of its crystal lattice, asso-

Table 18.2: Physical properties of the elements of IIIA group.

Element	T_m (°C)	T_b (°C)	ρ (g/cm)^3	R_{at} (Å)	$E°$ (V), $E_{p-p}{}^{3+}$/E	$E°$ (V), $E_{p-p}{}^{+}$/E
B	2040	2,550	2.48	0.91	–	–
Al	660.2	2,270	2.70	1.43	−1.622	–
Ga	29.8	2070	5.90	1.39	−0.65	–
In	156.2	2075	7.31	1.69	−0.343	−0.25
Tl	302.5	1457	11.85	1.71	+0.71	−0.336

ciated with weak forces of intermolecular interaction, which leads to the fact that when melting, the density of gallium increases. The metals of the group have different crystal lattices. Of interest is the wide temperature interval corresponding to the liquid state of In and Ga – about 2,000 °C. They are soft metals – In and Tl are easy to cut with a knife. The hardest of them is Al, which belongs to the light metals.

18.2 Boron and its compounds

18.2.1 Physical and chemical properties

Boron is a solid with a high melting point. In boron, there is no metallic bonding and it is a weak electrical conductor. Boron has a high ionization potential and electronegativity, as a result of which its electrons do not move throughout the crystal and it rather exhibits nonmetallic properties. Boron has a unique structure with a complex crystal lattice, in which the main structural unit is *icosahedrons,* consisting of 12 atoms (B_{12}). They are bonded by covalent bonds both within the individual structural units and between adjacent B_{12} units. Therefore, when melting, a very high melting temperature is required to break these covalent bonds. Another similar highly fusible element is carbon. Its allotropic form, graphite sublimates at above 4,000 °C. This nonmetal is made up of layers in which carbon atoms are bonded by multiple bonds and, as in the melting of boron, a network of very strong covalent bonds is broken. Similarly, boron exists in amorphous and crystalline states. Amorphous boron is a dark brown powder, and its crystals have a dark red color. In the crystalline state, the boron lattice is very strong, which manifests itself in high hardness, approaching the hardness of diamond, low entropy (7 J/mol K), and high melting point (2,300 °C).

The high ionization potential of boron practically excludes the existence of B^{3+}. Its oxidative state is related to the formation of three hybrid covalent bonds sp^2 and the planar geometry of the molecules. It should also be noted that boron is more similar to silicon than to its analogs from the group (diagonal similarity). Under normal conditions, boron is weakly active. At room temperature, it reacts only with fluorine up to BF_3. At 400–700 °C, boron interacts with other halogen elements, oxygen (to B_2O_3), sulfur (to B_2S_3), and at 1,200 °C it reacts with water:

$$2B + 3H_2O = B_2O_3 + 3H_2$$

Hydrohalic acids do not interact with boron. Boron is slowly oxidized only by hot concentrated nitric and sulfuric acids, aqua regia, and base melts in the presence of oxidizing agents:

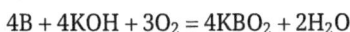

$$B + 3HNO_3 = H_3BO_3 + 3NO_2$$

$$4B + 4KOH + 3O_2 = 4KBO_2 + 2H_2O$$

18.2.2 Synthesis

Boron with technical purity is obtained from borax, from which boric acid first precipitates. Boric acid decomposes when heated to water and boron oxide, from which boron is reduced with magnesium or sodium at high temperature:

$$Na_2B_4O_7 + H_2SO_4 + 5H_2O = Na_2SO_4 + 4H_3BO_3\downarrow$$

$$2H_3BO_3 = B_2O_3 + 3H_2O$$

$$B_2O_3 + 3Mg = 3MgO + 2B$$

Other methods are used to obtain boron with high purity:

$$2BX_3 + 3H_2 \rightarrow 2B + 6HX$$

By each of the listed methods, boron is obtained in an amorphous state.

18.2.3 Chemical compounds

With metals, when heated, boron forms *borides*. All borides are compounds with an indeterminate composition (berthollides) that do not correspond to the usual valence. For example, magnesium boride has a composition of MgB_2, and chromium has the borides Cr_4B, Cr_2B, Cr_5B_3, CrB, Cr_3B_4, and CrB_2. The borides of the d-elements are solid chemically inert substances. The borides of the s-elements are more active and are decomposed by acids.

Boron in compounds with more electronegative elements is found in the oxidation state of +3. The compounds are formed by the excitation of boron atoms, as a result of which the three valence electrons become single electrons ($2s^12p^2$). In this case, two types of hybridization are possible: sp^2 (trigonal molecules or ions BF_3, BCl_3, BO_3^{3-}) and sp^3 (tetrahedral complex ions BF_4^-, BH_4^-).

Several compounds of boron with hydrogen (*boranes*) are known with the general formulas B_nH_{n+6} and B_nH_{n+4}, where $n > 2$: B_2H_6, B_4H_{10}, etc. They are highly poisonous gases and easily volatile liquids formed by the interaction of borides of s-elements with acids. The best studied is diborane B_2H_6, in the molecules of which, along with the two-electron bonds, there are also single-electron "bridge" bonds. The bonding in B_2H_6 can be described by the concept of hybridization. According to this concept, the four bonds separated by almost equal angles correspond to sp^3 hybridization. This bonding means an orbital that spans three atoms (three-center bond) and contains two electrons. The bond on the other B-H-B bridge is identical.

The simplest hydrogen compound of boron BH_3 (borane) does not exist under ordinary conditions, which is explained by the coordination unsaturation of the boron atom and the incomplete shielding of electrons from the positive charges of the nuclei.

Boron is second only to carbon in terms of the number of hydrides it forms. More than 50 neutral boranes, B_nH_m, and an even larger number of boron anions, $B_nH_m^{x-}$ are known. The molecules of boranes are different in shape from those of other known hydrides. The chemistry of boranes is very similar to organic chemistry despite the interesting differences between them. All boranes have positive ΔG values; therefore they are thermodynamically unstable and easily decompose into their constituent components.

The industrial synthesis of diborane takes place by reacting BF_3 with sodium hydride:

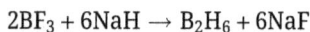

$$2BF_3 + 6NaH \rightarrow B_2H_6 + 6NaF$$

Boranes possess partially negatively charged hydrogen ions due to the low electronegativity of boron, resulting in the high chemical activity characteristic of these compounds. For example, diborane, like most of the neutral boranes, burns when mixed with pure oxygen. The products of this highly exothermic reaction are diboron trioxide and water vapor:

$$B_2H_6 + 3O_2 \rightarrow B_2O_3 + 3H_2O$$

The reaction with water to H_3BO_3 (B(OH)$_3$) and hydrogen is also highly exothermic:

$$B_2H_6 + 6H_2O \rightarrow 2H_3BO_3 + 6H_2$$

Most of the other boranes are synthesized from diborane. For example, tetraborane is obtained by the condensation of two diborane molecules, and pentaborane by subsequent interaction with another diborane molecule at high pressure:

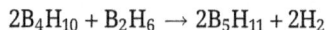

$$2B_2H_6 \rightarrow B_4H_{10} + H_2$$

$$2B_4H_{10} + B_2H_6 \rightarrow 2B_5H_{11} + 2H_2$$

Diborane is an important reagent in organic chemistry. This gas reacts with unsaturated hydrocarbons (containing double or triple carbon-carbon bonds) and forms alkylboranes:

$$B_2H_6 + 6CH_2 = CHCH_3 \rightarrow 2B(CH_2CH_2CH_3)_3$$

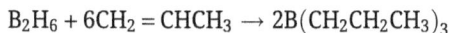

Of the borane anions, the tetrahydroborate ion BH_4^- is of interest. The crystal structure of sodium tetrahydroborate resembles that of NaCl, with BH_4^- ion appearing to be identical to Cl^- ion.

Sodium tetrahydroborate is of essential practical importance for organic chemistry as a moderate reducer, especially in the reduction of aldehydes to primary alcohols and ketones to secondary alcohols, without reducing other functional groups such as carboxylic ones. Sodium tetrahydroborate is produced by the reaction of diborane with sodium hydride:

$$2NaH + B_2H_6 \rightarrow 2NaBH_4$$

With oxygen, boron forms the *oxide* B_2O_3, which reacts with water to form *orthoboric acid* (H_3BO_3). In its solid state, it has a layered structure in which the molecules are bonded by strong hydrogen bonds. When heated, it releases water and transitions into *metaboric acid* (HBO_2) at 200 °C and boric oxide at 300 °C as shown in Figure 18.1.

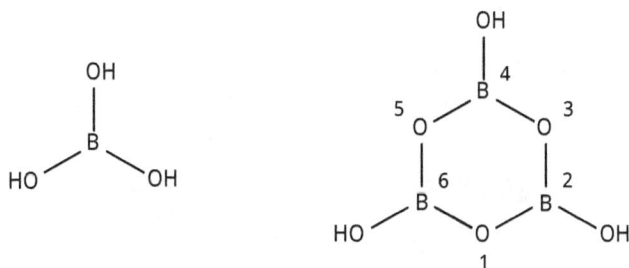

Figure 18.1: Structures of metaboric and orthoboric acids.

The formula of orthoboric acid is often written as $B(OH)_3$. This is due to the fact that when dissolved in water, it does not dissociate like other acids, but attaches hydroxide ions from water by a donor-acceptor mechanism, forming hydrogen cations (hydroxonium cations H_3O^+) and tetrahydroxyborate complex ions:

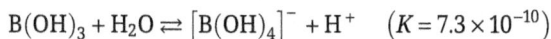

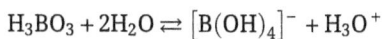

$$B(OH)_3 + H_2O \rightleftharpoons \left[B(OH)_4\right]^- + H^+ \quad (K = 7.3 \times 10^{-10})$$

$$H_3BO_3 + 2H_2O \rightleftharpoons \left[B(OH)_4\right]^- + H_3O^+$$

For this reason, orthoboric acid is a weak monobasic acid. The acidic properties of orthoboric acid are determined not by its electrolyte dissociation but by the acceptance of OH^-, that is, it behaves like a Lewis acid.

Of its salts, sodium tetraborate ($Na_2B_4O_7$) and its crystal hydrate ($Na_2B_4O_7 \cdot 10H_2O$) (borax) are of great practical importance. Borax melts at 750 °C and in the molten state dissolves the oxides of practically all metals. This property of borax is used to clean metal surfaces from oxide layers.

Boron forms compounds with all halogen elements: gaseous BF_3 and BCl_3, liquid BBr_3, and crystalline BI_3. Their molecules have a triangular structure due to the sp^2-hybridization of boron orbitals. Due to the presence of a free orbital at the boron atom, these compounds are strong electron pair acceptors (prototypes of Lewis acids). For example, BF_3 attaches an ammonia molecule to form an $H_3N.BF_3$, with the fluoride ion forms BF_4^-, etc.

$$BF_3 + :NH_3 \rightarrow F_3B:NH_3$$

Boron halides undergo complete hydrolysis with the formation of two acids:

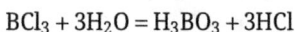

$$BCl_3 + 3H_2O = H_3BO_3 + 3HCl$$

The hydrolysis of boron trifluoride is accompanied by a secondary process of formation of the strong complex tetrafluoroboric acid HBF_4:

$$BF_3 + 3H_2O = H_3BO_3 + 3HF$$

$$BF_3 + HF = HBF_4$$

The salts of this acid tetrafluoroborates are known such as KBF_4 and $Ba(BF_4)_2$.

When boron reacts with sulfur, *boron sulfide* (B_2S_3) is formed, which, when interacting with water, undergoes complete hydrolysis:

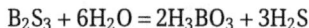

$$B_2S_3 + 6H_2O = 2H_3BO_3 + 3H_2S$$

At high temperatures, boron interacts with N_2 to form *boron nitride* (BN). Its diamond-like modification has several names: elbor, cubonite, and borazon. This substance is close to diamond in hardness, but is more heat-resistant, eithstanding heating up to 2,000 °C, while diamond burns at 1,000 °C.

The most important of the binary compounds of boron is *boron carbide*, which has the formula B_4C. Its actual structure is more correctly represented by the formula $B_{12}C_3$ since it consists of *icosahedrons* B_{12}, like boron itself, with carbon atoms binding all adjacent icosahedrons. One method for obtaining B_4C is the reduction of diboron trioxide with carbon:

$$2B_2O_3 + 7C \rightarrow B_4C + 6CO$$

Boron carbide is one of the hardest substances known. Its fibers are extremely strong and are used to make impenetrable clothes. Boron carbide is used as a starting material to obtain other hard materials such as titanium boride:

$$2TiO_2 + B_4C + 3C \rightarrow 2TiB_2 + 4CO$$

Titanium boride belongs to a different structural class of *borides*. These borides consist of hexagonal layers of boron ions, which are isoelectronic and isostructural with the allotropic form of carbon, graphite. Metal ions are located between the boron layers. Each boron atom has a charge of -1, and stoichiometrically this class of borides corresponds to metals in $+2$ oxidation state, as for example in MgB_2.

18.2.3.1 Usage and biological role

Boron is added to steel and various alloys to increase their heat resistance and durability. Pure boron absorbs neutrons, so it is used to protect against neutron radiation.

Boron halides are used as catalysts and hydridoborates as reducers in organic synthesis. Boron compounds with metals (borides) exhibit high hardness and heat resistance, which is why they are used to obtain special superhard and heat-resistant alloys and for the manufacture of various refractory products. Of these compounds, MgB_2 is very cheap, readily available, and it also exhibits superconducting properties at low temperatures. MgB_2 retains its superconductivity to 39 K, the highest measured value for such a compound. Boron carbide and nitride exhibit great heat resistance. Graphite-like boron nitride is used as a high-temperature lubricant, and diamond-like BN is used as a superhard material in the drilling of hard ores and in the processing of metals. The crystal hydrate of sodium tetraborate ($Na_2B_4O_7.10H_2O$) (borax) has a constant composition and its solutions are used in analytical chemistry to determine the concentrations of acidic solutions. The reaction of borax with acids proceeds according to the equation:

$$Na_2B_4O_7 + 2HCl + 5H_2O = 2NaCl + 4H_3BO_3$$

Boron is an important trace element that regulates a number of biochemical processes in plant and animal organisms. The average content of boron in the body is $10^{-5}\%$, and it is concentrated mainly in the lungs and liver, thyroid gland, brain, and heart muscle. An excess of boron is harmful to the human body, as it reduces the activity of adrenaline and disrupts the metabolism of hydrocarbons and proteins in the body. Boron compounds can also have a toxic effect on the human body and can penetrate even through healthy skin. Boron is a vital trace element for plants, affecting the metabolism and transport of carbohydrates. Along with Mn, Cu, Mo, and Zn, boron is among the five most important trace elements for plants, affecting the proper development and growth of plants. With a shortage of boron in plants, the yield and the amount of grain are strongly reduced. In addition to boric acid, which is a recognized mild antiseptic, the boron-containing preparation tolboxane is also known as a mild narcotic tranquilizer. Boron compounds are associated with the actions of some enzymes, but the mechanisms remain unclear.

18.3 Aluminum and its compounds

18.3.1 Physical properties

Aluminum is a silvery-white light metal melting at 660 °C. It is very ductile, malleable, and easily drawn into wires and sheets. It can be used to make a foil with a thickness of less than 0.01 mm. Aluminum exhibits very high thermal and electrical conductivity, which is why it is especially valuable as a structural material. Its alloys with various metals are light and strong. In the air it is quickly coated with a thin oxide shell (Al_2O_3), which is water-insoluble and protects it from further oxidation. The only limitation for the application of Al is its high cost, which, combined with its good qualities and wide field of application, defines it as an important product in the metallurgy of any country.

18.3.2 Preparation

For the first time, aluminum was obtained by reducing aluminum chloride with metallic sodium:

$$AlCl_3 + 3Na = 3NaCl + Al$$

The technical raw material for the production of Al is the mineral bauxite $Al_2O_3.xH_2O$. In the first stage of production, pure $Al(OH)_3$ is obtained from the ore, which is electrolyzed at 1,000 °C using graphite electrodes. In this process, Al is released at the cathode, and O_2 is released at the anode. This technology is associated with very high energy costs, which affects the production cost of the metal.

Currently, Al is obtained by electrolysis of molten salts. The melt contains 85–90% cryolite $3\ NaF.AlF_3$ (Na_3AlF_6) and 10–15% Al_2O_3. This mixture melts at 1,000 °C. When dissolved in molten cryolite, alumina behaves like a salt of aluminum and aluminum acid and dissociates into aluminum cations and anions of the acid residue of aluminum acid:

$$AlAlO_3 \rightleftharpoons Al^{3+} + AlO_3^{3-}$$

Cryolite also dissociates:

$$Na_3AlF_6 \rightleftharpoons 3Na^+ + AlF_6^{3-}$$

When an electric current passes through the melt, Al^{3+} and Na^+ cations move to the cathode. Aluminum, being less active than sodium, is reduced first. The reduced aluminum in the molten state is collected at the bottom of the bathtub, from where it is periodically removed. The anions AlO_3^{3-} and AlF_6^{3-} move to the anode (graphite electrode), and AlO_3^{3-} ions oxidize:

On the anode: $\qquad$ $4AlO_3{}^{3-} - 12e^- = 2Al_2O_3 + 3O_2$

On the cathode: $\qquad$ $4Al^{3+} + 12e = 4Al$

Overall equation: $\qquad$ $2Al_2O_3 = 4Al + 3O_2$

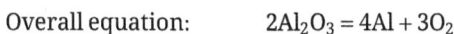

The alumina consumed in the process is constantly replenished. The amount of cryolite practically does not change. The production of aluminum by electrolysis requires large losses of electrical energy, so aluminum plants are usually built near power plants.

18.3.3 Chemical properties

Aluminum is a very active metal. In the row of metal activity, it stands after the alkali and alkaline earth metals. However, it is quite resistant to air because its surface is covered with a very dense layer of oxide, which protects the metal from contact with air. If the protective oxide layer is removed, it begins to react energetically with oxygen and water vapor from the air turning into a brittle mass – aluminum hydroxide:

$$4Al + 3O_2 + 6H_2O = 4Al(OH)_3$$

This reaction is accompanied by the release of heat. Purified from the protective oxide layer, aluminum interacts with water to release hydrogen:

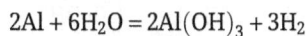

$$2Al + 6H_2O = 2Al(OH)_3 + 3H_2$$

When aluminum powder is put into a flame, it burns with blinding light:

$$4Al + 3O_2 = 2Al_2O_3 + Q$$

The high thermal effect of the Al oxidation process indicates the strong chemical affinity of aluminum for oxygen. These pronounced reduction properties of aluminum are used to extract some highly meltable metals from their oxides (as well as their salts) – the method is called *aluminothermy*. The reaction is accompanied by the release of a large amount of heat:

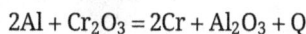

$$2Al + Cr_2O_3 = 2Cr + Al_2O_3 + Q$$

A mixture of 25% Fe_3O_4 and 75% Al, called thermite, is used for welding metal products. If a mixture of finely powdered aluminum is ignited with magnetite Fe_3O_4, then the reaction proceeds spontaneously by heating the mixture to 3,500 °C. Fe at this temperature is molten.

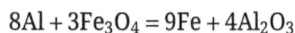

$$8Al + 3Fe_3O_4 = 9Fe + 4Al_2O_3$$

Aluminothermy is used to obtain some rare metals that form strong bonds with oxygen (niobium, tantalum, molybdenum, tungsten, etc.).

Aluminum also reduces oxides of nonmetals:

$$4Al + 3SO_2 \rightarrow 2Al_2O_3 + 3S$$

As a strong reducer, aluminum dissolves well in acids with nonoxidative action (dilute sulfuric and hydrochloric acids), releasing hydrogen:

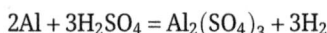

$$2Al + 6HCl = 2AlCl_3 + 3H_2$$

$$2Al + 3H_2SO_4 = Al_2(SO_4)_3 + 3H_2$$

$$2Al + 6H^+ \rightarrow 2Al^{3+} + 3H_2$$

Dilute nitric acid, when cold, passivates aluminum, but when heated, it dissolves in it with the release of nitric monoxide, nitrous oxide, free nitrogen, or ammonia, for example:

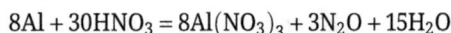

$$8Al + 30HNO_3 = 8Al(NO_3)_3 + 3N_2O + 15H_2O$$

Concentrated nitric acid passivates aluminum.

Since aluminum oxide and hydroxide exhibit amphoteric properties, the metal dissolves easily in aqueous solutions of bases, with the exception of ammonium hydroxide:

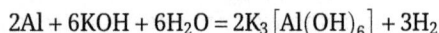

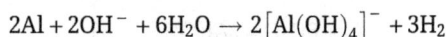

$$2Al + 6KOH + 6H_2O = 2K_3\left[Al(OH)_6\right] + 3H_2$$

$$2Al + 2OH^- + 6H_2O \rightarrow 2\left[Al(OH)_4\right]^- + 3H_2$$

Concentrated solutions of alkaline bases easily dissolve the metal, resulting in a hydroxyaluminate complex ion. When melting solid sodium base with aluminum, the metal dissolves to form sodium metaaluminate $NaAlO_2$.

Powdered aluminum easily interacts with almost all nonmetals. At the beginning of the reaction, heating is necessary, after which it proceeds very intensively and is accompanied by the release of a large amount of heat:

$$2Al + N_2 = 2AlN$$

$$4Al + 3C = Al_4C_3$$

With carbon, Al reacts directly at high temperature (2,000 °C). Aluminum carbide is obtained, which, when interacting with water, releases methane:

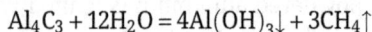

$$Al_4C_3 + 12H_2O = 4Al(OH)_3\downarrow + 3CH_4\uparrow$$

Aluminum sulfide can only exist in a solid state:

$$2Al + 3S = Al_2S_3$$

In aqueous solutions, it undergoes complete hydrolysis with the formation of aluminum hydroxide and hydrogen sulfide:

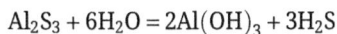

$$Al_2S_3 + 6H_2O = 2Al(OH)_3 + 3H_2S$$

AlF_3 is produced by the interaction of Al_2O_3 with HF at high temperature:

$$Al_2O_3 + 6HF = 2AlF_3 + 3H_2O$$

AlF_3 is a solid crystalline substance that forms very stable hexafluoride complex ion:

$$AlF_3 + 3F^- = [AlF_6]^{3-}$$

The other halides (AlX_3) are produced by direct interaction and are compounds with covalent bonds:

$$2Al + 3Br_2 = 2AlBr_3$$

Many of the halogenides of the p-elements are characterized by the formation of complex compounds with the halides of alkali, alkaline earth metals, and ammonia, especially fluorides. Thus, fluoride complexes are very common in boron and aluminum ($H[BF_4]$, $M_3[AlF_6]$). The stability of halogenide complexes decreases in the order F–Cl–Br–I.

18.3.4 Chemical compounds

Aluminum hydride (AlH_3) (alan) forms bridge bonds analogous to the bonds in boranes. It is obtained indirectly in an ether solution:

$$AlCl_3 + 3LiH = AlH_3 + 3LiCl$$

Aluminum hydride is a white solid polymeric substance $(AlH_3)_x$, which, when heated above 100 °C, decomposes into simple elements. When reacting with alkali metal hydrides, AlH_3 forms hydridoaluminates ($LiAlH_4$), which are strong reducers and are used in organic synthesis:

$$AlH_3 + LiH = LiAlH_4$$

Aluminum oxide (Al_2O_3) is a white substance with a melting point of 2,050 °C. It exists in an amorphous state and in several crystalline modifications. Amorphous Al_2O_3 is formed by heating and decomposition of aluminum hydroxide or aluminum salts at 400–600 °C:

$$2Al(OH)_3 \rightarrow Al_2O_3 + 3H_2O$$

As an amphoteric substance, it also interacts with acids and bases:

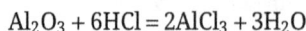

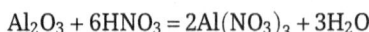

$$Al_2O_3 + 6HCl = 2AlCl_3 + 3H_2O$$

$$Al_2O_3 + 6HNO_3 = 2Al(NO_3)_3 + 3H_2O$$

In concentrated solutions of alkaline bases, the dissolution of aluminum oxide leads to the formation of a hexahydroxyaluminate complex ion:

$$Al_2O_3 + 6KOH + 3H_2O = 2K_3[Al(OH)_6]$$

When melted with alkaline bases, aluminum oxide forms metaaluminates:

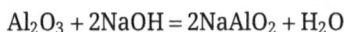

$$Al_2O_3 + 2NaOH = 2NaAlO_2 + H_2O$$

At a temperature above 600 °C, crystalline g–Al_2O_3 is formed, and at 1,000–1,200 °C – the l-form of the oxide, the so-called corundum. In nature, aluminum oxide occurs in the form of corundum and alumina. Natural corundum is a very hard and chemically inert compound (T_m = 2,072 °C). Corundum colored by red-colored chromium compounds is called ruby, and colored by titanium and iron compounds in blue color is called sapphire. Currently, these gemstones are easily and artificially obtained. Alumina is used as a raw material to obtain aluminum. Dehydrated aluminum oxide serves as an adsorbent in the purification and separation of organic substances.

Crystalline aluminum oxide also exhibits amphoteric properties, but does not dissolve in water, acids, and bases. When boiled in a concentrated solution of a base, it only partially passes into solution. Aluminum oxide is brought to a soluble state by melting with bases or potassium pyrosulfate:

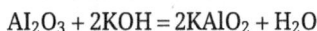

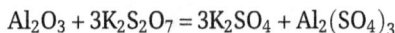

$$AI_2O_3 + 2KOH = 2KAlO_2 + H_2O$$

$$Al_2O_3 + 3K_2S_2O_7 = 3K_2SO_4 + Al_2(SO_4)_3$$

When aluminum oxide is melted with K_2CO_3 or soda, aluminates are formed, which are easily dissolved in water:

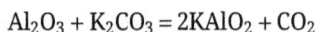

$$Al_2O_3 + K_2CO_3 = 2KAlO_2 + CO_2$$

Aluminum hydroxide (Al(OH)₃) is formed by the interaction of bases, ammonium hydroxide, and alkaline carbonates with solutions of aluminum salts:

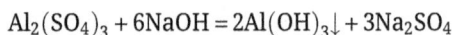

$$Al_2(SO_4)_3 + 6NaOH = 2Al(OH)_3\downarrow + 3Na_2SO_4$$

or by hydrolysis of salts, in which the process proceeds to the end:

$$Al_2S_3 + 6H_2O = 2Al(OH)_3\downarrow + 3H_2S\uparrow$$

Another method for obtaining $Al(OH)_3$ is to mix aqueous solutions of salts that mutually enhance their hydrolysis:

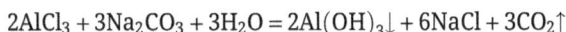

$$2AlCl_3 + 3Na_2CO_3 + 3H_2O = 2Al(OH)_3\downarrow + 6NaCl + 3CO_2\uparrow$$

The formula $Al(OH)_3$ is uncertain. In fact, the amorphous precipitate (gel) of aluminum hydroxide contains a large amount of bound water and has a variable composition $Al(OH)_3 \cdot xH_2O$. It is a white amorphous substance that, when heated, loses water, turning into aluminum oxide.

Aluminum hydroxide exhibits amphoteric properties. Freshly precipitated aluminum hydroxide easily dissolves in acids and bases (without ammonium hydroxide):

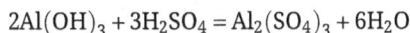

$$2Al(OH)_3 + 3H_2SO_4 = Al_2(SO_4)_3 + 6H_2O$$

$$Al(OH)_3 + 3KOH = K_3\left[Al(OH)_6\right]$$

Hydroxyaluminates, formed by the interaction of aluminum hydroxide with solutions of bases, are characterized by a variety of corresponding anions $[Al(OH)_6]^{3-}$, $[Al(OH)_4]^-$, $[AlO(OH)_4]^{3-}$, $[AlO_2(OH)_2]^{3-}$, $[Al(H_2O)_2(OH)_4]^-$, $[Al(H_2O)(OH)_5]^{2-}$, and the composition and content of anions depend on the conditions of synthesis.

$Al(OH)_3$ is a weak base and an even weaker acid, so aluminum salts are found in solution only in excess of acids, and aluminates – only in excess of bases. When diluting solutions with water, these compounds are strongly hydrolyzed. Dried aluminum hydroxide, which has lost some of its water, does not dissolve either in acids or in bases, analogous to aluminum oxide.

When a gel of $Al(OH)_3$ is stored, it crystallizes and turns into aluminum hydroxide-oxide (methahydroxide, boehmite) $AlO(OH)$ or $Al_2O_3 \cdot H_2O$. This process is observed in practically all insoluble hydroxides and accelerates when heated. When freshly precipitated gel is dehydrated in a vacuum, a highly porous amorphous aluminum oxide – *alumogel* is formed, which, like silica gel, is used as a sorbent. $Al(OH)_3$ absorbs various substances, so it is used to purify water.

In its *salts*, aluminum is found in the form of simple cations Al^{3+} or in the composition of oxyanions or complex anions. In addition to simple and complex salts, it also forms double salts.

In aqueous solution, aluminum ions Al^{3+} are in the form of hexaaquaaluminate ions $[Al(H_2O)_6]^{3+}$, which undergo hydrolysis to produce pentaaquahydroxyaluminate ions $[Al(H_2O)_5(OH)]^{2+}$ and then to tetraaquadihydroxyaluminate ions:

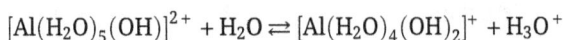

$$[Al(H_2O)_6]^{3+} + H_2O \rightleftharpoons [Al(H_2O)_5(OH)]^{2+} + H_3O^+$$

$$[Al(H_2O)_5(OH)]^{2+} + H_2O \rightleftharpoons [Al(H_2O)_4(OH)_2]^+ + H_3O^+$$

Thus, the solutions of aluminum salts are acidic, with almost the same acidity as that of acctic acid. The addition of hydroxide ions gives a gel-like precipitate of aluminum

hydroxide, but this product dissolves in excess of hydroxide ions to form an alumi-
nate ion:

$$[Al(H_2O)_6]^{3+} \rightarrow Al(OH)_3\downarrow \rightarrow [Al(OH)_4]^-$$

As a result, Al^{3+} is soluble at low and high pH, but insoluble under neutral conditions.

The most important of the simple salts of Al is aluminum sulfate, which is used in
large quantities in the purification of drinking and wastewater. Its coagulating action
is related to hydrolysis processes, which produce a precipitate of $Al(OH)_3$. Due to its
developed surface, $Al(OH)_3$ has the ability to retain particles of other substances and
drag them to the bottom of the vessel. For this purpose, $Al_2(SO_4)_3$ dissolves in water
together with calcium hydrogen carbonate. In this case, as a result of the exchange
interaction, aluminum carbonate is formed, which is completely hydrolyzed to form
aluminum hydroxide, which, with its developed surface, adsorbs impurities from the
water and precipitates them:

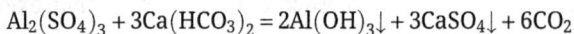

$$Al_2(SO_4)_3 + 3Ca(HCO_3)_2 = 2Al(OH)_3\downarrow + 3CaSO_4\downarrow + 6CO_2$$

There are many aluminum salts of oxygen-containing acids. Their characteristic feature is that their crystal lattice most often includes a specific number of water molecules as crystallization water. Such salts are called crystal hydrates, for example, aluminum sulfate crystallizes with nine water molecules $Al_2(SO_4)_3.9H_2O$. It is used as a
coagulant for natural water purification.

Aluminum halides exhibit peculiar properties (Figure 18.2). They "smoke" in the
air due to evaporation and interaction with moisture from the air, resulting in the
formation of crystal hydrates. Their dissolution in water is accompanied by the release of a large amount of heat. Crystal hydrates $AlX_3.6H_2O$, which are aqua-complex
compounds, are released from aqueous solutions. When they are heated, anhydrous
halides are not formed because hydrolysis takes place:

$$AlCl_3 \cdot 6H_2O = Al(OH)_3 + 3HCl + 3H_2O$$

Aluminum halides represent an interesting series of compounds: aluminum fluoride
melts at 1,290 °C, aluminum chloride sublimates at 180 °C, and aluminum bromide
and iodide melt at 97.5 and 190 °C, respectively. Therefore, fluoride is characterized
by a high melting point, as in ionic compounds, while the melting of bromide and iodide is similar to typical covalent compounds. Aluminum ions have a charge density
of 364 C/mm^3, so all anions, except the small fluoride ion, are expected to polarize by
Al^{3+} to form covalent bonds with aluminum. In fact, aluminum fluoride has an ionic
crystal structure, while bromide and iodide exist as covalently bonded dimers (Al_2Br_6
and Al_2I_6), analogous to diborane, with two bridging halogen atoms. Chloride has an
ionic crystal lattice in solid state, while in liquid and vapor phase it forms dimers.

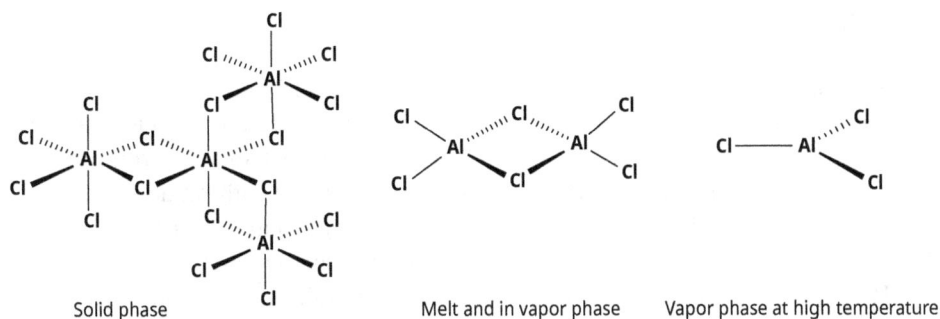

Figure 18.2: Structures of Al(III) chlorides.

An example of a double salt is potassium-aluminum sulfate $KAl(SO_4)_2 \cdot 12H_2O$ (potassium-aluminum alum or common alum), which is used as a filler in paper production. It is accepted that all double salts – sulfates of metals with +1 and +3 oxidation state, crystallizing with 12 molecules of water, are called *alum*. Their general formula can be expressed as $M_1M_2(SO_4)_2 \cdot 12H_2O$. In their dissociation, double salts dissociate completely to simple ions:

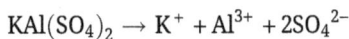

$$KAl(SO_4)_2 \rightarrow K^+ + Al^{3+} + 2SO_4^{2-}$$

Aqueous solutions of alum have an acidic character, due to their hydrolysis, which produces a precipitate of the weak base $Al(OH)_3\downarrow$.

One of the important complex salts of aluminum is cryolite (Na_3AlF_6). Industrial production of cryolite occurs by the reaction:

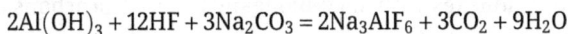

$$2Al(OH)_3 + 12HF + 3Na_2CO_3 = 2Na_3AlF_6 + 3CO_2 + 9H_2O$$

In the composition of oxyanions and hydroxyaminos, aluminum is found in *aluminates*. Anhydrous oxo-aluminates can be obtained by heating Al_2O_3 or $Al(OH)_3$ with metal oxides and hydroxides. Metaaluminates ($NaAlO_2$), orthoaluminates (K_3AlO_3), and various polyaluminates of the type $2CaO \cdot Al_2O_3$ and $5CaO \cdot 3Al_2O_3$ are formed. The aluminates of the alkali metals dissolve in water, but are stable in a strongly alkaline environment, whereas in a neutral one they are completely hydrolyzed:

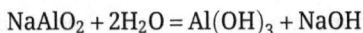

$$NaAlO_2 + 2H_2O = Al(OH)_3 + NaOH$$

Recently, a new class of materials has gained popularity: *aluminides* – these are strong, low-density materials. These intermetallic compounds usually have a defined stoichiometry, which is why they are often referred to as compounds rather than alloys (although the limit is very artificial). Magnesium aluminides with stoichiometry Mg_2Al_3, $Mg_{17}Al_{12}$, as well as Ni_3Al, $TaAl_3$, and $TiAl$ are of interest. They are used for the production of engines for aircraft.

18.3.5 Application

The metal aluminum is widely used in practice. In terms of its application in technology, aluminum ranks second after iron. It is used to make foil for radio engineering, as well as for food packaging. Aluminum is used to coat steel and cast-iron products in order to protect them from corrosion. The coated products withstand heating up to 1,000 °C. Aluminum alloys, characterized by great strength and lightness, are used in the production of heat exchangers, in aircraft construction and mechanical engineering. The most common of these alloys is duralumin (dural), which, in addition to aluminum, also contains copper, magnesium, and manganese.

Aluminum oxide is used for the preparation of refractory and chemically resistant ceramics, aluminum chloride – for the production of organic substances as a catalyst, and aluminum alum – in the leather and paper industries. $KAl(SO_4)_2.12H_2O$ is sometimes used to stop bleeding from small bleeding wounds (e.g., when shaving) because it causes the proteins to coagulate.

Aluminum compounds in combination with MgO and SiO_2 oxides have long been known as antacid agents. Most often, aluminum hydroxide is used in a number of antacid forms. Like other antacids, the compound is an insoluble base that neutralizes excess stomach acid:

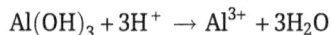

$$Al(OH)_3 + 3H^+ \rightarrow Al^{3+} + 3H_2O$$

Interestingly, aluminum, which is widely distributed in the geosphere, is practically not absorbed by living organisms. The reason for this is the poor solubility of Al hydroxide and phosphate, which does not allow these compounds to accumulate in the body. However, Al complexes with such bioligands as hydrocarbons and fats containing a large number of oxygen donor atoms are neurotoxic and, deposited in the tissues of the brain, contribute to the development of Alzheimer's disease. Therefore, it is currently recommended to avoid aluminum cookware.

18.4 Gallium, indium, and thallium

These are rare elements that do not have minerals of their own. They are found in the natural compounds of zinc, aluminum, lead, and other metals and are extracted from waste products after the processing of polymetallic ores. These three elements were detected by spectral methods.

Gallium, indium, and thallium in their free state are silvery-white soft metals with low melting points. The structure of gallium differs significantly from that of typical metals: its crystal lattice consists of Ga_2. The metallic bonds between the individual atomic pairs are fragile, so gallium melts very easily (29.8 °C). At the transition to a gaseous state, the strong covalent bonds in the atomic pairs are broken, which is why gallium boils at a very high temperature (2,237 °C) and, in this respect, shows a certain uniqueness – this is the element that has the widest temperature interval of existence in a liquid state.

The electrode potentials of gallium, indium, and thallium define them as metals of medium activity, and with acids these elements interact as such. Amphoteric gallium interacts with solutions and melts of bases similarly to aluminum, weakly amphoteric indium reacts with bases in the presence of oxidizing agents, while thallium does not interact with bases. In the elements of the gallium subgroup, there are three electrons in the outer electron shell, forming a configuration s^2p^1, which corresponds to an oxidation state of +3. Tl, like Pb and Bi, are elements of the VI period and in their electron configurations there is a filled 4f-subshell. As a result of the lanthanide contraction, the electron pair of the $6s^2$ subshell is more firmly bound to the nucleus and becomes "inert." Therefore, in thallium, the characteristic oxidation state is +1, for lead +2, and for bismuth +3. From the electron configurations of these elements, it can be seen that their oxidation states coincide with the number of p-electrons, that is, only p-electrons are involved in the formation of bonds, and the electron pair of $6s^2$ remains neutral and does not participate in chemical bonding, the so-called "inert pair effect." The energies of the E-X bonds of these elements are also important here (Table 18.3). The energies of the E-Cl bonds in the subgroup decrease, therefore, the energy required to promote the electron from the $6s^2$ subshell to 6p is not compensated by the energy released during the formation of two additional bonds ($6s^26p^1 \rightarrow 6s^16p^2$). Therefore, in thallium, +3 oxidation state is unstable and manifests itself only when interacting with strong electronegative elements. Similar phenomena are observed in the other p-elements of the sixth period.

Table 18.3: Ionization potentials and energies of the E-Cl bonds.

Elements	$I_2 + I_3$ (eV)	Energy of the bond E–Cl (kJ/mol)
Ga	51.0	241.9
In	46.7	205.9
Tl	50.0	152.7

18.4.1 Chemical properties

The hydrides of boron, aluminum and gallium differ to some extent from the other hydrogen compounds of the p-elements. Gallium hydrides, as well as those of boron, are highly volatile, gaseous, or liquid substances. Hydrides B_2H_6 and Ga_2H_6 are dimers in a gaseous state that are easily decomposed by water.

Similar to the behavior of aluminum, gallium, and indium without a protective layer react with water. When heated, the activity of the elements increases. Gallium and indium burn in oxygen with oxide formation at temperatures above 200 °C:

$$4Ga + 3O_2 \rightarrow 2Ga_2O_3$$

$$4In + 3O_2 \rightarrow 2In_2O_3$$

In general, the p-elements of IIIA group Al, Ga, and In show a great affinity for oxygen, especially characteristic for aluminum, which is used to obtain metals by aluminothermy (Table 18.4).

Table 18.4: Melting points of oxides.

Oxide	T_m (°C)	ΔH_{form} (kJ/mol)
B_2O_3	290	−1,254
Al_2O_3	2,053	−1,675.7
Ga_2O_3	1,725	−1,089.1
In_2O_3	1,910	−925.9
Tl_2O_3	717	−390.4

Gallium and indium, when heated, interact with sulfur, nitrogen, carbon, and halogen elements, forming respectively sulfides, nitrides, carbides, halides, etc. The chlorides of the metals of group IIIA have layered structures (Figure 18.3), without $GaCl_3$ and $AlCl_3$, which have a molecular structure consisting of dimers Ga_2Cl_6 and Al_2Cl_6.

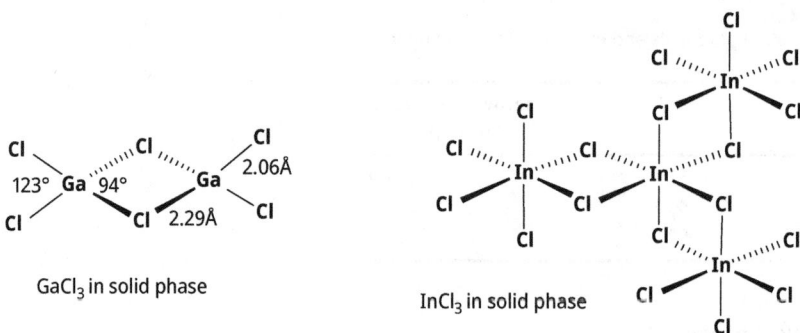

Figure 18.3: Structures of Ga(III) and In(III) chlorides.

But even in these compounds there is a significant covalence in the chemical bond, as evidenced by the amphoteric nature of the given halides and their tendency to hydrolyze. Fluoride is resistant to hydrolysis, while the rest of the metal halides are hydrolyzed more vigorously, especially the chlorides and bromides of gallium and indium:

$$GaCl_3 + 3H_2O \rightarrow 3HCl + Ga(OH)_3$$

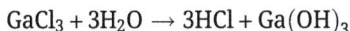

Such halides smoke in the air due to hydrolysis with water vapor in the air.

Thallium is more active than its counterparts. In air in the presence of moisture, it oxidizes to form TlOH, and with oxygen it gives oxides, Tl_2O_3 and Tl_2O. With sulfur, it forms sulfide (Tl_2S), and with the halogen elements – halides of the type TlX.

Pure gallium and indium are very resistant to the action of acids. They do not usually react with hydrochloric acid containing traces of CuCl.

In concentrated solutions of nitric and sulfuric acids, gallium, and indium, analogous to aluminum, are passivated. With dilute nitric acid, gallium and indium react as active metals, reducing it to NH_3, especially when heated. Gallium has a similar behavior to aluminum and dissolves in hot solutions of bases:

$$Ga + NaOH + 5H_2O \rightarrow Na\left[Ga(OH)_4(H_2O)_2\right] + 1 1/2 H_2$$

Indium also dissolves in bases, but significantly less than aluminum and gallium.

Thallium dissolves slowly in hydrochloric and sulfuric acids because it is covered with insoluble layers of chloride and sulfate and does not dissolve in bases.

p-Elements of the IIIA group react very actively with metals. With most metals, they form intermetallic compounds. Many of these compounds have higher melting points than the metals themselves.

18.4.2 Chemical compounds

In most of their compounds, gallium and indium are found in the oxidation state of +3. These are the oxides Ga_2O_3 and In_2O_3, the hydroxides $Ga(OH)_3$ and $In(OH)_3$, and their numerous salts: chlorides, sulfates, nitrates, etc. Oxides and hydroxides of gallium and indium are amphoteric and, when interacting with solutions of bases, form hydroxy-complex salts:

$$Ga_2O_3 + 6HCl \rightarrow 2GaCl_3 + H_2O$$

$$Ga_2O_3 + 6NaOH + 3H_2O \rightarrow 2Na_3[Ga(OH)_6]$$

A small number of compounds of gallium and indium are known in +1 oxidation state, including the oxides Ga_2O and In_2O, the sulfides Ga_2S and In_2S, and some halides. These compounds are unstable and disproportionate when heated:

$$3Ga_2O = 4Ga + Ga_2O_3$$

$$3InCl = 2In + InCl_3$$

In accordance with the regularities of the periodic table, at thallium, the number and stability of compounds in the oxidation state of +1 increase. The increased resistance to the lower oxidation states in the order Ga, In, Tl can be illustrated by the following regularity: Ga_2O_3 melts at 1,800 °C with no decomposition, In_2O_3 decomposes into In_2O and oxygen at 850 °C, and Tl_2O_3 decomposes into Tl_2O and oxygen at 90 °C.

In terms of density, plasticity, melting point, and toxicological action, thallium is similar to lead, and in terms of the properties of its compounds in an oxidation state of +1 it is similar to alkali metals and silver. For example, thallium hydroxide (TlOH) is a base, thallium and potassium nitrates are isomorphic, and thallium carbonate (Tl_2CO_3) in solution, like sodium and potassium carbonates, has an alkaline reaction due to hydrolysis, etc. Thallium halides TlCl, TlBr, and TlI, like silver halides, decompose under the action of light and are insoluble in water. Thallium hydroxide (TlOH) decomposes at temperatures higher than 100 °C similarly to silver hydroxide.

18.4.3 Use and biological role

The elements of the gallium subgroup (gallium, indium, thallium) find a variety of applications due to their valuable properties. Liquid gallium is used in quartz thermometers, allowing the measurement of temperatures above 1,000 °C, which is impossible with ordinary thermometers. Gallium successfully replaces toxic mercury in the preparation of compositions for dental fillings (gallium amalgam). Indium is distinguished by a high light reflectance and is used for reflector coatings. Indium coatings protect metals from corrosion. Indium is part of some alloys used in dentistry, as well as in some low-melting alloys (an alloy of indium, bismuth, tin, lead, and cadmium melts at 47 °C). Additives of indium to copper alloys increase their resistance to the action of sea water. The addition of this metal to silver increases the luster of silver and prevents its darkening in the air. The alloy of thallium (10%) with tin (20%) and lead (70%) exhibits very high acid resistance, withstanding the action of a mixture of sulfuric, hydrochloric, and nitric acids. Thallium increases the sensitivity of photoelements to infrared radiation. The compounds of Ga, In, and Tl are used as night vision devices. Compounds of Ga with elements of group VI A (sulfur, selenium, tellurium), as well as indium with various nonmetals, exhibit semiconductor properties. Applications of indium compounds include liquid crystal, flat panel and plasma displays, touch screen technology of mobile phones and ATMs, electronic ink, light-emitting diodes, solar cells, antistatic coatings, defrosting aircraft windows, and infrared reflecting coatings (oven doors).

Although trace amounts of these elements have been found in animal organisms, their biological role has not yet been clarified. Gallium nitrate is known as the first "non-platinum" carcinostatic agent, that is, a drug that stops the growth of tumor cells. The mechanism of action of the drug is not clear, but it is assumed that Ga^{3+} ions can partially inhibit DNA and RNA-dependent polymerase as well as inhibit the binding of calcium ions to ATP molecules. The compounds of indium and thallium are quite poisonous. Thallium causes hair loss, while indium has the opposite stimulating effect. Some of the thallium complexes are used to fight rodents. Some of the major toxicological effects of thallium are (a) disruption of potassium-dependent processes,

(b) interference with cysteine residues, (c) riboflavin sequestration, and (d) ribosomal inhibition and injury to myelin sheath. Thallium mimics the nutritionally essential element potassium and easily passes through the stomach wall into the blood stream. The body, after a while realizes that it is not potassium and sends it back to the intestines. However, further down the intestines it is again mistaken for potassium and reabsorbed. This cycle needs to be broken if thallium is to be removed from the body. Elimination of thallium from the body is slow with a half-life for elimination of 3–30 days. Because of this reason, thallium may act as a cumulative poison. Many complexing agents and chelating ligands were attempted to selectively remove thallium without success and thallium poisoning was therefore thought to be incurable. The isotope ^{111}In in the form of a complex with organic acids is used in radiopharmacy in various MRI medical examinations. The isotope ^{201}Tl is used in myocardial perfusion scintigraphy. The thallium scanning not only diagnoses ischemia but can also differentiate ischemia from a heart attack.

Chapter 19
Fourth main group of the periodic table

19.1 General characteristics of the elements of IVA group

The p-elements of the fourth group include carbon (C), silicon (Si), germanium (Ge), tin (Sn), and lead (Pb). The characteristics of the atoms of these chemical elements are presented in Table 19.1.

Table 19.1: Characteristics of atoms of elements of IVA group.

Element	C	Si	Ge	Sn	Pb
Valence electrons	$2s^2p^2$	$3s^2p^2$	$4s^2p^2$	$5s^2p^2$	$6s^2p^2$
Radius of the atom (nm)	0.077	0.117	0.139	0.158	0.175
Ionization potential (eV)	11.3	8.2	7.9	7.3	7.4
Electronegativity	2.5	1.8	1.8	1.8	1.8
Oxidation states	−4, +4, +2	−4, +4	+ 4, +2	+4, +2	+4, +2

When considering the main characteristics of their atoms, it can be seen that with an increase in their number, the nonmetallic properties of the elements weaken, and the metallic ones increase. The nonmetallic properties are characteristic of carbon and partly of silicon, while the metallic properties are characteristic of tin and lead. In germanium, metallic properties slightly prevail over nonmetallic ones.

The electron configuration of the last electron shell of these elements is ns^2np^2. In their penultimate electron shell, C and Si have two and eight electrons, respectively, which is the reason why these elements differ in properties from the others, which have 18 electrons in their penultimate shell and are complete electron analogs. The relatively large number of electrons in the outer electron shell determines, in principle, the nonmetallic character of the elements. If the p-elements of group III, without boron, are metals, then the metallic properties of group IVA are significantly less pronounced. During the transition from C to Pb, the atomic radius increases and the force of attraction of electrons from the nucleus decreases, which leads to a weakening of the nonmetallic character. Carbon is a nonmetal, silicon and germanium are semiconductors, α-modification of tin is a semiconductor, and β-tin and lead are metals (Table 19.2).

For silicon under ordinary conditions, the diamond-like crystal structure (cubic) is stable. These are hard and brittle crystals of steel-gray color. Since the sizes of silicon atoms are larger than those of carbon, the parameters of the crystal lattice of silicon are larger than those of diamond. Therefore, silicon has a lower melting point and hardness. Germanium is a substance with a grayish-white color and metallic luster, but with a diamond-like crystal lattice. Tin is polymorphic. β-Tin is resistant to temperatures higher

https://doi.org/10.1515/9783111712246-019

Table 19.2: Physical properties of p-elements of IVA group.

Element	T_m (°C)	T_b (°C)	R_{at} (Å)	ρ (g/cm³)	Hardness (kg/mm)	$E°$ (V) E^{4+}/E	$E°$ (V) E^{2+}/E
C	3,900	4,347	0.77	3.51	10,000	–	–
Si	1,414	2,630	1.18	2.33	980	–	–
Ge	958.5	2,690	1.39	5.32	385	–	0.2
α-Sn	231.9	2,337	1.58	5.85	30.2	0.009	−0.136
β-Sn				7.29			
Pb	237.4	1,753	1.75	11.34	3.9	0.80	−0.126

than 13.2 °C. It is a silvery-white metal with a tetragonal structure. At a temperature lower than 13.2 °C, α-tin (gray tin) with a diamond-type structure is formed. The transition from β-tin to α-tin is accompanied by a 25.6% increase in volume, whereby the tin turns into dust. Lead is a dark gray soft metal with a face-centered cubic lattice typical for metals. Lead has no polymorphic modifications.

The first three elements of the group have very high melting temperatures, while the last two metals of the group have low melting points and a wide temperature interval of existence in a liquid state, characteristic of metals. All elements of the group form compounds in which atoms bond with each other and form chains (catenation).

Carbon and silicon, with the active metals, form compounds in a negative oxidation state (−4), while the other elements of the group do not form such compounds. In their compounds with more electronegative elements, carbon, silicon, germanium, tin, and lead are found in oxidation states of + 2 and + 4. The first corresponds to the participation of only s-electrons in chemical bonds, and the second to all valence electrons. In accordance with the regularities in the structure of atoms and the periodic law, the maximum oxidation state + 4 is characteristic of carbon and silicon, the minimum + 2 for lead. In germanium and tin, both values of oxidation states are characteristic.

19.2 Carbon

Carbon has two properties that allow it to form such a wide range of compounds. These are *catenation* (the ability to form chains) and *multiple bonds* (the ability to form double and triple bonds). In terms of the number and variety of compounds it forms, carbon is second only to the element hydrogen. Carbon, compared to all other elements, shows the greatest tendency to catenate, for which three conditions are required: first, a valence greater than or equal to 2; secondly, the ability of the atoms of the element to bond with each other, with these bonds being as strong as the bonds with other elements; and thirdly, the kinetic inertness of the compounds to other molecules and ions. In fact, two chemical phenomena make life possible on Earth: hydrogen bonding and carbon catenation. Without these two phenomena, there can be no life in any form that we can imagine. The ability of carbon to form stable chains is primarily due to the fact that only in this element do the number of valence electrons coincide with the number of valence orbitals.

Carbon is widely distributed in nature in the form of carbon dioxide in the atmosphere, carbonates of calcium, magnesium, zinc, and iron in the earth's crust, and hydrocarbons in oil, natural gas, peat, and plants. In the free state, carbon occurs in nature in two allotropic modifications: graphite and diamond. There is also a synthetic modification carbine, obtained in 1963, but despite its variety of compounds, carbon is not one of the most common elements.

19.2.1 Allotropy and properties of carbon

The significant difference in the physicochemical properties of the polymorphic forms of carbon is noteworthy. Until the nineteenth century, graphite and diamond were thought to be two different substances. The crystal lattice of *graphite* consists of flat layers of carbon atoms that are in sp^2-hybrid state, linked to each other in regular hexagons (Figure 19.1). There is also an unpaired p-electron, the orbital of which is oriented perpendicular to the plane of the hexagons. These p-electrons form delocalized π bonds that cover the layer of hexagons.

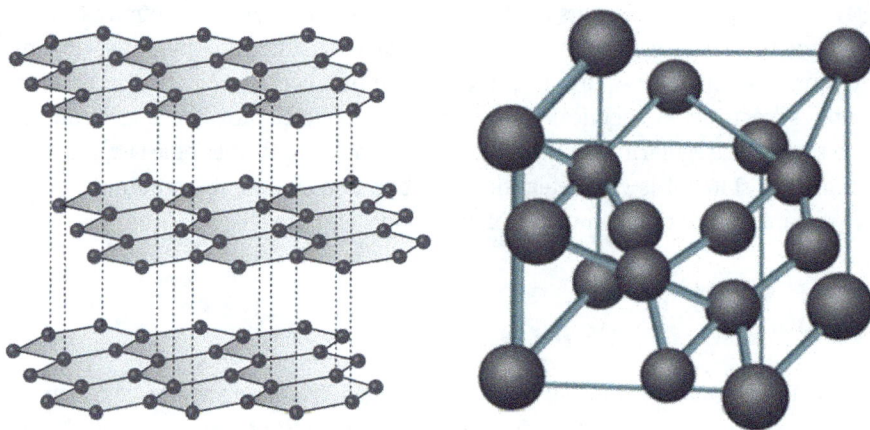

Figure 19.1: Structures of graphite and diamond.

The length of the carbon-carbon bonds in graphite is 141 pm, equal to those in benzene C_6H_6. This similarity between benzene and graphite is also manifested in the delocalized π bonds resulting from the overlap of the 2pz orbitals perpendicular to the plane of the rings. This results in a bond order between adjacent carbon atoms of 1⅓. The distance between the carbon layers is very large (335 pm) and is more than twice the value of the van der Waals radius of the carbon atom. Therefore, the attraction between the layers is very weak, and the adjacent layers of carbon atoms in graphite are bonded to each other by weak intermolecular interactions. The layers are easily separated from each other, which is why graphite leaves a mark on paper or wood.

The layered structure of graphite is the reason for its high electrical conductivity. Graphite is an excellent lubricant. It adsorbs gas molecules between its layers. Although graphite is thermodynamically more stable than diamond, it is kinetically more active. A wide range of substances react with graphite from alkali metals to halogen elements through metal halides.

Carbon atoms in *diamond* are sp³-hybridized and are bonded to each other in a three-dimensional tetrahedral network. This is an atomic crystal lattice with strong covalent bonds. Among all solids, diamond has the highest number of atoms per unit volume, which explains its high hardness (the hardest known substance) and a minimum value of standard entropy (2.4 J/mol K). Diamond is an electrical insulator, but it has excellent thermal conductivity, which is about five times better than that of copper. The high thermal conductivity is directly dependent on the structure of the diamond, which is a giant molecule consisting of a continuous network of covalent bonds. The thermal energy is transferred directly through its overall structure. Diamond is in a solid state up to 4,000 °C, as it takes a huge amount of energy to break the strong covalent bonds.

In nature, diamonds are found in the form of individual small crystals, and the largest of them are used as precious stones after grinding. Due to the value of diamonds, many attempts have been made to obtain them artificially. The synthesis of diamonds from graphite is carried out at very high pressure (10^{10} Pa) and temperature (3,000 °C). The density of diamond (3.5 g/cm³) is much greater than that of graphite (2.2 g/cm³), so an elementary application of Le Chatelier's principle shows that diamond is obtained from graphite, preferably at high pressure. In addition, in order to overcome the significant barrier of activation energy associated with the rearrangement of covalent bonds, a high temperature is also needed. The free energy of diamond is 2.9 kJ mol^{-1} and is higher than that of graphite, which practically does not allow diamond to decay into graphite. Artificial diamonds are mainly used for technical purposes.

Various carbon-containing substances are produced in industry such as coke, activated carbon, charcoal, and fibrous carbon. All of them have a deformed graphite structure, high porosity, and a large relative surface area. These substances are conventionally called amorphous carbon.

Graphite, and especially diamond, are chemically very inert: they do not interact with HNO_3 and aqua regia; they are stable in bases, oxygen, and sulfur. They interact with halogen elements and metals only at high temperatures. Amorphous carbon is more active and is oxidized by nitric and concentrated sulfuric acid according to the equations:

$$C + 4HNO_3 = CO_2 + 4NO_2 + 2H_2O$$
$$C + 2H_2SO_4 = CO_2 + 2SO_2 + 2H_2O$$

Carbon is chemically active only at high temperatures, with soot being the most active due to its large surface area. It binds with hydrogen directly at high temperature (1,200 °C) to form methane:

$$C + 2H_2 \rightleftharpoons CH_4 + Q$$

Since the straight reaction is exothermic, equilibrium can be directed to the right at low temperatures, but under this condition, the reaction rate is very low. In the presence of a Ni catalyst and increased pressure, equilibrium can be drawn to methane at moderately high temperatures, but the yield is very small.

Carbon reacts with metals at high temperatures and forms carbides (Be_2C, Al_4C_3, and SiC). Compared to oxygen and other more electronegative elements, carbon is referred to as a reducer. Carbon reacts with all halogens and forms compounds with the general formula CX_4. At high temperature, C also interacts with S to produce carbon disulfide, a widely used organic solvent.

19.2.2 Usage and biological role of carbon

Carbon is widely used in human economic activity. In the form of coke, it is used in metallurgy for the reduction of metals from their oxides. Soot is used in the production of rubber and black paints, and activated carbon is used as a sorbent in the purification of water, solutions, and air from impurities. Graphite is used to make pencils, refractory crucibles, and high-temperature products. Due to the high electrical conductivity of graphite, it is used to make electrodes for electrolysis and galvanic cells, and due to its chemical resistance, it is used for the preparation of special chemical equipment. Neutron-delaying blocks of pure graphite are used in atomic reactors. Around 80% of the world's diamond production and virtually all artificial diamonds are used in the metalworking industry. Large and transparent diamonds are used to make jewelry.

In terms of its importance and content (21%) in living organisms, carbon is the most biologically important element. Its compounds are formed and decomposed during all processes related to the beginning, development, and death of living organisms. Nature is designed in such a way that living organisms are formed by the compounds of these elements, which are capable of forming sufficiently strong, but at the same time, labile bonds. These bonds must be easily subjected to both homolytic and heterolytic breaking, as well as cyclization. That is why carbon is the most important organogen. Most of these compounds are subject to bioorganic chemistry.

The simplest inorganic carbon compounds (free carbon and CO) are toxic to humans. Prolonged contact with soot or dust causes changes in the structure of the lungs and disorders of their functions, leading to the formation of skin cancer. CO is also highly toxic, the poisonous effect of which is caused by the fact that CO binds to blood hemoglobin ~10^3 times more easily than oxygen and thus causes suffocation. Carbon dioxide (CO_2) is present in the biosphere as a product of respiration and oxidation. The annual emissions of CO and CO_2 into the atmosphere are 2×10^8 and 9×10^9 tons, respectively (for comparison, the emission of hydrocarbons is equal to 8×10^7 tons per year). CO_2 is poorly soluble in water, so its presence in biological fluids is negligible. However, an important fermentative reaction takes place in the stomach:

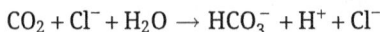

$$CO_2 + Cl^- + H_2O \rightarrow HCO_3^- + H^+ + Cl^-$$

As a result, proteins are broken down in an acidic environment. It should be noted that without the necessary enzymes, this reaction proceeds in the opposite direction.

Natural carbon contains three *isotopes*: carbon-12, which is the most abundant isotope (98.89%); a small fraction of carbon-13 (1.11%) and traces of carbon-14. Carbon-14 is a radioactive isotope with a half-life of 5.7×10^3 years. With such a short half-life, it is expected to have small traces of this isotope on the Earth. It is also found in living organisms. This isotope is also produced in reactions between neutrons from cosmic rays and N atoms in the upper atmosphere. In nature, this carbon reacts with oxygen to form radioactive carbon dioxide molecules. The latter is absorbed by plants during photosynthesis. Living organisms also absorb it when eating, and they all contain an identical ratio of radioactive carbon. After the death of the organism, further carbon intake is stopped, and carbon-14 remains present for a long time during the breakdown of the body. In this way, the age of an object can be determined by measuring the amount of carbon-14 present in the sample. This method provides an absolute scale of the age of objects between 1,000 and 20,000 years old, the so-called radioactive dating.

The repeated transition of carbon from one part of the bio-geosphere to another is called the carbon cycle, which is considered to be the most intense of all biogeochemical cycles. The existence of carbon in nature is mainly in limestones ($CaCO_3$) of biological origin and carbon dioxide (CO_2), which is the circulating form of inorganic carbon. The atmosphere contains only 0.03% CO_2. Although this concentration is low, it is sufficient for the primary bioproduction of autotrophic plants on land. CO_2 emissions into the atmosphere are regulated by the dissolution of CO_2 in the ocean, which contains 50 times more CO_2 than that in the atmosphere. The exchange of CO_2 between water, air, and land is expressed by the equation:

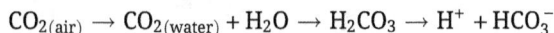

$$CO_{2(air)} \rightarrow CO_{2(water)} + H_2O \rightarrow H_2CO_3 \rightarrow H^+ + HCO_3^-$$

Although low in content and low in acidity (H_2CO_3), dissolved CO_2 in continental waters reacts with minerals. In carbonates, such as limestone, calcium in the form of water-soluble hydrogen carbonate is also involved in the circulation:

$$H_2CO_3 + CaCO_3 \rightarrow Ca(HCO_3)_2$$

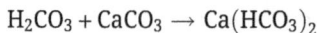

Hydrogen carbonate is exported with river water to the ocean, and invertebrate marine animals convert it to calcite or aragonite in their shells or external skeletons. Thus, in nature, the powerful layered deposits of limestone are created.

The carbon cycle in the biosphere takes place between self-nourishing and non-self-feeding organisms. Autotrophic organisms use CO_2 from the atmosphere as a source of carbon, while heterotrophic organisms obtain carbon in the form of organic compounds. The carbon produced by heterotrophs returns to the atmosphere as CO_2 through their breathing. The two fundamental processes (photosynthesis and respiration) actually determine the circulation of carbon in the biosphere. They are combined as two reversible processes in the equation:

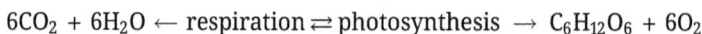

$$6CO_2 + 6H_2O \leftarrow \text{respiration} \rightleftharpoons \text{photosynthesis} \rightarrow C_6H_{12}O_6 + 6O_2$$

During photosynthesis, the formation of all biochemical compounds that make up living cells takes place. In them, light energy is accumulated in a chemically transformed form. This energy is consumed by autotrophic and heterotrophic organisms for chemical, osmotic, electrical (nerve cells), and mechanical (movement) work necessary for their vital activity, growth, and reproduction. At the same time, the opposite biochemical process of equimolecular consumption of O_2 and CO_2 release – respiration — takes place. In the past geological conditions, the organic biomass created by photosynthesis was not fully mineralized aerobically to CO_2. It was divided from the cycle and accumulated in deposits such as peat, coal, and oil, the burning of which enriches the atmosphere with CO_2. It should be noted that from the quaternary to the beginning of industrial society, this cycle was practically perfect. Most of the primary annual production was absorbed in the respiration of autotrophic and heterotrophic organisms, whereby the exhaled CO_2 fully compensated for the consumption of CO_2 in photosynthesis. Currently, the use of fossil fuels (coal, natural gas, and oil) leads to emissions of significant amounts of CO_2 into the biosphere and to undesirable effects (climate warming, etc.).

Carbon is used in medicine in the form of activated carbon, *Carbo medicinalis*, which is used as an adsorbent for intoxications. Its high adsorption capacity is due to its large surface area (1 g of substance has a surface area of 700 m^2). Sodium hydrogen carbonate ($NaHCO_3$) is used to a limited extent as an antacid agent in the case of increased acidity of the stomach and heartburn. It is also part of effervescent tablets.

19.2.3 Chemical compounds of carbon

19.2.3.1 Simple hydrocarbons and carbides

In the carbon atom, the number of valence electrons coincides with the number of valence orbitals. In this regard, carbon exhibits the most pronounced ability, compared to other elements, to form compounds with chain and cyclic bonds between atoms. These compounds, called *organic hydrocarbons*, are studied in organic chemistry. The simplest organic compounds (methane, ethylene, acetylene, acetic, and oxalic acids) are also of interest to inorganic chemistry.

Methane (CH_4) is widely distributed in nature. It is a major component of natural gas. The methane molecule has a tetrahedral structure, which is determined by the sp^3 hybridization of the valence orbitals. Methane is used to produce hydrogen, soot, and as a reducer in some chemical-technological processes. Hybridization is different in molecules of gaseous substances ethane (C_2H_6, sp^3), ethylene (C_2H_4, sp^2), and acetylene (C_2H_2, sp), which increases the multiplicity and bond energy between carbon atoms from 347 kJ/mol for ethane to 811 kJ/mol for acetylene.

Acetic acid (CH_3COOH) is a monobasic organic carboxylic acid, dissociating as a weak acid:

$$CH_3COOH \rightleftharpoons H^+ + CH_3COO^-, \qquad K = 1.74 \times 10^{-5}$$

All its salts (acetates) dissolve in water and, in solution, undergo reversible hydrolysis.

Oxalic acid $H_2C_2O_4$ is a dibasic carboxylic acid. Its salts, that is, oxalates, are insoluble except for alkali metal salts. Acetic and oxalic acids and their salts are used in inorganic synthesis or analysis.

Carbides are binary compounds of carbon with metals as well as with nonmetals with low electronegativity (except hydrogen). Carbides are solids with high melting points. According to the type of chemical bond, they are divided into ionic, covalent, and metal-like carbides.

Ionic carbides are compounds of C with metals from the first, second, and third (Al) main groups of the periodic table, obtained by direct interaction of carbon with metals at high temperatures:

$$4Al + 3C = Al_4C_3$$

Most of these ionic compounds contain dicarbide C_2^{2-} (Na_2C_2, CaC_2) or C^{4-} (Na_4C and Al_4C_3) ions. Ionic carbides are the only carbides with high chemical activity. They are divided into methanides and acetylenides. Methanides can be considered as products of the substitution of hydrogen for metal in methane: Na_4C, Mg_2C, Al_4C_3, etc. In hot water and acids, they decompose by releasing methane:

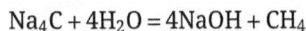

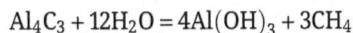

$$Na_4C + 4H_2O = 4NaOH + CH_4$$

$$Al_4C_3 + 12H_2O = 4Al(OH)_3 + 3CH_4$$

Acetylenides can be considered as products of the substitution of hydrogen for metal in acetylene: Na_2C_2, CaC_2, ZnC_2, etc. In hot water and acids, they decompose, releasing acetylene:

$$ZnC_2 + 2HCl = ZnCl_2 + C_2H_2$$

Since most nonmetals are more electronegative than carbon, there are only a few known *covalent carbides*. These include silicon carbide (SiC) (carborundum) and boron carbide $B_{12}C_3$. They are known for their high hardness, heat resistance, and chemical inertness. There is a huge interest in silicon carbide, which is used in the production of materials that can withstand much higher temperatures than metals. It is obtained by the reaction:

$$SiO_2 + 3C \rightarrow SiC + 2CO$$

Metallic carbides are compounds of carbon with most of the d-elements. Their composition varies widely and does not correspond to the usual notions of valence. The crystal structures of carbides of this type are lattices of metal atoms in which C atoms are placed. In some properties (metallic luster and electrical conductivity) they are close to metals. These carbides are characterized by high hardness and a high melting point. They are the most highly meltable of all substances.

Carbides are usually obtained by melting mixtures of metals with carbon in an inert atmosphere. Of the ionic carbides, calcium carbide is widely used to produce acetylene, and Al_4C_3 is used to produce methane. Other carbides are used as abrasive, highly melting solid materials in the processing of hard alloys and steel.

19.2.3.2 Carbon(II) compounds

In the +2 oxidation state, carbon is part of carbon(II) oxide, carbonyls, hydrogen cyanide, and cyanides. *Carbon(II) oxide* (CO) is a gaseous substance with a very strong bond in the molecule (1,070 kJ/mol). The set of molecular orbitals and their number of electrons in this molecule are similar to those in the nitrogen molecule, so these substances are quite close in many properties (bond energy, diamagnetism, mechanism of intermolecular interaction, etc.). The multiplicity of bonds in the molecules of CO and N_2 is equal to three. The formation of the triple bond in the molecule of CO can also be explained by the method of valence bonds, in which two of the bonds are formed by an exchange mechanism (by the unpaired electrons in each of the interacting atoms), and the third by a donor-acceptor mechanism (Figure 19.2).

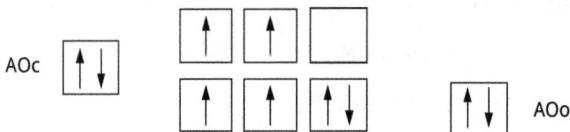

Figure 19.2: The triple bond in the molecule of CO.

Carbon(II) oxide is very toxic and especially dangerous because it is an odorless gas, so poisoning with it can go unnoticed. Its toxic effect is explained by the fact that CO easily combines with the hemoglobin in the blood, which makes it impossible to absorb oxygen.

This oxide belongs to the non-salt-forming category, that is, it does not interact with water and bases and does not form salts. When heated, it interacts reversibly with water vapor:

$$CO + H_2O \rightleftharpoons CO_2 + H_2$$

The equilibrium up to 830 °C is shifted in the direction of the products, and at a higher temperature in the direction of the reactants.

The main chemical function of CO is reduction. It interacts with oxygen and is converted into CO_2. Reactions involving its participation usually take place at high temperatures because they require a lot of activation energy:

$$2CO + O_2 = 2CO_2$$

This oxide is widely used as a reducer in metallurgy to obtain metals from their oxides:

$$Fe_2O_3 + 3CO = 2Fe + 3CO_2$$

It also interacts with other oxidizing agents, such as Cl_2, which produces carbonyl chloride:

$$CO + Cl_2 \rightleftharpoons COCl_2$$

Carbonyl chloride has the trivial name "phosgene" and is a chemical warfare substance.

Carbon(II) oxide is obtained by the incomplete combustion of carbon. It can be obtained industrially by converting solid fuel into gas, a process called "coal gasification." Carbon gasification is carried out in special devices called gas generators; therefore, the resulting gases are referred to as generator gases. The following processes take place:

$$C + O_2 = CO_2$$
$$C + CO_2 = 2CO$$
$$2C + O_2 = 2CO$$
$$2CO + O_2 = 2CO_2$$

The resulting gas consists of CO and nitrogen and is called air gas.

More efficient use of coal leads to the production of the so-called "water gas." It is produced by blowing heated water vapor over hot coals:

$$C + H_2O = CO + H_2$$

Water gas has a higher calorific value due to two "flammable" components: CO and H_2.

In industry, large quantities of CO are mainly obtained from methane (methane conversion). The conversion of methane is carried out with water vapor:

$$CH_4 + H_2O = CO + 3H_2$$

The reaction takes place at 700–800 °C in the presence of a nickel catalyst and in an excess of water vapor in order to shift the equilibrium in the direction of the products.

Carbon(II) oxide serves as a feedstock in the production of a wide variety of products, mainly organic. In inorganic chemistry and metallurgy, CO is used as a reducer in the production of many metals from their oxides and as a fuel.

Due to the presence of lone electron pairs, carbon(II) oxide enters into bonding reactions with metals, forming complex compounds called *carbonyls*: $Ni(CO)_4$, $Fe(CO)_5$, etc. The chemical bonds in carbonyls are formed by a donor-acceptor mechanism, whereby CO molecules are donors, and metal atoms are acceptors of electron pairs. The composition of carbonyls depends on the number of free orbitals in metal atoms. For example, there are five free valence orbitals in an excited iron atom, so its composition corresponds to the formula $Fe(CO)_5$. There are four free orbitals in the nickel atom, so the composition of the carbonyl is $Ni(CO)_4$. The formula of a cobalt carbonyl $Co_2(CO)_8$ is determined by the presence of an unpaired electron in the cobalt atom, by which a bond is formed between the two metal atoms, and four free orbitals that attach four CO molecules each. Metal carbonyls are liquids with low boiling points, decomposing when metal and carbon(II) oxide are heated. They are used to obtain particularly pure metals and as catalysts.

The triple chemical bond, similar to that in the molecules of CO and N_2, is also present in the cyanide ion (CN^-). *Hydrogen cyanide* (HCN) is a volatile liquid that is highly soluble in water. The aqueous solution of this compound is the weak monobasic hydrocyanic acid. The salts of this acid, the *cyanides*, are hydrolyzed reversibly in an aqueous solution:

$$KCN + H_2O \rightleftharpoons KOH + HCN$$

Cyanide ions form numerous complex compounds as ligands through their lone electron pairs. The formation of stable complexes is facilitated by the reduced resistance of metals to the action of oxidizing agents (in the presence of CN^-):

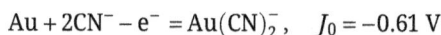

$$Au - e^- = Au^+, \qquad J_0 = 1.69 \text{ V}$$
$$Au + 2CN^- - e^- = Au(CN)_2^-, \quad J_0 = -0.61 \text{ V}$$

This is used in gold extraction technology:

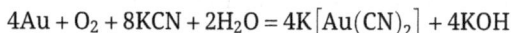

$$4Au + O_2 + 8KCN + 2H_2O = 4K\left[Au(CN)_2\right] + 4KOH$$

19.2.3.3 Carbon(IV) compounds
In the +4 oxidation state, carbon is part of carbon dioxide, carbonic acid and its salts, and compounds with halogen elements, sulfur, and others.

Carbon dioxide (CO_2) is constantly formed in nature during the combustion of fuels, the respiration of humans and animals, and the decay of plant residues. The CO_2 content in the atmosphere is 0.03% and tends to increase.

When carbon is completely burned, CO_2 is produced by releasing a large amount of heat:

$$C + O_2 = CO_2$$

In industry, CO_2 is obtained mainly by the thermal decomposition of limestone (at 900 °C):

$$CaCO_3 = CaO + CO_2$$

And in laboratory conditions, when hydrochloric acid reacts with $CaCO_3$:

$$CaCO_3 + 2HCl = CaCl_2 + CO_2 + H_2O$$

The carbon in the CO_2 molecule is in a state of sp hybridization, so the molecule is linear and nonpolar with strong double ($\sigma + \pi$) bonds between the carbon and oxygen atoms.

Carbon dioxide does not support combustion. It is heavier than air, which is why it is used in fire-fighting equipment. Small amounts of CO_2 dissolved in the blood stimulate the work of the respiratory center located in the brain stem. Therefore, minimal amounts of this gas must be present in the air. Larger amounts of it have a suffocating effect. Due to its significant density, CO_2 accumulates in the lower layers of the atmosphere and interferes with the free flow of heat exchange between the Earth and outer space. This is the so-called greenhouse effect, which leads to a slow increase in the temperature of the Earth, associated with fatal environmental risks.

The solubility of carbon dioxide in water is small – about 900 mL of CO_2 is dissolved in 1 L of water at 20 °C. The solution has a slightly acidic reaction due to the formation of a weak dibasic *carbonic acid* by the reversible reaction:

$$CO_2 + H_2O \rightleftharpoons H_2CO_3$$

The equilibrium of the above reaction is strongly shifted to the left, that is, a large part of the dissolved carbon dioxide is in the form of CO_2, and not H_2CO_3. For this reason, the dissociation constants of carbonic acid are very small ($K_1 = 4.5 \times 10^{-7}$; $K_2 = 4.7 \times 10^{-11}$) because the entire amount of carbon dioxide in the solution is taken into account when determining them.

Carbonic acid dissociates in two stages:

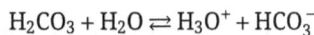

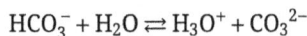

$$H_2CO_3 + H_2O \rightleftharpoons H_3O^+ + HCO_3^-$$

$$HCO_3^- + H_2O \rightleftharpoons H_3O^+ + CO_3^{2-}$$

and forms two types of salts: *normal carbonates* and acidic *hydrogen carbonates*. Their methods of preparation are trivial: interaction of CO_2 with basic oxides and bases or ionic reactions:

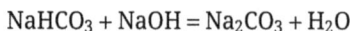

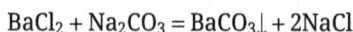

$$NaOH + CO_2 = NaHCO_3$$

$$CaO + CO_2 = CaCO_3$$

$$NaHCO_3 + NaOH = Na_2CO_3 + H_2O$$

$$BaCl_2 + Na_2CO_3 = BaCO_3\downarrow + 2NaCl$$

Alkali metal carbonates (and ammonium carbonate) are soluble in water, and their solutions have an alkaline reaction due to reversible hydrolysis:

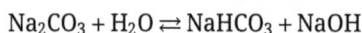

$$Na_2CO_3 + H_2O \rightleftharpoons NaHCO_3 + NaOH$$

When heated, carbonates decompose into oxides. The temperature at which decomposition takes place depends on the polarizing action of the cation. For example, the small size of Be^{2+} cation exhibits a highly polarizing effect. Therefore, $BaCO_3$ decomposes into BaO and CO_2 at 100 °C. Sodium and potassium carbonates decompose at 900–1,000 °C. Hydrogen carbonates of alkali metals turn into carbonates at 100 °C:

$$2NaHCO_3 = Na_2CO_3 + CO_2 + H_2O$$

All carbonates decompose by releasing CO_2 when reacting with acids. This reaction is used to detect carbonates because the release of CO_2 is accompanied by a characteristic noise.

Carbon dioxide and carbonates are widely used by humans. The main consumer of CO_2 is the food industry: the production of sugar, beer, carbonated waters, and its use in the form of "dry ice." In addition, CO_2 is used in extinguishing fires, transporting flammable liquids, and in the production of sodium carbonate and urea. Of the carbonates, sodium carbonate (used in the production of glass, soap, and paper), sodium hydrogen carbonate (in the food industry), and potassium carbonate (in the preparation of refractory glass and in photo processing) are of interest. Many natural carbonates of zinc, manganese, copper, lead, and iron are the starting materials for the production of the respective metals. $CaCO_3$ is very common in nature in the form of limestone, chalk, and marble. It is insoluble in water, so lime water ($Ca(OH)_2$ solution) becomes cloudy when CO_2 is passed through it:

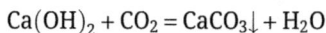

$$Ca(OH)_2 + CO_2 = CaCO_3\downarrow + H_2O$$

If carbon dioxide is passed through for a longer time, then the solution gradually becomes clearer because calcium carbonate turns into soluble hydrogen carbonate:

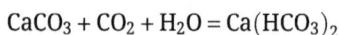

$$CaCO_3 + CO_2 + H_2O = Ca(HCO_3)_2$$

Calcium hydrogen carbonate, at the boiling point of the solution or during prolonged standing, decomposes with the release of CO_2 and the precipitation of $CaCO_3$. All these processes explain the cycle of calcium carbonate in nature. Soil and groundwater containing carbon dioxide dissolve calcium carbonate and excrete it in the form of hydrogen carbonate in rivers and seas. From there, it enters the organisms of marine animals to build their skeletons or, by releasing CO_2, it again turns into calcium carbonate, thus forming its accumulations and deposits.

Carbon forms compounds with all halogen elements – *halides* with a composition of CF_4, CCl_4, CBr_4, and CI_4. In the direction from fluorine to iodine, the stability of these compounds decreases. Especially important are carbon tetrachloride CCl_4 (a good noncombustible solvent of organic substances) and tetrafluoromethane (CF_4) (for the production of polymerized tetrafluoroethene $(C_2F_4)_n$, the so-called Teflon). The gaseous compounds CF_4, CF_3Cl, CF_2Cl_2, and $CFCl_3$, the so-called chlorofluorocarbons, are used in refrigeration equipment.

At 750–1,000 °C, carbon reacts with sulfur to form *carbon disulfide* CS_2. This liquid dissolves substances with a molecular crystal lattice well, but it is poisonous and, in the presence of oxygen, easily ignited:

$$CS_2 + 3O_2 = CO_2 + 2SO_2$$

Carbon disulfide is a sulfoanhydride, that is, with the sulfides of alkali metals, it forms sulfosalts (*sulfocarbonates*):

$$Na_2S + CS_2 = Na_2CS_3$$

Unlike other sulfosalts, which are soluble in water, sulfocarbonates are insoluble. When interacting with acids, these salts form sulfocarbonic acid, the only one of all sulfonic acids that does not decompose instantly at the moment of preparation, but rather slowly and at low temperatures is separated from the solution in the form of a yellow oily liquid:

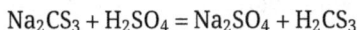

$$Na_2CS_3 + H_2SO_4 = Na_2SO_4 + H_2CS_3$$

Carbon in the oxidation state of $+4$ forms a compound that can be considered as carbonic acid, in the molecule of which the three oxygen atoms in the oxidation state of -2 are replaced by two nitrogen atoms in the oxidation state of -3. This is *cyanamide* H_2CN_2 – a solid, water-soluble substance, which is a weak acid in solution. Of great importance is the salt of this acid – calcium cyanamide ($CaCN_2$), used as an artificial fertilizer.

Other carbon(IV) compounds are also known, such as *cyanic acid,* HCNO ($K = 1.2 \times 10^{-4}$) and *thiocyanic acid* (HSCN) ($K = 0.14$). In terms of the structure of their acid residues, they are close to the cyanamide ion, as shown in Figure 19.3.

Cyanates, with a general structural formula –NCO, exist in the form of two isomeric modifications: $-N=C=O$ and $-O-C\equiv N$. Cyanates and cyanic acid itself in the liquid state contain a tautomeric mixture of the two isomers, and in the solid state, the isomer $-N=C=O$ predominates. The soluble ammonium rhodanide (NH_4SCN) and alkali metal rhodanides, used in qualitative analysis, as well as some rhodanides used as paints, are of practical importance from the salts.

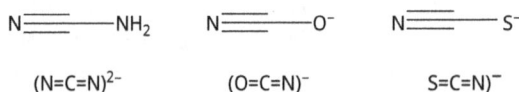

$$N\!\!=\!\!=\!\!-NH_2 \qquad N\!\!=\!\!=\!\!-O^- \qquad N\!\!=\!\!=\!\!-S^-$$

$$(N{=}C{=}N)^{2-} \qquad\qquad (O{=}C{=}N)^- \qquad\qquad S{=}C{=}N)^-$$

Figure 19.3: Structures of cyanamide, cyanate, and thiocyanate ions.

19.3 Silicon

19.3.1 General characteristics

The silicon atom has an electron configuration of $1s^2 2s^2 2p^6 3s^2 3p^2$. It is the only one of the group with a complete eight-electron configuration of the penultimate electron shell. Its atomic and ionic radii are significantly larger than those of carbon. Despite its larger atomic radius, silicon forms strong covalent bonds with other chemical elements, which is due to the presence of free d-orbitals in its outer electron shell. Unlike carbon, silicon is not characterized by sp and sp^2 hybridization, that is, the formation of double and triple bonds between silicon atoms is less likely than between silicon and atoms of other chemical elements. The most common configuration in its compounds is tetrahedral (sp^3). The presence of free d-orbitals also impacts the tendency of silicon to complex formation and the achievement of the maximum coordination number. In multiple cases, $sp^3 d^2$ hybridization and coordination number 6 are present. Additional strengthening of bonds is achieved by the so-called d_π–p_π bonding. A donor-acceptor interaction takes place, and a π bond is formed with the participation of a free electron pair of the atom bound to silicon. The bonds with the same type of atoms (silicon chains) and with hydrogen are weaker than those with carbon due to the impossibility of realizing such a donor-acceptor bond in carbon. The most characteristic oxidation state for silicon is +4. Compared to more electropositive elements, it also exhibits an oxidation state of −4.

19.3.2 Occurrence

Silicon is the second most abundant chemical element on the Earth. It does not exist in a free state but is part of about 3,000 minerals, which make up mountain ores and the Earth's crust. Quartz SiO_2 (sand) is widespread, with several varieties: quartz, rock crystal, agate, jasper, opal, and chalcedony. Crystals of pure silica sometimes reach huge sizes (mountain crystals up to 70 tons). The number and variety of silicates and aluminosilicates, as well as the products of their erosion, are also significant. Silicon compounds have entered the everyday life of humans since ancient times. Flint, clay, and various silicates were widely used to make tools, weapons, and vessels. Silicon was obtained and isolated for the first time by Berzelius in 1823. Its name derives from the Latin names *silex, silicis,* and means flint. There are three natural isotopes of silicon: ^{28}Si (92.18%), ^{29}Si (4.71%), and ^{30}Si (3.12%), and several artificially obtained ^{25}Si, ^{26}Si, ^{27}Si, ^{31}Si, and ^{32}Si.

19.3.3 Usage and biological role

At present, silicon is the basis of many new materials used in metallurgy, electrical engineering, construction, and instrumentation. Silicon is a useful additive to various alloys in metallurgy. Added to steel in quantities of 2–4%, it greatly increases its magnetic permeability. In this way, steel is prepared for transformers, electric motors, and generators, the so-called transformer steel. It is used in the production of acid-resistant cast iron, as well as for the removal of oxygen from steel and alloys during their melting. Ferrosilicon (alloy composed of iron and silicon) is a reducer in metallurgy and a useful additive to steels, helping to separate oxygen from them. Pure silicon is used in modern microelectronics and in semiconductor technology (for the production of solar batteries, integrated circuits, etc.). Integrated circuits are made from thin plates of silicon cut from a pure single crystal, on which thousands of circuit elements are mounted.

Silicon belongs to the trace elements. It is involved in the absorption of calcium, magnesium, phosphorus, potassium, sodium, sulfur, aluminum, cobalt, and many other elements. In the absence of silicon, 76 out of 104 elements are not absorbed by the body or are absorbed improperly. With silicon deficiency, the development of various diseases is possible. If calcium is the main element for the formation of hard bone tissues, Si is the element that determines the properties of flexible structures: connective tissues of tendons, walls of blood vessels and gastrointestinal tract, endocrine glands, cartilage, and valves of the cardiovascular system. Silicon is necessary for the normal functioning and formation of all these structures, giving them strength and elasticity. With silicon deficiency, the aging process develops faster. Plants containing silicon include cereals, horsetails, and palm trees. In animals, silicon is found in insignificant quantities in connective tissue, kidneys, and pancreas. It has been found that in diseases such as tuberculosis and cancer, the excretion of silicon from the kidneys decreases. The entrance of silicon compounds into the body causes leukocytosis, and inhalation of SiO_2 dust is the cause of silicosis (sclerosis of lung tissues). Organic compounds of Si, as an analog of C, are used to lower cholesterol levels.

19.3.4 Preparation

Silicon of technical purity is obtained from SiO_2 by reduction with carbon (industrial), magnesium (laboratory), or aluminum:

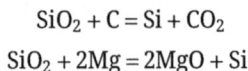

$$SiO_2 + C = Si + CO_2$$

$$SiO_2 + 2Mg = 2MgO + Si$$

Silicon with high semiconductor purity is obtained by thermal dissociation or reduction of its volatile compounds, which do not contain impurities:

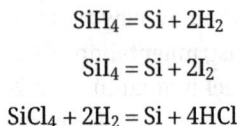

$$SiH_4 = Si + 2H_2$$

$$SiI_4 = Si + 2I_2$$

$$SiCl_4 + 2H_2 = Si + 4HCl$$

19.3.5 Physical and chemical properties of silicon

The most characteristic modification of silicon is crystalline silicon with an atomic crystal lattice similar to diamond. Silicon is in a state of sp^3-hybridization with a tetrahedral geometry. The simple substance has a gray-black color and metallic luster and is characterized by high hardness and a high melting point (1,428 °C). However, in terms of hardness, silicon is inferior to diamond because the bond energy between silicon atoms is less than the bond energy between carbon atoms. Pure silicon has negligible conductivity, but small amounts of impurities turn it into a semiconductor. It exhibits insulator properties more often. The so-called amorphous silicon is also known. In fact, this is not another allotropic modification but is highly dispersed crystals of the crystalline form. Its graphite-like modification is unstable.

Silicon is a chemically inert element. It does not interact with acids and aqua regia. Only a mixture of nitric and hydrofluoric acid dissolves silicon, turning it into a complex compound:

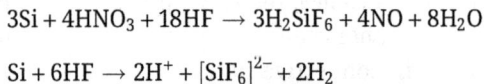

$$3Si + 4HNO_3 + 18HF \rightarrow 3H_2SiF_6 + 4NO + 8H_2O$$

$$Si + 6HF \rightarrow 2H^+ + [SiF_6]^{2-} + 2H_2$$

Under ordinary conditions, Si reacts only with fluorine and with melts and solutions of alkaline bases, whereby, unlike other nonmetals, it does not disproportionate but only oxidizes:

$$Si + 2NaOH + H_2O \rightarrow Na_2SiO_3 + 2H_2$$

$$Si + 4KOH = K_4SiO_4 + 2H_2$$

$$Si + 4KOH \rightarrow 4K^+ + [SiO_4]4^- + 2H_2$$

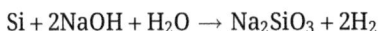

At elevated temperatures, it also interacts with other halogen elements to form EX_4, as well as with oxygen, other nonmetals, and water. Reactions with chlorine and oxygen occur at a noticeable speed only at 400 and 600 °C, respectively.

In relation to active metals, Si acts as an oxidizing agent, whereby silicides are obtained:

$$Si + 2Mg \rightarrow Mg_2Si$$

The chemical activity of silicon depends on the state of its surface. Thus, highly dispersed silicon obtained by the reaction

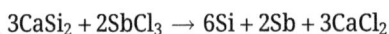

$$3CaSi_2 + 2SbCl_3 \rightarrow 6Si + 2Sb + 3CaCl_2$$

reacts with water to form SiO_2 and hydrogen:

$$Si + 2H_2O \rightarrow SiO_2 + 2H_2$$

Usually, the surface of silicon crystals is covered with a thin layer of silica (SiO_2), which prevents the material beneath this layer from reacting with air at temperatures up to 900 °C. At higher temperatures, a reaction with oxygen is observed to form silica, and at temperatures above 1,400 °C, silicon reacts with nitrogen from the air and forms silicon nitrides, SiN and Si_3N_4:

$$Si + O_2 \rightarrow SiO_2$$

$$2Si + N_2 \rightarrow 2SiN$$

$$3Si + 2N_2 \rightarrow Si_3N_4$$

19.3.6 Chemical compounds of silicon

19.3.6.1 Hydrides
Silicon does not react directly with hydrogen. Its hydrogen compounds are obtained by treating metallic silicides with mineral acids. A mixture of hydrogen compounds, *silanes*, is obtained, which are members of a common homologous series with a composition of Si_nH_{2n+2} (n = from 1 to 6), namely SiH_4 (monosilane), Si_2H_6 (disilane), Si_3H_6 (trisilane), etc., to Si_6H_{14} (hexasilane). The mixture is dominated by monosilane:

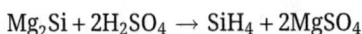

$$Mg_2Si + 2H_2SO_4 \rightarrow SiH_4 + 2MgSO_4$$

Monosilane is a colorless gas that spontaneously ignites and burns in air:

$$SiH_4 + 2O_2 \rightarrow SiO_2 + 2H_2O$$

The first members of the homologous row of silanes are gases, and the rest are easily volatile liquids, all of which are toxic substances. Compared to hydrocarbons, silanes are much more reactive, and the possibilities for silicon to form long chains are more limited. The latter is due to the significantly lower strength of the Si–Si bonds (222.5 kJ/mol) and Si–H bonds compared to the analogous C–C (348.4 kJ/mol) and C–H bonds. Silanes and hydrocarbons are associated with the different electronegativities of C and Si, resulting in a different distribution of electric charges:

$$C^{\delta^-} - H^{\delta^+} \qquad Si^{\delta^+} - H^{\delta^-}$$

In the air, silanes spontaneously ignite. They easily hydrolyze and interact with bases:

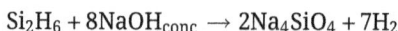

$$Si_2H_6 + (4 + 2x)H_2O \rightarrow 2SiO_2.xH_2O + 7H_2$$

$$Si_2H_6 + 8NaOH_{conc} \rightarrow 2Na_4SiO_4 + 7H_2$$

Unlike methane, monosilane interacts with water to produce metasilicic acid:

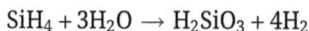

$$SiH_4 + 3H_2O \rightarrow H_2SiO_3 + 4H_2$$

Compared to other elements and compounds, silanes exhibit strong reducing properties:

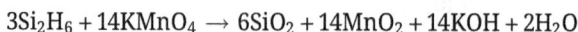

$$3Si_2H_6 + 14KMnO_4 \rightarrow 6SiO_2 + 14MnO_2 + 14KOH + 2H_2O$$

19.3.6.2 Silicides

Silicon at high temperatures interacts with numerous metals to form silicides. They are obtained by direct interaction between the reacting elements, for example:

$$2Ca + Si \rightarrow Ca_2Si$$

They are also obtained by reducing metal oxides with Si, by reacting Si with metal hydrides, as well as by reacting excess metal with Si(IV) oxide:

$$2MeO + 2Si = Me_2Si + SiO_2$$

$$2CaH_2 + Si = Ca_2Si + H_2\uparrow$$

$$SiO_2 + 4Mg = Mg_2Si + 2MgO$$

Silicides are crystalline substances with a metallic luster, usually silvery-white or gray in color. According to the type of chemical bond, silicides are generally divided into ionic-covalent (silicides of alkali and alkaline earth metals and Mg) and metal-like (silicides of transition metals). In their structures, there are M–Si, Si–Si, and M–M bonds.

The silicides of the metals of the I and II main groups of the periodic table are characterized by the combination of an ionic bond between metal atoms and Si with a covalent bond between Si atoms. They are chemically unstable and are easily decomposed by water and acids, releasing a mixture of silanes. The composition of the silicides of alkali and alkaline earth metals is constant, corresponding to the oxidation state of silicon (−4): Na_4Si, Mg_2Si, etc.

Metal-like silicones are characterized by a combination of a metallic bond between metal atoms with a covalent bond between Si atoms, as well as to a significant extent with the covalent bond M–Si, which increases with a decrease in the donor capacity of metals. The silicides of d- and f-elements are significantly more stable than the ionic covalent ones. They are metal-like and conduct electric current. They are notable for their hardness, fire resistance, chemical inertness, and high melting temperatures. Their composition does not always correspond to a certain stoichiometry or to the usual oxidation states of the elements ($FeSi$, Cr_3Si, Mn_5Si_3, etc.).

19.3.6.3 Halogenides and other compounds with more electronegative elements

With the halogen elements, silicon forms *tetrahalides*: gaseous SiF_4, liquid $SiCl_4$ and $SiBr_4$, and solid SiI_4. In general, molecular halides are characterized by an increase in melting and boiling temperatures with an increase in the atomic number of the halogen element (Figure 19.3).

Table 19.3: Melting and boiling points of Si(IV) halides.

Halogenide	T_m (°C)	T_b (°C)
SiF_4	−90 (pressure)	−95
$SiCl_4$	−68	+57
$SiBr_4$	+5	+153
SiI_4	+122	+290

Silicon halides are compounds with covalent bonds that are obtained by direct interaction. For practical purposes, they can be obtained by treating silica with halogens or calcium halides:

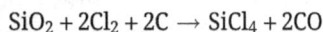

$$SiO_2 + 2Cl_2 + 2C \rightarrow SiCl_4 + 2CO$$

Silicon in the halides is in the sp³-hybrid state. Halogenides do not show a tendency for polymerization. The presence of free d-orbitals in the outermost electron shell determines the acceptor properties of these compounds and their tendency for complexation:

$$SiF_4 + 2HF \rightarrow H_2[SiF_6]$$

Tetrahalides are easily hydrolyzed:

$$\equiv Si - X + H_2O \rightarrow \equiv Si - OH + HX$$

Their hydrolysis leads to the production of polymeric oxygen-containing acids. The production of solid, white products in this process is the reason for the appearance of white smoke above the liquid and gaseous halides when they come into contact with the air and moisture.

In hydrolysis, these compounds exhibit the properties of halogen-anhydrides; that is, they are completely hydrolyzed to form two acids:

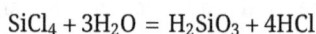

$$SiCl_4 + 3H_2O = H_2SiO_3 + 4HCl$$

Silicon tetrafluoride is also hydrolyzed, but the hydrofluoric acid binds to SiF_4, forming a complex compound. Thus, the general equation for hydrolysis has another form:

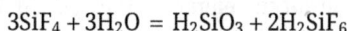

$$3SiF_4 + 3H_2O = H_2SiO_3 + 2H_2SiF_6$$

Hexafluorosilicic acid (H_2SiF_6) is stable in solution, but in its free state, it decomposes to a significant extent into HF and SiF_4. It is a strong dibasic acid, close in strength to sulfuric acid. Its salts are known as *fluorosilicates*: Na_2SiF_6, $BaSiF_6$, etc. A large part of these salts is well soluble in water. Interestingly, it is the fluorosilicates of alkali metals that are insoluble, which is practically very important since almost all salts of these metals with other acids are well soluble in water.

Tetrahalides are starting substances in the production of pure silicon.

With sulfur, when heated, silicon forms *silicon disulfide* (SiS_2). Similar to oxide, it is a solid whose main building blocks are SiS_4 tetrahedra. In this case, however, the binding of the tetrahedra occurs not only through one of the vertices but also through their common edges. Like most binary compounds of silicon, disulfide is completely hydrolyzed by the release of silicic acids and hydrogen sulfide, and when interacting with sulfides of alkali metals and with ammonium sulfide, it exhibits sulfo-anhydride properties:

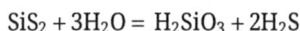

$$SiS_2 + 3H_2O = H_2SiO_3 + 2H_2S$$

$$Na_2S + SiS_2 = Na_2SiS_3$$

Silicon interacts with nitrogen at high temperatures (above 1,300 °C). *Silicon nitride* (Si_3N_4) is obtained, which interacts with alkaline bases with the release of ammonia:

$$Si_3N_4 + 12KOH \rightarrow 3K_4SiO_4 + 4NH_3$$

When sand and coal are heated in electric furnaces above 2,300 °C, *silicon carbide* – SiC (carborundum) – is obtained. It is not the only compound of silicon with carbon.

When silicon tetrahalides react with a Grignard reagent, alkyl and aryl derivatives of the halides $R_{4-n}SiX_n$ are obtained:

$$SiCl_4 + CH_3MgCl \rightarrow CH_3SiCl_3 + MgCl_2$$

The hydrolysis of these substances results in the production of organosilicon hydroxides. These compounds, called *silanols*, are involved in polycondensation processes, resulting in compounds containing polymer chains. These are called *silicones*. For example, the hydrolysis of trialkyl monochlorosilane (R_3SiCl) results in the production of hexaalkylsiloxane (Figure 19.4).

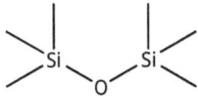

Figure 19.4: Structure of hexaalkylsiloxane.

The hydrolysis of alkyl trichlorosilane leads to the production of a polymer with a ring structure. In addition to the treatment of silicon tetrahalides with Grignard's reagent, alkyl-substituted chlorosilanes can also be obtained through the treatment of silicon and chlorinated hydrocarbons in the presence of a Cu catalyst:

$$Si + 2CH_3Cl \rightarrow (CH_3)_2SiCl_2$$

Oxides. Two silicon oxides are named SiO and SiO_2. The dioxide is produced by direct interaction between the elements and when heated to 800 °C. This is a very common thermodynamically stable refractory substance (sand) in nature. The crystal lattice of SiO_2 consists of silicon atoms, each of them surrounded by four oxygen atoms, the so-called silicon-oxygen tetrahedra, connected to each other by common vertices. Depending on the mutual orientation of the tetrahedra, there are several polymorphic modifications of SiO_2: α-quartz (up to 573 °C), β-quartz (573–867 °C), tridymite (867–1,470 °C), and cristobalite (from 1,470 °C to the melting point of 1,725 °C). In its most characteristic form (α-quartz), the basic building block in the crystal is a regular tetrahedron, in which the silicon atom occupies a central place. It is in sp^3 hybridization and forms four bonds with the four oxygen atoms located at the vertices of the tetrahedron. Many such tetrahedra are connected in a network, and the binding of two adjacent tetrahedra occurs by the common oxygen atom.

When molten SiO_2 is cooled, a glassy mass is formed, which is an amorphous form of SiO_2 called quartz glass. It is very valuable because it transmits ultraviolet radiation, is resistant to acids, and withstands sudden temperature changes. In nature, there are well-formed colorless and transparent quartz crystals, often in the form of hexagonal prisms, commonly called mountain crystals. It can dissolve oxides of other chemical elements in itself. This is how different types of silicate glasses are obtained. Crystals colored by impurities in violet are called amethyst, and those with a brownish color are called smoky topaz. Varieties of quartz also include flint and agate. Quartz is a component of clay and gneiss. Ordinary sand consists of fine quartz crystals. Pure sand is white, but due to impurities, it is yellowish.

Silica (SiO_2) is insoluble in water and resistant to acids. It dissolves only in hydrogen fluoride and hydrofluoric acid, forming volatile silicon tetrafluoride (SiF_4) or H_2SiF_6:

$$SiO_2 + 4HF \rightarrow SiF_4 + 2H_2O$$

$$SiO_2 + 6HF = H_2SiF_6 + 2H_2O$$

As an acidic oxide, SiO_2 is unstable in the presence of alkaline reagents. Under the action of strong bases, Si–O–Si bonds are destroyed, the basic building blocks (SiO_4 tetrahedra) pass into solution, and silicic acid salts are obtained:

$$SiO_2 + 2NaOH = Na_2SiO_3 + H_2O$$

In the melt, SiO_2 also reacts with alkaline carbonates and sulfates:

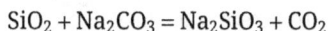

$$SiO_2 + Na_2CO_3 = Na_2SiO_3 + CO_2$$

When silicon dioxide reacts with silicon, the silicon suboxide (SiO) is obtained:

$$SiO_2 + Si \rightleftarrows 2SiO$$

The process is endothermic, and the equilibrium is shifted to the right only at high temperatures. At room temperature, monoxide is unstable. Therefore, the equilibrium is "frozen" by rapidly cooling the products obtained at high temperature. SiO is a brown solid with no particular practical application. It does not interact with water and is not an acidic anhydride.

Silica (SiO_2) is an anhydride of the silicic acids: *metasilicic* (H_2SiO_3), *orthosilicic* ($Si(OH)_4 \equiv H_4SiO_4$), and *polysilicic* ($nSiO_2 \cdot mH_2O$). In general, the composition of polysilicic acids is represented by the latter formula, but the variety of compositions and structures is huge. The group of metasilicic acids includes $SiO_2.H_2O = H_2SiO_3$ (metasilicic acid); $2SiO_2.H_2O = H_2Si_2O_5$ (dimetasilicic acid); $3SiO_2.H_2O = H_2Si_3O_7$ (trimetasilicic acid); $4SiO_2.H_2O = H_2Si_4O_9$ (tetrametasilicic acid). The group of orthosilicic acids includes $SiO_2.H_2O = H_4SiO_4$ (orthosilicic acid); $2SiO_2.3H_2O = H_6Si_2O_7$ (diorthosilicic acid); and $3SiO_2.4H_2O = H_8Si_3O_{10}$ (triorthosilicic acid).

A mixture of these slightly soluble acids is formed as a puffy precipitate (gel) when soluble silicates react with acids:

$$Na_2SiO_3 + HCl = NaCl + (H_2SiO_3, H_4SiO_4, nSiO_2 \cdot mH_2O)\downarrow$$

Silicic acids are weak acids, weaker than carbonic acid. Therefore, when storing solutions of their salts in loosely closed containers, decomposition of silicates by CO_2 (anhydride of a stronger acid) occurs:

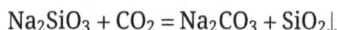

$$Na_2SiO_3 + CO_2 = Na_2CO_3 + SiO_2\downarrow$$

When dewatering a gel of silicic acids (under low heating in a vacuum), a highly porous substance, *silica gel*, is obtained, which is used as an adsorbent, carrier, and desiccant. In general, it corresponds to the formula $SiO_2 \cdot xH_2O$.

Salts of silicic acids, which are stable only in strongly alkaline solutions, are called *silicates*. When alkaline solutions of silicates are acidified and conditions are created for the production of the acid, it quickly polymerizes. During this polymerization process, particles of colloidal size are obtained, and finally, a product with an extremely developed surface, silica gel, is formed.

Of practical importance among the salts of silicic acids is sodium metasilicate, which is obtained by the interaction of quartz sand with a sodium base in a melt:

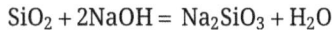

$$SiO_2 + 2NaOH = Na_2SiO_3 + H_2O$$

The aqueous solution of sodium silicate ("soluble glass") has a highly alkaline reaction due to the hydrolysis process:

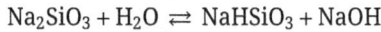

$$Na_2SiO_3 + H_2O \rightleftarrows NaHSiO_3 + NaOH$$

There is a very large number of complex silicates that are derivatives of various polysilicic acids. However, some grouping of the types of silicates is possible:
a) *Simple silicates with isolated tetrahedra SiO_4^{4-}*. Such are orthosilicates, in which there are no oxygen bridges between the individual tetrahedra.
b) *Silicates with a limited number of tetrahedra are linked to each other by Si–O–Si bonds.* Such compounds include tortveite – $Sc_2(Si_2O_7)$, beryl – $Al_2Be_3(Si_6O_{18})$, and wollastonite – $Ca_3(Si_3O_9)$.
c) *Chain silicates.* These are silicates with a unidirectional infinite structure composed of many interconnected tetrahedra. Compounds with chain structures include $Mg_2(Si_2O_6)$ and $Fe_2(Si_2O_6)$.
d) *Silicates with two-dimensional infinite structures.* With this network-like bonding of tetrahedra, the basic building blocks are $(Si_2O_5)^{2-}$. Minerals of this type include talc – $Mg_3(OH)_2(Si_4O_{10})$ and kaolinite – $Al_2(OH)_4(Si_2O_5)$.
e) *Silicates with a three-dimensional infinite structure.* In silicates with a three-dimensional infinite structure, tetrahedra form bonds in the three directions of space, creating spatial structures with an infinite number of elements. Such compounds include albite – $Na[AlSi_3O_8]$, anorthite – $Ca[Al_2Si_2O_8]$, and orthoclase – $K[AlSi_3O_8]$. Three-dimensional silicates are often characterized by a variety of gaps in the lattice where cations or water molecules can be located. These structures are particularly suitable for adsorbents and catalysts. Natural zeolites have such structures. Their composition can be expressed by the general formula $M_{x/n}[(Al_2O_3)_x(SiO_2)_y \cdot zH_2O]$, where n is the charge of the metal cation M^{n+} (usually sodium, calcium, or potassium), and z is the number of molecules of crystallization water.

Silicates of all metals except alkaline ones are insoluble in water and are widespread in nature, especially calcium and magnesium silicates. Natural silicates are salts of polysilicic acids. Their composition is usually expressed in the form of a ratio of oxides: $2MgO.SiO_2$ (olivine), $3MgO\cdot2SiO_2\cdot2H_2O$ (asbestos), and $2CaO\cdot MgO\cdot8SiO_2\cdot2H_2O$ (tremolite). The composition of natural silicates is very diverse because they crystallize isomorphically with aluminates, forming *aluminosilicates*: $Na_2O\cdot Al_2O_3\cdot6SiO_2$ (albite), $6Na_2O\cdot6Al_2O_3\cdot12SiO_2\cdot27H_2O$ (zeolite), $K_2O\cdot Al_2O_3\cdot6SiO_2$ (orthoclase), etc. Under the action of water and air over millions of years, the natural aluminosilicates forming mountain ores decompose to form quartz (sand) and kaolin (clay). For albite, this can be expressed by the equation:

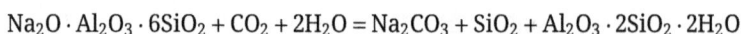

$$Na_2O \cdot Al_2O_3 \cdot 6SiO_2 + CO_2 + 2H_2O = Na_2CO_3 + SiO_2 + Al_2O_3 \cdot 2SiO_2 \cdot 2H_2O$$

This is how the destruction of ores and the formation of soil take place.

One of the most important artificially produced silicates is *glass*. Ordinary glass has a composition of $Na_2O\cdot CaO\cdot6SiO_2$. For special needs, various types of glass are obtained, resistant to chemical, mechanical, or thermal influences. They include boron, aluminum, arsenic, lead, and tin.

The glass is obtained by the joint melting of sand, limestone, and soda. Instead of soda, a mixture of sodium sulfate and coal is often used. Melting is associated with the decomposition of carbonates and the release of CO_2 according to the equation:

$$Na_2CO_3 + CaCO_3 + 6SiO_2 \rightarrow Na_2O \cdot CaO \cdot 6SiO_2 + 2CO_2$$

The resulting colorless viscous melt solidifies when cooled into a transparent vitreous mass. The vitreous state differs from the crystalline state and is not even considered a form of the solid state of matter. Unlike crystals, which have a repetition of the same structural units at significant distances along the lattice, here the amorphous structure is typical. Amorphous substances do not have a specific melting temperature, but a softening interval since not all bonds are equally strong. Therefore, glasses are considered not as solids, but as supercooled liquids. Glasses are also able to dissolve oxides of other elements, and they are often colored in characteristic colors. For example, the presence of CoO gives the glass a blue color and Cr_2O_3 a green color. The color of the glass can also be due to colloidal particles dispersed in it.

19.3.6.4 Usage

Both silicon and its compounds are widely used in human economic activities. Silica, in the form of sand, clay, and other silicon compounds, is widely used in the production of cement, glass, porcelain, faience, in the ceramic industry, etc.

The salts of silicic acids are among the most diverse and the most important industrially among all the salts of the chemical elements. On the one hand, they make up the Earth's crust. On the other hand, the importance of silicates used in the manufacture of glass, ceramics, porcelain, and cement is enormous.

Quartz is used both independently (for laboratory chemical vessels, fiber optics, etc.) and in the manufacture of glass. Ordinary glass retains ultraviolet rays, but quartz glass made of pure SiO_2 does not have this disadvantage. It is highly meltable, with a negligible coefficient of thermal expansion, so it can withstand sudden temperature changes. When working with quartz glass, one should take into account its low mechanical strength and its instability to alkaline substrates.

Glass has become an integral part of people's everyday lives. Along with its application as window glass, it serves as the basis for obtaining new thermal insulation and composite materials – foam glass, fibers, and fiberglass – characterized by extraordinary strength, elasticity, and lightness. The replacement of CaO with PbO in ordinary glass results in a glass with a high refractive index of light, the so-called crystal glass. Chemically resistant and fireproof glasses have an increased content of SiO_2, B_2O_3, and BaO.

Sodium silicate is also used in the woodworking industry as an adhesive that protects against decay, fire, etc., and sodium hexafluorosilicate is used as an insecticide for pest control in agriculture and in the production of cement and enamel. Fluorosilicates of magnesium, zinc, and aluminum are used in construction, giving waterproofness to building materials.

Cement also belongs to the silicate materials. Ceramic materials have long been known and used by humans. These are refractory materials made of clay, carbides, and oxides of some metals. Building materials include bricks, tiles, cladding slabs, and pipes. Refractory ceramics are used for lining kilns, and chemically resistant ones are used for lining apparatus in the chemical industry. Ceramics for domestic purposes include sanitary porcelain and earthenware as well as food containers. Technical ceramics are used to make capacitors, high-temperature crucibles and plates, parts of motors, etc. The use of ceramics in aircraft and rocket engineering is also increasing.

Many other silicon compounds are used in practice. Silicides $MoSi_2$, WSi_2, etc. are used in the production of refractory products. Silane (SiH_4) and silicon tetrachloride are starting substances in the synthesis of silicon-organic compounds, which are becoming increasingly used.

Silicon-organic compounds are used to produce heat-resistant rubber-like polymers, adhesives, refractory varnishes and enamels, water-repellent substances in the preparation of fabrics, heat-resistant lubricants, electrical insulation materials, etc. From the group of silicon-organic compounds, silicones have very valuable qualities that in many respects surpass the qualities of the corresponding organic carbon compounds and find a wide variety of applications – for artificial prostheses, heart valves, high-temperature oils, rubbers, etc. Their main qualities and great advantages are their chemical inertness and temperature resistance. The number of silicones is extremely large. A new branch of chemistry is emerging – organosilicon chemistry. Polymer products with molecular masses of up to 3 million atomic units are obtained.

19.4 Germanium, tin, and lead

Occurrence
Germanium does not form its own minerals and is found as an impurity in the sulfides of zinc, copper, and silver as well as in coal. For this reason, it was not known for a long time. Its existence was predicted by Mendeleev on the basis of the periodic law in 1871 and was discovered in 1886. The properties of germanium were remarkably close to those predicted by Mendeleev.

The main mineral of tin is cassiterite (SnO_2) and of lead is galena (PbS). Tin and lead, along with copper, silver, and gold, have been known since ancient times. The alloy of tin with copper (bronze) was used by humans at the dawn of the development of civilization (Bronze Age).

Synthesis
Germanium with technical purity is obtained by reducing its oxide with H_2:

$$GeO_2 + 2H_2 = Ge + 2H_2O$$

To obtain germanium with high purity, the zone melting method is used. Zone melting is based on the different solubility of impurities in the solid and liquid states; in the solid phase, it is significantly less.

Metals are obtained from their most common ores through reduction processes. Tin is obtained from cassiterite by reduction:

$$SnO_2 + C = Sn + CO_2$$

Lead is obtained from galena (PbS). Galena is first oxidized:

$$2PbS + 3O_2 = 2PbO + 2SO_2$$

And then it undergoes a reduction:

$$PbO + C = Pb + CO$$

The pure metals, tin and lead, are obtained by electrolysis.

19.4.1 Physical and chemical properties

Germanium is a brittle substance with a metallic luster and a diamond-type crystal lattice. It is a semiconductor with a melting point of 937 °C. Tin and lead are metals of silvery-white color, ductile, and low-melting. Their melting points are 232 and 334 °C, respectively. Tin has three allotropic modifications: gray, white, and rhombic (brittle) tin. White and rhombic tin have a metallic crystal lattice and metallic properties, while gray tin has a covalent-type lattice. The latter is a semiconductor with a diamond-like structure. The transition to gray tin occurs at low temperatures, at which the metal turns into a gray powder. Tin products are made of white tin, and interestingly, the presence of even traces of gray tin leads to the spontaneous transformation of white tin into gray. This rapid process leads to the dusting of tin products, known as "tin plague." Tin plague is a phenomenon in which the chemical element tin passes from its tetragonal β-allotropic form to its α-form. Lead has only one modification and is a grayish-white, soft, and malleable metal.

Germanium, tin, and lead form compounds with ionic and covalent bonds. The tendency to give up electrons in these elements is less pronounced than in the p-elements of group III. Tin and lead in the row of standard potentials are immediately before hydrogen, and germanium is after it, between copper and silver. Germanium does not react with nonoxidizing acids (HCl and dilute H_2SO_4), but tin reacts slowly when heated:

$$Sn + 2HCl = SnCl_2 + H_2$$

$$Sn + H_2SO_{4dilut} = SnSO_4 + H_2$$

and quickly in concentrated HCl with hydrogen release:

$$Sn + 4HCl_{conc} \rightarrow H_2[SnCl_4] + H_2$$

Lead does not interact with dilute hydrochloric and sulfuric acids due to the formation of poorly soluble $PbCl_2$ and $PbSO_4$. With concentrated hydrochloric acid, lead forms $H_2[PbCl_4]$:

$$Pb + 4HCl_{conc} \rightarrow H_2[PbCl_4] + H_2$$

With concentrated sulfuric acid, all three metals react slowly when heated, whereby germanium is oxidized to the maximum oxidation state to form an acid, tin forms tin(II) sulfate, and lead passes into solution in the form of an acidic salt:

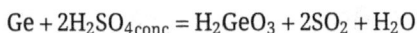

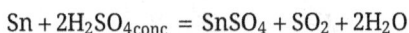

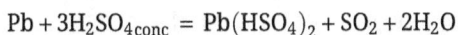

$$Ge + 2H_2SO_{4conc} = H_2GeO_3 + 2SO_2 + H_2O$$

$$Sn + 2H_2SO_{4conc} = SnSO_4 + SO_2 + 2H_2O$$

$$Pb + 3H_2SO_{4conc} = Pb(HSO_4)_2 + SO_2 + 2H_2O$$

In dilute nitric acid, germanium is stable, while tin and lead are converted to nitrates, reducing HNO_3 to various products (NH_4NO_3, N_2, N_2O, and NO):

$$4Sn + 10HNO_{3dilut} \rightarrow 4Sn(NO_3)_2 + NH_4NO_3 + 3H_2O$$

$$3Sn + 8HNO_{3dilut} = 3Sn(NO_3)_2 + 2NO + 4H_2O$$

$$3Sn + 16HNO_3 \rightarrow 3Sn(NO_3)_4 + 4NO + 8H_2O$$

$$3Pb + 8HNO_{3dilut} = 3Pb(NO_3)_2 + 2NO + 4H_2O$$

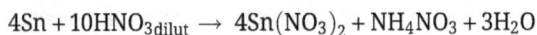

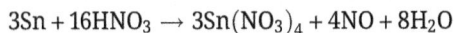

In concentrated nitric acid, under normal conditions, germanium, tin, and lead are passivated, but when heated, they slowly oxidize. Germanium and tin are oxidized by nitric acid to the maximum oxidation state with the formation of germanic acid (H_2GeO_3) and tin acid (H_2SnO_3) (in fact, hydrated oxides $EO_2 \cdot H_2O$ are formed), while lead is oxidized to the oxidation state +2 with the formation of salts. Concentrated nitric acid has a weak effect on lead because the resulting lead nitrate is difficult to dissolve in concentrated HNO_3:

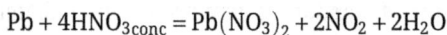

$$Ge + 4HNO_{3conc} = H_2GeO_3 + 4NO_2 + H_2O \rightarrow xGeO_2 \cdot yH_2O + NO_2$$

$$Sn + 4HNO_{3conc} = H_2SnO_3 + 4NO_2 + H_2O \rightarrow xSnO_2 \cdot yH_2O + NO_2$$

$$Pb + 4HNO_{3conc} = Pb(NO_3)_2 + 2NO_2 + 2H_2O$$

Organic acids react with tin and lead in the presence of oxygen to form soluble compounds:

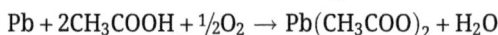

$$Pb + 2CH_3COOH + \tfrac{1}{2}O_2 \rightarrow Pb(CH_3COO)_2 + H_2O$$

Germanium and tin interact with aqua regia:

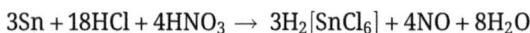

$$3Ge + 12HCl + 4HNO_3 \rightarrow 3GeCl_4 + 4NO + 8H_2O$$

$$3Sn + 18HCl + 4HNO_3 \rightarrow 3H_2[SnCl_6] + 4NO + 8H_2O$$

Germanium, tin, and lead are dissolved in concentrated solutions and base melts in the presence of oxidizing agents, whereby hydroxy-complex salts are formed in the solutions, and oxysalts are formed in the melt:

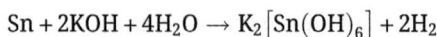

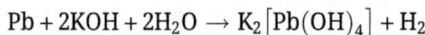

$$Sn + 2KOH + 4H_2O \rightarrow K_2[Sn(OH)_6] + 2H_2$$

$$Pb + 2KOH + 2H_2O \rightarrow K_2[Pb(OH)_4] + H_2$$

Germanium dissolves better in the presence of oxygen or oxidizing agents:

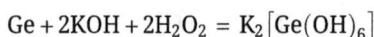

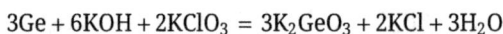

$$Ge + 2KOH + 2H_2O_2 = K_2[Ge(OH)_6]$$

$$3Ge + 6KOH + 2KClO_3 = 3K_2GeO_3 + 2KCl + 3H_2O$$

At ordinary temperatures, germanium and tin are resistant to oxygen, while lead becomes covered with a thin layer of oxide and loses its metallic luster. When heated, the activity of the elements in the germanium subgroup increases.

With water, germanium, tin, and lead do not interact under ordinary conditions. Tin does not react with water in the cold, but when heated, it reacts with the formation of SnO_2 and H_2:

$$Sn + 2H_2O \rightarrow SnO_2 + 2H_2$$

In the cold, water does not interact with lead, but in the presence of O_2, a reaction occurs:

$$2Pb + 2H_2O + O_2 \rightarrow 2Pb(OH)_2$$

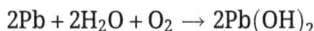

Hot water vapor also reacts with lead:

$$Pb + H_2O \rightarrow PbO + H_2$$

If water contains a small amount of dissolved CO_2, the surface of lead is covered with a protective layer of $PbCO_3$:

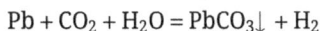

$$Pb + CO_2 + H_2O = PbCO_3\downarrow + H_2$$

and lead does not dissolve further in water. At a higher CO_2 content, $PbCO_3$ dissolves with the formation of lead hydrogen carbonate:

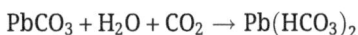

$$PbCO_3 + H_2O + CO_2 \rightarrow Pb(HCO_3)_2$$

which was the cause of lead poisoning in the population of ancient Rome due to the lead water pipes used at that time.

19.4.2 Chemical compounds

Germanium, tin, and lead form compounds in oxidation states of +2 and +4.

19.4.2.1 Oxides

Oxides of the type MO and MO_2 are known. Lead combines directly with oxygen to form PbO, while germanium and tin form dioxides. The oxides of the heavier members of the IV A group can be considered ionic compounds. Tin(IV) oxide (SnO_2) is the stable oxide of tin, while lead(II) oxide (PbO) is the stable oxide of lead. In lead, two more mixed oxides are known: $PbO.PbO_2$ (Pb_2O_3) and $2PbO.PbO_2$ (Pb_3O_4), as well as multiple nonstoichiometric oxides located between Pb_3O_4 and PbO_2.

Lead(II) oxide exists in two crystalline forms: yellow and red. α-PbO is tetragonal and red in color. It is thermodynamically stable at low temperatures. β-PbO is rhombic and yellow in color. It is thermodynamically stable at high temperatures and metastable at room temperature.

In addition to direct synthesis, PbO can also be obtained by the thermal decomposition of Pb(II) salts. PbO dissolves in acids, with the exception of HCl and H_2SO_4 due to the low solubility of the respective salts. In the presence of strong oxidizing agents, PbO_2 can be produced from PbO. Brown lead(IV) oxide (PbO_2) is quite stable and appears to be a good oxidizing agent.

The Pb(IV) ion is a highly polarizing ion that exists permanently in an aqueous solution. Pb(IV) oxide is an insoluble solid in which Pb^{4+} ions are stabilized in the high-energy crystal lattice. Even in this state, it can be argued that there is covalent bonding in the structure. The addition of an acid, e.g., nitric acid, leads to the formation of Pb(II) and the release of oxygen gas:

$$2PbO_2 + 4HNO_3 \rightarrow 2Pb(NO_3)_2 + 2H_2O + O_2$$

In the cold, with concentrated hydrochloric acid, the oxide of Pb(IV) undergoes a substitution reaction with an intermediate product – the covalently bonded Pb(IV) chloride. When heated, the unstable Pb(IV) chloride decomposes into Pb(II) chloride and chlorine gas:

$$PbO_2 + 4HCl \rightarrow PbCl_4 + 2H_2O$$

$$PbCl_4 \rightarrow PbCl_2 + Cl_2$$

The practical significance of PbO_2 is its use as an oxidizing agent:

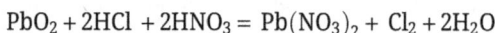

$$PbO_2 + 2HCl + 2HNO_3 = Pb(NO_3)_2 + Cl_2 + 2H_2O$$

$$3PbO_2 + 2Cr(OH)_3 + 10KOH = 2K_2CrO_4 + 3K_2\left[Pb(OH)_4\right] + 2H_2O$$

19.4.2.2 Hydroxides and oxosalts

In oxidation state +2, oxides GeO, SnO, PbO, and their corresponding hydroxides $Ge(OH)_2$, $Sn(OH)_2$, and $Pb(OH)_2$ are amphoteric. Therefore, when they react with acids, cationic salts are formed – $GeCl_2$, $SnSO_4$, $Pb(NO_3)_2$, and when interacting with solutions and melts of bases – anionic salts: $K_2[Ge(OH)_4]$, Na_2SnO_2, $K_2[Pb(OH)_4]$, etc. Cationic salts are stored in solutions in a strongly acidic environment and anionic salts in a basic medium in order to suppress hydrolysis. All these compounds are reducers, with their reduction properties increasing from the compounds of Pb(+2) to the compounds of Ge(+2).

In the oxidation state +4, in oxides (GeO_2, SnO_2, and PbO_2) and hydroxides (H_2GeO_3, H_2SnO_3, and $Pb(OH)_4$), acidic properties predominate, so most of their salts are of the anionic type: K_2GeO_3, $K_2[Ge(OH)_6]$, Na_2SnO_3, $Na_2[Sn(OH)_6]$, K_2PbO_3, K_4PbO_4, and $K_2[Pb(OH)_6]$. In lead and tin, some salts of the cationic type are also known: $SnCl_4$, $Sn(SO_4)_2$, and $PbCl_4$. Anionic-type salts are hydrolyzed reversibly:

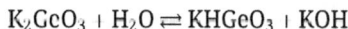

$$K_2GeO_3 + H_2O \rightleftharpoons KHGeO_3 + KOH$$

but of the cationic type completely:

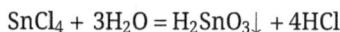

$$SnCl_4 + 3H_2O = H_2SnO_3\downarrow + 4HCl$$

All compounds in which Ge, Sn, and Pb are in the +4 oxidation state are good oxidizing agents.

19.4.2.3 Hydrides

With hydrogen, Ge, Sn, and Pb form hydrides: GeH_4 (germane), SnH_4 (stannane), and PbH_4 (plumbane). These are thermodynamically unstable compounds that decompose under low heating. The hydrides of Ge(IV), Sn(IV), and Pb(IV) are produced indirectly.

19.4.2.4 Halides

Germanium and tin form all types of halogenides (fluorides, chlorides, bromides, iodides) in oxidation states +2 and +4, and lead forms all halides in oxidation state +2 and only two compounds in oxidation state +4 (PbF_4 and $PbCl_4$). Tetrabromide and tetraiodide of lead do not exist because, at the moment of production, they decompose in an intramolecular redox type:

$$PbBr_4 = PbBr_2 + Br_2$$

$$PbI_4 = PbI_2 + I_2$$

Sn(IV) chloride is a typical covalent metal chloride. It is an oily liquid that smokes in moist air, forming hydrogen chloride gas and gel-like Sn(IV) hydroxide, represented below as $Sn(OH)_4$ (although it is actually a hydrated oxide):

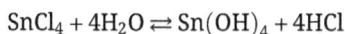

$$SnCl_4 + 4H_2O \rightleftharpoons Sn(OH)_4 + 4HCl$$

Pb(IV) chloride is a yellow, oily substance that, like its tin counterpart, decomposes in the presence of moisture and explodes when heated.

Pb(II) chloride, bromide, and iodide are water-insoluble substances. Bright yellow crystals of Pb(II) iodide are formed if a soluble iodide is added to a colorless solution of a Pb(II) salt:

$$Pb^{2+} + 2I^- \rightarrow PbI_2$$

The addition of a large excess of iodide causes the precipitate to dissolve and form a solution containing tetraiodoplumbate ions:

$$PbI_2 + 2I^- \rightleftharpoons [PbI_4]^{2-}$$

Tin(II) chloride is used quite often as a reducer:

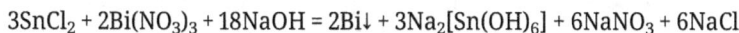

$$SnCl_2 + I_2 + 2HCl = SnCl_4 + 2HI$$

$$3SnCl_2 + 2Bi(NO_3)_3 + 18NaOH = 2Bi\downarrow + 3Na_2[Sn(OH)_6] + 6NaNO_3 + 6NaCl$$

19.4.2.5 Sulfides

Germanium, tin, and lead form stable sulfides with sulfur: GeS, SnS, PbS, as well as GeS_2 and SnS_2. Lead disulfide decomposes at the moment of production:

$$PbS_2 = PbS + S$$

Of the sulfides, the most stable are those in which the elements are in the oxidation state (+2), namely SnS (dark brown) and PbS (black). PbS is the least soluble compound of lead. It is obtained by precipitating soluble lead salts with hydrogen sulfide or soluble alkaline sulfides:

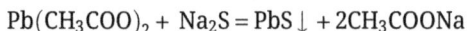

$$Pb(CH_3COO)_2 + Na_2S = PbS \downarrow + 2CH_3COONa$$

All sulfides interact with nitric acid:

$$3GeS + 10HNO_3 + H_2O = 3H_2GeO_3 + 3H_2SO_4 + 10NO$$

$$3PbS + 8HNO_3 = 3Pb(NO_3)_2 + 2NO + 3S + 4H_2O$$

The sulfides SnS and SnS$_2$ dissolve in concentrated hydrochloric acid. The monosulfide dissolves due to the removal of sulfide ions in the weak electrolyte, hydrogen sulfide:

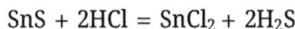

$$SnS + 2HCl = SnCl_2 + 2H_2S$$

When the disulfide is dissolved, disulfide ions are removed from the precipitate-saturated solution equilibrium. A process of disproportionation takes place:

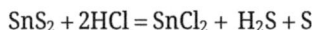

$$SnS_2 + 2HCl = SnCl_2 + H_2S + S$$

The GeS$_2$ and SnS$_2$ sulfides belong to the sulfo-anhydrides. Of interest is their dissolution in ammonium sulfide (NH$_4$)$_2$S and yellow ammonium sulfide (NH$_4$)$_2$S$_2$. A soluble salt of H$_2$SnS$_3$ acid, ammonium thiostannate, is obtained:

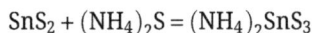

$$SnS_2 + (NH_4)_2S = (NH_4)_2SnS_3$$

19.4.3 Usage and biological role of germanium, tin, and lead

Germanium is used in microelectronics due to its semiconductor properties. The addition of germanium dioxide to glass makes it particularly transparent (optical glass). Tin is used as an anticorrosion coating. In significant quantities, tin is used to make food cans as well as in the production of various alloys: with copper (bronze), with zinc (brass), with antimony (babbitt), with zirconium (for nuclear reactors), with niobium (superconductors), etc. Tin dioxide is part of white enamels and glazes, tin dichloride is used as a reducer, and tin disulfide as a pigment. In industry, large quantities of lead are used for the preparation of various alloys, lead batteries, and for protection against radioactive radiation. The addition of lead(II) oxide to glass gives it the property of strongly refracting light (crystal glass). The double lead oxide Pb$_3$O$_4$, which is an orange powder, is used on a large scale as a surface coating on iron and steel, protecting them from corrosion. Mixed metal oxides, such as calcium-lead(IV) oxide $CaPbO_3$, are also used as effective protection of steel structures against salt water. White $PbCO_3.Pb(OH)_2$ is used in the production of white synthetic pigments. Tetraethyl lead $Pb(C_2H_5)_4$ is applied as an additive to gasoline, lead dioxide is used as an oxidizing agent, lead chromate $PbCrO_4$ as a pigment in yellow paints, and PbF_2 as a solid electrolyte.

Although it has not been sufficiently studied, germanium turns out to be an element with a wide spectrum of biological activity. It carries oxygen to the tissues of the body, increases the immune status, and exhibits antitumor activity. At present, interest in this element is growing. Numerous studies have found that various compounds of germanium have analgesic, hypotensive, antianemia, immunomodulatory, and bactericidal effects.

Tin is a micro-biogenic element that participates in the synthesis of some nucleic acids. Metallic tin is not toxic, and its soluble compounds show low toxicity. The exceptions are SnH_4, $Sn(CH_3)_4$, and $Sn(C_2H_5)_4$, which affect the nervous system.

Lead is one of the most dangerous elements for living organisms. Once in the body, lead is difficult to emit, accumulates in it, and its toxic dose is very low. It accumulates mainly in the skeletal system, where it displaces calcium from the bones. Both lead and its compounds are toxic. In case of poisoning, lead affects the nervous, digestive, and cardiovascular systems. Chronic poisoning is known as the disease "saturnism." Some lead compounds (PbO and $Pb(CH_3COO)_2$) have been used since ancient times. They were applied externally as compresses and patches with astringent action in purulent-inflammatory skin diseases.

The elements of the IVA group differ significantly from each other in terms of their content in the body and their bioactivity, starting from the first organogen, carbon, which plays a central role in living organisms, through the vital biological functions of the microelements silicon and germanium, to the serious toxicity of tin and lead. As a rule, the toxicity of elements and their inorganic and organometallic compounds rises with an increase in the atomic mass of the elements.

Chapter 20
Fifth main group of the periodic table

20.1 General characteristics of the elements of VA group

The fifth main group includes the p-elements nitrogen (N), phosphorus (P), arsenic (As), antimony (Sb), and bismuth (Bi). The main characteristics of atoms of the elements of VA group are presented in Table 20.1.

Table 20.1: Characteristics of atoms of elements of IA group.

Element	N	P	As	Sb	Bi
Valence electrons	$2s^2 2p^3$	$3s^2 3p^3$	$4s^2 4p^3$	$5s^2 5p^3$	$6s^2 6p^3$
Radius of the atom (nm)	0.071	0.130	0.148	0.161	0.182
Ionization potential (eV)	14.5	10.5	9.8	8.6	7.3
Electronegativity	3.0	2.1	2.0	1.9	1.9

As can be seen from the data, the radii of the atoms of the elements in the group increase, and the ionization potentials and electronegativity decrease. Therefore, the nonmetallic properties are weakened and the metallic properties are enhanced in the elements of the group: nitrogen, phosphorus, and arsenic are nonmetals; antimony is a highly amphoteric metal; and bismuth is a metal with very weak amphoteric characteristics. The chemical properties of the p-elements of group V gradually confirm their covalent character in comparison with the previous subgroups. The tendency to form anionic complexes is much more pronounced. The ability to form cations is characteristic mainly for bismuth, but is less pronounced in antimony and practically does not manifest itself in other p-elements of the group. The chemical bonds in nitrogen and phosphorus compounds are covalent. They are of significant polarity in arsenic and antimony, and are ionic in bismuth. Despite the fact that these elements are complete electronic analogues, nitrogen and phosphorus differ significantly from other elements in the properties of the elements and their chemical compounds. It should also be noted that two of the most different nonmetals in terms of reactivity are located in the same group: active phosphorus and inactive nitrogen. Therefore, nitrogen and phosphorus are usually considered separately, and arsenic, antimony, and bismuth are considered together.

Nitrogen is a gas, and the rest are solids. Phosphorus is soft, without metallic luster. As and Sb have an increased hardness, and Sb also has a metallic luster. N and P do not conduct electric current and form acidic oxides, so they are definitely classified as nonmetals. However, in the case of As, there are differences in its allotropic forms, one of which has typical metallic characteristics, while the second exhibits nonmetal-

https://doi.org/10.1515/9783111712246-020

lic ratios quite similar to those of phosphorus. Sb and Bi are also almost on the metal-nonmetal border, as is As. Their electrical resistances are much higher than those of metals, such as Al (2.8 $\mu\Omega$ cm) and even higher than those of typical "weak" metals, such as lead (22 $\mu\Omega$ cm). In general, these two elements are categorized as metals. Almost entirely, all three boundary elements form covalent compounds. Melting and boiling points are a good indicator for defining the boundary between metallic and nonmetallic relations (Table 20.2). In the fifth group, these parameters increase with the increase in the sequence number in the group, with the exception of some discrepancies in the melting temperatures between Sb and Bi, showing that the light members of V group follow the trend of typical nonmetals, and only Bi exhibits metallic properties.

Table 20.2: Melting and boiling points of elements of IA group.

Element	T_m (°C)	T_b (°C)
N	−210	−196
P	44	281
As	615 subl	615 subl
Sb	631	1,387
Bi	271	1,564

20.2 Nitrogen

20.2.1 Brief description

The nitrogen atom has an electronic configuration of $[He]2s^2 2p^3$. Nitrogen exhibits all oxidation states from −3 to +5. It is a typical nonmetal and has only one allotropic form: N_2. It is found in the Earth's atmosphere and, together with oxygen, is a major component of air, which consists of 78% N_2 and 21% O_2. It is a colorless, odorless gas that does not support combustion. Apart from its role in the nitrogen cycle, nitrogen has a very important place as an inert buffer for the chemically active gas O_2 in the atmosphere. Without N_2, any spark in the atmosphere would lead to a huge fire. Nitrogen is used as an inert medium in many chemical processes. Liquid nitrogen is used to maintain low temperatures in cryostatic and vacuum installations, a property due to its difficult liquefaction (at −196 °C). However, the main area of application of N_2 is in the synthesis of NH_3, nitric acid, artificial fertilizers, and explosives.

Nitrogen is one of the six main biogenic elements that make up living organisms. It is mainly part of proteins. It is vital for the growth of living organisms, as it participates in the composition of some nutrients. Under normal conditions, nitrogen does not have a physiological effect on the human body. However, at increased pressure (e.g., in divers), nitrogen penetrates into the tissues and the so-called nitrogen narcosis occurs. That is why divers' spacesuits today are not supplied with air, but a helium-oxygen mixture.

The diatomic molecule N_2 has a triple bond. The high strength of molecular nitrogen is the reason for its weak chemical activity. Only with some active metals, for example, with lithium, nitrogen interacts as an oxidizing agent at low temperatures, forming nitrides:

$$6Li + N_2 \rightarrow 2Li_3N$$

Other metals and hydrogen are oxidized by nitrogen at high temperatures. In terms of electronegativity, nitrogen is second only to fluorine and oxygen. In the reaction with oxygen, nitrogen is a reducer. The interaction of these elements with the formation of NO occurs noticeably only at 4,000 °C. The activating energy of this process is the highest known: 540 kJ/mol.

It is interesting to trace the differences between the first two members of the group: nitrogen and phosphorus (Table 20.3). Nitrogen in its highest oxidation state is a strong oxidizing agent in an acidic solution, while the highest oxidation state of phosphorus appears to be quite stable. In fact, unlike nitrogen, in its highest oxidation state, phosphorus is the most stable, while in its lowest oxidation state it is thermodynamically the most unstable. When reviewing the energies of the bonds, it can be seen why different types of bonds are preferred for the two elements. The nitrogen molecule N_2 is the stable form of the element, and this is manifested in both chemical reactions and nitrogen-containing compounds. For this reason, triple bonds are formed much more often than single (or double) bonds. With phosphorus, the difference in the energies of the single and triple bonds is much smaller. Thus, phosphorus contains groups of singly bonded phosphorus atoms, and the strong P–O bond becomes dominant for phosphorus chemistry. Elemental P reacts energetically with oxygen and forms oxides, while nitrogen is very stable in oxidation.

Table 20.3: Differences in chemical bonds of nitrogen and phosphorus.

Nitrogen bonds	E_{bond} (kJ/mol)	Phosphorus bonds	E_{bond} (kJ/mol)
N–N	247	P–P	200
N≡N	942	P≡P	481
N–O	201	P–O	335

The energy of the nitrogen-nitrogen triple bond is even greater than that of the carbon-carbon triple bond. The large difference between the energies of the N≡N and N–N bonds (about 700 kJ/mol) causes the N≡N bonds to be preferred in nitrogen reactions rather than the formation of chains of single nitrogen-nitrogen bonds, as observed in the chemical compounds of carbon. In addition, the fact that N_2 is a gas means that the entropy factor also favors the formation of the N_2 molecule in chemical reactions. This difference in the behavior of nitrogen and carbon can be traced by comparing the combustion of hydrazine (N_2H_4) with that of ethene (C_2H_4). When hydrazine is burned, N_2 is produced, and when ethene is burned, carbon dioxide is released:

$$N_2H_4 + O_2 \rightarrow N_2 + 2H_2O$$

$$C_2H_4 + 3O_2 \rightarrow 2CO_2 + 2H_2O$$

In the fifth and sixth main groups, the second members, P and S, are prone to catenation.

20.2.2 Synthesis

In industry, nitrogen is obtained by fractional distillation of liquefied air, while in laboratory practice, thermal decomposition reactions of some substances such as ammonium salts (dichromate and nitrite) or sodium azide are used:

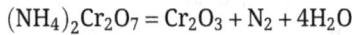

$$(NH_4)_2Cr_2O_7 = Cr_2O_3 + N_2 + 4H_2O$$

$$NH_4NO_2 = N_2 + 2H_2O$$

$$2NaN_3 = 2Na + 3N_2$$

20.2.3 Biological role

Nitrogen is present in living organisms in the form of a variety of organic compounds such as amino acids, peptides, and purine bases, as well as in the form of free N_2 coming in with the inhaled air. The nitrogen cycle in nature closely connects the geosphere and the biosphere, confirming their unity. There are many bacteria capable of easily converting some nitrogen compounds into others, with a change in the oxidation state of nitrogen. If, in industry, the synthesis of NH_3 is carried out under difficult conditions, in the biosphere the binding of atmospheric N_2 and its conversion to NH_3 is much easier – enzymatically with the participation of nitrogenase:

$$N_2 + 16ATP + 8e + 8H^+ \rightarrow 2NH_3 + 16ADP + 16[P \text{ in inorganic phosphates}] + H_2$$

where ATP and ADP are adenosine triphosphate and adenosine diphosphate, respectively, and the initial ATP is considered to be in the form of a complex with Mg.

The microorganisms involved in this reaction are present in some root crops and also in blue-green algae. The enzyme nitrogenase, which contains proteins, as well as Mo and Fe, is active only under anaerobic conditions. Studies have shown that when N_2 is reduced to NH_3, NH=NH and NH_2–NH_2 are not formed. This suggests that the enzyme probably has two active centers: on one of them the nitrogen molecule breaks down, and on the other the H atom is coordinated.

Other mutual transformations of nitrogen compounds also occur in nature: nitrification or oxidation of NH_3 to NO_2, as well as reduction of nitrate ions from nitrogen fertilizers under the action of plant enzymes or anaerobic bacteria to NO_2 or even to NH_3.

Inorganic nitrogen compounds are generally toxic, with the exception of N_2 and N_2O in small quantities. A huge amount of various oxides (NO_x) and other nitrogen compounds are released into the atmosphere annually. The NO molecule, despite its difficulty in forming from the corresponding simple substances, is present in the atmosphere in enormous quantities. It is estimated that 7×10^7 tons of atmospheric N_2 react with O_2 annually as a result of high-temperature processes, such as the combustion of fuels in industry and transport. It has been proven that nitrogen oxides, as well as ozone, are capable of interacting with the products of incomplete combustion of fuel to form highly toxic peroxonitrates. Under the action of solar radiation in the upper layers of the atmosphere, photochemical reactions take place with the participation of NO_x, which are catalyzed by the solid dust particles contained there. In the human body, an amount of ~100 mg/day of NO is formed by arginine. NO molecules are able to penetrate the cell walls of blood vessels and regulate blood circulation. In addition, NO controls insulin secretion, renal filtration, reparative processes in tissues, etc. Therefore, NO has a double effect, manifesting itself both as toxic and as undoubtedly beneficial. When taking the common cardiological drug nitroglycerin, it is hydrolyzed with the formation of a nitrate ion, which is reduced by the iron of hemoglobin to NO, and then it is NO that causes relaxation of the smooth muscles of the vessels. The other nitrogen oxides such as NO_2 and N_2O_3 are highly toxic and cause suffocation and inflammation of the lungs. Nitrite ion (NO_2^-) is especially toxic, as it oxidizes methemoglobin and disrupts the process of O_2 transport in the body. In addition, the nitrite ion forms the carcinogenic nitrosamine in the stomach. $NaNO_2$ has been used in the past as a vasodilator for angina pectoris and spasms of the brain vessels. Recently, $NaNO_2$ has not been recommended due to its undoubted toxicity, and has been replaced by nitroglycerin or nitrosorbitol, which do not have such side effects. Inhalation of ammonia (NH_3) vapors in large quantities is harmful, because it creates a strong base environment on the surface of mucous membranes of the larynx and lungs, which causes their irritation and inflammation. In addition, small molecules of NH_3 easily penetrate cell membranes and become competitors to many ligands when coordinating with metal ions in the body.

20.2.4 Chemical compounds

20.2.4.1 Hydrides

The most important nitrogen hydride is ammonia, but in addition there are two other hydrides such as hydrazine (N_2H_4) and hydronitric acid (HN_3), as well as hydroxyl-amine (NH_2OH).

Ammonia (NH_3) is produced by the interaction of N_2 and H_2 in the equation:

$$N_2 + 3H_2 \rightleftarrows 2NH_3, \quad \Delta H^\circ = -46 \text{ kJ/mol}$$

The reaction is reversible and exothermic. In accordance with the Le Chatelier principle, the straight reaction is favored at high pressure and low temperature. The yield of ammonia increases with an increase in pressure; therefore, the process is carried out in steel columns under a pressure of 150–1,000 atm. With an increase in temperature, the yield of ammonia decreases. At low temperatures, however, the reaction rate is low, and very high pressure is dangerous for installations. Thus, in order to maintain a high reaction rate, industrial synthesis of ammonia is carried out under optimal conditions: a temperature of about 450 °C and a pressure of 300 atm in the presence of a catalyst (iron is activated with potassium and aluminum oxides). The practical yield under these conditions is about 22%.

Ammonia is obtained in laboratory conditions by mixing ammonium salts with concentrated solutions of bases; for example, ammonium sulfate and calcium hydroxide:

$$(NH_4)_2SO_4 + Ca(OH)_2 \rightarrow CaSO_4 + 2H_2O + 2NH_3$$

Other laboratory methods for obtaining ammonia are thermal dissociation of ammonium salts or interaction of nitrides with water:

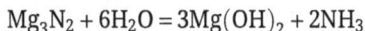

$$NH_4Cl = NH_3 + HCl$$

$$Mg_3N_2 + 6H_2O = 3Mg(OH)_2 + 2NH_3$$

Ammonia is a colorless poisonous gas with a very strong characteristic odor. It is the only gas that exhibits basic properties. In the ammonia molecule, nitrogen is in sp^3 hybridization, in which one hybrid orbital is nonbinding. Therefore, the ammonia molecule has a pyramidal shape and is polar. Ammonia condenses to a liquid state at −35 °C. This boiling point is much higher than that of phosphine, PH_3 (−134 °C), and is precisely due to the fact that ammonia molecules form strong hydrogen bonds with each other. Liquid ammonia is a good polar solvent.

Due to the presence of a free electron pair in the molecule, ammonia behaves like a strong Lewis base and combines by a donor-acceptor mechanism with the cations of many metals to form complex compounds: $[Zn(NH_3)_4]Cl_2$, $[Cu(NH_3)_4]SO_4$, and $[Cd(NH_3)_6]$ $(OH)_2$. For example, NH_3 replaces water molecules as a stronger Lewis base than water:

$$[Ni(OH_2)_6]^{2+} + 6NH_3 \rightarrow [Ni(NH_3)_6]^{2+} + 6H_2O$$

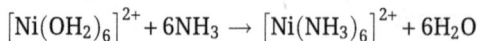

One of the "classic" Lewis acid-base reactions is that between the gaseous electron-deficient boron trifluoride and ammonia to obtain a white solid, in whose molecule the ammonia electron pair binds by a donor-acceptor mechanism to the free atomic orbital in boron:

$$NH_3 + BF_3 \rightarrow F_3B{:}NH_3$$

In redox reactions, ammonia appears as a reducer, because nitrogen is in its extreme negative oxidation state of −3. It burns in an oxygen environment:

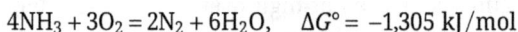

$$4NH_3 + 3O_2 = 2N_2 + 6H_2O, \quad \Delta G° = -1{,}305 \text{ kJ/mol}$$

Complete combustion is thermodynamically less favorable, but in the presence of a platinum catalyst, this mechanism is kinetically preferred, that is, the activating energy of this alternative mechanism is lower than that of combustion to nitrogen:

$$4NH_3 + 5O_2 \rightarrow 4NO + 6H_2O, \quad \Delta G° = -1{,}132 \text{ kJ/mol}$$

Ammonia reduces metals from their oxides when heated and discolors a solution of potassium permanganate:

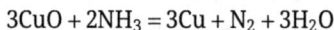

$$3CuO + 2NH_3 = 3Cu + N_2 + 3H_2O$$

$$10NH_3 + 6KMnO_4 + 9H_2SO_4 = 5N_2 + 6MnSO_4 + 3K_2SO_4 + 24H_2O$$

Ammonia also acts as a reducing agent in reactions with chlorine. In an excess of ammonia, nitrogen is formed, and the excess of ammonia further reacts with HCl to white solid NH_4Cl:

$$2NH_3 + 3Cl_2 \rightarrow N_2 + 6HCl$$

$$HCl + NH_3 \rightarrow NH_4Cl$$

In excess of chlorine, a very different reaction occurs. In this case, the product is nitrogen trichloride – a colorless, explosive, oily liquid:

$$NH_3 + 3Cl_2 \rightarrow 3HCl + NCl_3$$

Ammonia dissolves easily in water: at room temperature, above 50 g of ammonia dissolves in 100 g of water, forming a solution with a density of 0.880 g/mL. When dissolved in water, the NH_3 and H_2O molecules join with hydrogen bonds to ammonia hydrate ($NH_3{\cdot}H_2O$), which is partially converted to ammonium hydroxide dissociating as a weak base:

$$NH_3 + H_2O \rightleftarrows NH_3.H_2O \rightleftarrows NH_4OH \rightleftarrows NH_4^+ + OH^- \quad K = 1{,}8.10^{-5}$$

However, in practice, whole NH_4OH molecules have not been isolated in aqueous solutions, which does not mean that NH_4OH is a strong base. The relatively low concentration of hydroxide ions (respectively, pH) of its aqueous solutions (compared to that of strong alkali and alkaline earth hydroxides) is due to the lower degree of interaction of ammonia with water.

In a basic medium, the equilibrium of the above reactions shifts to the left – ammonia practically does not dissolve in bases. In an acidic environment, on the contrary, the equilibrium shifts to the right, due to the binding of OH^- with hydrogen cations – ammonia dissolves in acids with the formation of salts containing ammonium cation (NH_4^+):

$$NH_3 + HCl = NH_4Cl$$

$$2NH_3 + H_2SO_4 = (NH_4)_2SO_4$$

Ammonia forms ammonium salts with almost all acids. All ammonium salts are well soluble in water. When they are dissolved, partial (reversible) hydrolysis takes place with the formation of acidic solutions:

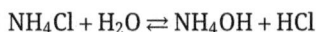

$$NH_4Cl + H_2O \rightleftharpoons NH_4OH + HCl$$

The decomposition of ammonium salts when heated can proceed without change in oxidation states:

$$NH_4Cl = NH_3 + HCl$$

$$(NH_4)_2CO_3 = 2NH_3 + CO_2 + H_2O$$

or by an intramolecular redox mechanism:

$$(NH_4)_2Cr_2O_7 = N_2 + Cr_2O_3 + 4H_2O$$

$$NH_4NO_3 = N_2O + 2H_2O$$

Liquid ammonia is a good solvent of alkali, alkaline earth, and other metals, forming blue-colored electrically conductive solutions with them. The autodissociation of ammonia is similar to that of water, but occurs to a negligible extent:

$$2NH_3 \rightleftharpoons NH_4^+ + NH_2^-$$

Ammonia is used to produce nitric acid, artificial fertilizers, and sodium carbonate. Aqueous solution of ammonia (ammonia water) is used as a liquid fertilizer, as well as in various syntheses and in medicine. Liquid ammonia is used as a coolant in industrial refrigerators. Ammonium salts such as NH_4NO_3, $(NH_4)_2SO_4$, and $(NH_4)_2HPO_4$ are artificial fertilizers.

When ammonia reacts with metals, amides, imides, and nitrides are formed. *Amides* are products of the substitution of one hydrogen atom in the ammonia molecule ($NaNH_2$), *imides* – of two (Na_2NH), and *nitrides* – of three (Na_3N). In the reactions of formation of these substances, ammonia exhibits oxidation ability due to hydrogen in the oxidation state of +1:

$$2Na + 2NH_3 = 2NaNH_2 + H_2$$

Nitrides, like carbides, are binary compounds of nitrogen with metals, as well as with nonmetals with low electronegativity (with the exception of hydrogen). According to the type of chemical bond, they are divided into ionic, covalent, and metal-like.

Ionic nitrides are nitrogen compounds with metals of the first and second groups of the periodic table. They have an ion crystal lattice and relatively low thermal stability. They interact with acids and hydrolyze in aqueous solution:

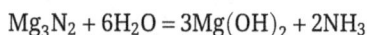

$$Mg_3N_2 + 6H_2O = 3Mg(OH)_2 + 2NH_3$$

The nitrides of Li (IA group) and Mg (IIA group) can be obtained at ordinary temperature:

$$3Mg + N_2 = Mg_3N_2$$

The thermal effects of the chemical interactions of N_2 with H_2, O_2, and metals show that nitrogen has greater chemical activity toward hydrogen and metals than toward oxygen.

The nitrides of the p- and d-elements are of the greatest importance in the technique. They are produced by the interaction of metals with nitrogen when heated.

There are several known *covalent nitrides* of p-elements. Covalent nitrides include nitrides of Si, C, and B. They are distinguished by high hardness, heat resistance, and chemical inertness.

Metal nitrides are nitrogen compounds with most of the d-elements. Their composition varies widely and does not correspond to the usual concepts of valence. In some properties (metallic luster and electrical conductivity), they are close to metals. These nitrides are characterized by high hardness and high melting points, similar to the corresponding carbides.

Nitrides are compounds with great hardness, durability, and low chemical activity. In general, they are stable in water, acids, and alkalis. Therefore, to increase the hardness and stability of metal products, their surface is treated with nitrogen. In engineering, the formation of nitrides in the form of a thin layer on the surface of metals and alloys is called nitriding.

Hydrazine (N_2H_4) is a colorless liquid substance with high toxicity. The oxidation state of nitrogen in hydrazine is intermediate, equal to −2. Its molecules have an asymmetrical structure, which explains their polarity. The presence of free electron pairs in nitrogen atoms explains its donor properties.

Hydrazine dissolves in water forming hydrates $N_2H_4 \cdot H_2O$ and $N_2H_4 \cdot 2H_2O$ (with hydrogen bonds), which partially turn into the very weak base – hydrazinium hydroxide:

$$N_2H_4 + H_2O \rightleftharpoons N_2H_4.H_2O \rightleftharpoons N_2H_5OH \rightleftharpoons N_2H_5^+ + OH^- \quad K = 8.5.10^{-7}$$

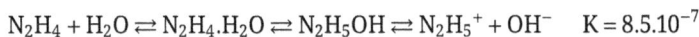

When reacting with acids, hydrazine forms hydrazinium salts. There are two types of hydrazinium salts, in which hydrazine is either monoprotonated or diprotonated:

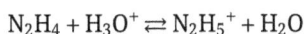

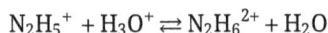

$$N_2H_4 + H_3O^+ \rightleftharpoons N_2H_5^+ + H_2O$$

$$N_2H_5^+ + H_3O^+ \rightleftharpoons N_2H_6^{2+} + H_2O$$

Hydrazine is a stronger reducer than ammonia. It burns in the air, and in solutions it is oxidized even by weak oxidizing agents:

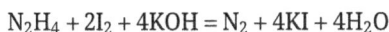

$$N_2H_4 + O_2 = N_2 + 2H_2O$$

$$N_2H_4 + 2I_2 + 4KOH = N_2 + 4KI + 4H_2O$$

Due to the fact that the oxidation state of nitrogen in hydrazine is intermediate, hydrazine can also occur as an oxidizing agent:

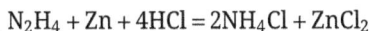

$$N_2H_4 + Zn + 4HCl = 2NH_4Cl + ZnCl_2$$

It is assumed that in this reaction, hydrazine is reduced not by zinc, but by atomic hydrogen at the moment of its separation when zinc reacts with acid.

Hydrazine is obtained by the oxidation of dissolved ammonia with sodium hypochlorite (Raschig method):

$$2NH_3 + NaClO = N_2H_4 + NaCl + H_2O$$

It is used as rocket fuel, in organic syntheses, in the production of plastics, rubber, explosives, and insecticides. Hydrazine and all its derivatives are highly toxic substances.

Hydronitric acid (HN_3) is a colorless liquid that is quite different from other nitrogen hydrides. It exhibits acidic properties, with a pK_a value similar to that of acetic acid:

$$HN_3 + H_2O \rightleftharpoons H_3O^+ + N_3^-$$

The compound has an irritating odor and is extremely toxic. It is a highly explosive liquid, easily decomposing into hydrogen and nitrogen:

$$2HN_3 \rightarrow H_2 + 3N_2$$

The three nitrogen atoms in the hydronitric acid molecule are linearly arranged, and with hydrogen they form an angle of 110°. The order of the two nitrogen-nitrogen bonds is about 1½ and 2½, respectively. The bonding can be represented by two resonant structures, one of which contains two double N=N bonds, and the other one N–N bond and one N≡N bond.

Hydronitric acid is produced by interaction of hydrazine with nitric acid:

$$N_2H_4 + HNO_2 = HN_3 + 2H_2O$$

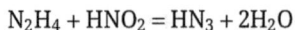

The salts of hydronitric acid are called *azides*. The azide ion (N_3^-) is isoelectronic with carbon dioxide and is assumed to have the same electronic structure. Chemically, the azide ion behaves like a pseudo-halide ion. When mixing a soluble azide with a solution containing silver ions, a precipitate of silver azide (AgN_3) is obtained, analogous to silver chloride (AgCl). Azide ions form complex ions, similar to those of halide ions, such as $[Sn(N_3)_6]^{2-}$, analogous to $[SnCl_6]^{2-}$.

Of the known azides, lead(II) azide has important applications as a detonator. It is a relatively safe compound, but on impact it decomposes explosively like dynamite:

$$Pb(N_3)_2 \rightarrow Pb + 3N_2$$

Hydroxylamine (NH_2OH) is a solid with a molecular crystal lattice. Its molecules can be considered as derivatives of NH_3, in which one hydrogen atom is replaced by a hydroxyl group.

Hydroxylamine dissolves in water, forming a monohydrate that dissociates as a weak base:

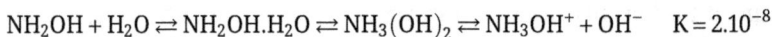

$$NH_2OH + H_2O \rightleftharpoons NH_2OH.H_2O \rightleftharpoons NH_3(OH)_2 \rightleftharpoons NH_3OH^+ + OH^- \quad K = 2.10^{-8}$$

With acids, hydroxylamine forms salts such as hydroxylamine chloride, $(NH_3OH)Cl$, and hydroxylamine hydroxysulfate, $(NH_3OH)HSO_4$. Their formulas can be also written as follows: $NH_2OH \cdot HCl$ and $NH_2OH \cdot H_2SO_4$.

Hydroxylamine and its conjugate acid, hydroxylammonium ion (NH_3OH^+), can easily disproportionate. Hydroxylamine disproportionates to N_2 and NH_3, while hydroxylammonium ion to nitrous oxide and ammonium cation:

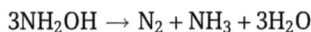

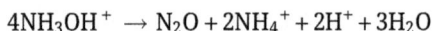

$$3NH_2OH \rightarrow N_2 + NH_3 + 3H_2O$$

$$4NH_3OH^+ \rightarrow N_2O + 2NH_4^+ + 2H^+ + 3H_2O$$

Hydroxylamine is a very strong reducer (stronger than NH_3 and N_2H_4), especially in alkaline environments. At the same time, in an acidic environment, it manifests itself as an oxidizing agent. The reduction properties of hydroxylamine can be illustrated by the reactions:

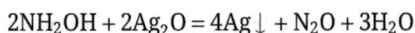

$$2NH_2OH + I_2 + 2KOH = N_2 + 2KI + 4H_2O$$

$$2NH_2OH + 2Ag_2O = 4Ag \downarrow + N_2O + 3H_2O$$

and oxidizing – with the reaction:

$$2NH_2OH + 4FeSO_4 + 3H_2SO_4 = (NH_4)_2SO_4 + 2Fe_2(SO_4)_3 + 2H_2O$$

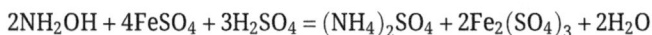

Hydroxylamine is obtained by various methods, one of which is the reduction of $NaNO_2$ with sulfur dioxide:

$$NaNO_2 + 2SO_2 + 3H_2O = NH_2OH + NaHSO_4 + H_2SO_4$$

It is used in the production of caprolactam, from which artificial fibers are produced, as well as in analytical chemistry.

20.2.4.2 Halides

Nitrogen forms many compounds with the halogen elements. All halides of the NX_3 type are known, of which chloride and fluoride are the most important. Nitrogen tri-chloride is a typical covalent chloride, a yellow viscous liquid. In its pure form, the compound is highly explosive due to its positive free energy of formation. It reacts with water forming NH_3 and $HClO$:

$$NCl_3 + 3H_2O \rightarrow NH_3 + 3HClO$$

Nitrogen trifluoride is thermodynamically stable with low chemical activity. It is a colorless, odorless gas. Nitrogen trifluoride does not react with water. Such stability and low reactivity are quite common in covalent fluorides. Despite the free electron pair of nitrogen, similar to that of ammonia, nitrogen trifluoride is a weak Lewis base. The F–N–F angle in nitrogen trifluoride (102°) is significantly smaller than a typical tetrahedral angle. One explanation for its behavior as a weak Lewis basis and a smaller than 109½° angle is that the bond is predominantly p-orbital in nature (for which 90° would be the most optimal angle), and that the free electron pair in nitrogen is located on the s-orbital rather than on a hybrid sp^3-orbital.

There is an unusual reaction in which nitrogen trifluoride acts as a Lewis base: at a very low temperature with oxygen, it forms the stable compound nitrogen fluoride oxide (NF_3O) under conditions of electrical discharge:

$$2NF_3 + O_2 \rightarrow 2NF_3O$$

Nitrogen fluoride oxide is often used as a classic example of a compound with a coordination covalent bond between the nitrogen and oxygen atoms.

In the group of nitrogen fluorides, in addition to NF_3, there are other compounds of different composition, such as N_2F_2 and N_2F_4.

Nitrogen bromide and iodide are crystalline substances that exist as $NBr_3 \cdot 6NH_3$ and $NI_3 \cdot NH_3$. The stability of nitrogen halides decreases toward iodine.

20.2.4.3 Oxides

Nitrogen forms oxides with oxygen in all positive oxidation states from +1 to +5: N_2O, NO, N_2O_3, NO_2, and N_2O_5. In addition, there is also nitrogen trioxide (NO_3), commonly referred to as the nitrate radical, which is found in small amounts in the atmosphere. Each of the oxides is actually thermodynamically unstable in terms of its decomposition into starting elements, but they are all kinetically stable. With an increase in molecular mass, the physical state of oxides changes: N_2O and NO are gases, NO_2 and N_2O_3 are liquids, and N_2O_5 is a solid. For all oxides, the standard Gibbs energy of formation is positive, so they are obtained indirectly.

Nitrous(I) oxide (N_2O, nitrous oxide) is formed by thermal dissociation of ammonium nitrate, which occurs when heated smoothly to 170–230 °C:

$$NH_4NO_3 = N_2O + 2H_2O$$

The linear structure of nitrous oxide is explained by the sp hybridization of the central atom. It is isoelectronic with the carbon dioxide and the azide ion in hydronitric acid, with the difference that in nitrous oxide the atoms are arranged asymmetrically, with bond lengths N–N and N–O of 113 pm and 119 pm, respectively. The order of the two N–N and N–O bonds is 2½ and 1½, respectively.

Like hydronitric acid, nitrous oxide can be represented by two resonance structures, one containing N=O and N=N bonds, and the other containing N–O and N≡N bonds.

Nitrous(I) oxide is a colorless gas with a characteristic pleasant odor, the so-called. "laughing gas." It has a narcotic effect and is used in medicine for anesthesia in painful operations.

Nitrous oxide is a neutral oxide and does not form hydroxide with water. It is thermodynamically unstable, and at temperatures above 500 °C it decomposes into nitrogen and oxygen. It does not interact with acids, bases, and halogen elements. It is quite inert, although it is almost the only known gas, apart from oxygen, which supports combustion.

It exhibits the oxidizing ability when heated:

$$N_2O + H_2 = N_2 + H_2O$$

Magnesium burns with nitrous oxide to magnesium oxide and nitrogen:

$$N_2O + Mg \rightarrow MgO + N_2$$

Nitrous(II) oxide (NO) is produced by reacting 30% HNO_3 with Cu or other weak reducers:

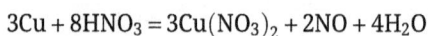

$$3Cu + 8HNO_3 = 3Cu(NO_3)_2 + 2NO + 4H_2O$$
$$3Ag + 4HNO_3 = 3AgNO_3 + NO + 2H_2O$$

Nitrous(II) oxide is also formed by the oxidation of ammonia with a platinum catalyst:

$$4NH_3 + 5O_2 = 4NO + 6H_2O$$

The direct interaction

$$N_2 + O_2 = 2NO$$

under normal conditions is thermodynamically impossible because the reaction is exothermic and is characterized by positive values of the Gibbs energy (86.6 kJ/mol). However, this reaction is accompanied by an increase in entropy ($\Delta S° = 34.7$ kJ/mol K), so when the temperature rises, it becomes reversible. For example, at 3,000 K, the equilibrium mixture already contains about 4% NO. The rapid cooling of the equilibrium mixture prevents the decay of NO. It forms in the atmosphere during storms and lightning. Small amounts of NO are formed in the cylinders of cars, which, together with the release of CO, determines the harmful emissions of exhaust gases.

One of the most interesting simple molecules (interpreted by the MOM) is that of nitric oxide, exhibiting paramagnetic properties. Its energy diagram of molecular orbitals resembles that of carbon monoxide, but with one extra unpaired electron occupying the antibonding orbital. Therefore, the bond order in the NO molecule is 2½. Molecules containing unpaired electrons are usually very active. When cooled, nitric oxide turns into a colorless liquid due to its tendency to form N_2O_2 dimers, in which the two nitrogen atoms are bonded by a single bond.

According to the MOM diagram, nitric oxide easily loses its electron from the antibonding orbital and forms the nitrosyl ion, NO^+, which is diamagnetic and has a shorter N–O bond length (106 pm) than that in the NO molecule (115 pm). This ion is isoelectronic with nitrogen and carbon monoxide molecules with triple bonds and it forms many analogous metal complexes.

Colorless nitric(II) oxide, like N_2O, is a non-salt-forming (neutral) oxide that does not interact with water and bases, and does not form acids and salts. Despite its low solubility in water, NO does not interact with it and is not an anhydride of any acid. In the air, it is spontaneously oxidized to nitrogen dioxide (NO_2):

$$2NO + O_2 \rightleftharpoons 2NO_2$$

When nitric acid reacts with most of the metals, it is difficult to define whether NO or NO_2 is released, since NO is spontaneously oxidized by oxygen in the air to NO_2. In this case, the concentration of the acid is also important. As a rule, when interacting with metals, more dilute solutions of nitric acid (less than 30–40%) release NO, and more concentrated ones release NO_2.

Nitric(II) oxide exhibits oxidizing ability when heated:

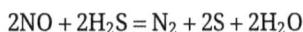

$$2NO + 2H_2S = N_2 + 2S + 2H_2O$$

There are known reactions in which nitric oxide exhibits reducing properties. For example, it is instantly oxidized by the oxygen of the air and by halogen elements:

$$2NO + O_2 = 2NO_2$$

$$2NO + Cl_2 = 2NOCl$$

The biological role of nitric oxide has been gaining more and more popularity in recent years. It is currently known that this small molecule plays a vital role in the organisms of all mammals. Organic compounds, such as nitroglycerin, were known to relieve angina, low blood pressure, and relax smooth muscle tissue. It has been found that nitric oxide plays an important role in dilating blood vessels; therefore, it is crucial for controlling blood pressure. An enzyme (nitric oxide synthase) is known, whose task is to produce nitric oxide. Nitric oxide deficiency is associated with high blood pressure, while septic shock, a common cause of death in intensive care units, is due to an excess of nitric oxide. Currently, numerous biochemical studies are being conducted to study the role of NO in the body. The influence of nitric oxide on other organs and functions (memory, stomach, kidneys, etc.) is also studied.

Dinitrogen trioxide (N_2O_3, nitrous trioxide), the most unstable of the nitrogen oxides, is a dark blue liquid that decomposes at temperatures above 230 °C. It is obtained by cooling an equivalent stoichiometric mixture of nitrous oxide and nitrogen dioxide:

$$NO + NO_2 \rightleftharpoons N_2O_3$$

Although dinitrogen trioxide contains two nitrogen atoms in +3 oxidation state, its structure is asymmetric. This indicates that this structure is built as a simple combination of two molecules with unpaired electrons (NO and NO_2). In confirmation of this, the length of N–N bond in N_2O_3 is unusually large (186 pm) compared to the length of the single bond in hydrazine (145 pm).

The bond length shows that one of the oxygen atoms is bonded to a nitrogen atom with a double bond, while each of the other two oxygen-nitrogen bonds has an order of about 1½. This value is an average between single and double bonds.

Nitrous trioxide is one of the acidic oxides of nitrogen. In fact, it is an acidic anhydride of nitric acid. Thus, when nitrous trioxide is mixed with water, nitrous acid (HNO_2) is formed, and in an excess of hydroxide ions in an alkaline medium, a nitrite ion is obtained:

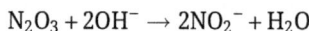

$$N_2O_3 + H_2O = 2HNO_2$$

$$N_2O_3 + 2OH^- \rightarrow 2NO_2^- + H_2O$$

Like any acid oxide, N_2O_3 interacts with alkaline bases and forms nitrites, salts of HNO_2:

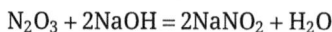

$$N_2O_3 + 2NaOH = 2NaNO_2 + H_2O$$

Nitrogen(IV) oxide (NO_2, nitrogen dioxide) is produced by the decomposition of $Cu(NO_3)_2$, $Pb(NO_3)_2$, and nitrates of other heavy metals when heated, a reaction in which a mixture of nitrogen dioxide and oxygen is obtained:

$$Cu(NO_3)_2 \rightarrow CuO + 2NO_2 + \tfrac{1}{2}O_2$$

$$2Pb(NO_3)_2 = 2PbO + 4NO_2 + O_2$$

Nitrogen dioxide is also produced by reactions of metals with concentrated nitric acid:

$$Cu + 4HNO_3 \rightarrow Cu(NO_3)_2 + 2H_2O + 2NO_2$$

As mentioned, NO_2 is also formed during the oxidation of nitric oxide:

$$2NO + O_2 \rightleftarrows 2NO_2$$

In NO_2 molecule, there is one unpaired electron at the nitrogen atom; therefore, when the temperature decreases, dimerization occurs, which presupposes the existence of the oxide N_2O_4. These two toxic oxides (NO_2 and N_2O_4) exist in a state of dynamic equilibrium. At low temperatures, the formation of colorless N_2O_4 is favored, while high temperatures lead to the formation of the brown gas nitrogen dioxide. At normal boiling point (-1 °C), this mixture contains 16% NO_2, which increases to 99% at 135 °C:

$$N_2O_4 \rightleftarrows 2NO_2$$

The nitrogen dioxide molecule is V-shaped with an O–N–O angle equal to 134°. This angle is slightly larger than the true trigonal-planar angle of 120°. Since nitrogen has one electron left, not an electron pair, the bonding angle is more open. The length of the oxygen-nitrogen bond shows an order of 1½, like that in the NO_2 portion of nitrous trioxide. The O–N–O angle in N_2O_4 molecule is almost identical to that in the nitrogen dioxide molecule.

Nitrogen dioxide has acidic properties. It manifests itself as a mixed anhydride and forms two acids at the same time with water: nitrous HNO_2 and nitric HNO_3:

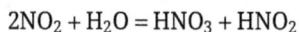

$$2NO_2 + H_2O = HNO_3 + HNO_2$$

This powerful corrosive mixture of oxoacids is formed when nitrogen dioxide from car gases reacts with rain or moisture. This is one of the main harmful components in urban rainfall.

Nitrogen dioxide is a powerful oxidizing agent. When heated, the next reaction occurs:

$$2NO_2 \rightarrow 2NO + O_2$$

It exhibits its oxidizing ability at high temperatures:

$$2NO_2 + 2C = N_2 + 2CO_2$$

NO and NO_2 are intermediates in the production of nitric acid.

Nitric(V) oxide (N_2O_5, nitrogen pentoxide), as nitric anhydride, is most commonly produced by dehydration of nitric acid using P_2O_5 or by reacting chlorine with silver nitrate:

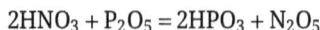

$$2HNO_3 + P_2O_5 = 2HPO_3 + N_2O_5$$

$$4AgNO_3 + 2Cl_2 = 2N_2O_5 + 4AgCl + O_2$$

In the liquid and gas phases, the molecule has a structure similar to those of the other oxides such as N_2O_3 and N_2O_4. The difference here is that one oxygen atom binds the two NO_2 units.

Solid N_2O_5 is an ionic substance, whose crystal structure actually consists of alternating nitrile cations NO_2^+ and nitrate anions NO_3^-.

N_2O_5 oxide is a volatile colorless substance in a solid state. At room temperature, it decomposes into NO_2 and O_2. As an acidic oxide, it reacts with water to form nitric acid:

$$N_2O_5 + H_2O \rightarrow 2HNO_3$$

All nitrogen oxides are oxidizing agents. The most powerful oxidizing agent of these is N_2O_5, which at room temperature ignites organic substances and oxidizes metals and other reducing agents:

$$N_2O_5 + SO_2 = N_2 + SO_3 + 2O_2$$

20.2.4.4 Nitrogen oxoacids and their salts

Nitrogen forms several acids, the most important being nitrous (HNO_2) and nitric (HNO_3) acids.

Nitrous acid is produced by the passage of nitrogen(III) oxide (N_2O_3) or an equimolar mixture of gases such as NO and NO_2 through water:

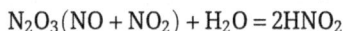

$$N_2O_3(NO + NO_2) + H_2O = 2HNO_2$$

and also, from its salts – *nitrites*:

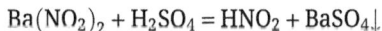

$$Ba(NO_2)_2 + H_2SO_4 = HNO_2 + BaSO_4\downarrow$$

To obtain nitrites, the reaction of the mixture of NO and NO_2 with solutions of bases and carbonates is used:

$$Ca(OH)_2 + NO + NO_2 = Ca(NO_2)_2 + H_2O$$

$$Na_2CO_3 + NO + NO_2 = 2NaNO_2 + CO_2$$

Nitrites are also formed when alkali metal nitrates are heated or reduced:

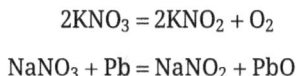

$$2KNO_3 = 2KNO_2 + O_2$$

$$NaNO_3 + Pb = NaNO_2 + PbO$$

The nitrite ion is V-shaped due to the free electron pair at the central nitrogen atom, with an angle of 115°, smaller than that of nitrogen dioxide (134°). The length of N–O bond is 124 pm, longer than that of NO_2 (120 pm), but still much shorter than the single N–O bond (143 pm).

Nitrous acid is very unstable, and in an anhydrous state, it is not obtained. It exists only in dilute aqueous solutions, where it dissociates as a weak acid:

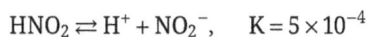

$$HNO_2 \rightleftharpoons H^+ + NO_2^-, \quad K = 5 \times 10^{-4}$$

Along with its acid dissociation, HNO_2 can also dissociate as a base to nitrosyl ions (NO^+):

$$HNO_2 \rightleftharpoons NO^+ + OH^-$$

Dissociation by basic type is intensified in the presence of strong acids, as evidenced by the formation of nitrosyl cation derivatives, for example, nitrosyl hydrogen sulfate ($NOHSO_4$), called nitrosyl-sulfuric acid.

Nitrous acid and its salts, in which the nitrogen atom is in an intermediate oxidation state (+3), exhibit redox duality. For example, in the reaction:

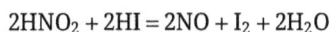

$$2HNO_2 + 2HI = 2NO + I_2 + 2H_2O$$

nitrous acid is an oxidizing agent, and in the reaction:

$$NaNO_2 + K_2Cr_2O_7 + 4H_2SO_4 = 3NaNO_3 + Cr_2(SO_4)_3 + K_2SO_4 + 4H_2O$$

its salt sodium nitrite is a reducer. Substances with redox duality are prone to disproportionation reactions. Nitrous acid and nitrites also disproportionate: HNO_2 – when its concentration in the solution increases, and nitrites – when heated:

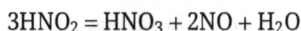

$$3HNO_2 = HNO_3 + 2NO + H_2O$$

$$4KNO_2 = 2KNO_3 + N_2O + K_2O$$

Nitrous acid is used as a reagent in organic chemistry. For example, when mixing nitrous acid with an organic amine (in this case, aniline $C_6H_5NH_2$), diazonium salts are obtained:

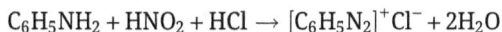

$$C_6H_5NH_2 + HNO_2 + HCl \rightarrow [C_6H_5N_2]^+Cl^- + 2H_2O$$

Diazonium salts are used for the synthesis of a wide range of organic compounds.

Nitric acid is obtained in industry according to the scheme:

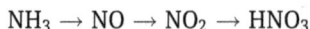

$$NH_3 \rightarrow NO \rightarrow NO_2 \rightarrow HNO_3$$

The Ostwald process for nitric acid synthesis is carried out in three stages. First, a mixture of NH_3 and O_2 (or air) passes through a platinum metal net. This is a very efficient exothermic process in which the contact time with the catalyst is limited to reduce unwanted side effects. It is performed at low pressure due to the entropy effect in order to shift the equilibrium to the products:

$$4NH_3 + 5O_2 \rightarrow 4NO + 6H_2O$$

In the second stage, additional oxygen is added to oxidize NO to nitrogen dioxide. It is carried out at low temperature and under pressure:

$$2NO + O_2 \rightarrow 2NO_2$$

The last stage of the process is the reaction of nitrogen dioxide with water. Under normal conditions, nitrogen dioxide reacts with water to form not only HNO_3 but also nitrous acid, since NO_2 is a mixed anhydride of the two acids:

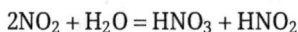

$$2NO_2 + H_2O = HNO_3 + HNO_2$$

It is possible that the process also proceeds to nitrogen(II) oxide:

$$3NO_2 + H_2O = 2HNO_3 + NO$$

Therefore, in industry, the process takes place in excess of air and high pressure (3–12 atm):

$$4NO_2 + O_2 + 2H_2O = 4HNO_3$$

The structure of nitric acid shows that the side O–N bonds are much shorter in length (121 pm) than the O–N bond associated with the hydrogen atom (141 pm). These small bond lengths correspond to double bonds between the nitrogen and the two oxygen atoms. Here, in addition to the σ-electrons, there are also four electrons involved in the O–N–O π-system (two on binding and two on antibonding orbitals), determining an order of the nitrogen-oxygen bond equal to 1½.

The structure of the nitrate ion is planar-trigonal with nitrogen-oxygen bonds of short length (122 pm) – slightly shorter than those in the nitrite ion. The nitrate ion is isoelectronic with the carbonate ion with a bond order equal to 1⅓.

Nitric acid (HNO_3) is produced and used in large quantities. In the anhydrous state, it is a liquid freezing at −41 °C; in the solid state, it is an ionic compound $NO_2^+OH^-$ – nitronium hydroxide:

$$4HNO_3 = 4NO_2 + O_2 + 2H_2O$$

Under normal conditions, nitric acid is a colorless liquid. The nitrogen dioxide released during its decomposition dissolves in the acid, giving it a yellow tint. The decomposition of concentrated HNO_3 is accelerated in the light, so it is stored in laboratories in dark glass containers.

HNO_3 dissolves indefinitely in water and dissociates almost irreversibly as a strong acid:

$$HNO_3 = H^+ + NO_3^-, \quad K = 23$$

In the presence of very strong acids (H_2SO_4 and $HClO_4$), nitric acid exhibits basic properties: it dissociates with the formation of the nitronium cation NO_2^+ and forms compounds such as NO_2HSO_4 (nitronium hydrogen sulfate) and NO_2ClO_4 (nitronium perchlorate):

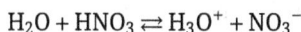

$$2HNO_3 \rightleftharpoons H_2NO_3^+ + NO_3^-$$

$$H_2NO_3^+ \rightleftharpoons H_2O + NO_2^+$$

$$H_2O + HNO_3 \rightleftharpoons H_3O^+ + NO_3^-$$

The general equation of the reaction is

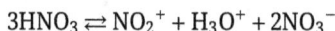

$$3HNO_3 \rightleftharpoons NO_2^+ + H_3O^+ + 2NO_3^-$$

Nitric acid, especially concentrated acid, is a strong oxidizing agent. It oxidizes all metals, except gold and platinum, as well as many nonmetals to their highest oxidation states. Concentrated nitric acid is actually an aqueous solution with a content of about 69% HNO_3 (corresponding to a concentration of 16 mol/L). This is the so-called "fuming nitric acid," which has strong oxidizing properties due to the oxygen released during its decomposition, and is a solution of nitric dioxide in nitric acid. Even when dilute, nitric acid is such a powerful oxidizing agent that it is rarely reduced to hydrogen when reacting with metals. In these interactions, nitric acid is usually reduced to NH_4^+, N_2, HNO_2, and nitrogen oxides (N_2O, NO, and NO_2). The formation of one or another product depends on the concentration of the acid, its purity, temperature, as well as the type of reducer. It is generally considered that during the oxidation of active metals, dilute nitric acid is reduced to NH_4^+ ions:

$$8Na + 10HNO_{3\,dil} = 8NaNO_3 + NH_4NO_3 + 3H_2O$$

When it interacts with metals of medium activity, it is reduced to N_2 and N_2O:

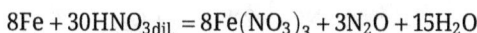

$$5Zn + 12HNO_{3\,dil.} = 5Zn(NO_3)_2 + N_2 + 6H_2O$$

$$8Fe + 30HNO_{3\,dil.} = 8Fe(NO_3)_3 + 3N_2O + 15H_2O$$

The dilute nitric acid when reacting with weakly active metals is reduced to NO:

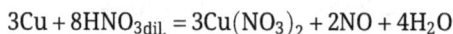

$$3Cu + 8HNO_{3\,dil.} = 3Cu(NO_3)_2 + 2NO + 4H_2O$$

In reactions of concentrated nitric acid with most metals and nonmetals, nitrogen dioxide is considered to be the main product of its reduction:

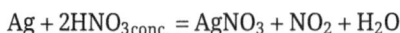

$$Ag + 2HNO_{3\,conc.} = AgNO_3 + NO_2 + H_2O$$

$$S + 6HNO_{3\,conc.} = H_2SO_4 + 6NO_2 + 2H_2O$$

It should be borne in mind that all these equations are conditional, because in reality mixtures of the products of the reduction are formed. The mechanism of reactions of nitric acid with metals and nonmetals is complex and has not been fully studied. A number of metals in concentrated nitric acid are passivated under normal conditions (Al, Cr, Ni, and Ti), but interact when heated or when the acid is diluted.

For the oxidation of gold, platinum, and other precious metals, a mixture of concentrated nitric and hydrochloric acid is used in a ratio of 1:3, called aqua regia. In this mixture, a very strong oxidizing agent is formed – atomic chlorine:

$$HNO_3 + 3HCl = NOCl + Cl_2 + 2H_2O$$

$$NOCl = NO + Cl^{\bullet}$$

The reduction activity of metals increases due to the formation of complexes, so the oxidation of precious metals occurs with the formation of complex acids:

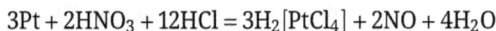

$$Au + HNO_3 + 4HCl = H[AuCl_4] + NO + 2H_2O$$

$$3Pt + 2HNO_3 + 12HCl = 3H_2[PtCl_4] + 2NO + 4H_2O$$

Nitric acid is produced in large quantities and is widely used as an oxidizing agent in rocket fuels and explosives, for metal processing, in organic synthesis, and for the production of nitrates.

Nitrates are formed when HNO_3 reacts with metals, oxides, hydroxides, and carbonates. All nitrates are soluble in water. Nitrates in solutions are weak oxidizing agents, but in melts or in mixtures with molten bases, their oxidation activity increases:

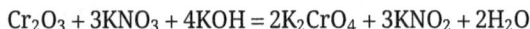

$$Cr_2O_3 + 3KNO_3 + 4KOH = 2K_2CrO_4 + 3KNO_2 + 2H_2O$$

When heated, nitrates of alkali and alkaline earth metals release O_2 and turn into nitrites:

$$2NaNO_3 = 2NaNO_2 + O_2$$

Nitrates of almost all other metals decompose into oxides:

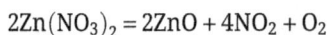

$$2Zn(NO_3)_2 = 2ZnO + 4NO_2 + O_2$$

and nitrates of some heavy metals (silver, mercury, etc.) decompose when heated to metals:

$$2AgNO_3 = 2Ag + 2NO_2 + O_2$$

A special position is occupied by ammonium nitrate, which up to a melting point (170 °C) is partially decomposed according to the equation:

$$NH_4NO_3 = NH_3 + HNO_3$$

at a temperature of 170–230 °C:

$$NH_4NO_3 = N_2O + 2H_2O$$

and at temperatures higher than 230 °C, it decomposes with an explosion according to the equation:

$$2NH_4NO_3 = 2N_2 + O_2 + 4H_2O$$

Nitrates are used in the chemical and technological processes of processing ores and minerals, in a number of chemical syntheses, in the production of explosives, etc. Many of the nitrates are used as nitrogen fertilizers.

20.3 Phosphorus

20.3.1 Brief description

Phosphorus is an element of great importance for living nature, due to the participation of its compounds in a number of life processes. It owes its name to the ability to emit light, that is, to phosphorescent (from the Greek words "phos" – light, and "foros" – bringing, or phosphorescent shine). This happens in the dark when white phosphorus is exposed to air. Phosphorus is a typical nonmetal, but its nonmetallic properties are less pronounced than those of nitrogen. Its most characteristic oxidation state is +5, but there are also compounds in which phosphorus is in +3 and +1 as well as in −3 and −2 oxidation states.

The chemistry of phosphorus is characterized by the stability of tetrahedral P_4 molecules. This tetrahedral unit is found not only in white phosphorus but also in the oxides P_4O_6 and P_4O_{10}, in the sulfide P_4S_{10}, as well as in the mixed oxide-sulfide compounds $P_4O_6S_4$ and $P_4S_6O_4$. Only at $T > 1{,}000$ °C phosphorus has diatomic molecules P_2, similar to N_2 molecules. In the solid state, several allotropic forms are characteristic, all of which give tetrahedral molecules, P_4, when melted.

20.3.2 Occurrence and synthesis

Phosphorus is part of many natural minerals – phosphorite, $Ca_3(PO_4)_2$; apatite, $3Ca_3$ $(PO_4)_2 \cdot CaX_2$ (X = F, Cl, Br, I); and also in the bones of living organisms.

Phosphorus is obtained by reducing phosphorite with coal (C) in the presence of sand (SiO_2) at 1,400–1,500 °C in electric furnaces and in an inert atmosphere by the reaction:

$$2Ca_3(PO_4)_2 + 10C + 6SiO_2 = P_4 + 6CaSiO_3 + 10CO$$

The process is carried out at a high temperature, and when the vapors condense, white phosphorus is obtained. The resulting white phosphorus is processed into red, which is used in the preparation of matches, for the production of semiconductor materials, phosphoric(V) oxide and phosphoric acid.

20.3.3 Forms of elemental phosphorus

In the free state, phosphorus forms more than 10 allotropic modifications, among which the most studied are white, red, and black phosphorus.

White phosphorus is a white, soft, waxy substance with a molecular crystal lattice consisting of tetrahedral P_4 molecules at the nodes.

It melts at 44 °C and boils at 257 °C. At 800 °C, the dissociation of tetraatomic molecules to diatomic P_2 begins, and at 1,200 °C, it dissociates to atoms. It is obtained by the condensation of phosphorus vapors. White phosphorus is a highly reactive substance, probably due to its structure, characterized by strongly tense bonds and weak dispersion forces between neighboring molecules. In air at room temperature, it oxidizes to P_4O_{10}, and at 40 °C, spontaneous ignition and powerful combustion of phosphorus occur:

$$P_4 + 5O_2 = P_4O_{10}$$

The study of the oxidation of white phosphorus led for the first time to the discovery of chain reactions by the Russian physicist Semenov, a Nobel Prize laureate.

Due to its high affinity for oxygen, white phosphorus must be stored under water. Although it is insoluble in hydrogen bond-forming solvents (such as water), it is extremely highly soluble in nonpolar solvents such as carbon disulfide (CS_2) and organic solvents. It burns in an environment of halogen elements, nitrogen dioxide, and CO_2, oxidized by nitric and concentrated sulfuric acid to H_3PO_4:

$$P_4 + 10Cl_2 = 4PCl_5$$
$$2P_4 + 10NO_2 = 2P_4O_{10} + 5N_2$$
$$P_4 + 10CO_2 = P_4O_{10} + CO$$
$$3P_4 + 20HNO_3 + 8H_2O = 12H_3PO_4 + 20NO$$

When heated in basic solutions, white phosphorus disproportionates:

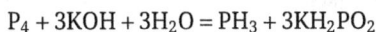

$$P_4 + 3KOH + 3H_2O = PH_3 + 3KH_2PO_2$$

White phosphorus is highly toxic, both when taken internally and when inhaled or through the skin. Its vapor is especially toxic.

Red phosphorus is formed when white phosphorus is heated to 300 °C in the absence of air. When exposed to ultraviolet radiation (from fluorescent lamps), white phosphorus slowly turns red. Red phosphorus is nontoxic, stable in air, and ignites above 400 °C. It is more inert than white phosphorus and is not as reactive. It is a red amorphous powder in which P_4 tetrahedra are chained together. Thus, red phosphorus is a polymer with less tight bonds than those in the white version. When heated, a crystalline modification of long polymer chains of phosphorus atoms (P_n) is obtained. The melting point of red phosphorus is about 600 °C, and at this temperature the polymer chains break to P_4, which are also contained in white phosphorus. Like any covalently bonded polymer, red phosphorus is insoluble in any type of solvent.

Black phosphorus is the most stable form under ordinary conditions. It is obtained from white phosphorus when heated to about 200 °C and at high pressure. It is curious that thermodynamically the most stable form of phosphorus – black phosphorus, is the most difficult to be synthesized – from white phosphorus at a pressure of about 1.2 GPa. Black phosphorus has a layered structure (analogous to that of graphite). It is the chemically most stable and inert modification of phosphorus. It is not toxic, exhibits the weakest activity, and strongly resembles graphite in appearance and properties. Black phosphorus burns at 290 °C.

20.3.4 Chemical properties

Phosphorus is an active element. The chemical properties of phosphorus are highly dependent on its modifications. Particularly chemically active is white phosphorus, which easily interacts with metals, hydrogen, oxygen, and halogen elements. Phosphorus does not interact with water and mineral acids. With metals, it forms compounds called phosphides, which are characterized by high hardness (Na_3P and Ca_3P_2). Some of them, such as those of Ga and In, are semiconductors. With oxygen it forms oxides such as P_2O_3 (P_4O_6) and P_2O_5 (P_4O_{10}), and with the halogen elements it forms halides with a composition of PX_3 and PX_5. With sulfur, it forms compounds with different compositions: P_4S_3, P_4S_5, and P_4S_7. Phosphorus does not react directly with carbon, but many compounds have been indirectly obtained, mainly with organic substances.

Phosphorus reacts with acids with oxidizing properties:

$$3P + 5HNO_3 + 2H_2O \rightarrow 3H_3PO_4 + 5NO$$

It dissolves into bases:

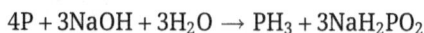

$$4P + 3NaOH + 3H_2O \rightarrow PH_3 + 3NaH_2PO_2$$

Phosphorus interacts with salts and exhibits reducing properties:

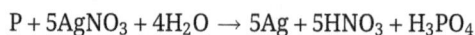

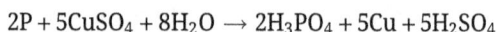

$$P + 5AgNO_3 + 4H_2O \rightarrow 5Ag + 5HNO_3 + H_3PO_4$$
$$2P + 5CuSO_4 + 8H_2O \rightarrow 2H_3PO_4 + 5Cu + 5H_2SO_4$$

20.3.5 Use and biological role

Phosphorus is an important vital element for animals and plants. All green parts of plants contain complex phosphorus compounds. In animals and humans, P is concentrated in the teeth, bones, nerves, and brain tissues, where a part of complex organic compounds play an important biological role. According to its content in the human body (~1%), it occupies an intermediate position between macro- and microelements. Phosphorus is one of the organogenic elements and plays an extremely important role in metabolism. More than 86% of phosphorus is included in the composition of hard tissues in animals in the form of calcium orthophosphate ($Ca_3(PO_4)_2$), hydroxyapatite ($3Ca_3(PO_4)_2 \cdot Ca(OH)_2$), and fluorapatite.

Inorganic phosphorus is a component of the buffer system of the blood, as well as other biological fluids. It ensures the maintenance of the alkaline-acid balance. The phosphate buffer system is contained both in the blood and in the cellular fluid of other tissues, especially in kidneys. In cells, it is represented by KH_2PO_4 and K_2HPO_4, and in the blood plasma and intercellular space by NaH_2PO_4 and Na_2HPO_4 with pH = 7.4. The main role in the mechanism of action of this system is played by the $H_2PO_4^-$ ion. The addition of a strong acid or a strong base causes a protective reaction of the buffer system by maintaining a constant pH value of the medium, which is explained by the binding of H^+ and OH^- and the formation of weakly dissociating electrolytes:

$$H_2PO_4^- \rightleftharpoons H^+ + HPO_4^{2-}$$

The increased concentration of H^+ leads to a shift of equilibrium to the left, that is, to the formation of acid. The phosphate buffer of the blood is closely related to the hydrocarbonate buffer:

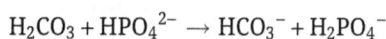

$$H_2CO_3 + HPO_4^{2-} \rightarrow HCO_3^- + H_2PO_4^-$$

It should be noted that hydrocarbons and fatty acids can be used by the body as a source of energy only when pre-phosphorylated, that is, when H_xPO_4 groups are added. The metabolism of phosphorus in the body is closely related to the metabolism of calcium, and these two biocycles together represent an important component of metabolism as a whole.

Organic phosphorus compounds are part of nucleic acids and thus participate in the processes of cell growth and division. Some phosphorus compounds play a key role in the storage and use of genetic information. Another important function of organic phosphorus compounds is to help convert nutrients into energy and to make full use of vitamins. They participate in enzymatic processes, providing the biochemical functions of a number of vitamins, participate in the regulation of metabolic processes, in the transmission of nerve impulses and muscle contraction. The most intense phosphorus metabolism takes place in the muscle tissue.

In the form of phosphate, phosphorus is a necessary component of adenosine triphosphoric acid (ATP). Phosphorus is part of proteins, nucleic acids, nucleotides, and other biologically active compounds. ATP hydrolysis transmits to the body the energy reserve necessary for its vital activity. Adenosine phosphate and creatine phosphate, which are macroergic compounds of phosphorus, accumulate the energy released in the process of glycolysis and oxidative phosphorylation. The accumulated energy is further used in the movement of muscles, the synthesis of various compounds in the body, the transport of substances through cell membranes, etc. Phosphorus compounds are involved in the construction of many enzymes and catalysts for the processes of metabolism of organic substances, creating conditions for the use of potential energy. Thus, the acid residue of phosphoric acid is part of a large number of coenzymes, and phosphorylation is one of the most important ways to convert vitamins into active forms.

The human body normally contains between 600 and 900 g of phosphorus in the form of inorganic phosphate and organic compounds. In the blood, the concentration of organic phosphorus compounds varies widely, while the amount of inorganic phosphorus is sufficiently stable. The excretion of phosphorus from the body occurs mainly in the urine. The recommended daily doses are within very wide limits and depend on the age of the person: in infants and children up to 1 year old it is 300–500 mg/day, up to 10 years old it is 1,650 mg/day, and in adults it is 1,200 mg/day. The phosphorus, necessary for humans, is taken with some foods, such as dairy products, egg yolk, and fish. Insufficient intake of phosphorus in the body causes growth retardation, softening of bone tissue, metabolic disorders, etc. If the phosphates themselves are practically harmless, the elemental phosphorus (especially white), igniting in the air, can cause severe burns. Phosphorus forms P_2O_5 on the surface of the skin, which is hydrolyzed by releasing a large amount of heat, which further worsens burning.

Phosphorus is used in the production of matches, in pyrotechnics, in military affairs, for processing metal surfaces, etc. Some phosphorus compounds are highly poisonous and are used to control pests in agriculture.

20.3.6 Chemical compounds

20.3.6.1 Hydrides
Phosphorus reacts very weakly with hydrogen and forms the compound phosphine
(PH_3), which is an analogue of ammonia in composition:

$$2P + 3H_2 \rightarrow 2PH_3$$

Unlike NH_3, this compound is thermodynamically unstable ($\Delta G°_{form} > 0$), that is, it is
difficult to form directly, so it is obtained indirectly from metallic phosphides:

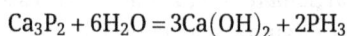

$$Ca_3P_2 + 6H_2O = 3Ca(OH)_2 + 2PH_3$$

or when phosphorus disproportionates with bases.

In phosphine, the H–P–H angle is only 93°, not 107°, as is the H–N–H angle in the
ammonia molecule. Such an angle in phosphine suggests that the phosphorus atom
uses p-orbitals rather than sp^3-hybrid orbitals to bind.

On the other hand, PH_3 differs significantly from ammonia, since the P–H bond is
much less polar than the N–H bond. The chemical bonds P–H in the PH_3 molecule are
practically covalent-nonpolar, because the electronegativity of phosphorus (2.13) and
hydrogen (2.10) are almost the same. PH_3 molecules are nonpolar and therefore phosphine is slightly soluble in water, and its hydrate is quite unstable and almost does
not dissociate:

$$PH_3 + H_2O \rightleftharpoons PH_3.H_2O \rightleftharpoons PH_4OH \rightleftharpoons PH_4^+ + OH^- \quad K = 4.10^{-29}$$

When phosphine reacts with very strong acids, the formation of phosphonic salts is
observed, such as PH_4Cl, PH_4Br, and PH_4I, which are completely hydrolyzed in aqueous solution.

PH_3 is a stronger reducer than NH_3. Its oxidation usually occurs up to +5 oxidation
state:

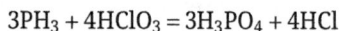

$$3PH_3 + 4HClO_3 = 3H_3PO_4 + 4HCl$$

In air at $T > 150$ °C, phosphine ignites spontaneously as a result of the exothermic reaction:

$$PH_3 + 2O_2 = H_3PO_4$$

PH_3 molecules and their derivatives are good donors and, as ligands, they form complex compounds, such as $[Ni(PH_3)_4]$ and $[Co(N_2)(PR_3)_3]$.

Phosphine is an extremely toxic gas, so work with substances that emit PH_3
(metal phosphides and phosphonium salts) should be carried out in a fireplace.

Although phosphine itself has minor uses, substituted phosphines are important
reagents in organometallic chemistry. The most commonly used one is triphenylphos-
phine, $P(C_6H_5)_3$, commonly abbreviated in the literature as PPh_3.

Other hydrogen compounds of phosphorus are also known, such as P_2H_4 (diphosphine), an analogue of hydrazine, and P_4H_2 (tetraphosphine), for which there is evidence that it is triple-polymerized ($P_{12}H_6$). Unlike phosphine, P_2H_4 is a liquid, and P_4H_2 is a solid. They manifest themselves as very strong reducers.

20.3.6.2 Phosphides

At temperatures of 600–1,200 °C, phosphorus oxidizes metals to form phosphides of the type Ca_3P_2 and Na_3P. Phosphorus also forms phosphides with more electropositive nonmetals (boron, silicon, and arsenic).

Phosphides of alkali and alkaline earth metals have a stoichiometric composition corresponding to the oxidation state of phosphorus −3: Na_3P, Ca_3P_2, etc. They interact intensively with water and acids with the release of PH_3 and decompose even by air moisture:

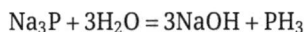

$$Na_3P + 3H_2O = 3NaOH + PH_3$$

The phosphides of d-elements of the first and second groups also have a defined stoichiometric composition, but are more inert and are decomposed only by acids:

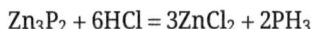

$$Zn_3P_2 + 6HCl = 3ZnCl_2 + 2PH_3$$

The phosphides of the other d-elements have an unstable composition, depending on how they are obtained. For example, iron phosphides show a composition of Fe_3P, Fe_2P, FeP, and FeP_2. These phosphides exhibit metallic or semiconductor properties. They are resistant to the action of oxygen, water, bases, and acids. The most important is the application of phosphides as semiconductor materials.

20.3.6.3 Halide

Phosphorus reacts with halogen elements to form halides with a composition of PX_3 and PX_5, and the most studied of which are chlorides.

Phosphorus trichloride (PCl_3) is a colorless liquid, and phosphorus pentachloride (PCl_5) is a white solid. Phosphorus trichloride is obtained by reacting chlorine in excess phosphorus:

$$P_4 + 6Cl_2 \rightarrow 4PCl_3$$

An excess of chlorine leads to the production of phosphorus pentachloride:

$$P_4 + 10Cl_2 \rightarrow 4PCl_5$$

Phosphorus trichloride reacts with water to H_3PO_3 and hydrogen chloride:

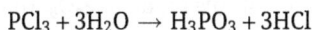

$$PCl_3 + 3H_2O \rightarrow H_3PO_3 + 3HCl$$

This behavior contrasts with that of nitric trichloride, which is hydrolyzed to ammonia and hypochlorous acid:

$$NCl_3 + 3H_2O \rightarrow NH_3 + 3HClO$$

Phosphorus trichloride has a trigonal structure. The compound is an important reagent in organic chemistry. It is used to convert alcohols into chlorine-derived compounds. Thus, 1-propanol in the presence of phosphorus trichloride is converted to 1-chloropropane:

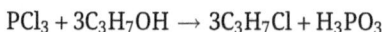

$$PCl_3 + 3C_3H_7OH \rightarrow 3C_3H_7Cl + H_3PO_3$$

The structure of phosphorus pentachloride is interesting because it is different in the solid phase at room temperature and in the gas phase at high temperatures. In the gas and liquid phases, phosphorus pentachloride is a trigonal bipyramidal covalent molecule, while in the solid phase, it adopts an ionic structure, $(PCl_4)^+(PCl_6)^-$.

Phosphorus pentachloride is also used as an organic reagent, but with less significant application. Like phosphorus trichloride, it reacts with water. This is a two-step process, with the first stage producing phosphoryl chloride ($POCl_3$), also known as phosphoryl oxychloride:

$$PCl_5 + H_2O \rightarrow POCl_3 + 2HCl$$

$$POCl_3 + 3H_2O \rightarrow H_3PO_4 + 3HCl$$

Phosphoryl chloride is one of the most important industrial phosphorus compounds. This thick, toxic liquid, which smokes in humid air, is produced industrially by the catalytic oxidation of phosphorous trichloride:

$$2PCl_3 + O_2 \rightarrow 2POCl_3$$

A wide range of chemical reagents are obtained from phosphoryl chloride. For example, tri-*n*-butyl phosphate $(C_5H_{11}O)_3PO$ is an important selective solvent in the separation of uranium and plutonium compounds.

Phosphorus trihalides are good donors due to the presence of an electron pair at the phosphorus atom, while pentahalogenides exhibit acceptor properties because of the free d-orbital:

$$PF_5 + KF \rightarrow K[PF_6]$$

20.3.6.4 Oxides

Phosphorus forms several oxides, and the most important of which are P_2O_3 and P_2O_5.

Phosphorus(III) oxide (P_2O_3, anhydride of phosphorous acid) is formed by the interaction of phosphorus in oxygen deficiency. The formula P_2O_3 is empirical, and the true composition of the molecule of this oxide is P_4O_6. Under normal conditions, it is a white crystalline mass. When reacting with water, phosphorus(III) oxide forms phosphoric acid (H_3PO_3).

Phosphorous(V) oxide (P_2O_5, anhydride of phosphoric acid) is a white, brittle, snow-like mass. The true composition of the molecule corresponds to the formula P_4O_{10}. It is characterized by an extremely high affinity for water. In terms of hygroscopicity, it surpasses all known substances. When interacting with water, depending on the amount of water, it forms different phosphoric acids (HPO_3, H_3PO_4, and $H_4P_2O_7$):

$$P_4O_{10} + 6H_2O \rightarrow 4H_3PO_4$$

P_4O_{10} takes chemically bound water from strong oxoacids, releasing their anhydrides:

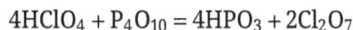

$$4HClO_4 + P_4O_{10} = 4HPO_3 + 2Cl_2O_7$$

Many compounds are dehydrated in this way by P_4O_{10}; for example, nitric acid is dehydrated to pentoxide and organic amides $RCONH_2$ to nitriles RCN. Phosphorous(V) oxide is used as a desiccant of gases as well as in organic synthesis during dehydration reactions.

In complex structures of both oxides, the tetrahedral arrangement of phosphorus atoms (a characteristic of P_4 molecules) is preserved.

20.3.6.5 Oxoacids of phosphorus and their salts

Phosphorus forms many acids, and five of them are of greatest importance, which can be divided into two groups: those containing a P–H bond (H_3PO_2 and H_3PO_3) and not containing a P–H bond (HPO_3, $H_4P_2O_7$, and H_3PO_4). In general, in oxoacids, hydrogen atoms are mainly bonded to oxygen atoms, and not to the central atom of the anion. It is these hydrogen atoms that are separated during the dissociation of the acid in an aqueous solution with the formation of hydroxonium cations (H_3O^+) and take part in neutralization reactions, that is, they determine the basicity of the acid. Usually, in the case of oxoacids, it is assumed that in order to exhibit significant acidity, they must be richer in oxygen, and the hydrogen atom must be bonded to an oxygen atom. For example, nitric acid ($HO)NO_2$ is stronger than nitrous acid ($HO)NO$, and in this order the oxidation state of the central atom also decreases. In inorganic acids, as a rule, the total number of hydrogen atoms in the molecule corresponds to the basicity of the acid, but this is not always the case. In some acids, hydrogen atoms are present directly connected to the central atom. These hydrogen atoms are not replaced by a metal ion, that is, they do not determine the basicity of the acid. Phosphorus shows such uniqueness in that in some of its acids the hydrogen atoms are directly bonded to the central phosphoric atom. Thus, phosphoric acid has three hydrogen cations released during dissociation, phosphorous acid (H_3PO_3) has two, and hypophosphorous acid (H_3PO_2) has only one. Orthophosphoric acid ($H_3P^{5+}O_4$) is tribasic, phosphorous acid ($H_3P^{3+}O_3$) is dibasic, and hypophosphorous acid ($H_3P^{1+}O_2$) is monobasic (Figure 20.1).

Hypophosphorous acid (H_3PO_2) does not have its own anhydride. In the molecule of this acid, two hydrogen atoms are directly bonded to the phosphorus atom: $H(H_2PO_2)$ or

HPO_2H_2. The formal oxidation state of phosphorus (+1) in this acid does not coincide with the number of bonds formed by the phosphorus atom itself. Hypophosphorous acid is a solid substance that is highly soluble in water. It dissociates to one degree as a monobasic weak acid:

$$H_3PO_2 \rightleftharpoons H^+ + H_2PO_2^-, \quad K = 8.5 \times 10^{-2}$$

The acidic residue of this acid is the hypophosphite ion ($H_2PO_2^-$), and its salts are called *hypophosphites*. Hypophosphites of alkali and alkaline earth metals are known. They are obtained by disproportionating phosphorus with bases:

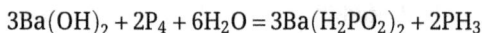

$$3Ba(OH)_2 + 2P_4 + 6H_2O = 3Ba(H_2PO_2)_2 + 2PH_3$$

Hypophosphorous acid is obtained from its salts through ionic reactions:

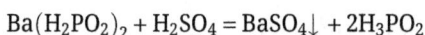

$$Ba(H_2PO_2)_2 + H_2SO_4 = BaSO_4\downarrow + 2H_3PO_2$$

Hypophosphorous acid and hypophosphites are active reducers. They reduce metals from the corresponding salts by precipitating the metal in the form of a thin shiny layer on the surface of nonconductive materials (glass, ceramics, and plastic):

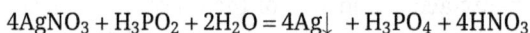

$$4AgNO_3 + H_3PO_2 + 2H_2O = 4Ag\downarrow + H_3PO_4 + 4HNO_3$$

This interaction is the basis of chemical methods for applying metal coatings in cases where electrolysis is not applicable.

Phosphorous acid (H_3PO_3) has one hydrogen atom directly bonded to the phosphoric atom: $H_2(HPO_3)$ or H_2PO_3H. Therefore, in aqueous solution, this medium-strength acid dissociates into two degrees, that is, it is dibasic:

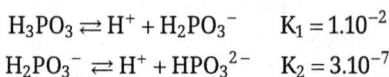

$$H_3PO_3 \rightleftharpoons H^+ + H_2PO_3^- \quad K_1 = 1.10^{-2}$$
$$H_2PO_3^- \rightleftharpoons H^+ + HPO_3^{2-} \quad K_2 = 3.10^{-7}$$

and forms two types of salts: monosubstituted (NaH_2PO_3) and disubstituted (Na_2HPO_3, $MgHPO_3$, etc.) *phosphites*.

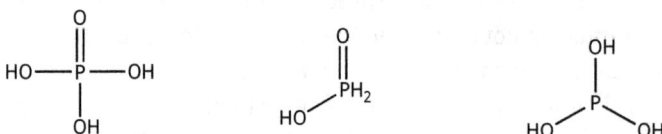

Figure 20.1: Structures of orthophosphoric, hypophosphorous, and phosphorous acids.

The structure of phosphorous acid conforms to the formula $PHO(OH)_2$. Two tautomeric forms have been found in aqueous solution:

$$P(OH)_3 \rightleftharpoons PHO(OH)_2$$

and the equilibrium is practically shifted completely to the right toward one form. From the other form, only some organic derivatives are known.

Phosphorous acid and phosphites are strong reducers: they reduce the weakly active metals from their salts, sulfuric acid to SO_2, etc.:

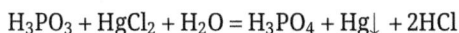

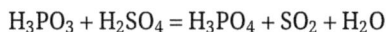

$$H_3PO_3 + HgCl_2 + H_2O = H_3PO_4 + Hg\downarrow + 2HCl$$
$$H_3PO_3 + H_2SO_4 = H_3PO_4 + SO_2 + H_2O$$

Phosphorous acid is produced by the reaction of P_4O_6 with water, by hydrolysis of PCl_3, or by the interaction of sulfuric acid with its salts:

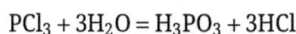

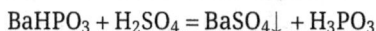

$$P_4O_6 + 6H_2O = 4H_3PO_3$$
$$PCl_3 + 3H_2O = H_3PO_3 + 3HCl$$
$$BaHPO_3 + H_2SO_4 = BaSO_4\downarrow + H_3PO_3$$

Phosphorous acid and its salts phosphites are used as reducers in inorganic synthesis.

According to the water-acidic oxide ratio, acids are divided into ortho-, pyro-, and meta-acids and acids with variable composition. Ortho-acids include acids in which the water-acidic oxide ratio exceeds 1. These acids include the orthophosphoric H_3PO_4 [$(H_2O):(P_2O_5)$ = 3:1]. In meta-acids, this ratio is equal to 1; for example, metaphosphoric acid, HPO_3 [$(H_2O):(P_2O_5)$ = 1:1]. These acids also include nitric, sulfuric, and many others. Pyro-acids (they got their names from the Greek word "pyr" – fire) are obtained from ortho-acids as a result of the release of water during heating:

$$2H_3PO_4 \longrightarrow H_4P_2O_7 + H_2O\uparrow$$

From the above, it follows that in the oxidation state (+5), phosphorus forms three acids: *metaphosphoric* HPO_3, *pyrophosphoric* $H_4P_2O_7$, and *orthophosphoric acid* H_3PO_4 (Figure 20.2).

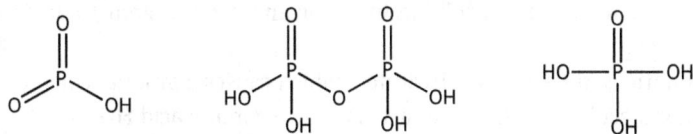

Figure 20.2: Structures of metaphosphoric, pyrophosphoric, and orthophosphoric acids.

They are considered as products of the interaction of P_4O_{10} with water in different ratios of reactants:

$$P_4O_{10} + 2H_2O = 4HPO_3$$
$$P_4O_{10} + 4H_2O = 2H_4P_2O_7$$
$$P_4O_{10} + 6H_2O = 4H_3PO_4$$

Of the greatest practical importance is the orthophosphoric acid (H_3PO_4), known simply as phosphoric acid. In the anhydrous state, it is a solid with a melting point of 42 °C. In water, it dissolves indefinitely and dissociates into three degrees ($K_1 = 7 \times 10^{-3}$, $K_2 = 8 \times 10^{-8}$, $K_3 = 5 \times 10^{-13}$):

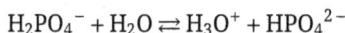

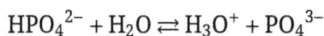

$$H_3PO_4 + H_2O \rightleftharpoons H_3O^+ + H_2PO_4^-$$
$$H_2PO_4^- + H_2O \rightleftharpoons H_3O^+ + HPO_4^{2-}$$
$$HPO_4^{2-} + H_2O \rightleftharpoons H_3O^+ + PO_4^{3-}$$

At ordinary temperature, phosphoric acid is chemically inert, but when heated, it reacts with most metals, metal oxides, and bases. Orthophosphoric acid is obtained not only by direct interaction of P_4O_{10} with H_2O (thermal method) but also by sulfuric acid (extraction) method. According to the sulfuric acid method, H_3PO_4 is obtained from phosphorite:

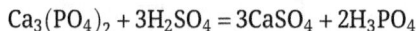

$$Ca_3(PO_4)_2 + 3H_2SO_4 = 3CaSO_4 + 2H_3PO_4$$

The thermal method consists in burning phosphorus to P_4O_{10} with subsequent interaction with water. Thermal phosphoric acid is characterized by high purity.

For laboratory purposes, pure orthophosphoric acid can be obtained by the oxidation of phosphorus with 30–40% nitric acid as follows:

$$3P_4 + 20HNO_3 + 8H_2O = 12H_3PO_4 + 20NO$$

Orthophosphoric acid is tribasic and forms three types of salts: *normal phosphates* Na_3PO_4, *monohydrogen phosphates* Na_2HPO_4, and *dihydrogen phosphates* NaH_2PO_4. Acidic phosphates are soluble; in normal phosphates, only alkali metal salts are well soluble. The phosphates of other metals are insoluble in water and are resistant to acids, bases, and heating.

When heated, phosphoric acid becomes dehydrated, which causes a gradual condensation of phosphoric acid molecules. The first product is pyrophosphoric acid $H_4P_2O_7$:

$$2H_3PO_4 \rightarrow H_4P_2O_7 + H_2O$$

Pyrophosphoric acid is tetrabasic and superior in strength to phosphoric acid. Like phosphoric acid, here each phosphorus atom has a tetrahedral configuration (see below).

The next product of condensation is triphosphoric acid ($H_5P_3O_{10}$):

$$3H_4P_2O_7 \rightarrow 2H_5P_3O_{10} + H_2O$$

The subsequent condensation gives products with an even greater degree of polymerization. These polyacids have a linear structure and a general formula $H_{n+2}P_nO_{3n+1}$.

At temperatures above 300 °C, the dehydration process continues, and pyrophosphoric acid turns into metaphosphoric acid (HPO_3), which exists as a polymer product with a ring structure $(HPO_3)_n$.

Orthophosphoric acid is used to obtain phosphorous fertilizers and phosphates. Natural deposits of phosphorites and apatites are the main sources for industrial extraction of highly concentrated phosphorus fertilizers. Phosphorus fertilizers are of different types: superphosphate $Ca(H_2PO_4)_2$, precipitate $CaHPO_4 \cdot 2H_2O$, ammophos $NH_4H_2PO_4 + (NH_4)_2HPO_4$, and phosphorite flour, which is a crushed phosphorite $Ca_3(PO_4)_2$. In addition to the above, the very valuable one for plants is mixed fertilizer containing phosphorus, nitrogen, and potassium – nitrophoska $(NH_4)_2HPO_4 + NH_4NO_3 + KCl$, in practice, which has good nourishing properties. Phosphorus fertilizers have a beneficial effect on the growth and development of plants: they increase the starch content in potatoes, sugar in beets, oil in sunflowers, etc.

Phosphates also occupy a serious place in everyday life. Trisodium phosphate is used in the household as a cleaning agent. Other sodium phosphates, for example, sodium pyrophosphate ($Na_4P_2O_7$) and sodium tripolyphosphate ($Na_5P_3O_{10}$), are often added to detergents because they react with calcium and magnesium ions from drinking water to form soluble compounds and prevent scale deposition. Phosphates are also added to detergents as fillers and aggregates. Disodium hydrogen phosphate is used in the production of pasteurized processed cheese. Ammonium salts (diammonium hydrogen phosphate and ammonium dihydrogen phosphate), in addition to the production of nitrogen-phosphorus fertilizers, are also useful in the preparation of excellent fire-resistant materials. Calcium phosphates are used as mild abrasives and polishing agents in toothpastes and also in the household. For example, baking powder contains calcium dihydrogen phosphate and sodium hydrogen carbonate:

$$Ca(H_2PO_4)_2 + NaHCO_3 \rightleftharpoons CaHPO_4 + Na_2HPO_4 + CO_2 + H_2O$$

Phosphoric acid is used as an additive to soft drinks. Its low acidity regulates pH and prevents bacterial growth in bottled solutions. It is also suitable for use in metal containers. Phosphate ions react with metal ions to inert phosphate compounds, which prevent any potential metal poisoning. Phosphoric acid is used both in industry and for interior car repairs, to create protective coatings on metal and steel surfaces, removing rust from them.

20.4 Arsenic, antimony, and bismuth

20.4.1 Brief description

As, Sb, and Bi under ordinary conditions have layered crystalline modifications, and the layers are connected by forces of intermolecular interaction (α-forms). These are their stable modifications. The substances in α-modifications are crystals with a metallic luster: As and Sb with a silvery-white color, and Bi with a reddish tint. Down the group, the volume of atoms increases, and the multiplicity of the covalent bond and the distance between the atoms equalize. Molecular structures are more typical for light elements, and coordination structures are more characteristic for heavy ones. This leads to the fact that bismuth has a coordination lattice, that is, the bonds are equal. In the bismuth lattice, the metal bond predominates. In As and Sb, there are allotropic modifications, similar to white P, containing the molecules As_4 and Sb_4. These modifications are unstable. The yellow modification of arsenic becomes α-rhombic during the stage of formation of black As, similar to black P. Yellow Sb undergoes similar transformations. The black modifications of As and Sb have semiconductor properties.

20.4.2 Occurrence and synthesis

As, Sb, and Bi are poorly distributed in nature and are found mainly in the form of sulfide minerals: FeAsS (arsenopyrite), As_2S_3 (auripigment), Sb_2S_3 (antimonite), and Bi_2S_3 (bismuthinite). To obtain the metals from natural sulfides, a pyrometallurgical method is applied: first, the sulfides are burned to oxides, and then the oxides are reduced with C:

$$2E_2S_3 + 9O_2 = 2E_2O_3 + 6SO_2$$
$$2E_2O_3 + 3C = 4E + 3CO_2$$

20.4.3 Properties

Under normal conditions, As and Sb are fragile crystalline substances with semiconductor properties. Bi is a metal usually coated with a dark gray oxide layer. The elements of the arsenic subgroup in the order of standard potentials are located after hydrogen, so they do not interact with HCl and dilute sulfuric acid. In acids with oxidizing properties, they dissolve:

$$2Sb + 6HNO_{3dil} \rightarrow Sb_2O_3 + 6NO_2 + 3H_2O$$
$$Bi + 4HNO_{3dil} \rightarrow Bi(NO_3) + NO + 2H_2O$$

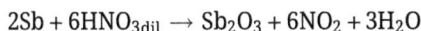

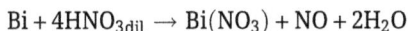

Concentrated nitric acid oxidizes As and Sb to an oxidation state of +5, and bismuth to an oxidation state of +3, because in bismuth the maximum oxidation state is not a characteristic:

$$3As + 5HNO_3 + 2H_2O \rightarrow 3H_3AsO_4 + 5NO$$
$$2Sb + 10HNO_3 = Sb_2O_5 + 10NO_2 + 5H_2O$$
$$Bi + 6HNO_3 = Bi(NO_3)_3 + 3NO_2 + 3H_2O$$

With concentrated sulfuric acid, antimony and bismuth react in the following reactions:

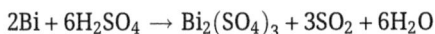

$$2Sb + 6H_2SO_{4conc} \rightarrow Sb_2(SO_4)_3 + 3SO_2 + 6H_2O$$
$$2Bi + 6H_2SO_4 \rightarrow Bi_2(SO_4)_3 + 3SO_2 + 6H_2O$$

They dissolve in aqua regia:

$$3Sb + 15HCl + 5HNO_3 \rightarrow 3SbCl_5 + 5NO + 10H_2O$$
$$Bi + 3HCl + HNO_3 \rightarrow BiCl_3 + NO + 2H_2O$$

Antimony and bismuth are resistant to basic solutions and melts, while arsenic in basic melts forms salts of arsenic acid:

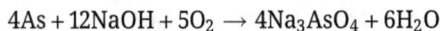

$$4As + 12NaOH + 5O_2 \rightarrow 4Na_3AsO_4 + 6H_2O$$

The interaction with bases takes place in the presence of oxidizing agents.

Antimony (when heated) and bismuth interact with water vapor and hydrogen sulfide:

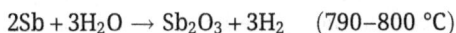

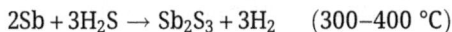

$$2Sb + 3H_2O \rightarrow Sb_2O_3 + 3H_2 \quad (790\text{--}800\ °C)$$
$$2Sb + 3H_2S \rightarrow Sb_2S_3 + 3H_2 \quad (300\text{--}400\ °C)$$
$$2Bi + 3H_2O \rightarrow Bi_2O_3 + 3H_2$$
$$2Bi + 3H_2S \rightarrow Bi_2S_3 + 3H_2$$

Metallic bismuth is a good reducer in terms of some chlorides and oxides:

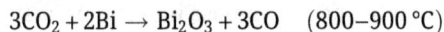

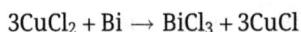

$$PCl_3 + Bi \rightarrow BiCl_3 + P$$
$$3PbO + 2Bi \rightarrow BiO_3 + 3Pb$$
$$AsCl_3 + Bi \rightarrow BiCl_3 + As$$
$$3CO_2 + 2Bi \rightarrow Bi_2O_3 + 3CO \quad (800\text{--}900\ °C)$$
$$3CuCl_2 + Bi \rightarrow BiCl_3 + 3CuCl$$

As can be seen from the examples given, bismuth in these cases forms stable compounds.

Metal modifications of the elements of the arsenic subgroup are more inert under ordinary conditions, but when heated, their activity increases. They burn in the air with the formation of E_2O_3. With the halogen elements, they form compounds of the type EX_3 and EX_5. They react with sulfur to form sulfides with a composition of E_2S_3 and E_2S_5. With metals, they form arsenides of the type Mg_3As_2, antimonides Me_3Sb_2, and bismuthides Ca_3Bi_2. Their compounds with alkali and alkaline earth metals have a salt-like character, but in the presence of water they decompose to hydrides EH_3. With other metals, the elements of the arsenic subgroup form compounds of the type GaAs, InSb, and AlAs, which are used in semiconductor technology. These persistent compounds do not react with nitrogen, carbon, and hydrogen.

20.4.4 Use and biological role

The main application of the elements of the arsenic subgroup is as components in some alloys that are used in fire-fighting equipment, laboratory practice, topography, etc. In general, these elements are toxic. Arsenic and partly antimony are trace elements. In the human body, the content of As ranges from 0.008 to 0.02 mg per 100 g of tissue. The high toxicity of some of the compounds of arsenic is known, but its positive role is also undoubted. Of the elements in the group, drugs with the participation of As, despite its toxicity, are of interest. Small amounts of As have a beneficial effect on the hematosis process and contribute to the compaction of bone tissue, the accumulation of protein substances and fats, and weight gain. As_2O_3 is used for necrotization of the nerve tissue in dentistry. Some organic preparations with the participation of As, which are coordination compounds, are used in the treatment of acute intestinal infections, as well as syphilis, malaria, typhus, etc. Medicinal preparations containing antimony are used to combat leprosy, tropical fever, sleeping sickness, etc. Preparations based on bismuth compounds have been known for about 200 years and are used to treat gastrointestinal diseases. The most effective modern preparation of this kind is the complex citrate of bismuth, $K_3(NH_4)_2Bi_6O_3(OH)_5(Hcit)_4$.

20.4.5 Chemical compounds

20.4.5.1 Hydrides
Arsenic, antimony, and bismuth do not react directly with hydrogen. The hydrides AsH_3 (arsine), SbH_3 (stibine), and BiH_3 (bismuthine) are toxic gases. They are thermodynamically unstable and decompose into H_2 and metals when slightly heated:

$$2EH_3 = 2E + 3H_2$$

The hydrides of these elements are very strong reducers. They are obtained by the action of acids on compounds of arsenic, antimony, and bismuth with active metals:

$$Ca_3As_2 + 6HCl = 3CaCl_2 + 2AsH_3$$

Free As and its compounds are reduced from nascent H_2 (Zn in acidic medium) to AsH_3:

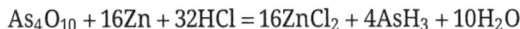

$$As_4O_{10} + 16Zn + 32HCl = 16ZnCl_2 + 4AsH_3 + 10H_2O$$

The resulting AsH_3 at high temperature decomposes thermally to metallic arsenic (Marsh-Liebig sample). This sample is used in medicine to prove arsenic intoxication.

20.4.5.2 Arsenides, antimonides, and bismuthides
These compounds are produced by direct interaction with more electropositive elements. Ionic compounds with alkali and alkaline earth metals (K_3E, Ca_3E_2, etc.) have stoichiometric composition and ionic chemical bonds. They easily hydrolyze, and under the action of mineral acids, they release the corresponding hydrogen compounds of elements. Covalent arsenides, bismuthides, and antimonides of p-elements exhibit good semiconductor properties. Metallic arsenides, bismuthides, and antimonides of transition metals have an unstable composition, depending on the method of their preparation. They exhibit metallic or semiconductor properties.

20.4.5.3 Halide
Arsenic, antimony, and bismuth interact with the halogen elements. In the oxidation state +3, the compounds of all three elements with all the halogen elements are known, while in the oxidation state +5, only fluorides and chlorides are known. The halogenides of the three elements of type EX_3 are obtained by direct interaction between the elements. The higher halides of EX_5 can be obtained from the trihalides under appropriate conditions:

$$SbCl_3 + Cl_2 = SbCl_5$$

All halides are hydrolyzed in aqueous solution: arsenic halides with the formation of two acids, and antimony and bismuth halides with the formation of oxysalts:

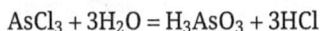

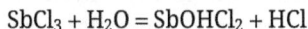

$$AsCl_3 + 3H_2O = H_3AsO_3 + 3HCl$$

$$SbCl_3 + H_2O = SbOHCl_2 + HCl$$

In both halogenides and hydrides, donor properties are weakly expressed and decrease toward bismuth. Pentahalogenides, like phosphorus pentahalogenides, possess trigonal bipyramidal covalent molecules and exhibit acceptor properties because of the free d-orbitals:

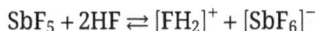

$$SbF_5 + 2HF \rightleftharpoons [FH_2]^+ + [SbF_6]^-$$

20.4.5.4 Sulfides

Arsenic, antimony, and bismuth react with sulfur to form slightly water-soluble sulfides. Most often they are produced by precipitation reactions from acidic solutions of their salts and soluble sulfides. Trisulfides of all three elements are known, and arsenic and antimony also form pentasulfides. Bismuth(V) sulfide has not been obtained. Compounds of arsenic and antimony with sulfur (As_2S_3, As_2S_5, Sb_2S_3, and Sb_2S_5) are dissolved in alkaline bases, alkaline sulfides, and ammonium sulfide until soluble salts of the respective acids are obtained:

$$As_2S_3 + 6NaOH = Na_3AsO_3 + Na_3AsS_3 + 3H_2O$$

In the above reaction, salts of arsenic and thioarsenic acids are obtained. Arsenic and antimony sulfides are sulfoanhydrides, that is, they dissolve in solutions of alkaline sulfides and ammonium sulfides to form sulfosalts:

$$As_2S_5 + 3K_2S = 2K_3AsS_4$$

$$Sb_2S_3 + 3Na_2S = 2Na_3AsS_3$$

Sulfides are insoluble in water and acids that do not exhibit oxidative action. They dissolve when reacting with oxidative-acting acids, for example, concentrated nitric acid:

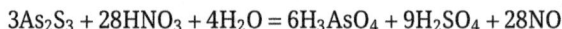

$$3As_2S_3 + 28HNO_3 + 4H_2O = 6H_3AsO_4 + 9H_2SO_4 + 28NO$$

Bi_2S_3 does not belong to sulfoanhydrides but interacts with concentrated HNO_3:

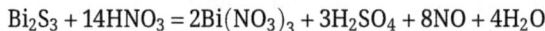

$$Bi_2S_3 + 14HNO_3 = 2Bi(NO_3)_3 + 3H_2SO_4 + 8NO + 4H_2O$$

20.4.5.4.1 Oxides, hydroxides, and oxoacids

Oxides with a composition of E_2O_3 and E_2O_5 are known for the three elements.

Compounds in oxidation state +3

When As, Sb, and Bi react with oxygen, *oxides* such as As_2O_3, Sb_2O_3, and Bi_2O_3 are formed, and their basic properties are enhanced in this order. As_2O_3 is characterized by a molecular structure, Sb_2O_3 has a pseudomolecular chain structure with rhombic syngony, and Bi_2O_3 has a coordination structure with a bismuth coordination number of 6.

The acidic oxide As_2O_3 reacts with water to form *arsenic acids* such as $HAsO_2$ (metaarsenous) and H_3AsO_3 (orthoarsenous). Both acids are weak. Metaarsenous acid, when dissolved, passes into orthoarsenous, which dissociates both in acidic and basic types, i.e. it is ampholyte. When solutions of arsenous acids are neutralized, *metaarsenites* of the type $NaAsO_2$ and *orthoarsenites* are formed, which are divided into normal (Ag_3AsO_3) and acidic (NaH_2AsO_3 and Na_2HAsO_3).

Amphoteric oxide Sb_2O_3 does not react with water. Its corresponding amphoteric *hydroxide* $Sb(OH)_3$ is obtained by the reaction of bases or ammonium hydroxide with salt solutions:

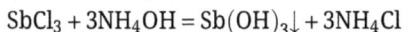

$$SbCl_3 + 3NH_4OH = Sb(OH)_3{\downarrow} + 3NH_4Cl$$

The amphoteric nature of $Sb(OH)_3$ is manifested in its interaction with acids and bases:

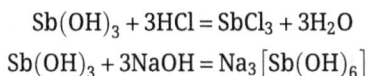

$$Sb(OH)_3 + 3HCl = SbCl_3 + 3H_2O$$
$$Sb(OH)_3 + 3NaOH = Na_3\left[Sb(OH)_6\right]$$

For this reason, antimony occurs in the composition of its salts in the form of a simple cation Sb^{3+}, an oxyanion SbO_2^- and a hydroxocomplex $Sb(OH)_6^{3-}$. Antimony (+3) salts, when dissolved in water, are hydrolyzed along the cation to form insoluble oxysalts, which is explained by the strong polarizing effect of the cation Sb^{3+}.

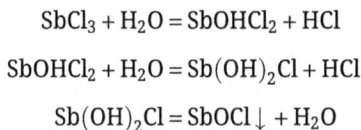

$$SbCl_3 + H_2O = SbOHCl_2 + HCl$$
$$SbOHCl_2 + H_2O = Sb(OH)_2Cl + HCl$$
$$Sb(OH)_2Cl = SbOCl{\downarrow} + H_2O$$

Bismuth(III) oxide Bi_2O_3 and *hydroxide* $Bi(OH)_3$ are practically non-amphoteric. In its salts, trivalent bismuth is found in the form of a simple cation Bi^{3+}. When dissolved, bismuth(III) salts are hydrolyzed, similar to antimony(III) salts, to form oxysalts:

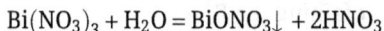

$$Bi(NO_3)_3 + H_2O = BiONO_3{\downarrow} + 2HNO_3$$

Basic salts, containing antimonyl and bismuthyl ions, are released as precipitates.

The compounds of As(III), Sb(III), and Bi(III) are reducers, and their reduction ability in this order decreases: for example, $HAsO_2$ is oxidized by iodine, antimony(III) hydroxide by bromine, and bismuth(III) hydroxide is only partially oxidized by chlorine:

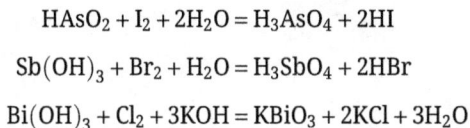

$$HAsO_2 + I_2 + 2H_2O = H_3AsO_4 + 2HI$$
$$Sb(OH)_3 + Br_2 + H_2O = H_3SbO_4 + 2HBr$$
$$Bi(OH)_3 + Cl_2 + 3KOH = KBiO_3 + 2KCl + 3H_2O$$

Oxidizing properties are exhibited only by Bi(III) compounds. They are reduced to metallic Bi with phosphorous acids and Bi(III) salts, with compounds of divalent tin, for example:

$$2Bi(NO_3)_3 + 3K_2\left[Sn(OH)_4\right] + 6KOH = 2Bi{\downarrow} + 3K_2\left[Sn(OH)_6\right] + 6KNO_3$$

Compounds in oxidation state +5

The higher *oxides* As_2O_5, Sb_2O_5, and Bi_2O_5 exhibit acidic properties and are produced indirectly.

Arsenic(V) oxide (As_2O_5, arsenic anhydride) exhibits a great affinity for water and, when interacting with it, forms *orthoarsenic acid* (H_3AsO_4), which is similar in properties to orthophosphoric acid (H_3PO_4). It forms normal salts, *orthoarsenates*, of the type Na_3AsO_4 and acidic salts such as *hydrogen arsenates* (Na_2HAsO_4) and *dihydrogen arsenates* (NaH_2AsO_4). *Metaarsenates* of the type $NaAsO_3$ are also known, although the corresponding acid in free form has not been obtained. Orthoarsenic acid is produced by the interaction of HNO_3 with arsenic or As_2O_3:

$$3As_2O_3 + 4HNO_3 + 7H_2O = 6H_3AsO_4 + 4NO$$

Antimony(V) oxide (Sb_2O_5) in hydrated form $Sb_2O_5 \cdot xH_2O$ is formed by the reaction of Sb with nitric acid. It is insoluble in water, but interacts with bases to form hydroxo-complex salts:

$$Sb_2O_5 + 2NaOH + 5H_2O = 2Na\left[Sb(OH)_6\right]$$

Antimonic acid (H_3SbO_4) is a weak acid. Salts of this acid are called *stibates* or *antimonates*.

Bismuth(V) oxide (Bi_2O_5) is formed by the reaction of Bi_2O_3 with ozone. Its corresponding acid does not exist, and *bismuthates* are produced by the interaction of strong oxidizing agents with Bi(III) compounds.

In the hydroxides of the elements, acidic properties prevail in the oxidation state of +5. Therefore, no salts containing the E^{5+} cations are known for them.

The compounds of As, Sb, and Bi in oxidation state +5 are oxidizing agents, with strongest oxidizing properties exhibited by bismuth compounds:

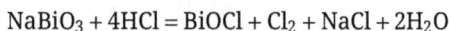

$$NaBiO_3 + 4HCl = BiOCl + Cl_2 + NaCl + 2H_2O$$

Chapter 21
Sixth main group of the periodic table

21.1 General characteristics of the elements of VIA group

The sixth main group contains the elements oxygen (O), sulfur (S), selenium (Se), tellurium (Te), and polonium (Po). The atoms of these elements contain six valence electrons each (Table 21.1). The first four elements are found in the composition of copper-containing minerals, so in the past they received the group name chalcogens, which means "related to copper ores."

Table 21.1: Characteristics of atoms of elements of VIA group.

Elements	O	S	Se	Te	Po
Valence electrons	$2s^2 2p^4$	$3s^2 3p^4$	$4s^2 4p^4$	$5s^2 5p^4$	$6s^2 6p^4$
Radius of the atom (nm)	0.073	0.102	0.117	0.136	0.146
Ionization potential (eV)	13.6	10.4	9.8	9.0	8.4
Electronegativity	3.5	2.5	2.4	2.1	2.0

From O to Po, there is an increase in atomic radii and a decrease in ionization potentials and electronegativity. Therefore, in the elements in the group, nonmetallic properties are weakened and metallic ones are enhanced: oxygen, sulfur, and selenium are nonmetals; tellurium has weak metallic characteristics; and polonium is a radioactive unstable metal. All elements of the sixth main group form more than one allotropic forms, for example, O_2 (oxygen) and O_3 (ozone), S – rhombic, monoclinic, etc. Oxygen and ozone are gases, and allotropic forms of S are solids. They are nonmetallic in nature. The allotropic forms of Se and Te are solid, but have a metallic luster, which grows when heated, that is, they exhibit the qualities of semiconductors (Table 21.2).

Table 21.2: Physical properties of elements of VIA group.

Element	T_m (°C)	T_b (°C)	R_{at} (Å)	ρ (g/cm^3)	Electrical resistance (mΩ cm)	$E°$ (V), E_{p-p}^{2+}/E
O	−218	−182	0.66	1.27	–	–
S_{monocl}	119.3	444.6	1.02	1.96	2×10^{33}	−0.48
S_{rhomb}	112.8	684.8	1.16	2.07	2×10^{11}	−0.48
S_{hexag}	220.4	1,087	1.35	4.82	2×10^5	−0.92
Se	452.0	962		6.25	43	−1.14
Te	254					
Po						

https://doi.org/10.1515/9783111712246-021

The electronic configuration of the last electron shell in these elements is ns^2np^4, which determines the most typical oxidation state of −2. Oxygen differs from other elements in which there are free d-orbitals and respectively the possibility of two excited states, with 4 and 6 valence electrons. Unlike oxygen and sulfur, selenium, tellurium, and polonium each have 18 electrons in their penultimate electron shell and are complete electron analogues. That said, only oxygen exhibits a second valence and the most common oxidation state −2 with some minor exceptions. The other elements exhibit variable valence and oxidation states: −2, +2, +4, +6. The stability of compounds with −2 and +6 oxidation states decreases from the top to the bottom in the group, while the stability of those in +4 oxidation states increases in the same direction.

Chemical elements of group VIA can interact with hydrogen, resulting in hydrogen compounds with a molecular structure, for example, H_2S. In the p-elements of VIA group, the ability to form cationic forms is weakly expressed. Only polonium forms compounds in which it is found in the form of a cation: $Po(SO_4)_2$ and $Po(NO_3)_4$. In its free state, sulfur and elements of the selenium subgroup are moderately stable substances. When heated, they burn in the air with the formation of dioxides such as SO_2, SeO_2, TeO_2, and PoO_2. Under normal conditions, they react only with fluorine to form fluoride of different compositions: EF_6, EF_4, and E_2F_{10}. With the rest of the halogen elements, without iodine, they react when heated to form halides such as EX_4, S_2Cl_2, and SCl_2. A regularity in the group is the decrease in chemical activity in relation to hydrogen and nonmetals with an increase in the atomic number, that is, chemical elements with smaller atomic numbers are more active. The elements of group VIA react with carbon, phosphorus, and metals, forming substances called sulfides, selenides, and tellurides: CS_2, P_2S_5, P_2Se_5, HgS, and FeS. Sulfur, selenium, and tellurium are dissolved in solutions containing their anions to form polyanions such as S_n^{2-}, Se_n^{2-}, and Te_n^{2-}. They also react with many organic compounds, such as saturated hydrocarbons, and dehydrate them. The reaction of sulfur with olefins is used in the vulcanization of rubber.

21.2 Oxygen

21.2.1 Brief description

Oxygen is a colorless, odorless, and tasteless gas that condenses into a pale blue liquid. Its density is 1.4 g/cm^3. Because it has a small molar mass and forms a nonpolar molecule, oxygen has very low melting and boiling points. The boiling point of oxygen is about −180 °C, and its melting point is −218 °C. It is a gas that does not burn, but maintains combustion. Almost all elements of the periodic table react with oxygen at room temperature or when heated, with the exception of precious metals and noble gases.

Oxygen is the second electronegative element in the periodic table after fluorine; hence, its typical oxidation state is –2. Compounds in which oxygen is in oxidation state –2 (oxides, bases, acids, and salts) are considered in the chemistry of each element. In some of its compounds, oxygen also manifests itself in other oxidation states, such as –1 (peroxides), –½ (superoxides), –⅓ (ozonides), +2, and +1. In oxidation states +2 and +1, oxygen is found in its compounds with fluorine (OF_2 and O_2F_2), which are highly reactive as oxidizing and fluorinating agents.

The diatomic molecule O_2 is paramagnetic. Its formation is explained by the method of molecular orbitals. The energy of the double bond in the molecule O_2 is quite high (498 kJ/mol), so the breakdown of the molecule into atoms begins at a temperature higher than 2,000 K. Oxygen is a strong oxidizing agent, but due to the strength of the bond in its molecule, oxidation (combustion) reactions usually take place at high temperatures.

The order of the bond in the oxygen molecule is 2 due to the six binding and the two antibonding electrons with parallel spins. This is the ground state of O_2 molecule (triplet oxygen). Therefore, in normal triplet form, molecules are paramagnetic, that is, they become magnetic in the presence of a magnetic field, due to the spin magnetic moments of the unpaired electrons in the molecule and the negative exchange energy between neighboring O_2 molecules. Liquid oxygen is attracted by magnets to such an extent that, in laboratory demonstrations, a bridge of liquid oxygen can be formed, in which its own weight is maintained between the poles of a strong magnet. However, an energy of the order of 95 kJ/mol is sufficient to induce pairing of single antibonding electrons (singlet oxygen). This diamagnetic form of oxygen becomes paramagnetic in seconds or minutes, depending on the concentration and surroundings of the molecule. Singlet oxygen is a high-energy state of molecular O_2 in which all electron spins are paired. It reacts much more easily with ordinary organic compounds. In nature, it is formed from water in the process of photosynthesis by the energy from sunlight. It is also formed in the troposphere by photolysis of ozone from short-wave light, as well as in the immune system of organisms as a source of active oxygen. It is used in the treatment of skin cancer. The diamagnetic form of oxygen can be obtained in laboratory conditions by reacting hydrogen peroxide with sodium hypochlorite:

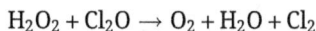

$$H_2O_2 + Cl_2O \rightarrow O_2 + H_2O + Cl_2$$

21.2.2 Occurrence and synthesis

Oxygen is the most abundant element in nature. If hydrogen predominates in the universe and nitrogen in the Earth's atmosphere, in the lithosphere (the Earth's crust and upper layer of the mantle) bound oxygen predominates. Oxygen makes up 47% of the mass of the Earth's solid shell, but is even more abundant in the water shell of our planet (hydrosphere) – almost 86%. It is part of most rocks and more than 1,000 minerals, of which mainly in the composition of sand, clays, mountain ores, and water. Many chemical elements exist in nature in the form of oxygen-containing compounds. Oxygen is present in the air and makes up 21% of the Earth's atmosphere. Usually, oxidizing gases, such as oxygen, do not occur as natural components in the atmospheres of planets. The "normal" atmosphere of a planet contains mainly hydrogen, methane, ammonia, and carbon dioxide. The unusually high concentration of oxygen as a component of the Earth's atmosphere is the result of the oxygen cycle. This biogeochemical cycle describes the movement of oxygen between the atmosphere, the biosphere, and the lithosphere. The current oxygen content and the appearance of signs of life were achieved nearly 5×10^8 years ago.

In nature, oxygen occurs in the form of three isotopes – ^{16}O, ^{17}O, and ^{18}O, with ^{16}O being the most common (99.76%). Fourteen radioisotopes of oxygen have been identified, the most stable of which are ^{15}O and ^{14}O with half-lives of 122.24 s and 70.606 s, respectively. The rest of the radioisotopes have significantly shorter half-lives.

Oxygen in industry is obtained by fractional distillation of liquefied air (in parallel with the production of nitrogen), as well as by electrolysis of water. In laboratories, it is obtained by decomposition of some salts, oxides, and peroxides when heated:

$$NaClO_4 = NaCl + 2O_2$$

$$2Ag_2O = 4Ag + O_2$$

$$2BaO_2 = 2BaO + O_2$$

Most often in laboratory conditions, O_2 is obtained by decomposition of potassium chlorate or hydrogen peroxide; in both cases, manganese(IV) oxide is used as a catalyst:

$$2KClO_3 \longrightarrow 2KCl + 3O_2$$

$$2H_2O_2 \longrightarrow 2H_2O + O_2$$

21.2.3 Biological action of oxygen

Oxygen is the element that sustains life on the Earth. The main components of atmospheric air are: N_2 (78.08%), O_2 (20.8%), Ar (0.93%), CO_2 (0.02–0.04%), Ne (1.92×10^{-3}%), He (5.24×10^{-4}%), Kr (1.14×10^{-4}%), H_2 (5.0×10^{-5}%), and Xe (8.7×10^{-6}%). It is noteworthy that the oxygen content of the atmosphere is remarkably constant, despite the diversity in the oxidative processes of respiration and combustion taking place on the Earth. The main factor maintaining the constancy of oxygen content in the Earth's atmosphere is photosynthesis. In this process, it is not terrestrial green plants that play a major role, but plankton and algae in the world's oceans, which account for about 80% of the oxygen released. In fact, life on the Earth is possible only in a fairly narrow range of oxygen content in the atmosphere – from 13% to 30%. At an oxygen content of less than 13%, aerobic organisms (using O_2 for their existence) would die, and at a higher than 30% content, the oxidation and combustion processes would proceed so intensely that they would ignite everything on the Earth.

Without oxygen, it is impossible to imagine the numerous and extremely important life processes, especially breathing. Only some plants and the simplest animals can survive without oxygen. They are called anaerobic. In most living organisms, an important part of metabolism is the respiratory cycle, which leads to the rapid formation of many substances. In the exhaled air, in addition to CO_2, there are certain amounts of hydrocarbons, alcohols, ammonia, formic acid (HCOOH), acetic acid (CH_3COOH), formaldehyde (HCHO), acetone (($CH_3)_2CO$), etc. When a person breathes in rarefied air at an altitude of 10 km, due to a lack of oxygen in the exhaled air, the content of ammonia, amines, phenol, acetone, and hydrogen sulfide increases sharply. In living organisms, oxygen is used for the oxidation of various substances, the main being the process of interaction of oxygen with hydrogen and the formation of water, as a result of which a significant amount of energy is released. Aerobic organisms also receive energy due to the oxidation of nutrients in cells and tissues to CO_2, H_2O, and $(NH_2)_2CO$. In the process of normal breathing, the molecular oxygen entering the lungs is reduced to water:

$$O_2 + 4H^+ + 4e^- \rightarrow 2H_2O$$

Molecular oxygen (O_2) is not toxic to living organisms, unlike its other forms, such as ozone (O_3), atomic O, the excited molecule $O_2\bullet$, and the radicals OH•, $HO_2\bullet$, $O_2^-\bullet$. In pathological conditions, the above process proceeds as an incomplete reduction:

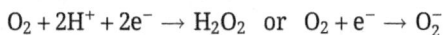

$$O_2 + 2H^+ + 2e^- \rightarrow H_2O_2 \quad \text{or} \quad O_2 + e^- \rightarrow O_2^-$$

The resulting O_2^- is called a superoxide radical. It can be useful when it destroys uncontrolled growing cells, but it can also be very toxic when it destroys the cell membranes of healthy cells necessary for the body. In addition, the harmful effect of the superoxide radical is that it deactivates enzymes, depolymerizes polysaccharides, and causes single breaks in the structure of DNA. In the intermediate, slow single-electron reduction of O_2 up to a superoxide radical, various substances in the body with appropriate potential can participate. In this reaction, H_2O_2 is formed, which, in the next stage of single-electron reduction, gives a hydroxide radical (OH•) with high reactivity, rapidly oxidizing any substance in the cell. The hydrophobic molecule O_2 passes easily into the cell through the hydrophobic lipid membranes and begins to oxidize organic substances to $O_2^-\bullet$ and OH• radicals. These polar radicals are "closed" in the cell, because they cannot escape back through the cell membranes. Specific enzymatic and non-enzymatic antioxidants – superoxidismutase, catalase and peroxidase can neutralize already formed free radicals. In addition, there are low molecular weight substances – antioxidants (e.g., vitamins C and E), which neutralize these particles nonenzymatically. Sometimes the release of superoxide radicals is beneficial. For example, antitumor antibiotics (bleomycin) form complexes with metal ions M^{n+}, catalyzing the rapid reduction of $O_2•$ to a superoxide radical that destroys DNA in tumor formations.

Oxygen is often used in medicine in the form of therapeutic inhalation mixtures, and the composition of which depends on the purpose of treatment. For example, mixtures of O_2 with air containing 40–60% O_2 are used to saturate tissues with oxygen. In case of poisoning with CO or acidic gases, mixtures of O_2 with CO_2 containing up to 5% CO_2 are used, the so-called "carbogen." In the treatment of burns, strokes and trophic diabetic ulcers, hyperbaric oxidation is applied in hyperbaric chambers with increased partial pressure of oxygen. With the help of oxygen enemas, intestinal anaerobic parasites are destroyed.

The role of oxygen in modern industry is also indisputable. Oxygen is widely used in metallurgy, as an oxidizing agent in rocket fuel, and so on.

21.3 Ozone

The allotropic modification of oxygen is ozone (O_3). Ozone is a gas with a pungent, specific odor. It consists of three oxygen atoms bonded in a polar V-shaped molecule (sp^2 hybridization of the oxygen atom). The angle between the O–O–O atoms is 117°. The bonds of σ- and π-type between O atoms are of the same length and have an order of about 1½. The angles and lengths of the bonds in the ozone molecule are very similar to those of the isoelectronic nitrite ion.

Ozone forms compounds with the alkali and alkaline earth metals. These compounds contain the ozonide ion (O_3^-). This ion is large in size, so it forms the most stable compounds with large cations, such as cesium. The ion O_3^- has also been shown to be V-shaped.

Ozone is one of the strongest oxidizing agents, and at room temperature, it oxidizes simple and complex substances to maximum oxidation states:

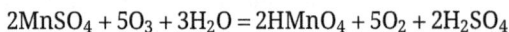

$$Cl_2 + 3O_3 = Cl_2O_7 + O_2$$

$$2MnSO_4 + 5O_3 + 3H_2O = 2HMnO_4 + 5O_2 + 2H_2SO_4$$

The high oxidizing activity of ozone is explained by its decomposition by the reaction

$$2O_3 = 3O_2$$

taking place in two stages:

$$1)\ O_3 = O_2 + O$$

$$2)\ O + O_3 = 2O_2$$

The release of atomic O in the first stage is the reason for the high oxidation capacity of O_3. The wide range of oxidative processes involving ozone is illustrated by the following reactions: the first in the gas phase, the second in aqueous solution, and the third in the solid state:

$$2NO_2 + O_3 \rightarrow N_2O_5 + O_2$$

$$CN^- + O_3 \rightarrow OCN^- + O_2$$

$$PbS + 4O_3 \rightarrow PbSO_4 + 4O_2$$

Ozone is obtained from oxygen in special devices, the so-called ozonators. Its content is determined by the iodine released during the following reaction:

$$O_3 + 2KI + H_2O = I_2 + 2KOH + O_2$$

Ozone is formed in small quantities in juniper forests and in the atmospheric air during storms. In the atmosphere, it is formed mainly by the photochemical reaction (hv):

$$O_2 + O \rightarrow O_3$$

Active atomic oxygen is formed due to the process:

$$NO + O_2 \rightarrow NO_2 + O\bullet$$

At an altitude of 10–30 km, the ozone content in the atmosphere increases sharply under the influence of the otherwise harmful ultraviolet radiation of the Sun. Thus, the formation of an ozone layer in the atmosphere protects life on the Earth. The beneficial effect of ozone in the atmosphere lies in the fact that ozone not only absorbs the biologically active and thus dangerous part of the ultraviolet radiation of the Sun, but also takes part in the formation of the thermal regime of the Earth's surface. It retains the heat released from the Earth in such spectral intervals in which this heat is absorbed, but quite weak by CO_2 and H_2O. In the modern world, emissions of industrial gases into the atmosphere lead to the gradual destruction of the ozone layer, which has now become one of the most pressing environmental problems.

The ozone layer, which is located high in the atmosphere, has protective functions, while in the ground layer, ozone can have an adverse effect. It is highly toxic to humans. Its maximum permissible concentration in the air is 0.5 mg/m^3. Ozone affects the structure of the lungs, inhibiting their function, and thus reduces tolerance to respiratory diseases. Ozone intensively oxidizes sulfur-containing amino acids and enzymes (cysteine, methionine, as well as tryptophan, histidine, and tyrosine). The drugs used for ozone poisoning and related radicals are primarily effective antioxidants, such as vitamins C and E.

The bactericidal action of ozone is very effective in disinfecting and removing unpleasant odors in drinking water, although this method is significantly more expensive than the usual chlorination. Its decontaminating effect is associated with the formation of superoxide radicals, which destroy the cell membranes of the microorganisms.

Ozone is used industrially as a fuel oxidizer, for bleaching paper, for discoloring some food products (flour, butter, ctc.).

21.4 Chemical compounds of oxygen

The chemical compounds of oxygen are considered in the individual chemical elements. Here is a brief overview of its binary compounds – oxides, peroxides, superoxides, and ozonides.

Oxides are compounds formed by atoms of two elements, one of which is oxygen in the oxidation state of -2. Oxides include all compounds of the elements with oxygen (Fe_2O_3 and P_4O_{10}), except those containing oxygen atoms linked to each other by a chemical bond, such as peroxides, superoxides, and ozonides (Na_2O_2, KO_2, and KO_3), as well as the compounds of oxygen with fluorine (OF_2 and O_2F_2). The latter are called oxygen fluorides, because the oxidation state of oxygen in them is positive.

The melting and boiling temperatures of oxides vary over a very wide range. At room temperature, depending on the type of crystal lattice, they can be located in different aggregation states. This is determined by the nature of the chemical bond in the oxides, which can be ionic or covalent polar. In the gaseous and liquid states at room temperature are the oxides that form molecular crystal lattices. As the polarity of the molecules increases, the melting and boiling temperatures increase. Oxides forming ionic crystal lattices, for example, CaO and BaO, are solids with very high melting points ($>1{,}000$ °C). In some oxides, polar covalent bonds are present. They form crystal lattices in which the atoms of the element are connected by several "bridge" oxygen atoms, forming infinite three-dimensional networks, for example, Al_2O_3, SiO_2, TiO_2, and BeO. These oxides also have very high melting points.

According to chemical properties, oxides are divided into the following types: non-salt-forming, salt-like, and salt-forming, the latter being divided into basic, amphoteric, and acidic. *Non-salt-forming* oxides are those oxides to which neither acids nor bases correspond (CO, N_2O, and NO).

Salt-like oxides are called double oxides, which consist of atoms of the same metal in different oxidation states. Most often, double or salt-like oxides are formed by metals that exhibit different oxidation states in their compounds: Pb_3O_4, Fe_3O_4, and Mn_3O_4. The formulas of these oxides can be represented as $2PbO{\cdot}PbO_2$, $FeO{\cdot}Fe_2O_3$, and $MnO{\cdot}Mn_2O_3$. For example, Fe_3O_4 is a basic oxide of FeO, chemically bonded to amphoteric oxide Fe_2O_3, which in this case exhibits the acidic properties. Thus, Fe_3O_4 can be formally considered as a salt $Fe(FeO_2)_2$, formed by the reaction of the base $Fe(OH)_2$ with the acid $HFeO_2$, which does not exist in nature:

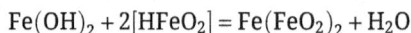

$$Fe(OH)_2 + 2[HFeO_2] = Fe(FeO_2)_2 + H_2O$$

From the acid $Pb(OH)_4$ and the base $Pb(OH)_2$, two double oxides can be obtained, Pb_2O_3 and Pb_3O_4, which are considered as salts. The first is the salt of H_2PbO_3, and the second is of H_4PbO_4.

In the group of oxides (especially in the oxides of d-elements), there are many compounds with a variable stoichiometric composition (berthollides), in which the oxygen content varies within fairly wide limits; for example, the composition of titanium(II) oxide (TiO) varies within the limits $TiO_{0.65}$–$TiO_{1.25}$.

Salt-forming oxides are the oxides that form salts. Oxides of this type are divided into three classes: basic, amphoteric, and acidic. *Basic oxides* are the oxides whose element is a cation in the corresponding salt or base. *Acidic oxides* are the oxides whose element is part of the anion in the formation of the corresponding salt or acid. *Amphoteric oxides* are called oxides that, depending on the conditions of the reaction, can exhibit both acidic and basic properties.

There are some regularities in the change in the properties of oxides (Table 21.3). For example, with an increase in the oxidation state of the element and a decrease in the ionic radius (respectively, a decrease in the effective negative charge of the oxygen atom – δ_O), the acidic properties of the oxide are enhanced. This explains the regular change in the properties of the oxides of the elements of the periodic table from basic to amphoteric and acidic. In a period, when the sequence number of the elements increases, the acidic properties of the oxides are enhanced, as well as the strength of the corresponding acids.

Table 21.3: Properties of oxides in a period.

Oxides	Na_2O	MgO	Al_2O_3	SiO_2	P_4O_{10}	SO_3	Cl_2O_7
Effective charge, δ_O	−0.81	−0.42	−0.31	−0.23	−0.13	−0.06	−0.01
Properties of oxides	Basic	Basic	Amphoteric	Acidic	Acidic	Acidic	Acidic

In the main groups of the periodic table, from the top to the bottom in the group, an increase in the basic properties of oxides is observed; for example, BeO is amphoteric, while MgO, CaO, SrO, BaO, and RaO are basic, and in this order their basic properties are enhanced.

For example, Cr_2O_3 has a melting point of 2,266 °C, a value characteristic of ionic compounds, while CrO_3 has a melting point of 196 °C, a value characteristic of covalent compounds. As the oxidation state of the element increases, the acidic properties of the oxides are enhanced and the basic ones are weakened. For example, MnO is a typical basic oxide, Mn_2O_3 is an amphoteric oxide with predominantly basic properties, MnO_2 is an amphoteric oxide with predominantly acidic properties, and MnO_3 and Mn_2O_7 are typical acidic oxides.

The basic oxides include the oxides of all metals of the first and second main groups, as well as the oxides of the transition metals in their lower oxidation states. The oxides of the most active metals (alkali and alkaline) under normal conditions interact directly with water, forming hydroxides, which are strong and water-soluble bases.

Most nonmetal oxides are acidic (CO_2, SO_3, P_4O_{10}, etc.). Oxides of nonmetals always contain covalent bonds. Those in which the element is in its lowest oxidation state are neutral, while those with an element in higher oxidation states are acidic.

For example, nitrous oxide (N_2O) is neutral, while dinitrogen pentoxide dissolves in water to nitric acid. Oxides of transition metals in their highest oxidation states exhibit mainly acidic properties; for example, CrO_3, Mn_2O_7, and V_2O_5. Acidic oxides react with water to form the corresponding acid. Some oxides, for example, SiO_2 and MoO_3, do not react directly with water, and their corresponding acids are produced indirectly. All acidic oxides, regardless of their relation to water, react with bases and basic oxides.

Amphotericity (from the Greek amphoteros – both) is the ability of chemical compounds (oxides, hydroxides, and amino acids) to exhibit both acidic and basic properties, depending on the properties of the second reactant involved in the reaction. The same substance (ZnO), reacting with a strong acid or acidic oxide, exhibits the properties of a base oxide, and when reacting with a strong base or basic oxide exhibits the properties of acidic oxide. Amphoteric oxides, as a rule, are stable not only in water but also in acids and bases. A typical example of such an oxide is Al_2O_3, which practically does not react with acids and interacts with bases only in melts. Some of the oxides participate in redox reactions, leading to changes in the oxidation state of the given element.

Oxides are produced either by direct interaction between elements and oxygen, or by thermal decomposition of hydroxides or salts of oxygen-containing acids.

21.4.1 Peroxides, superoxides, and ozonides

Oxygen is part of peroxides, superoxides, and ozonides having the oxidation states −1, −½, and −⅓, respectively. In these compounds, oxygen is in the form of O_2^{2-} (peroxide ion), O_2^- (superoxide ion), and O_3^- (ozonide ion). The first two are taken as derivatives of the oxygen molecule, which has accepted two electrons and one electron, respectively, and the third as a derivative of the ozone molecule that has accepted one electron.

An important compound of oxygen is *hydrogen peroxide* (H_2O_2), discussed in a previous chapter. *Metal peroxides* are compounds of the cations of alkali and alkaline earth metals with peroxide anions O_2^{2-}; therefore, they are considered as salts of the acid H_2O_2. They are obtained by the oxidation of molten metals in excess oxygen. Of the alkali metals, only lithium reacts with oxygen to form a normal oxide. The rest of the alkaline elements give up peroxides, superoxides, and ozonides. Sodium reacts with O_2 to form sodium peroxide:

$$2Na + O_2 \rightarrow Na_2O_2$$

Sodium peroxide is diamagnetic, with the length of the oxygen-oxygen bond being about 149 pm, much longer than that in the oxygen molecule (121 pm). Diamagnetism can be explained by the weaker bond due to the filled antibonding π_{2p} orbitals, according to the molecular orbital method. The bond order between oxygen atoms is 1, which determines the less stability of these compounds. The other three alkali metals react with excess O_2 to superoxides containing the paramagnetic superoxide ion O_2^-:

$$K + O_2 \rightarrow KO_2$$

The length of the oxygen-oxygen bond in these ions (133 pm) is less than that in peroxide, but slightly larger than that in O_2 itself, according to the method of molecular orbitals. The order of the bond between oxygen atoms becomes 1½, located between 1 in O_2^{2-} and 2 in the oxygen molecule. These large polarizing anions are usually stabilized by cations with low charge density.

All these compounds react energetically with water to metal hydroxides, with alkaline peroxide forming hydrogen peroxide and superoxide forming hydrogen peroxide and oxygen:

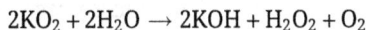

$$Li_2O + H_2O \rightarrow 2LiOH$$

$$Na_2O_2 + 2H_2O \rightarrow 2NaOH + H_2O_2$$

$$2KO_2 + 2H_2O \rightarrow 2KOH + H_2O_2 + O_2$$

When heated, they release oxygen:

$$2CaO_2 = 2CaO + O_2$$

Barium peroxide is used to produce H_2O_2. Sodium peroxide is used to bleach linen, woolen fabrics, etc. It is used to regenerate air in submarines and space stations, which is related to the fact that when interacting with CO_2, oxygen is released:

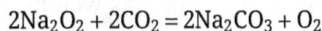

$$2Na_2O_2 + 2CO_2 = 2Na_2CO_3 + O_2$$

Potassium superoxide is used in some types of self-breathing equipment, as it strongly absorbs exhaled carbon dioxide and moisture and releases oxygen:

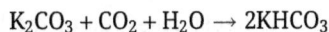

$$4KO_2 + 2CO_2 \rightarrow 2K_2CO_3 + 3O_2$$

$$K_2CO_3 + CO_2 + H_2O \rightarrow 2KHCO_3$$

Ozonides, as well as superoxides, are paramagnetic color compounds with a strong oxidative effect. When heated, they decompose and release oxygen:

$$2KO_3 = 2KO_2 + O_2$$

Ozonides energetically interact with water and acids:

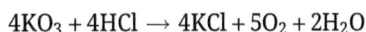

$$4KO_3 + 2H_2O \rightarrow 4KOH + 5O_2$$

$$4KO_3 + 4HCl \rightarrow 4KCl + 5O_2 + 2H_2O$$

21.5 Sulfur

21.5.1 Brief description

Sulfur is a typical nonmetal. Under normal conditions, sulfur is an insulator, forming yellow crystals with a density of 2.07 g/cm^3. Sulfur exists in the form of S_8 octa-atomic molecules, forming molecular crystals in three allotropic modifications: rhombic (α-sulfur), monoclinic (β-sulfur), and γ-sulfur. The most stable under normal conditions is rhombic sulfur. This modification changes to monoclinic at 95.5 °C. Both modifications have a crown-shaped form. Rhombic sulfur is yellow in color and monoclinic sulfur is pale yellow. The third sulfur modification has a hexagonal lattice. This is an unstable modification and becomes a mixture of rhombic and monoclinic modifications. Its crystals contain S_6 rings. Monoclinic sulfur melts at 119.3 °C, and rhombic sulfur at 112 °C. Sulfur boils at 445 °C. S_6 and S_4 molecules appear in vapor when the temperature rises. At 800–900 °C, S_2 molecules appear in, and these diatomic molecules break down into atoms at 1,000 °C. Sulfur dissolves well in carbon disulfide (CS_2) and organic solvents. It does not dissolve in water.

21.5.2 Occurrence and synthesis

Sulfur is a well-distributed chemical element and has been known to mankind since ancient times. It is found in large quantities in a free state (native sulfur), in the form of compounds such as sulfates and sulfides, as well as in the gaseous form such as hydrogen sulfide in the natural gas of some mineral sources. Natural sulfates are $Na_2SO_4 \cdot 10H_2O$ – mirabilite, $CaSO_4 \cdot 2H_2O$ – gypsum, and sulfides are PbS – galena, ZnS – sphalerite, Cu_2S – chalcosine, etc.

The production of sulfur is done by melting the ore in the deposits using hot water under pressure (the Frasch method). In addition, sulfur is obtained from hydrogen sulfide present in natural or exhaust gases (Claus method) by oxidation in case of oxygen deficiency (incomplete combustion of hydrogen sulfide):

$$2H_2S + O_2 = 2S \downarrow + 2H_2O$$

In the absence of deposits of native sulfur, it is most often obtained by the Claus method, which allows for the recovery of some of the waste gases:

$$2H_2S + SO_2 = 3S + 2H_2O$$

21.5.3 Chemical properties

By transition from oxygen to sulfur, the strength of the single σ bond increases due to the reduced interelectron repulsion. At the same time, the strength of the π bond decreases, which is associated with an increase in the radius and a reduced overlap of the p-atomic orbitals. Therefore, if oxygen is characterized by the formation of multiple (σ + π) bonds, sulfur and its analogues are characterized by the formation of single chain bonds –E–E– E–E–. Forming chain and cyclic compounds is one of the characteristic properties of the atoms of the chalcogenic elements, which is called catenation. Catenation is not limited to simple substances but also occurs in the chemical compounds of these elements.

Sulfur has a pronounced chemical activity. It interacts with almost all chemical elements, except for precious metals, inert gases, and iodine, and its activity toward metals is significantly greater than toward nonmetals. All sulfur reactions are associated with the opening of the S_8 ring and the formation of chains of sulfur atoms or compounds containing such chains. Therefore, chemical reactions with sulfur take place at an elevated temperature, facilitating the breaking of the S–S bonds. Even the simple inorganic reaction of sulfur with sodium sulfite is complicated:

$$S_8 + 8Na_2SO_3 \rightarrow 8Na_2S_2O_3$$

Sulfur manifests itself as an oxidizing agent when interacting with metals:

$$Zn + S = ZnS$$

$$2Al + 3S = Al_2S_3$$

In the row of standard potentials, p-elements of group VIA stand after hydrogen; therefore, they do not react with acids with nonoxidative action. They are oxidized only by acids with oxidative action:

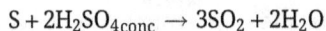

$$S + 4HNO_3 \rightarrow SO_2 + 4NO_2 + 2H_2O$$

$$S + 6HNO_3 = H_2SO_4 + 6NO_2 + 2H_2O$$

$$S + 2H_2SO_{4conc} \rightarrow 3SO_2 + 2H_2O$$

In fuming sulfuric acid (oleum), sulfur dissolves to form solutions colored from yellow to light blue, containing paramagnetic particles of unknown composition. Sulfur does not interact with water. In bases, when heated, it is disproportionated according to the equation:

$$3S + 6NaOH = 2Na_2S + Na_2SO_3 + 3H_2O$$

During prolonged boiling, sulfur reacts with bases according to the equation:

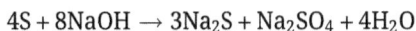

$$4S + 8NaOH \rightarrow 3Na_2S + Na_2SO_4 + 4H_2O$$

21.5.4 Use and biological role

Sulfur is widely used in industry. It is mainly used to obtain sulfuric acid and vulcanization of rubber. Its most important use is in the vulcanization of rubber (mixing rubber with sulfur when heated). Rubber with a higher sulfur content is called ebonite, which is used as an insulating material.

In the geosphere, sulfur is present in its native state, in the form of impurities in coal, in the form of compounds with the components of natural gas and oil, in the form of numerous ores of metal sulfides and polysulfides, and in the form of sulfates (mainly sodium, calcium, and magnesium, and less often zinc and iron). In the biosphere, sulfur forms compounds close to natural inorganic polysulfides. These are various proteins with sulfur bridges S–S; for example, methionine, lipoic acid, glutathione, thiamine, and coenzyme A. A fragment of the sulfur cycle in nature can be represented by the simplified scheme: plants $\rightleftarrows$ animals $\rightarrow$ metabolism $\rightarrow$ SO_4^{2-}.

Plants, unlike animals, are capable of synthesizing S-containing amino acids. Animals ingest the S-containing proteins stored in plants, which does not compensate for the need for sulfur intake during feeding. This is achieved by taking dietary supplements, such as the good source of sulfur and powerful antioxidant methionine, contained in dairy products, etc.

Interestingly, some microorganisms (the so-called thionic bacteria) absorb H_2S and CO_2 and convert them into hydrocarbons $C_x(H_2O)_y$ according to the equation:

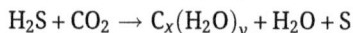

$$H_2S + CO_2 \rightarrow C_x(H_2O)_y + H_2O + S$$

Some thionic bacteria are capable of oxidizing sulfides and polysulfides (pyrite, FeS_2) not only to free sulfur but also to thiosulfate ions, and at a high rate. The specific metabolic characteristics of these bacteria allow for their wide practical use in various technologies.

Sulfur in the body is part of the sulfhydryl groups –SH of proteins, and is also present in the form of sulfates and hydrogen sulfide in the gastrointestinal tract. Protein tissues are able to absorb excess superoxide radicals and thus prevent further tissue destruction. In this case, the –SH groups are converted into –S–S–H groups. These groups play an important role in the body's self-defense against radiation. They are the first to react with free radicals H• and OH• and protect nucleic acids. In the body, the –SH groups are converted into SO_4^{2-}, $S_2O_3^{2-}$, $S_4O_6^{2-}$, and S_8 groups under the action of oxidative enzymes. Of all these processes of oxidation of sulfhydryl groups, the most important is the formation of endogenous sulfuric acid, which neutralizes the poisonous phenol, cresol, and indole, produced in the intestine by amino acids. Sulfuric acid also binds xenobiotics (foreign proteins) due to the formation of the corresponding sulfuric acid esters.

Sulfur is part of many medications. Sulfur is an effective remedy for sports injuries, joint diseases, spikes, eczema, acne, warts, hemorrhoids, stomach problems, and lack of appetite. It has an immunostimulant effect, regulates blood sugar levels, and improves the skin. Suspension of sulfur in water is used for cyanide poisoning (CN^- ions), as sulfur converts highly toxic cyanides into SCN^- rhodanides. Sulfur suspension in ether serves to treat some skin diseases.

21.6 Chemical compounds of sulfur

21.6.1 Hydrides

Hydrogen sulfide (H_2S) is the most important sulfur hydride. Hydrogen sulfide is a highly toxic gas with an unpleasant suffocating odor. Naturally, hydrogen sulfide is produced by anaerobic bacteria. In fact, this process, which occurs in degeneration, is the source of natural sulfur in the atmosphere. Under laboratory conditions, H_2S is formed by direct interaction of hydrogen and sulfur at 150 °C, by treating iron(II) sulfide with hydrochloric acid, or by hydrolysis of aluminum sulfide:

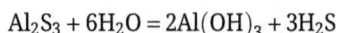

$$H_2 + S = H_2S$$

$$FeS + 2HCl = FeCl_2 + H_2S$$

$$Al_2S_3 + 6H_2O = 2Al(OH)_3 + 3H_2S$$

As an analogue of the water molecule, the hydrogen sulfide molecule has a V-shaped structure with a smaller angle between the bonds. Hydrogen sulfide molecules, like water molecules, are polar, but unlike water molecules, hydrogen bonds practically do not occur between them. Therefore, hydrogen sulfide is very volatile and slightly soluble in water. The concentration of its saturated aqueous solution at 20 °C is about 0.4%. In solution, hydrogen sulfide is a weak acid, dissociating into two degrees:

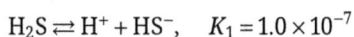

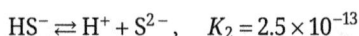

$$H_2S \rightleftharpoons H^+ + HS^-, \quad K_1 = 1.0 \times 10^{-7}$$

$$HS^- \rightleftharpoons H^+ + S^{2-}, \quad K_2 = 2.5 \times 10^{-13}$$

H_2S is a good reducer. Depending on the conditions, it can be oxidized by the reactions:

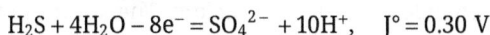

$$H_2S - 2e^- = S + 2H^+, \qquad J° = 0.14 \text{ V}$$

$$HS^- - 2e^- = S + H^+, \qquad J° = -0.06 \text{ V}$$

$$H_2S + 4H_2O - 8e^- = SO_4^{2-} + 10H^+, \quad J° = 0.30 \text{ V}$$

Hydrogen sulfide burns in the air to sulfur or SO_2, depending on the stoichiometric ratio:

$$2H_2S + O_2 \rightarrow 2H_2O + 2S$$

$$2H_2S + 3O_2 \rightarrow 2H_2O + 2SO_2$$

When comparing the standard redox potentials, it can be seen that the most likely oxidation of hydrogen sulfide is to sulfur, and further oxidation to sulfuric acid is not excluded:

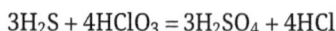

$$3H_2S + K_2Cr_2O_7 + 4H_2SO_4 = 3S\downarrow + Cr_2(SO_4)_3 + K_2SO_4 + 7H_2O$$

$$3H_2S + 4HClO_3 = 3H_2SO_4 + 4HCl$$

H_2S is used to produce sulfur and various sulfides, as well as in organic synthesis (thiophenes and mercaptans). It is highly toxic because it binds to the hemoglobin of the blood and paralyzes the respiratory tract. It is more toxic than hydrogen cyanide. When hydrogen sulfide is inhaled, it mainly attacks the central nervous system, as a result of which breathing is paralyzed.

Sulfur is characterized by the formation of *polysulfanes* (H_2S_n), which is related to the tendency of S to catenate. Compounds with $n = 2-8$ are isolated in an individual state, while higher homologues are isolated only in mixtures. Polysulfanes are synthesized in different ways:

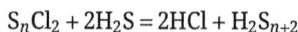

$$Na_2S_n + 2HCl = 2NaCl + H_2S_n \quad (n = 4-6)$$

$$S_nCl_2 + 2H_2S = 2HCl + H_2S_{n+2}$$

Polysulfanes are unstable, and easily disproportionate to H_2S and free sulfur. Colorless disulfan (H_2S_2) is an analogue of hydrogen peroxide (H_2O_2). It has a similar structure and exhibits oxidizing properties. With the help of H_2S_2, GeS, and SnS, sulfides are oxidized to thioderivatives of Ge(IV) and Sn(IV), and sulfides As_2S_3 and Sb_2S_3 to thioderivatives of As(V) and Sb(V):

$$GeS + (NH_4)_2S_2 \rightarrow (NH_4)_2GeS_3$$

$$As_2S_3 + 2(NH_4)_2S_2 = 2NH_4AsS_3 + (NH_4)_2S$$

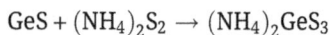

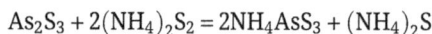

21.6.2 Sulfides

Compounds of sulfur, selenium, and tellurium with more electropositive elements (metals and some nonmetals) are called chalcogenides, among which sulfides are the most important. The properties of sulfides, formed by the elements in a given period of the periodic table, change from basic to acidic. The sulfides of the typical (alkali and alkaline earth) metals have basic properties, and those of the nonmetals – acidic properties. The former, when dissolved in water, form an alkaline medium due to reversible hydrolysis, and the latter are completely decomposed by water with the formation of two acids:

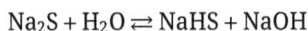

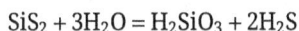

$$Na_2S + H_2O \rightleftharpoons NaHS + NaOH$$

$$SiS_2 + 3H_2O = H_2SiO_3 + 2H_2S$$

The sulfides of some amphoteric metals (Al_2S_3, Cr_2S_3, and Fe_2S_3) are irreversibly hydrolyzed to form poorly soluble hydroxides and hydrogen sulfide:

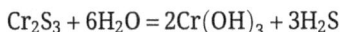

$$Cr_2S_3 + 6H_2O = 2Cr(OH)_3 + 3H_2S$$

When basic sulfides react with acidic sulfides, sulfosalts or thiosalts (e.g., sodium thiosilicate) are formed:

$$Na_2S + SiS_2 = Na_2SiS_3$$

Sulfides are divided into four groups depending on their relation to water and solutions:
1) Sulfides of alkali and alkaline earth metals, as well as ammonium sulfide, are *soluble in water.*
2) *Insoluble in water, but soluble in acids with low oxidizing ability* (HCl, dilute sulfur, etc.) are the following sulfides: ZnS, FeS, and MnS. Their solubility is explained by the following reaction:

$$ZnS + H_2SO_4 = ZnSO_4 + H_2S$$

3) Sulfides of heavy metals, including Cu_2S, CuS, PbS, HgS, Ag_2S, Bi_2S_3, and MoS_2, are *soluble in acids with oxidative action* (nitric acid, concentrated sulfuric acid, etc.). Their solubility is explained by the oxidation of the sulfide ion:

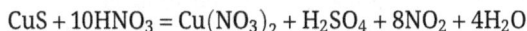

$$CuS + 10HNO_3 = Cu(NO_3)_2 + H_2SO_4 + 8NO_2 + 4H_2O$$

In some sulfides, along with the oxidation of the sulfide ion, the metal is also oxidized to its highest oxidation state:

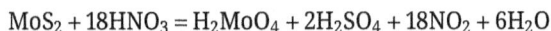

$$MoS_2 + 18HNO_3 = H_2MoO_4 + 2H_2SO_4 + 18NO_2 + 6H_2O$$

4) *Soluble in solutions of sulfides of the first group.* These are the sulfides of nonmetals and amphoteric metals: As_2S_3, As_2S_5, Sb_2S_3, Sb_2S_5, P_2S_3, P_2S_5, CS_2, SiS_2, GeS, GeS_2, SnS_2, and B_2S_3. Their common name is sulfoanhydrides. Their solubility is explained by the formation of soluble sulfosalts or thiosalts:

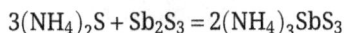

$$3(NH_4)_2S + Sb_2S_3 = 2(NH_4)_3SbS_3$$

If a solution of a sulfosalt reacts with an acid with low oxidizing ability, the sulfoacid, formed during the exchange reaction, instantly decomposes into sulfide and hydrogen sulfide:

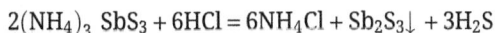

$$2(NH_4)_3 SbS_3 + 6HCl = 6NH_4Cl + Sb_2S_3\downarrow + 3H_2S$$

Like sulfanes, there are homologous series of polysulfides of the type M_2S_n ($n = 2–9$), of which the most famous are disulfides. These are solid compounds containing a disulfide ion (S_2^{2-}), analogous to O_2^{2-}. Thus, FeS_2 does not contain iron in a higher oxidation state, but a disulfide ion (S_2^{2-}). These solid structures have been found for the metals manganese, iron, cobalt, nickel, ruthenium, and osmium. The disulfide ion also forms compounds with alkali and alkaline earth metals. They are part of the family of polysulfides containing the S_n^{2-} ion, where n has values between 2 and 6. Unlike heavy metal disulfides, alkaline disulfides are soluble in water. Their different solubility is used to separate chemical elements. Sulfides are used as pigments, semiconductors, in chemical synthesis, leather production, pest control of agricultural crops, etc.

21.6.3 Halogenides and oxyhalides

Sulfur reacts with all halogen elements, except iodine, and forms compounds in oxidation states +6 (SF_6, SO_2F_2, and SO_2Cl_2), +4 (SF_4, SCl_4, $SOCl_2$, and $SOBr_2$), +2 (SCl_2), and +1 (S_2Cl_2). All of them, with the exception of sulfur hexafluoride, are completely hydrolyzed to form two acids:

$$SCl_4 + 3H_2O = H_2SO_3 + 4HCl$$

Compounds that interact with water to form two acids (one of which is hydrogen halide) are called halides. Therefore, most of the compounds of sulfur with halogen elements are halogen anhydrides.

The most important sulfur halide is sulfur hexafluoride (SF_6). The sulfur tetrafluoride (SF_4) is chemically more active than SF_6. Surprisingly, the most stable chlorides are those with a low oxidation state of sulfur: SCl_2 and S_2Cl_2. Sulfur hexafluoride (SF_6) with octahedron-shaped molecule is produced by direct combustion of molten sulfur into gaseous fluorine:

$$S + 3F_2 \rightarrow SF_6$$

Sulfur hexafluoride is an inert gaseous substance ($\Delta G°_{form} = -1,103$ kJ/mol), used as a gaseous insulator in high-voltage installations. Sulfur(I) chloride (S_2Cl_2) is used in the rubber industry for cold vulcanization of thin-walled rubber products, and sulfur tetrafluoride (SF_4) is used for fluorination of organic compounds.

The most important of the sulfur oxohalogenides, sulfuryl chloride (SO_2Cl_2), is synthesized in industry by direct chlorination of SO_2 in the presence of a $FeCl_3$ catalyst:

$$SO_2 + Cl_2 \rightarrow SO_2Cl_2$$

The compound is stable up to 300 °C, after which it dissociates into SO_2 and Cl_2. Due to this property, SO_2Cl_2 is used as a chlorinating and oxidizing reagent. SO_2Cl_2 is considered as sulfuric acid chloranhydride:

$$SO_2Cl_2 + 2H_2O = H_2SO_4 + 2HCl$$

The SO_2X_2 molecule is a deformed tetrahedron with O–S–O angle of the order of 120–126°.

21.6.4 Sulfur compounds with nitrogen

There are several nitrogen compounds of sulfur. Some of them are of interest because their structures and bond lengths cannot be explained in terms of known chemical bond theories. The classic example is *tetrasulfur tetranitride* (S_4N_4).

Of even greater interest is the polymer $(SN)_x$, called polythiazyl. This bronze-colored metal-like compound was synthesized in 1910, and only 50 years later, when studying its properties, it was proved that $(SN)_x$ is an excellent electrical conductor, and exhibits the properties of a superconductor at very low temperatures (0.26 K). Currently, there is an increased interest in nonmetallic compounds with metallic properties, both because of their potential use in everyday life and in view of the development of the modern theory of superconductivity.

21.6.5 Oxides

Sulfur oxides of different composition and structure have been obtained; some of which have a cyclic structure, and two of them are the most widely used – SO_2 and SO_3. A number of other unstable oxides are also known. For example, the oxide S_2O is produced by reacting thionyl chloride with silver sulfide.

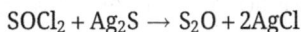

$$SOCl_2 + Ag_2S \rightarrow S_2O + 2AgCl$$

S_8O is obtained from H_2S_7 and $SOCl_2$ or by oxidation of S_8 with trifluoroacetic acid.

Sulfur dioxide (SO_2) is a colorless, toxic gas with a sour taste. As an anhydride of sulfurous acid, when dissolved in water, SO_2 forms mainly hydrates with an undetermined composition $SO_2 \cdot nH_2O$. It partially interacts with water to form the unstable sulfurous acid H_2SO_3, which belongs to the weak dibasic acids ($K_1 = 1.6 \times 10^{-2}$, $K_2 = 1.0 \times 10^{-7}$):

$$SO_2 + H_2O \rightleftharpoons H_2SO_3$$

The sulfur dioxide molecule is V-shaped with an S–O bond length of 143 pm and an O–S–O angle of 119°. This bond length is much shorter than the single sulfur-oxygen bond (163 pm) and very close to the typical sulfur-oxygen double bond (140 pm). The angle close to 120° indicates the similarity of sulfur dioxide to the trigonal sp^2-hybrid state.

Sulfur dioxide is obtained by frying pyrite, sulfur, and sulfides of nonferrous metals:

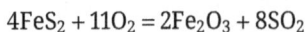

$$4FeS_2 + 11O_2 = 2Fe_2O_3 + 8SO_2$$
$$S + O_2 = SO_2$$
$$2NiS + 2O_2 = 2NiO + SO_2$$

In laboratory conditions, the following reactions are used to produce SO_2:

$$Na_2SO_3 + H_2SO_4 = Na_2SO_4 + SO_2 + H_2O$$
$$Cu + 2H_2SO_{4conc} = CuSO_4 + SO_2 + 2H_2O$$

For laboratory preparation, the reaction of soluble sulfite with dilute acid is most often used:

$$SO_3^{2-} + 2H^+ \rightarrow H_2O + SO_2$$
$$HSO_3^- + H^+ \rightarrow H_2O + SO_2$$

SO_2 is a typical acidic oxide and, as such, it reacts with basic oxides and bases.

Sulfur dioxide is one of the good reducing agents that is easily oxidized to a sulfate ion:

$$SO_2 + 2H_2O \rightarrow SO_4^{2-} + 4H^+ + 2e^-$$

It is oxidized by oxygen to sulfur trioxide. The reactions take place under heating and in the presence of catalysts – V_2O_5 with K_2O and Pt (heterogeneous contact catalysis) or NO (homogeneous catalysis), discussed below.

The oxidizing properties of SO_2 are manifested when interacting with strong reducers, H_2S:

$$2H_2S + SO_2 = 3S + 2H_2O$$

The presence of a free electron pair in the SO_2 molecule determines not only its reduction properties but also its tendency to complexation, in particular the formation of hydrates. The SO_2 molecule serves as a neutral ligand in numerous transition metal complexes; for example, $[RuCl(NH_3)_4(SO_2)]Cl$. SO_2 can coordinate to the metal either by sulfur or by oxygen atoms, manifesting itself as a monodentate or bidentate ligand.

Sulfur dioxide is an important intermediate in the production of sulfuric acid. It is also used as a bleaching agent and preservative in the food and wine industry (to stabilize white wines). In small quantities, it is used to disinfect fruit stores, to discolor wool, silk, and sugar, as well as in some organic productions. The maximum allowable sulfur dioxide concentration in humans is about 5 ppm, while plants begin to suffer at a concentration of 1 ppm.

Sulfur trioxide (SO_3, anhydride of sulfuric acid) is widely used in the production of sulfuric acid. It is produced by oxidation of dioxide in the presence of catalysts. There are two methods for producing SO_3: contact and nitrous methods. In the contact method, the oxidation of SO_2 with oxygen from the air occurs at 450–500 °C in the presence of a catalyst (V_2O_5 and Pt):

$$2SO_2 + O_2 = 2SO_3$$

In the nitrous method, the catalyst is a mixture of nitrogen oxides such as NO and NO_2. Catalytic oxidation takes place according to the scheme of a two-step process, in which the catalyst reacts with one of the substances until an intermediate is obtained, which in the second step reacts with the other substance, and the catalyst is reduced:

$$NO + \frac{1}{2}O_2 = NO_2$$
$$SO_2 + NO_2 = SO_3 + NO$$

SO_3 interacts energetically with water, and the reaction is accompanied by the release of heat:

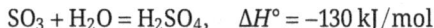

$$SO_3 + H_2O = H_2SO_4, \quad \Delta H° = -130 \text{ kJ/mol}$$

Due to this exothermic reaction, the resulting sulfuric acid begins to boil and sulfuric acid smoke (fuming H_2SO_4) is formed. To prevent this, the absorption of SO_3 is carried out not with water but with concentrated H_2SO_4, which dissolves SO_3 to form a product called *oleum* ("oil"):

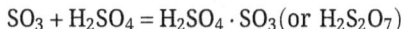

$$SO_3 + H_2SO_4 = H_2SO_4 \cdot SO_3 (\text{or } H_2S_2O_7)$$

To obtain sulfuric acid itself, the oleum is mixed with water:

$$H_2SO_4 \cdot SO_3 + H_2O = 2H_2SO_4$$

Sulfur trioxide is a colorless, low-boiling liquid at room temperature. The liquid and gaseous phases contain a mixture of SO_3 and S_3O_9. The liquid freezes at 16 °C to S_3O_9 crystals with a trimeric structure. The trimer is isoelectronic and isostructural with the polyphosphate $(P_3O_9)^{3-}$ and polysilicate $(Si_3O_9)^{6-}$ ions. In the presence of moisture, long-chain solid polymers with the structure $HO(SO_3)_nOH$ are formed, where n is about 10^5. When liquid SO_3 boils, the gas molecules that are formed are plane SO_3. As with sulfur dioxide, all sulfur-oxygen bond lengths are the same (142 pm) and in close proximity to the value of typical double bonds.

Sulfur trioxide is one of the compounds with the highest reactivity. It exhibits oxidizing properties. With sulfur and carbon, it is reduced to SO_2:

$$2SO_3 + C \rightarrow 2SO_2 + CO_2$$

At temperatures above 500 °C, SO_3 is also reduced by carbon monoxide:

$$SO_3 + CO \rightarrow SO_2 + CO_2$$

With metal sulfides, the reduction takes place to SO_2 or free sulfur.

When SO_3 reacts with gaseous H_2S, the products SO_2, H_2O, and S are formed. Sulfur trioxide acts as a strong Lewis acid, with which metal oxides form the corresponding sulfates:

$$Fe_2O_3 + 3SO_3 \rightarrow Fe_2(SO_4)_3$$

21.6.6 Oxoacids

Anhydrous *sulfuric acid* (H_2SO_4) is a dense liquid that solidifies at 10.3 °C and boils at 340 °C with partial decomposition to SO_3 and H_2O. It dissolves indefinitely in water, forming hydrogen bonds and hydrates, $H_2SO_4 \cdot xH_2O$. In solution, it dissociates into two degrees and is a strong acid ($K_1 = 10^3$, $K_2 = 1.3 \times 10^{-2}$). When sulfuric acid is dissolved, a large amount of heat is released, where the solution boils and splashes, which is very dangerous. Therefore, the dissolution of sulfuric acid should be carried out carefully and with continuous stirring to prevent local overheating. When diluted, the concentrated acid is poured in a thin stream into the water, and not vice versa. The reasons for this are that sulfuric acid is heavier than water and sinks immediately to the bottom of the container, preventing the acid from splashing. Dilute acid solutions left in the air concentrate slowly, which should be taken into account when working with H_2SO_4. The great thermal effect of dissolving sulfuric acid is explained by the formation of stable H_2SO_4 hydrates: $H_2SO_4 \cdot H_2O$, $H_2SO_4 \cdot 2H_2O$, and $H_2SO_4 \cdot 4H_2O$. Their formation also explains the water-taking and drying properties of concentrated sulfuric acid.

The sulfuric acid molecule is made up of a tetrahedral arrangement of oxygen atoms around the central sulfur atom. The lengths of the S–OH (1.54 Å) and S–O (1.43 Å) bonds in the H_2SO_4 molecule are such that the S–O bonds can be considered as double and the S–OH bonds as single. The small lengths of the bonds and their high energy suggest the dual nature of the sulfur-oxygen bonds in the sulfate ion.

SO_2 is used for industrial production of sulfuric acid. Sulfur dioxide is obtained by burning sulfur or "frying" sulfides (mainly pyrite, FeS_2). In the next step, SO_2 is oxidized to SO_3. The final stage of industrial production of sulfuric acid is the absorption of SO_3 by water to sulfuric acid:

$$SO_3 + H_2O = H_2SO_4$$

Most often this is done not with water (due to the highly exothermic nature of the process), but with a concentrated H_2SO_4 (97%). The resulting sulfuric acid is called "oleum."

Dilute sulfuric acid is a weak oxidizing agent. It only interacts with metals that are found before hydrogen in the row of relative activity. In this interaction, sulfates of metals are formed in their lower oxidation states and hydrogen is released:

$$Fe + H_2SO_4 = FeSO_4 + H_2$$

Concentrated sulfuric acid is a strong oxidizing agent. It oxidizes all metals, except gold and platinum, and many of the nonmetals – carbon, phosphorus, sulfur, etc. The reduction of sulfuric acid leads to the formation of SO_2, sulfur, or hydrogen sulfide:

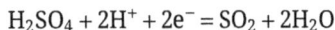

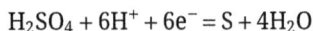

$$H_2SO_4 + 2H^+ + 2e^- = SO_2 + 2H_2O$$

$$H_2SO_4 + 6H^+ + 6e^- = S + 4H_2O$$

$$H_2SO_4 + 8H^+ + 8e^- = H_2S + 4H_2O$$

Usually, all three products are formed, with the predominant being SO_2:

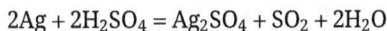

$$2Ag + 2H_2SO_4 = Ag_2SO_4 + SO_2 + 2H_2O$$

$$C + 2H_2SO_4 = CO_2 + 2SO_2 + 2H_2O$$

The released sulfur and hydrogen sulfide also interact with sulfuric acid:

$$3H_2S + H_2SO_4 = 4S + 4H_2O$$

$$S + 2H_2SO_4 = 3SO_2 + 2H_2O$$

Sulfuric acid with a concentration higher than 93% does not react with iron, which allows it to be transported by rail in tanks of ordinary steel. Lead is stable in dilute sulfuric acid due to the formation of insoluble $PbSO_4$ on its surface (which is important for its application in batteries). However, with concentrated sulfuric acid, lead is oxidized to form a soluble acidic salt:

$$Pb + 3H_2SO_4 = Pb(HSO_4)_2 + SO_2 + 2H_2O$$

Sulfuric acid is produced on a large scale, as it is an important raw material in the production of a number of other chemicals. The main consumer of sulfuric acid is the industrial production of mineral fertilizers. In addition, it is used in the production of many acids and salts, organic products, and explosives, for the purification of petroleum products, for cleaning iron and steel products, etc. Sulfuric acid is one of the most commonly used reagents in laboratory practice.

Sulfuric acid salts are *sulfates* (Na_2SO_4, $CaSO_4$, and $Al_2(SO_4)_3$) and *hydrogen sulfates* ($KHSO_4$ and $Mg(HSO_4)_2$). Since the dissociation of the second degree is significant, sulfuric acid gives mainly normal salts – sulfates. Sulfates in solid state are crystal hydrates: $CuSO_4 \cdot 5H_2O$ – bluestone, $FeSO_4 \cdot 7H_2O$ – green vitriol, etc. Double sulfates (*alum*) are relatively easy to synthesize; for example, $NaAl(SO_4)_2 \cdot 12H_2O$ and $KCr(SO_4)_2 \cdot 12H_2O$. Sulfates are widely used in industry. Ammonium sulfate is an artificial fertilizer; sodium sulfate is used in glass, paper, and other industries; aluminum sulfate in water purification; copper sulfate for pest control of agricultural crops; iron sulfate for the preparation of mineral paints, etc.

Sulfuric dioxide (anhydride) partially reacts with water to form the unstable sulfurous acid (H_2SO_3) (Figure 21.1), which belongs to the weak dibasic acids.

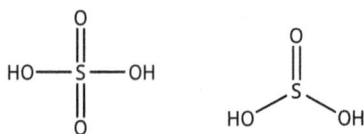

Figure 21.1: Structures of sulfuric and sulfurous acids.

There is no evidence that sulfurous acid exists in solution, but its molecule was found in the gas phase. The conjugate bases of this unstable acid are the common anions bisulfite (or hydrogen sulfite) and sulfite. The acid is found in acidic rain. Raman spectral studies of SO_2 solutions in water show only signals due to the SO_2 molecule and the hydrogen sulfite ion (HSO_3^-) in agreement with the following equilibrium:

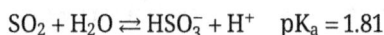

$$SO_2 + H_2O \rightleftharpoons HSO_3^- + H^+ \quad pK_a = 1.81$$

Other spectral studies (^{17}O NMR) provide evidence that solutions of sulfurous acid and protonated sulfites contain a mixture of isomers that are in equilibrium:

$$[H - OSO_2]^- \rightleftharpoons [H - SO_3]^-$$

The salts of sulfurous acid (*sulfites*), when dissolved in water, are partially hydrolyzed to form an alkaline medium:

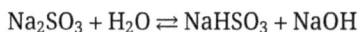

$$Na_2SO_3 + H_2O \rightleftharpoons NaHSO_3 + NaOH$$

S(IV) compounds exhibit redox duality, and their reduction properties are more typical. For example, in a solution, oxygen easily oxidizes sulfites to sulfates. Therefore, sulfites are stored dry and their solutions are prepared immediately before use. The interaction of sodium or potassium sulfite as reducers with potassium permanganate is the most common redox reaction. Depending on the acidity of the medium, different oxidation products are obtained:

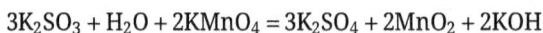

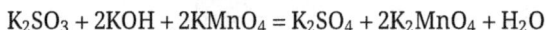

$$5Na_2SO_3 + 2KMnO_4 + 3H_2SO_4 = 5Na_2SO_4 + 2MnSO_4 + K_2SO_4 + 3H_2O$$

$$3K_2SO_3 + H_2O + 2KMnO_4 = 3K_2SO_4 + 2MnO_2 + 2KOH$$

$$K_2SO_3 + 2KOH + 2KMnO_4 = K_2SO_4 + 2K_2MnO_4 + H_2O$$

In cases where S(IV) compounds are oxidizing agents, they are usually reduced to sulfur:

$$H_2SO_3 + 2H_2S = 3S + 3H_2O$$

A consequence of the redox duality of S(IV) compounds is their disproportionation. For example, sodium sulfite turns into sulfate and sulfide when heated, and when an aqueous solution of SO_2 is heated to 150 °C in a soldered tube, H_2SO_4 and sulfur are formed:

$$4Na_2SO_3 = 3Na_2SO_4 + Na_2S$$

$$3SO_2 + 2H_2O = 2H_2SO_4 + S$$

In concentrated solutions of hydrogen sulfites, pyrosulfites are formed, in which the S–S bond is realized, and not S–O–S:

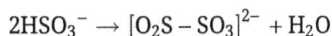

$$2HSO_3^- \rightarrow [O_2S - SO_3]^{2-} + H_2O$$

Aqueous solution of sulfur dioxide (sulfurous acid) as well as solutions of bisulfites and sulfites are used as reducing agents and disinfectants. They are used in materials that can be damaged by chlorine-containing bleach.

Due to its tendency to catenation and the diversity in its oxidation states, sulfur forms many oxoacids with different stability. The first group includes *polysulfuric acids*, with the general formula $H_2SO_4 \cdot nH_2O$ or $H_2S_nO_{3n+1}$, and the technical name "oleum," given above. At $n = 2$, the acid has the formula $H_2S_2O_7$ (*di sulfuric* and *pyrosulfuric*), and at $n = 3$, the formula of the acid is $H_2S_3O_{10}$ (*trisulfuric acid*). Sulfuric acid (H_2SO_4) is more correctly called orthosulfuric acid, because it contains the largest number of hydroxyl groups bonded to one central atom S(VI). When H_2SO_4 is dehydrated or when its aqueous solution is saturated with SO_3, disulfuric acid ($H_2S_2O_7$) is formed:

$$H_2SO_4 + SO_3 = H_2S_2O_7$$

In the molecule of disulfuric acid (Figure 21.2), the two SO_4^{2-} tetrahedra bind to a common oxygen atom.

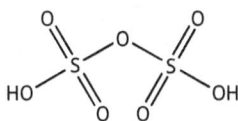

Figure 21.2: Structure of pyrosulfuric acid.

As can be seen, the pyrosulfate ion has a structure $[O_3S-O-SO_3]^{2-}$.

At boiling point, the solution of Na_2SO_3 with sulfur forms sodium thiosulfate, which corresponds to *thiosulfuric acid* ($H_2S_2O_3$) (Figure 21.3). Its composition can be expressed in two graphic formulas, corresponding to its resonance structures, from which it is seen that the molecule contains sulfur in two oxidation states: one in maximum positive (+6) and another in maximum negative (−2):

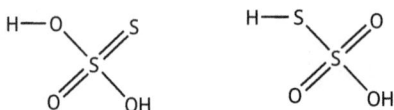

Figure 21.3: Structure of thiosulfuric acid.

When heated strongly, sodium thiosulfate is disproportionated to three different oxidizing states of sulfur: sodium sulfate, sodium sulfide, and sulfur:

$$4Na_2S_2O_3 \rightarrow 3Na_2SO_4 + Na_2S + 4S$$

It is for this reason that thiosulfuric acid is unstable and decomposes at the moment of production by an intramolecular redox mechanism:

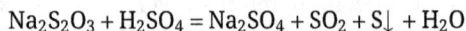

$$Na_2S_2O_3 + H_2SO_4 = Na_2SO_4 + SO_2 + S{\downarrow} + H_2O$$

The rate of the reaction is judged by the opalescence of the solution due to the formation of S.

Sodium thiosulfate is used to absorb chlorine:

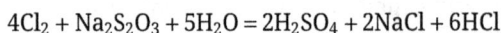

$$4Cl_2 + Na_2S_2O_3 + 5H_2O = 2H_2SO_4 + 2NaCl + 6HCl$$

and in photography as a binding fixative:

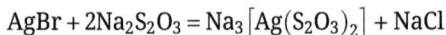

$$AgBr + 2Na_2S_2O_3 = Na_3\left[Ag(S_2O_3)_2\right] + NaCl$$

In very dilute aqueous solutions, sulfur dioxide (SO_2) reacts with hydrogen sulfide (H_2S) to form *polythionic acids* with the general formula $H_2S_nO_6$, where $n = 3$–6. In these acids, sulfur atoms are linked to each other in straight chains; for example, *trithionic acid* ($H_2S_3O_6$) and *tetrathionic acid* ($H_2S_4O_6$) (Figure 21.4). They are the conjugate acids of polythionates in which the bond −S−S− is realized.

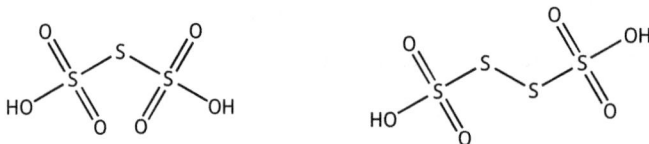

Figure 21.4: Structures of trithionic and tetrathionic acids.

The salt of tetrathionic acid (sodium tetrathionate) is formed when $Na_2S_2O_3$ reacts with I_2:

$$2Na_2S_2O_3 + I_2 = 2NaI + Na_2S_4O_6$$

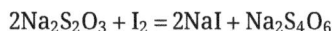

This reaction is the basis of one of the methods of quantitative analysis called iodometry.

In addition to the sulfur acids listed above, there is another group called *peroxysulfuric acids*. Peroxydisulfuric acid has the formula $H_2S_2O_8$. Although the sulfate ion contains sulfur in its highest oxidation state +6, it can be oxidized by electrolysis to a peroxydisulfate ion:

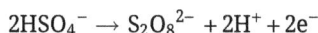

$$2HSO_4^- \rightarrow S_2O_8^{2-} + 2H^+ + 2e^-$$

The peroxydisulfate ion contains the bond –O–O– and has a similar structure to that of the tetrathionate ion (Figure 21.5).

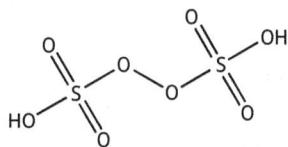

Figure 21.5: Structure of peroxydisulfuric acid.

The structure of the $S_2O_8^{2-}$ ion is made up of two SO_4 tetrahedra joined together by a peroxide bond –O–O–. Peroxysulfuric acids contain in their molecules the peroxide group O_2^{2-}. The most commonly used of them are *peroxymonosulfuric acid* (H_2SO_5) and *peroxydisulfuric acid* ($H_2S_2O_8$). These acids and their salts (peroxysulfates, $K_2S_2O_8$ and $(NH_4)_2S_2O_8$)) are strong oxidizing agents:

$$S_2O_8^{2-} + 2e^- \rightarrow 2SO_4^{2-}, \quad E° = +2.01\,V$$

When an oxygen atom in the hydroxide group of H_2SO_4 is replaced with a peroxide group, peroxymonosulfuric acid (H_2SO_5) is formed (by hydrolysis of peroxydisulfuric acid, $H_2S_2O_8$):

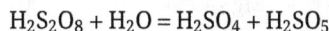

$$H_2S_2O_8 + H_2O = H_2SO_4 + H_2SO_5$$

Peroxymonosulfuric acid is monobasic, because the hydrogen atom of the peroxide group does not dissociate. In solid state, it is highly explosive. Its salts are thermally unstable and release oxygen when heated. Peroxysulfuric acids can be mixed with water in any proportions. For example, H_2SO_5 actively interacts with water:

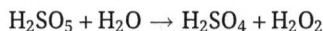

$$H_2SO_5 + H_2O \rightarrow H_2SO_4 + H_2O_2$$

The most important application of peroxysulfuric acids is in the production of H_2O_2:

$$H_2S_2O_8 + 2H_2O \rightarrow 2H_2SO_4 + H_2O_2$$

When the hydroxide group in H_2SO_4 is replaced by the isoelectronic groups F^- and Cl^-, the *halogen sulfonic acids* are formed, which can be represented by the formulas $F(OH)SO_2$ and $Cl(OH)SO_2$. In practice, they are synthesized by reacting liquid sulfur trioxide with hydrogen halide (HF and HCl). Fluorosulfonic acid (HSO_3F) is a very strong and stable acid in aqueous solution, dissociating according to the following equation:

$$HSO_3F = H^+ + SO_3F^-$$

Fluorosulfonic acid is practically not hydrolyzed, unlike chlorosulfonic acid. Fluorine enhances the polarization of the O–H bond, which leads to an increase in the strength of fluorosulfonic acid. Chlorosulfonic acid, similar to it, is completely hydrolyzed:

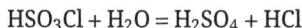

$$HSO_3Cl + H_2O = H_2SO_4 + HCl$$

21.6.7 Use and biological role of chemical compounds of sulfur

As a result of the vital activity of mankind, many sulfur compounds are released into the atmosphere, and the most active air pollutants are SO_2 and H_2S. The oxide SO_2 enters the atmosphere in large quantities during the combustion of coal and petroleum products (nearly 8×10^7 tons per year), and as a result of processes in metallurgy. The life of SO_2 is 4–7 days, after which it passes into the sulfate aerosols. Particularly harmful are the "acidic rains" resulting from intensive SO_2 emissions from industry.

Hydrogen sulfide is released in large quantities (~ 3.5×10^7 tons per year) from marine and ocean waters, as well as from underground sulfur sources. Industrial emissions of hydrogen sulfide are much smaller. H_2S is constantly formed at the bottom of large water basins when sulfates, dissolved in water, react with organic matter, for example, with methane:

$$MgSO_4 + CH_4 \rightarrow MgS + CO_2 + 2H_2O \rightarrow MgCO_3 + H_2S$$

The resulting H_2S rises from the bottom of the reservoir to a depth of ~150 m, as it meets the O_2 penetrating from above. In addition, up to a depth of 200 m, there are bacteria that oxidize H_2S to S. When the surface of the reservoir is polluted, the upper limit of the rise of H_2S from the bottom increases, and the released H_2S becomes dangerous for living organisms living in surface depths.

SO_2 and H_2S gases are highly toxic substances. The action of H_2S consists in the fact that it inhibits the enzyme cytochrome oxidase, responsible for the transfer of electrons in the respiratory chain. SO_2 interacts with moisture from the mucous membranes of the pharynx and larynx, causing irritation and inflammation of the tissues.

Many of the inorganic sulfur compounds are used in medicine. Sodium thiosulfate ($Na_2S_2O_3$) is the most important antidote to heavy metals that have entered the body, such as Hg, Pb, Bi, and Tl. It is also used for burns with liquid bromine. Metal sulfates have a wide range of applications: $Na_2SO_4 \cdot 10H_2O$ (Glauber's salt) is used as a laxative; $MgSO_4$ is used as a hypotensive agent (pressure reducer); $CaSO_4$ (gypsum) is used to fix bones in case of fracture; $BaSO_4$ is a radiopaque substance in studies of the gastrointestinal tract; and $CuSO_4 \cdot 5H_2O$ and $ZnSO_4 \cdot 7H_2O$ are antiseptics.

21.7 Selenium, tellurium, and polonium

21.7.1 Brief description

Sulfur and selenium are nonmetals. Tellurium is also a nonmetal, but it has metallic features, and in appearance it resembles a metal, but is fragile. Polonium is a metal that is close in physical properties to lead.

Elements exist in different polymorphic forms. Selenium, like sulfur, forms several modifications. The two crystalline modifications of selenium (rhombic and monoclinic) consist of octa-atomic rings, Se_8. Both modifications are thermodynamically unstable compared to the crystalline modifications of sulfur. They consist of infinite spiral chains of atoms. The presence of unbonded electron pairs leads to a certain delocalization of the electron density along the volume of the crystal and, as a consequence, to the appearance of some metallic features. Modification in appearance resembles a metal, but its metallic properties are very weakly expressed. Semiconductor properties are quite clearly manifested in this modification. One polymorphic modification is known for tellurium. It is the silvery-white semimetallic form, similar in structure to the gray form of selenium, which also exhibits semiconductor properties.

The most pronounced are the metallic properties of polonium. Polonium has two modifications. The low-temperature modification (stable to temperatures up to 100 °C) has a cubic structure, and the high-temperature one is rhombohedral. In both modifications, the coordination number of polonium is equal to 6. It is a soft metal, similar in physical properties to Bi and Pb.

While sulfur is an insulator with a relative resistance of 2×10^{33} mΩ/cm, selenium (2×10^{11} mΩ/cm) and tellurium (2×10^5 mΩ/cm) occupy an intermediate position in electrical conductivity and are semiconductors. Sulfur, selenium, and tellurium, as well as all nonmetals, are characterized by a negative temperature coefficient. In its two modifications, polonium has a resistance characteristic for real metals (43 mΩ/cm) and a positive temperature coefficient.

21.7.2 Occurrence, properties, and synthesis

Selenium and tellurium are permanent companions of sulfur and are always found in nature as a triad. Selenium and tellurium are rare scattered elements present in the form of impurities in sulfides. Polonium is present in trace quantities in uranium-containing minerals, as it is a product of the radioactive decay of uranium.

Selenium was discovered in 1817 by Berzelius in the industrial waste from the production of sulfuric acid. The color of selenium is red or gray. Its density is 4.8 g/cm³ and its melting point is 221 °C. Selenium boils at 685.3 °C. It exhibits some metallic properties and has weak electrical conductivity, but when irradiated and heated, it can increase significantly, and can be used as a semiconductor. Tellurium is a gray silvery substance in a solid aggregation state with a metallic luster. The density of this chemical element is 6.24 g/cm³, its melting point is 450 °C, and its boiling point is 990 °C. Tellurium, analogous to selenium, has a weak electrical conductivity, which can increase when irradiated and heated. This chemical element also exhibits semiconductor properties. Polonium is a radioactive chemical element that was discovered by Marie Curie in 1898, and is named after her homeland – Poland. It is about 300 times more radioactive than uranium and its distribution in nature is very weak. Polonium can be mined from uranium ore, with only 0.2 g of polonium produced from 1 ton of ore.

Chemically, selenium and tellurium are similar to sulfur, but with less activity. When heated, they react with a number of metals and nonmetals, forming the corresponding compounds – selenides, tellurides, oxides, halides, etc. Selenium and tellurium are analogues of sulfur in terms of their redox properties, but their oxidative activity is less pronounced, while the reduction activity is stronger. In the order of relative activity of metals, p-elements of group VIA are located after hydrogen; therefore, they do not react with acids with nonoxidizing action. Like sulfur, they are oxidized by acids with the following oxidizing action:

$$Se + 4HNO_3 \rightarrow SeO_2 + 4NO_2 + 2H_2O$$

Se and Te are obtained from waste products during the processing of sulfide ores of nonferrous metals. After proper processing, they are converted into oxides, which are reduced by SO_2:

$$EO_2 + 2SO_2 \rightarrow E + 2SO_3$$

Metallic Po is obtained by thermal decomposition of polonium sulfide in a vacuum (275 °C) or polonium dioxide (~ 500 °C). It can also be obtained by reducing PoO_2 with hydrogen (200 °C) or $PoBr_2$ with dry ammonia:

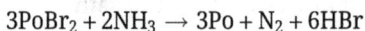

$$3PoBr_2 + 2NH_3 \rightarrow 3Po + N_2 + 6HBr$$

21.7.3 Use and biological role

Selenium, tellurium, and their compounds are mainly used in semiconductor technology. Selenium is used in copiers. Tellurium has the property of increasing the hardness and resistance of steel and lead and is used as a valuable additive in their composition.

Selenium has established itself relatively recently as a vital element essential for health due to its unique antioxidant qualities. It is found in enzymes and amino acids. The element is part of an enzyme that regulates fat metabolism in the body. It is concentrated mostly in the eyes, skeletal muscles, heart, and liver. Although it activates some enzymes, it is highly poisonous in large quantities. This element shows one of the narrowest tolerance ranges. Clinical deficiency is defined by levels around 0.05 ppm in food intake, while concentrations above 5 ppm cause chronic poisoning. Therefore, selenium as a dietary supplement should be approached with special caution. Sufferers of selenium poisoning excrete dimethylselenium ($(CH_3)_2Se$) with the smell of garlic. In humans, through a balanced diet, food usually provides an adequate level of Se. Selenium-rich foods are mushrooms, garlic, asparagus, fish, and animal liver. A correlation has been shown between higher levels of Se in water and a reduced incidence of breast and colon cancer. Selenium deficiency leads to reduced resistance to some viral infections. Se deficiency is also manifested in growth retardation, liver necrosis, pancreatic atrophy, etc. When talking about selenium deficiency, it primarily refers to Se(IV) compounds, specifically SeO_3^{2-} and $HSeO_3^-$. In pharmacy, selenium is added to various medicinal preparations to combat dandruff, hair loss, and eczema.

Microamounts of tellurium have been found in animals and humans, but so far, its biological role is practically unclear.

Polonium is a dangerous radioactive element. More than 27 isotopes of polonium with mass numbers from 192 to 218 are known. This heavy radioactive metal accumulates in the leaves of tobacco plants. Through tobacco smoke, it enters the lungs and leads to dangerous consequences for organs and blood. Polonium-210 is one of the most dangerous radionuclides for living organisms. Radioactive polonium is distributed in the body mainly by diffusion and is predominantly deposited in the lymphoid organs, such as lymph nodes, stomach, spleen, and kidneys. This determines its serious impact on immunity when ingested through food, water, and air. Increased susceptibility to exogenous and endogenous infections of the body (pneumonia, pleurisy, etc.) has been reported in animals.

21.7.4 Chemical compounds

21.7.4.1 Hydrides
Selenium and tellurium form hydrogen selenide (H_2Se) and hydrogen telluride (H_2Te) with hydrogen. These are very toxic gases with an unpleasant odor. The standard

Gibbs formation energy for H_2Se has a low positive value (19.7 kJ/mol), so hydrogen selenide can be obtained by the synthesis from H_2 and Se under conditions shifting the equilibrium to the right, while in H_2Te this is excluded ($\Delta G°_{form}$ = 85 kJ/mol). H_2Po decays at temperatures above 0 °C in moist air and light. H_2Pe is even more unstable and can only be produced in traces.

Selenium and tellurium hydrides are synthesized by treating the corresponding metal chalcogenides with water or acids:

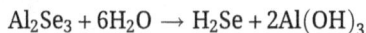

$$Al_2Se_3 + 6H_2O \rightarrow H_2Se + 2Al(OH)_3$$

$$Al_2Te_3 + 6HCl \rightarrow 2H_2Te + 2AlCl_3$$

H_2Te is also produced by cathodic reduction of tellurium at 0 °C. The anode is a platinum electrode, and 50% sulfuric acid is used as an electrolyte.

During the transition from H_2O to H_2S, H_2Se, and H_2Te, the valence angles H–E–H in the molecules decrease, which weaken the stability of the molecules H_2S, H_2Se, and H_2Te compared to that in the molecule of H_2O.

In aqueous solutions, hydrides (H_2E) are weak dibasic acids. The strength of acids and their reducing ability increase from S to Te due to an increase in atomic radii, and hence a decrease in the energy of E–H bond in the molecules and facilitation of the proteolytic interaction with water:

$$H_2E + H_2O \rightleftharpoons H_3O^+ + HE^-$$

Under standard conditions, hydrides (H_2E) are gases with toxic effects. As the mass and size of the molecules increase, the intermolecular interactions increase; therefore, the melting and boiling temperatures increase. The abnormally high temperatures of the phase transitions in water are explained by the increased intermolecular interactions in water by hydrogen bonds.

H_2Se and H_2Te exhibit a strong reducing effect and are oxidized by atmospheric oxygen:

$$2H_2E + O_2 \rightarrow 2H_2O + 2E$$

21.7.4.2 Halogenides and oxohalogenides

Among the different classes of chalcogenide compounds, their halides are the most numerous and diverse. Halogenide compounds containing chalcogenide atoms in different oxidation states have been synthesized and isolated – from the lowest to the highest; for example Te_3Cl_2, Se_2X_2 (X = F, Cl, Br), TeI, $PoCl_2$, SeX_4 (X = F, Cl), TeX_4 (X = F, Br), $PoBr_4$, PoI_4, SeF_6, and TeF_6. Fluorides are quite different from others in terms of stability, chemical activity, and composition. E_2X_2 is obtained by the direct interaction of simple substances or by reduction of higher halides with the corresponding chalcogens:

$$SeX_4 + Se \rightarrow Se_2X_2$$

Dihalides (EX_2) are obtained by halogenation of E_2X_2:

$$E_2Cl_2 + Cl_2 \rightleftharpoons 2ECl_2$$

EX_2 has low boiling points. EX_2 molecules have an angular shape, and their stability increases from Se to Po.

The most numerous are tetrahalides, which are synthesized in different ways: $E + X_2$, $EO_2 + X_2$, $CCl_4 + EO_2$, etc. The structures of gaseous homologs are trigonal bipyramids with a free electron pair; thus, their structures are strongly deformed. The solid halides, SeX_4 and TeX_4, are characterized by the formation of tetramers. The tetrahalides of polonium have a salt-like character and a predominantly ionic bond.

The tetrahalides of S, Se, Te, and Po are hydrolyzed easily to form the corresponding acids:

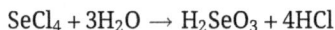

$$SeCl_4 + 3H_2O \rightarrow H_2SeO_3 + 4HCl$$

They interact with metal halides to form complex salts:

$$TeF_4 + KF \rightarrow K[TeF_5]$$
$$TeF_4 + 2KF \rightarrow K_2[TeF_6]$$
$$PoCl_4 + 2KCl \rightarrow K_2[PoCl_6]$$

The geometric shape of the complex ions $[SeCl_6]^{2-}$ and $[TeF_6]^{2-}$ corresponds to a regular octahedron, and the ion $[TeF_5]^-$ has a pyramidal structure.

From hexafluorides, SeF_6 and TeF_6 are obtained. Like SF_6, they are hydrolyzed.

In tellurium, there are some subhalogenides obtained by $Te + X_2$ reaction, which are characterized by the presence of zigzag chains of Te atoms, analogous to the chains in free Te.

Like sulfur oxohalogenides, Se and Te derivatives (SeO_2X_2 and TeO_2F_2) are chemically active compounds and exist as colorless volatile liquids or gases under normal conditions.

21.7.4.3 Selenides and tellurides

Selenides and tellurides of metals are very close analogues of sulfides. They usually have a variable composition. Many of them form isomorphic crystals with sulfides. Selenides and tellurides of zinc, lead, cadmium, and so on are mainly used as semiconductors. Diselenides of heavy metals (Mo, W, Ta, and Nb) have a layered structure that determines their low coefficient of friction. They are used as a component of lubricants for apparatus, operating in a vacuum in space and in conditions of different radiation.

21.7.4.4 Oxygen-containing compounds

Sulfur, selenium, and tellurium are part of a large number of oxygen-containing compounds (oxides, acids, and salts) in oxidation states of +4 and +6. If sulfur is more characterized by compounds in oxidation state +6, selenium and tellurium are more characterized by +4 oxidation states. Po is part of compounds in oxidation states of +2 and +4.

The compounds of the elements in the oxidation state of +4 are the oxides with the composition EO_2 and their oxoacids and salts. Selenium dioxide (SeO_2) is a solid substance that dissolves well in water with the formation of the relatively stable selenious acid (H_2SeO_3), which is lower in strength than sulfurous acid:

$$SeO_2 + H_2O = H_2SeO_3$$

In the SO_2–SeO_2–TeO_2–PoO_2 series, the acidic properties weaken. SeO_2 is acidic, TeO_2 is amphoteric, and PoO_2 is basic with weak amphotericity. TeO_2 and PoO_2 have high crystal lattice energy. Telluric dioxide (TeO_2) does not interact with water, but dissolves in bases to form tellurites:

$$TeO_2 + 2OH^- \rightarrow TeO_3{}^{2-} + H_2O$$

TeO_2 also dissolves in dilute HCl to H_2TeCl_6 according to the equation:

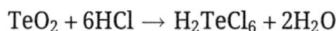

$$TeO_2 + 6HCl \rightarrow H_2TeCl_6 + 2H_2O$$

Analogous compounds of H_2TeCl_6 are also known in other elements of the periodic table, for example, H_2PbCl_6, H_2TiCl_6, and H_2PtCl_6.

Polonium dioxide (PoO_2) interacts with acids as a typical basic oxide:

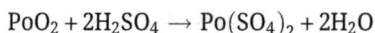

$$PoO_2 + 2H_2SO_4 \rightarrow Po(SO_4)_2 + 2H_2O$$

Se and Te dioxides are produced by reacting elements with NO_2 or concentrated HNO_3:

$$Se + 2NO_2 \rightarrow SeO_2 + 2NO$$

PoO_2 is produced directly by heating the metal in air (250 °C).

An increase in the size of atoms in the Se–Te–Po series affects the structure of solid dioxides. SeO_2 contains infinite chains in which each of the Se atoms is surrounded pyramidally by three oxygen atoms ($-SeO_3-$), one of which is finite. The two polymorphic modifications, α-TeO_2 and β-TeO_2, contain the groups (TeO_4) in the form of deformed trigonal bipyramids joined by their ribs or tips. PoO_2 is an ionic compound with a structure similar to CaF_2.

Of the acids of the elements in +4 oxidation state (H_2SeO_3, H_2SO_3, and H_2TeO_3), only H_2SeO_3 is released in the free state. The strength of the acids decreases in the order H_2SO_3–H_2SeO_3–H_2TeO_3. The increase in the atomic radius and the decrease in the electronegativity of the central atom leads to a weaker shift of the electron density from the oxygen atom of the hydroxyl group and to a weaker polarization of the O–H bond. The salts of H_2SeO_3 and H_2TeO_3, called selenites and tellurites, hydrolyze to form alkaline solutions.

The reduction properties of Se(IV) and Te(IV) compounds are weaker and the oxidizing properties are stronger than those of sulfur(IV) compounds. In selenious and tellurous acids, the reduction properties are less pronounced, that is, they are much more difficult to oxidize to derivatives of Se(VI) and Te(VI). The oxidative properties of H_2TeO_3 are more pronounced than in sulfuric acid, and Se(IV) compounds exhibit stronger oxidizing properties than S(IV) and Te(IV) compounds. For example, when sulfur is mixed with selenious acid, the second is reduced first:

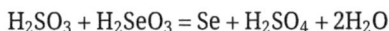

$$H_2SO_3 + H_2SeO_3 = Se + H_2SO_4 + 2H_2O$$

The compounds of the elements in oxidation state +6 are oxides with composition EO_3 (SeO_3 and TeO_3) and their oxoacids and salts. SeO_3 is obtained by dehydration of selenic acid (H_2SeO_4) at 150 °C using phosphorus anhydride. TeO_3 is synthesized by dehydration of orthotelluric acid (H_6TeO_6) at 350 °C. The two oxides SeO_3 and TeO_3, when heated, release oxygen and form dioxides. SeO_3 dissolves well in water to form H_2SeO_4. Solid SeO_3 consists of tetrahedra SeO_4^{2-} united in cyclic tetramers $Se_4O_{12}^{8-}$. Solid TeO_3 is made up of octahedra TeO_6^{6-}, united by their common vertices that form chains. Unlike SO_3 and SeO_3, TeO_3 is poorly hydrated.

SeO_3 is a strong oxidizer. It oxidizes HCl to Cl_2, as well as phosphorus to P_2O_5:

$$SeO_3 + 2HCl \rightarrow H_2SeO_3 + Cl_2$$

The oxidative properties of TeO_3 are less pronounced, and the reaction occurs when heated.

The oxoacids of the elements in +6 oxidation state, H_2SO_4, H_2SeO_4, and H_6TeO_6 (Figure 21.6), are synthesized by the oxidation of dioxides or their corresponding acids:

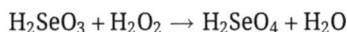

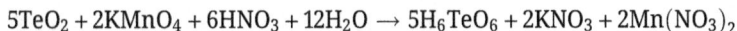

$$H_2SeO_3 + H_2O_2 \rightarrow H_2SeO_4 + H_2O$$
$$5TeO_2 + 2KMnO_4 + 6HNO_3 + 12H_2O \rightarrow 5H_6TeO_6 + 2KNO_3 + 2Mn(NO_3)_2$$

They are also obtained by the oxidation of simple substances with strong oxidizing agents:

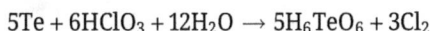

$$5Te + 6HClO_3 + 12H_2O \rightarrow 5H_6TeO_6 + 3Cl_2$$
$$BaTeO_4 + H_2SO_4 + 2H_2O \rightarrow H_6TeO_6 + BaSO_4$$

Selenic acid (H_2SeO_4) is lesser in strength than sulfuric acid, as evidenced by the reaction:

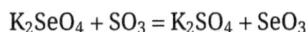

$$K_2SeO_4 + SO_3 = K_2SO_4 + SeO_3$$

in which the anhydride (SO_3) of the stronger acid (H_2SO_4) displaces the anhydride (SeO_3) of the weaker acid (H_2SeO_4) from its salt. In terms of oxidative activity, H_2SeO_4 is noticeably superior to H_2SO_4, because for Se the oxidation state of +4 is more characteristic than +6, and for S it is the opposite. Selenic acid with water forms the crystal hydrates $H_2SeO_4 \cdot H_2O$ and $H_2SeO_4 \cdot 4H_2O$. Selenic acid salts (selenates) are in many respects analogous to sulfates, for example, in solubility and crystalline forms. They also form double salts similar to alum. At high temperatures, they are less stable than sulfates. When heated on charcoal, they easily release O_2 and free Se.

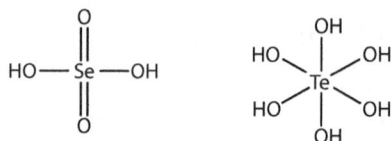

Figure 21.6: Structures of selenic and orthotelluric acids.

The two acids of Te(VI) – metatelluric (H_2TeO_4) and orthotelluric (H_6TeO_6) – are poorer in oxidative activity to selenic acid, but superior to sulfuric acid. The structure of orthotelluric acid (H_6TeO_6) is different from that of sulfuric and selenic acids. The crystal structure of solid H_6TeO_6 ($t_m = 136\ °C$) consists of molecules with a regular octahedral shape, which is also preserved in solution. Tellurates are not isomorphic to sulfates and selenates. Orthotelluric acid is titrated with a base as a monobasic acid to form salts of the type $MTeO(OH)_5$ and is weaker than carbonic acid. Products of complete (Ag_6TeO_6 and Na_6TeO_6) and partial (NaH_5TeO_6, $Na_2H_4TeO_6$, and $Na_4H_2TeO_6$) neutralization are obtained. Orthotelluronic acid is a stronger oxidizing agent than sulfuric acid. Most often, the corresponding metals are obtained as products of H_2SeO_4 and H_6TeO_6 reduction.

Chapter 22
Seventh main group of the periodic table

22.1 General characteristics and properties of halogen elements

The p-elements of the seventh group (Table 22.1) are collectively called halogen elements, that is, "salt-forming" (from halos – salt and genao – gives birth), because they
form too many salts.

Table 22.1: The main characteristics of atoms of elements of VIIA group.

Elements	F	Cl	Br	I	At
Valence electrons	$2s^2 2p^5$	$3s^2 sp^5$	$4s^2 4p^5$	$5s^2 5p^5$	$6s^2 6p^5$
Radius of the atom (nm)	0.071	0.099	0.114	0.133	–
Ionization potential (eV)	17.4	13.0	11.8	10.4	9.2
Electronegativity	4.0	3.0	2.8	2.5	–
Bond length in the molecule (nm)	0.142	0.200	0.229	0.267	–
Energy of the bond (kJ/mol)	155	239	190	149	117
$J°\ (X_2 + 2e^- = 2X^-)$ (V)	2.72	1.36	1.09	0.54	–

In halogen elements the radii of atoms increase, electronegativity decreases and the
bond length in the molecules increases. The bond energy during the transition from F_2
to Cl_2 increases, because in the chlorine molecule additional donor-acceptor bonds are
formed due to the free d-orbitals and filled p-orbitals. In the transition from chlorine to
bromine, iodine and astatium, the bond energy decreases, which is because of the increasing bond length.

The electron configuration of valence electrons in halogen atoms is $ns^2 np^5$. It also
determines their typical oxidation state of −1. The addition of an electron results in
the formation of halide ions with a stable eight-electron noble gas configuration. Halogen atoms vary in the number of their inner electron shells, but they all have seven
electrons in their outer electron shell.

The p-elements of group VII are characterized by nonmetallic properties. These
are the most active nonmetals (Table 22.2). In the free state, these elements have a
coordination number equal to 1 and molecular crystal lattices in solid and liquid
states. They are characterized by diatomic molecules. Under normal conditions, fluorine and chlorine are gases, bromine is a liquid substance, iodine is a crystalline, easily volatile substance with a molecular crystal lattice. Fluorine and chlorine, as well
as bromine and iodine vapors, are toxic.

Halogen elements do not conduct heat and electric current. Their physical properties change as their relative atomic mass increases: the density of the elements increases,
their melting and boiling points rise, their smell weakens, and the color becomes more

https://doi.org/10.1515/9783111712246-022

Table 22.2: Melting and boiling points of elements of VIIA group.

Elements	T_m (°C)	T_b (°C)	ρ (g/cm³)
F	−220.6	−187.7	1.3
Cl	−100.9	−34.2	1.9
Br	7.2	58.76	3.4
I	113.5	184.35	4.4

saturated. Metallic properties appear in iodine and are expressed in the metallic luster of its crystals. Fluorine is a light-yellow gas with a pungent specific odor. Solid fluorine has a rhombic lattice at a temperature of −228 °C, and at a temperature higher than −229 °C – a cubic molecular lattice. Chlorine is a yellow-green gas with a pungent irritating odor. Bromine is a mobile dark red liquid (the only liquid nonmetal) with moderate solubility in water (33.6 g/L at 25 °C). Iodine is a solid with a weak metallic sheen and a black-violet color. It sublimates easily, forming violet smoke, and dissolves slightly in water (0.33 g/L at 25 °C). The vapor pressures of bromine and iodine are quite high. Thus, when bromine is opened, toxic red-brown bromine vapors are released, and iodine crystals emit toxic violet vapors when slightly heated. Although iodine resembles a metal, it behaves like a typical nonmetal in most chemical reactions. Astatine is a radioactive element obtained in 1940. In its free state, it is very volatile and poorly soluble in water. All isotopes of astatine have a fairly short lifespan. Therefore, they emit high-intensity radiation. However, it has been proven that the chemistry of this element follows the trends observed in other group members. Astatine, formed as a product of uranium isotopes, is one of the rarest elements on Earth.

Fluorine forms compounds in only one oxidation state −1. The other halogen elements, except in this oxidation state, form compounds in positive oxidation states: +1, +3, +5, +7, and in chlorine +4 and +6 are also possible.

In chemical reactions, the halogen elements exhibit oxidizing properties, accepting a single electron and reaching the stable configuration of an inert gas. The oxidative activity of the halogen elements decreases down the group: fluorine is the strongest oxidizing agent, chlorine and bromine are also oxidizing agents, and iodine exhibits both oxidizing and reducing properties. Due to the attenuating oxidative activity of the halogen elements in the group from top to bottom, each higher standing halogen element displaces the lower one of their compounds with the metals:

$$Cl_2 + 2NaBr = 2NaCl + Br_2$$
$$Cl_2 + 2KI = 2KCl + I_2$$

Halogen elements are chemically very active and practically do not form cations. Only in iodine are known compounds in which it can be in the form of cations I^+ and IO^+ (INO_3 and $IONO_3$). Halogen elements react with most elements of the periodic table. Fluorine interacts even with inert gases, except for the light He, Ne, and Ar to form fluorides of different compositions.

Fluorine is the most active of all halogen elements, although its electronic affinity is less than that of chlorine (333.1 and 369.5 kJ/mol for fluorine and chlorine, respectively). The higher activity of fluorine compared to chlorine is explained by the significantly lower dissociation energy of the F_2 molecule (155 kJ/mol) than that of the Cl_2 molecule (239 kJ/mol). Unlike other analogs, it interacts directly with xenon:

$$Xe + F_2 \rightarrow XeF_2, \quad \Delta G = -161.2 \, kJ/mol$$
$$Xe + 2F_2 \rightarrow XeF_4, \quad \Delta G = -256.7 \, kJ/mol$$

It reacts with oxygen at low temperatures and an electric spark to form oxygen fluorides with a composition of OF_2, O_3F_2, and O_2F_2. Nonmetals and most metals in powder form burn in a fluorine atmosphere to form fluorides.

Chlorine also reacts directly with most nonmetals (except carbon, oxygen, nitrogen) and metals at low temperature. Under certain conditions, it forms compounds with xenon $XeCl_4$ and $XeCl_2$. In cold with water, it forms clathrates $Cl_2 \cdot H_2O$ and $Cl_2 \cdot 8H_2O$. The chemical activity of bromine is less than that of fluorine and chlorine. It does not react directly with oxygen, nitrogen, carbon, and inert gases. Even less pronounced chemical activity is observed in iodine.

In general, halogen elements are very active, so they are found mainly in the form of compounds. The most common of the halogen elements is chlorine, which in the form of sodium, potassium, calcium, and magnesium chlorides is found in the waters of oceans, seas, and salt lakes. In the solid state, chlorine is part of the minerals: halite $NaCl$, silvin KCl, and carnalite $KCl \cdot MgCl_2 \cdot 6H_2O$. The main fluorinated minerals are fluorite CaF_2, cryolite Na_3AlF_6, and fluorapatite $3Ca_3(PO_4)_2 \cdot CaF_2$. Bromine, iodine, and astatium refer to the scattered elements. The main source of their production is drilling waters from oil fields and some seaweed. Astatine is unstable and radioactive, which is why it is inaccessible and insufficiently studied.

22.2 Fluorine

22.2.1 Brief description

A special place among the halogen elements is occupied by fluorine, which, unlike other elements, does not have a d-subshell. It occurs only in oxidation state −1 and is a stronger oxidizing agent than even oxygen. Fluorine gas is extremely active in chemical reactions. It interacts with all metals and nonmetals and with most complex substances.

The high chemical activity of fluorine is explained by the following two reasons:
(1) The fluorine molecule is unstable, while the bonds that fluorine forms with other elements are very strong. That is why reactions involving fluorine are character-

ized by high negative values of the Gibbs energy, that is, such reactions are thermodynamically probable.

(2) Reactions involving fluorine are characterized by low values of activation energy, so they have a high rate (kinetic factor) even at low temperatures.

In all its reactions, F_2 is an oxidizer that oxidizes all elements to their maximum oxidation states:

$$S + 3F_2 = SF_6$$

$$P_4 + 10F_2 = 4PF_5$$

$$2H_2O + 2F_2 = 4HF + O_2$$

$$2NH_3 + 3F_2 = 6HF + N_2$$

22.2.2 Synthesis

Fluorine is produced by electrolysis in melts of potassium or sodium fluoride, in which the simple substance fluorine is released at the anode:

$$2F^- - 2e^- = F_2$$

Chemical oxidation is not possible because there is no stronger oxidizing agent than F_2.

22.2.3 Use and biological role

Fluorine is used in uranium production technology and in the separation of its isotopes, which is very important in the use of uranium in nuclear reactors. Fluorine is used in large quantities in the production of organofluorine compounds (Teflon), cooling agents (freons), electrical insulation materials, and insecticides. HF is also used in uranium technology, in glassmaking, and in chemical analyses. The complex compound Na_3AlF_6 is used in the production of aluminum. Fluorine is a biogenic trace element. In the form of fluoride ions (F^-) it increases the resistance of teeth to cavities, stimulates blood formation and immunity, participates in the construction of the skeleton, and protects against the development of osteoporosis. The required amount for an adult is 2–3 mg/day. It accumulates in bone tissue and teeth in the form of fluorapatite. However, the intake of large amounts of fluorine compounds causes the disease fluorosis. Fluoride ions replace iodide ions in tyrosine and weaken the activity of the thyroid gland, and also block the active centers of many enzymes. Fluorine enters the body through drinking water. In areas where drinking water is high in fluorine, people's teeth become stained, and where it is too low in fluorine, the risk of frequent tooth decay seriously increases. Therefore, fluoride tablets are given prophylactically in childhood, and almost all toothpastes contain fluoride.

22.2.4 Chemical compounds of fluorine

22.2.4.1 Hydrogen fluoride

Among the numerous fluorine compounds, hydrogen fluoride is of great importance, which is obtained by heating calcium fluoride (mineral fluorite or fluorspar) with concentrated sulfuric acid at 300 °C:

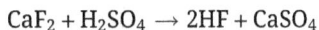

$$CaF_2 + H_2SO_4 \rightarrow 2HF + CaSO_4$$

Hydrogen fluoride is a colorless gas with a pungent suffocating odor that easily liquefies into a smoking liquid with a boiling point of about 20 °C, much higher than the boiling points of other hydrogen halides. This is the result of very strong hydrogen bonds between adjacent hydrofluoride molecules that form $(HF)_n$ associations, where $n = 2$–6. Fluorine shows the highest electronegativity of all elements, so the hydrogen bond formed with fluorine is the strongest of all possible. Hydrogen bonds are linear with respect to hydrogen atoms, but are oriented 120° with respect to fluorine atoms. In this way, the molecules form zigzag chains. Due to the presence of hydrogen bonds, hydrogen fluoride at 20 °C turns into a liquid state, and at −83 °C into a solid state. It dissolves indefinitely in water, forming hydrofluoric acid with water molecules, which dissociates into ions as a weak acid:

$$HF + H_2O \rightleftharpoons H_3O^+ + F^-$$

The weak HF has a $pK_a = 3.2$, unlike the other hydrohalic acids, which are much stronger. The relatively weak dissociation of hydrofluoric acid can be explained by the fact that the fluorine-hydrogen bond is much stronger than the other hydrogen-halogen bonds. In more concentrated solutions, hydrofluoric acid is further ionized to linear HF_2^- ions, unlike the behavior of other acids:

$$F^- + HF \rightleftharpoons HF_2^-$$

The HF_2^- ion is so stable that alkaline salts (such as KHF_2) crystallize in solution. There is evidence that in HF_2^- the hydrogen atom is centrally located between the two fluorine atoms.

HF has strong corrosive activity, although it is a weak acid. It is one of the few substances that attacks glass, and for this reason, hydrofluoric acid is always stored in plastic bottles. The reaction with glass results in the formation of gaseous SiF_4 or the hexafluorosilicate ion (SiF_6^{2-}):

$$SiO_2 + 4HF = SiF_4 + 2H_2O$$

$$SiO_2 + 6HF \rightarrow SiF_6^{2-} + 2H^+ + 2H_2O$$

Hydrofluoric acid reacts with oxides and bases to form fluorides:

$$NaOH + HF \rightarrow NaF + H_2O$$

$$KOH + 2HF \rightarrow KHF_2 + H_2O$$

22.2.4.2 Fluorides

Fluorides are binary compounds of fluorine (−1) with all other chemical elements. They are obtained directly or by using HF. Metal fluorides are crystalline substances with ionic bonds. Only alkali metal fluorides are soluble in water. When dissolved, they are partially hydrolyzed. Most metal fluorides are slightly soluble in contrast to other metal halides. Nonmetal fluorides are gases or liquids with molecular structures. They are very diverse in properties, some of them are inert (CF_4, NF_3, and SF_6), and others are easily decomposed by water:

$$SiF_4 + 3H_2O = H_2SiO_3 + 4HF$$

In fluorides formed by the elements of the same period (LiF, BeF_2, BF_3, CF_4, NF_3, and OF_2), the same regularity is observed in the change of properties from basic to acidic, as in oxides.

Fluorine forms several compounds with oxygen, of which only one is stable – oxygen difluoride OF_2. Oxygen difluoride is a colorless, highly toxic gas with a pungent specific odor, slightly soluble in water. The two elements interact directly when cooled under the action of an electric charge, which produces O_2F_2. Oxygen difluoride is perishable and gradually breaks down into its constituent elements. Of the halogen elements, only fluorine does not form acidic oxides. Therefore, oxygen-containing acids of fluorine have not been isolated so far. There is evidence of the existence of hypofluoric acid HOF. It is obtained as an intermediate compound in the oxidation of water with fluorine at −40 °C. HOF is released as a colorless crystalline substance that decomposes explosively at room temperature to HF and O_2:

$$F_2 + H_2O \rightarrow HOF + HF$$

$$2HOF \rightarrow 2HF + O_2$$

Complex compounds containing fluoride ions as ligands are widespread, for example, Na_3AlF_6, K_2WF_8, H_2SiF_6, and HPF_6. The fluoride ion is a strong field ligand that forms strong bonds with all complexing agents. Therefore, the addition of hydrofluoric acid to HNO_3 leads to the formation of a mixture that oxidizes substances such as Si, B, W, Ta, and Nb, which are stable even in royal water:

$$W + 2HNO_3 + 8HF = H_2WF_8 + 2NO + 4H_2O$$

$$3Si + 4HNO_3 + 18HF = 3H_2[SiF_6] + 4NO + 8H_2O$$

$$3Pt + 4HNO_3 + 18HF = 3H_2[PtF_6] + 4NO + 8H_2O$$

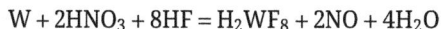

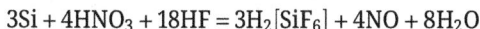

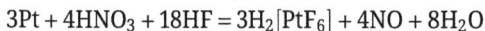

All fluorine compounds are toxic. They accumulate gradually in the body, which leads to serious poisoning.

22.3 Chlorine, bromine, and iodine

22.3.1 Synthesis

Chlorine is obtained in industry by electrolysis of an aqueous solution of NaCl (chlor-alkali electrolysis). For laboratory purposes, Cl_2 is obtained in small quantities by oxi-dation of Cl^- with strong oxidizing agents such as MnO_2 and $KMnO_4$:

$$4HCl + MnO_2 = MnCl_2 + Cl_2 + 2H_2O$$

$$10KCl + 2KMnO_4 + 8H_2SO_4 = 2MnSO_4 + 5Cl_2 + 6K_2SO_4 + 8H_2O$$

Bromine is obtained from the more affordable bromides, which react with oxidizing agents, including chlorine, which displaces bromine:

$$2KBr + Cl_2 = 2KCl + Br_2$$

$$2NaBr + MnO_2 + 2H_2SO_4 = Br_2 + MnSO_4 + Na_2SO_4 + 2H_2O$$

Iodine is obtained from seaweed, where it is present in the form of sodium and potas-sium iodides, when it is displaced by Cl_2. In laboratory conditions, it is obtained in the same way as Br_2:

$$2NaI + Cl_2 = 2NaCl + I_2$$

22.3.2 Properties

Chlorine is a yellow-green gas, heavier than air, and highly toxic (inhalation of chlo-rine leads to suffocation and death). Chlorine Cl_2 molecules are stronger than F_2, which is explained by the additional donor-acceptor interactions between chlorine atoms involving free d-orbitals, which are not present in fluorine atoms. Bromine is a red-brown liquid that turns into vapor at 59 °C. Bromine vapors are toxic, and like Cl_2, they cause suffocation. Iodine is a dark violet crystalline substance that easily subli-mates. Iodine vapors have a violet color and are toxic.

The solubility of halogen elements in water decreases toward the heavier mem-bers of the group. Chlorine and bromine are well soluble in water. Chlorine dissolves in water up to 1% (at 20 °C). This solution is called chlorine water. In chlorine water, chlorine partially disproportionates by the reversible reaction:

$$Cl_2 + H_2O \rightleftharpoons HCl + HClO$$

The resulting hypochlorous acid HClO is unstable and decomposes with the formation of atomic oxygen, thanks to which chlorine water discolors organic paints and oxidizes many inorganic substances:

$$HClO = HCl + O$$

In base solutions, the equilibrium of the disproportionation reaction is shifted to the right and it proceeds practically irreversibly. Depending on the conditions, different products are obtained, for example, in the cold, NaClO is obtained, and when heated NaClO$_3$ is obained:

$$Cl_2 + 2NaOH = NaCl + NaClO + H_2O$$

$$3Cl_2 + 6NaOH = 5NaCl + NaClO_3 + 3H_2O$$

Bromine dissolves in water better than chlorine (up to 4%). The resulting solution is called bromine water. In bromine water, bromine dissociates similarly to chlorine. Bromine solutions in water have a golden-yellow or red-brown color, depending on the concentration.

The solubility of iodine in water is very small (0.028 g in 100 g H$_2$O). Iodine dissolves better in saturated solutions of KI. In this case, a complex compound is formed:

$$KI + I_2 = K[I_3]$$

Halogen elements react actively with aqueous solutions of bases. When a solution of NaOH is saturated with chlorine, the so-called bleach is obtained, the bleaching effect of which is due to the reactions:

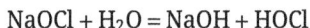

$$NaOCl + H_2O = NaOH + HOCl$$

$$HOCl = HCl + O$$

If dry slaked lime reacts with chlorine, the so-called chlorine lime is obtained, which has a strong oxidizing effect and is used as a disinfectant:

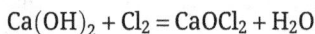

$$Ca(OH)_2 + Cl_2 = CaOCl_2 + H_2O$$

Chlorine lime is a white powder with a pungent odor of chlorine. In humid air and under the action of CO$_2$, it decomposes:

$$2CaOCl_2 + CO_2 + H_2O = CaCO_3\downarrow + CaCl_2 + 2HOCl$$

Halogen elements dissolve well in organic solvents. Of interest is the behavior of iodine in various organic solvents. In polar oxygen-containing solvents (methanol (CH$_3$OH), ethanol (C$_2$H$_5$OH), diethyl ether (C$_2$H$_5$OC$_2$H$_5$), and acetone (CH$_3$COCH$_3$)), iodine dissolves well to form dark brown-colored solutions, which is due to the solvation of iodine mole-cules. A solution of iodine in ethanol is used as a medicine (the so-called iodine tinc-ture). In nonpolar or weakly polar oxygen-free organic solvents (chloroform ($CHCl_3$), carbon tetrachloride (CCl_4), gasoline, benzene (C_6H_6)), iodine dissolves, in which case the violet color of iodine vapor is preserved. The violet color of the solutions indicates that the molecules are not solvated.

The halogen elements have extremely high chemical activity. Although bromine and iodine are not as active as fluorine and chlorine, their chemical activity is still quite significant.

Chlorine is a strong oxidizing agent, oxidizing most metals, nonmetals, and numer-ous complex compounds:

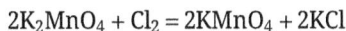

$$2Fe + 3Cl_2 = 2FeCl_3$$

$$2K_2MnO_4 + Cl_2 = 2KMnO_4 + 2KCl$$

Bromine also oxidizes many simple substances and destroys wood and rubber. In terms of oxidizing ability, iodine is inferior to other halogen elements, but neverthe-less, it oxidizes many metals and compounds and destroys rubber. In iodine, the re-duction ability is quite pronounced, for example, it is oxidized by nitric acid:

$$3I_2 + 10HNO_3 = 6HIO_3 + 10NO + 2H_2O$$

22.3.3 Use and biological role

Chlorine is used for the sterilization of drinking water and for bleaching fabrics and paper, in the production of organochlorine products, and as an oxidizing agent in chemical synthesis and analysis. Bromine is used to make many medicinal prepara-tions, and bromides are used in chemical synthesis and analysis. Silver bromide is used in photography. Iodine is widely used in medicine (in the form of iodides), as it is a vital element for humans. With iodine deficiency, the thyroid gland does not pro-duce iodine-containing hormone, which is why specific diseases develop, especially in mountainous areas where water is poor in iodine. Therefore, table salt NaCl is io-dized, that is, sodium iodide (NaI) is added to it. It is also used for disinfection in the form of iodine tincture and iodasept.

Chlorine when inhaled causes suffocation. It was used as a chemical warfare agent. Bromine vapor and, to a lesser extent, iodine have a similar effect. When inhaled, bro-mine vapors cause a runny nose, watery eyes, and conjunctivitis. Nosebleeds and aller-gic reactions are possible. Bromine and iodine that come in contact to the skin cause serious burns.

All halogen elements are characterized by high biological activity. Chlorine is one of the most important macronutrients in the body. Chlorine in the body is found mainly in the extracellular space. In the form of chloride ions, Cl^- participates in the formation of gastric juice, blood plasma, and together with sodium ions is the cause of the accumulation of water in the body. The role of chloride ions is very diverse. They take part in the formation of the blood buffer system, regulate the osmotic pressure in the water-salt metabolism, promote the deposition of glycogen in the liver, and maintain high acidity in the stomach. Hydrochloric acid in the stomach under the action of pepsin ensures the breakdown of peptides $R-CO-NH-R^1$ into carboxylic acids RCOOH and amines NH_2R^1. The Cl^- ion is also part of other enzyme systems, such as activating the enzyme amylase, secreted by the salivary glands. The required amount of chlorine is 2 g/day, and the harmless dose is 5–7 g/day. The need for chlorine, even in excess, is obtained from food and mainly from table salt (90%). The chlorine content in the human body is 0.25% and in the blood plasma −0.35%. The body contains more than 200 g of NaCl, of which 45 g is dissolved in the blood. Of the simple inorganic compounds of chlorine, the aqueous 0.9% solution of NaCl is widely used in medicine as a basis for the preparation of medicines (saline). For disinfection, chlorine lime $CaOCl_2$ is applied, which, under the action of CO_2, water and light, releases active oxygen, destroying the cell membranes of the simplest microorganisms. A strong remedy against aerobic bacteria is decaoxide $[(ClO_2)_2O = O(ClO_2)_2]$.

Bromine is a biogenic trace element. It is involved in the regulation of the central nervous system and affects the function of the gonads and thyroid gland. Bromine salts, KBr and NaBr, have a calming effect. They are part of a number of dosage forms that are used to treat mental disorders. The daily requirement for bromide ions is 0.5–2 mg/day. It accumulates in the brain and thyroid gland. It enters the body mainly through food – bread, dairy, and legumes. NaBr, used as a drug, replaces I^- in the thyroid gland with Br^-, enhances the activity of the adrenal gland. Especially sensitive to the action of Br^- ions is the CNS, which reacts to the action of bromide ions by balancing and slowing down the processes of excitation, which is manifested in the calming effect of NaBr. In case of an excess of Br^- in the body, a "salt diet" with a high content of NaCl is recommended, replacing Br^- ions in the body.

Iodine is a biogenic trace element. It is necessary for the normal functioning of the thyroid gland. The hormone of this gland, thyroxine, is an iodine compound that controls metabolism. When the thyroid gland is disturbed, goiter or Graves' disease develops. The need for iodine fluctuates in the range of 50–200 µg/day. The content of iodine in ordinary foods is not high. More iodine is contained in sea fish. Iodine deficiency causes hypothyroidism, in which endemic goiter grows. Thus, the thyroid gland, by increasing its size, tries to increase the amount of hormone produced. An excess of iodine leads to a rather active functioning of the thyroid gland (hyperthyroidism), which is expressed in an increase in metabolic processes, weight loss, excitability, and tachycardia. Molecular iodine I_2 exhibits weak antiseptic properties.

22.3.4 Chemical compounds

22.3.4.1 Hydrogen halides

The halogen elements react with H_2 to form the corresponding hydrogen halides. Chlorine reacts with hydrogen in the light, and the reaction takes place with an explosion by a chain mechanism, similar to the reaction between O_2 and H_2:

$$Cl_2 = 2Cl$$

$$Cl + H_2 = HCl + H$$

$$H + Cl_2 = HCl + Cl, \text{ etc.}$$

Bromine interacts with hydrogen at high temperature and the reaction also has a chain mechanism. The interaction of iodine with hydrogen (at 500 °C in the presence of a platinum catalyst) is a reversible process and the yield is too low:

$$H_2 + I_2 \rightleftarrows 2HI$$

To obtain hydrochloric acid under laboratory conditions (previously in industry), the reaction between concentrated sulfuric acid and NaCl is used:

$$2NaCl + H_2SO_4 = Na_2SO_4 + 2HCl$$

The same approach cannot be used for HBr and HI. To obtain hydrogen bromide and hydrogen iodide, the following reactions are mainly used:

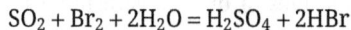

$$PX_3 + 3H_2O = H_3PO_3 + 3HX$$

$$SO_2 + Br_2 + 2H_2O = H_2SO_4 + 2HBr$$

Hydrogen chloride (HCl), hydrogen bromide (HBr), and hydrogen iodide (HI) are gases that are well soluble in water. Their solutions are strong acids that completely dissociate into hydrogen cations (H_3O^+) and halide anions:

$$HX + H_2O \rightarrow H_3O^+ + X^-$$

The most important of the hydrohalic acids is hydrochloric acid. At maximum saturation of water with hydrogen chloride, concentrated hydrochloric acid is obtained, in which the content of hydrogen chloride is 37%. It is also called "smoking" hydrochloric acid because when left in the air it constantly emits white smoke with a pungent suffocating odor (HCl). The mixture of three parts of concentrated hydrochloric acid and one part of concentrated nitric acid is called "aqua regia," which has a very strong oxidizing effect and dissolves precious metals – Au, Pt, etc. The dissolving effect of aqua regia is due to chlorine from nitrosyl chloride, released during the mixing of acids:

$$3HCl + HNO_3 = NOCl + Cl_2 + 2H_2O$$

In the HF–HCl–HBr–HI series, the strongest is HI, and the weakest is HF, which is explained by a decrease in the strength of the covalent bond H–X from fluorine to iodine. As strong acids, HI, HBr, and HCl interact with bases, basic oxides, salts, and with active metals.

The oxidative activity of HCl, HBr, and HI is small. They oxidize only those metals that are located before hydrogen in the row of relative activity (with the exception of lead, on the surface of which a protective layer of the insoluble $PbCl_2$, $PbBr_2$, and PbI_2 salts is formed):

$$Zn + 2HCl = ZnCl_2 + H_2$$

The reduction activity of hydrohalic acids increases from HCl to HI. The solution of HI darkens during long-term storage due to oxidation by O_2 in the air, resulting in the formation of iodine:

$$4HI + O_2 = 2I_2 + 2H_2O$$

Under the same conditions, hydrochloric and hydrobromic acid solutions do not change.

Concentrated sulfuric and nitric acids oxidize HBr and HI, but do not oxidize HCl acid:

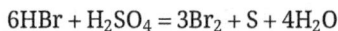

$$6HBr + H_2SO_4 = 3Br_2 + S + 4H_2O$$

$$HI + 6HNO_3 = HIO_3 + 6NO_2 + 3H_2O$$

Hydrogen chloride and hydrochloric acid are widely used in the processing of metals as well as in organic and inorganic synthesis. Hydrogen bromide is mainly used in the production of bromine derivatives of organic substances. Hydrogen iodide is used to produce iodides.

22.3.4.2 Halides

The type of chemical bond in the halides of the elements of a single period changes from ionic in the halides of typical metals to covalent in the halides of nonmetals. Halogenides are considered in all elements of the groups, which is why only a brief overview is provided here.

The halides of alkali and alkaline earth metals dissolve in water without hydrolyzing. Soluble halides of less active metals are hydrolyzed reversibly:

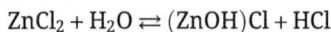

$$ZnCl_2 + H_2O \rightleftharpoons (ZnOH)Cl + HCl$$

The halides of nonmetals are completely decomposed by water with the formation of two acids:

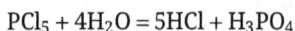

$$PCl_5 + 4H_2O = 5HCl + H_3PO_4$$

Metal halides are obtained by reacting hydrohalic acids with metals and their oxides, hydroxides, and carbonates:

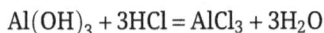

$$Mg + 2HCl = MgCl_2 + H_2$$

$$FeO + 2HCl = FeCl_2 + H_2O$$

$$Al(OH)_3 + 3HCl = AlCl_3 + 3H_2O$$

$$CaCO_3 + 2HCl = CaCl_2 + H_2O + CO_2$$

Basically, there are two possible ways to form metal halides: the interaction of a metal and a halogen element to form a metal halide with a higher oxidation state, and the interaction of a metal and a hydrogen halide to form a metal halide with a lower oxidation state of the metal. The production of Fe(III) chloride and Fe(II) chloride illustrates this pattern:

$$2Fe + 3Cl_2 \rightarrow 2FeCl_3$$

$$Fe + 2HCl \rightarrow FeCl_2 + H_2$$

In the first case, Cl_2 acts as a strong oxidizing agent, while in the second case, hydrogen chloride appears as a weak oxidizing agent.

To obtain insoluble halides, ion exchange reactions in salt solutions are convenient:

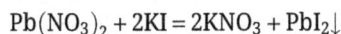

$$Pb(NO_3)_2 + 2KI = 2KNO_3 + PbI_2\downarrow$$

The halides of nonmetals are mainly obtained by direct interaction:

$$Si + 2Cl_2 = SiCl_4$$

$$P_4 + 10Cl_2 = 4PCl_5$$

Halogenides are used in a wide variety of fields of engineering. Sodium chloride is a starting substance in the production of NaOH, Cl_2, Na_2CO_3, and Na_2SO_4. Potassium chloride is an artificial fertilizer and raw material for the production of KOH and potassium salts. Aluminum chloride is a catalyst, CCl_4 is a solvent of fats and oils, $SiCl_4$ is a starting substance in the production of semiconductor silicon, and $NiCl_2$ is a catalyst. Silver bromide is used in photography, and sodium and potassium iodides are used in medicine.

22.3.4.3 Interhalogen compounds

Halogen elements can interact with each other to form interhalogen compounds. Especially numerous are the compounds of fluorine with the other halogen elements: with chlorine ClF, ClF_3, and ClF_5, with bromine BrF, BrF_3, and BrF_5, and with iodine IF, IF_3, IF_5, and IF_7. Chlorine forms the following compounds: BrCl, ICl, and ICl_3, and bromine – IBr. They also can contain oxygen: $ClOF_3$, ClO_2F, ClO_3F, ClO_2F_2, etc. All interhalogen compounds are toxic.

The structure of the molecules of these compounds is explained by the theory of the repulsion of valence electron pairs (Gillespie's method). For example, the ClF_5 molecule has the shape of a tetragonal pyramid (according to experimental data). This is possible if, at the chlorine atom ($3s^2p^5$), electrons from the 3p-subshell are divided and passed to the 3d-subshell ($3s^23p^33d^2$) with subsequent sp^3d^2-hybridization of the orbitals. In this case, six hybrid AO are formed: five single-electron bonding and one two-electron nonbonding AO. The bonding AO attach fluorine atoms by an exchange mechanism and a four-sided pyramid molecule of ClF_5 is formed.

Interhalogens interact with the halides of active metals that affirm their acidic properties:

$$KF + ClF_5 = K[ClF_6]$$

In water, these compounds are completely hydrolyzed, resulting in the formation of two acids:

$$ClF_5 + 3H_2O = 5HF + HClO_3$$

All interhalogen compounds are strong oxidizing agents.

22.3.4.4 Oxides, oxoacids, and oxysalts

Chlorine, bromine, and iodine form oxides in all even and odd oxidation states from +1 to +7 (without +3) as well as salts and acids in oxidation states +1, +3, +5, and +7 (see Table 22.3 and Figures 22.1 and 22.2). During the formation of these compounds in the atoms of the halogen elements, the rupture of the electron pairs, the transition of electrons of the free orbitals from the d-subshell, and the hybridization of the valence orbitals occur.

Oxides of chlorine, bromine, and iodine are thermodynamically unstable compounds with positive values of the Gibbs formation energy, so they are not obtained from simple substances, but by other methods, for example:

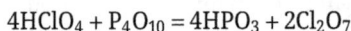

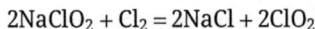

$$4HClO_4 + P_4O_{10} = 4HPO_3 + 2Cl_2O_7$$

$$2NaClO_2 + Cl_2 = 2NaCl + 2ClO_2$$

Oxides of halogen elements are easily detonated (by decomposition to a halogen element and oxygen) and are subject to the chemistry of explosives.

Oxides with odd oxidation state, when interacting with H_2O, form the corresponding acids:

$$Cl_2O + H_2O = 2HOCl$$

$$Cl_2O_7 + H_2O = 2HClO_4$$

Oxides with even oxidation states contain an unpaired electron in their molecules, so these compounds combine into dimers. The most important of these is chlorine dioxide

$[ClO_2 (Cl_2O_4)]$, which is more stable than other oxides and is used as an effective fabric bleacher.

At room temperature, *dichlorine monoxide* (Cl_2O) exists as a yellow-brown gas with a pungent odor and toxic effect. It is soluble in water and organic solvents. This oxide has a strong oxidizing effect and is a good chlorinating agent. The earliest method for the synthesis of Cl_2O is the reaction of Hg(II) oxide with chlorine. However, this method is expensive and very dangerous due to poisonous mercury:

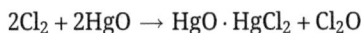

$$2Cl_2 + 2HgO \longrightarrow HgO \cdot HgCl_2 + Cl_2O$$

Table 22.3: Oxides and oxoacids of halogens.

Oxidation states	Oxides	Formulas and names of acids	Names and examples of salts
+1	Cl_2O Br_2O –	HClO – hypochlorous HBrO – hypobromous HIO – hypoiodous	Hypochlorite: $Ca(ClO)_2$ Hypobromite: nabro Hypoiodite: KIO
+2	ClO and Cl_2O_2 Br_2O_2 IO	– – –	– – –
+3	– – –	$HClO_2$ – chlorous	Chlorite: $NaClO_2$ Bromite: $NaBrO_2$ Iodite: $NaIO_2$
+4	ClO_2 and Cl_2O_4 BrO_2 –	– –	–
+5	Br_2O_5 I_2O_5	$HClO_3$ – chloric $HBrO_3$ – bromic HIO_3 – iodic	Chlorate: $KClO_3$ Bromate: $KBrO_3$ Iodate: KIO_3
+6	ClO_3 and Cl_2O_6 BrO_3 and Br_2O_6 I_2O_6	– –	– –
+7	Cl_2O_7 – – –	$HClO_4$ – perchloric $HBrO_4$ – perbromic HIO_4 – metaperiodic H_5IO_6 – orthoperiodic	Perchlorate: $KClO_4$ Perbromate: $KBrO_4$ Metaperiodate: KIO_4 Orthoperiodate: K_5IO_6

Chlorine dioxide (ClO_2) is a yellow gas that condenses into a dark red liquid at 11 °C. The compound is well soluble in water, forming relatively stable green solutions. Chlorine dioxide is paramagnetic, similar to nitrogen dioxide. However, unlike NO_2, it does not show a tendency to dimerization. The length of the Cl–O bond is only 140 pm – much less than 170 pm (single bond length), but very close to that of a typical chlorine-oxygen double bond. It should be noted that this oxide has nothing to do with the transient ClOO radical formed in the upper atmosphere.

The synthesis of ClO_2 involves the reduction of chlorine(V) (in the form of ClO_3^-) using Cl^- in an acidic environment to chlorine(IV) in ClO_2 and chlorine(0) in Cl_2:

$$2ClO_3^- + 4H^+ + 2Cl^- \rightarrow 2ClO_2 + Cl_2 + 2H_2O$$

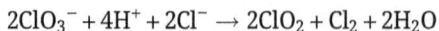

In an aqueous solution, chlorine dioxide dissolves by disproportionation:

$$6ClO_2 + 3H_2O \rightarrow 5HClO_3 + HCl$$

Chlorine dioxide is a very powerful oxidizing agent. It is used to bleach wood pulp in the paper industry. Its advantage is that it does not attack the structure of the pulp and preserves the mechanical strength of the paper, unlike Cl_2.

Dichlorine hexaoxide (Cl_2O_6) is a bright red liquid. In its solid state, it is a crystalline substance of orange color. The molecules of liquid Cl(VI) oxide have the structure $O_2Cl-O-ClO_3$ and the crystalline $[ClO_2]^+[ClO_4]^-$. It is obtained by the oxidation of ClO_2 with ozone:

$$2ClO_2 + 2O_3 \rightarrow 2O_2 + Cl_2O_6$$

Chlorine(VI) oxide is an unstable substance and decomposes at a temperature of 0–10 °C:

$$Cl_2O_6 \rightarrow O_2 + 2ClO_2$$

At temperatures above 20 °C, chlorine also appears in the decomposition products. With water, it reacts intensely causing an explosion:

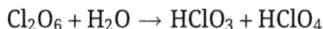

$$Cl_2O_6 + H_2O \rightarrow HClO_3 + HClO_4$$

It interacts with bases in solution, whereby disproportionation occurs:

$$Cl_2O_6 + 2KOH \rightarrow KClO_3 + KClO_4 + H_2O$$

It exhibits strong oxidizing properties. With $AlCl_3$ forms $ClO_2[Al(ClO_4)_4]$ and with $FeCl_3$ forms $ClO_2[Fe_2(ClO_4)_7]$.

Dichlorine heptoxide (Cl_2O_7) is a typical acidic oxide. The molecule of Cl_2O_7 is polar and has the structure $O_3Cl-O-ClO_3$. Dichlorine heptoxide is a colorless, oily liquid. It decomposes at a temperature of 120 °C and under mechanical stress, but is relatively more stable than other chlorine oxides. It dissolves slowly in cold water and forms perchloric acid:

$$Cl_2O_7 + H_2O \rightarrow 2HClO_4$$

Cl_2O_7 exhibits strong oxidizing properties.

Dichlorine heptoxide is obtained by heating perchloric acid with P_4O_{10} or oleum:

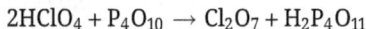

$$2HClO_4 + P_4O_{10} \rightarrow Cl_2O_7 + H_2P_4O_{11}$$

Another method for obtaining Cl_2O_7 is the electrolysis of a solution of $HClO_4$ with platinum electrodes at 0 °C. Pure Cl_2O_7 can be synthesized in a vacuum by heating some perchlorates, for example, $Nb(ClO_4)_5$ or $MoO_2(ClO_4)_2$.

Diiodine pentoxide (I_2O_5) is an anhydride of iodic acid. It forms white crystals that darken in the light due to its decomposition. I_2O_5 is used as an oxidizing agent. It dissolves well in water, is slightly soluble in ethanol and insoluble in diethyl ether, carbon disulfide, and chloroform. It is obtained by decomposing iodic acid at high temperature or with dehydrating agents:

$$2HIO_3 \rightarrow I_2O_5 + H_2O$$

The strength of acids of halogen elements increases with increasing oxidation state. For example, chlorine forms oxoacids and oxyanions in all positive odd oxidation states from +1 to +7. The shapes of the ions and the corresponding acids are tetrahedral, with the chlorine atom in the center. The small lengths of the chlorine-oxygen bonds indicate the presence of multiple bonds formed because of the filled p-orbitals of the oxygen atom and the free d-orbitals of the Cl atom.

The following pattern is observed in the *oxoacids* of chlorine: $HClO$ is very weak, $HClO_2$ is weak, $HClO_3$ is strong, and $HClO_4$ is a very strong acid.

Perchloric acid $HClO_4$ is a colorless compound and in aqueous solution it is stronger than sulfuric and nitric acids. In the order from $HClO$ to $HClO_4$, in addition to the strength of acids, their stability also increases, but in the same order their oxidative effect decreases due to the decay of acids (with oxygen release). As a result of the above, $HClO$ and $HClO_2$ are stable only in very dilute solutions, $HClO_3$ is stable in solutions up to 30–40%, and $HClO_4$ can also be obtained in an anhydrous state. The decomposition of unstable acids occurs in parallel with disproportionation and an intramolecular redox process:

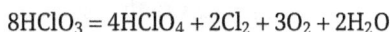

$$3HClO = 2HCl + HClO_3$$

$$2HClO = 2HCl + O_2$$

$$4HClO_3 = HCl + 3HClO_4$$

$$8HClO_3 = 4HClO_4 + 2Cl_2 + 3O_2 + 2H_2O$$

The salts of the oxygen-containing acids of the halogen elements are more stable than the acids themselves.

In the order of acids formed by different halogen elements in the same oxidation state, the strength of acids decreases: $HClO$ ($K = 5.10^{-8}$), $HBrO$ ($K = 2.10^{-9}$), and HIO ($K = 2.10^{-11}$).

Hypohalous acids ($HClO$, $HBrO$, and HIO) are formed when the respective halogen element is dissolved in cold water:

$$X_2 + H_2O \rightleftharpoons H^+ + X^- + HXO$$

They can be obtained by the reaction:

$$2X_2 + 2HgO + H_2O \rightarrow HgO \cdot HgX_2 + 2HXO$$

Hypohalogenous acids are very weak acids; that is why solutions of their salts *hypoha-logenites* show an alkaline reaction as a result of hydrolysis processes:

$$XO^- + H_2O \rightleftharpoons HXO + OH^-$$

As mentioned, all oxoacids of the halogen elements are oxidizing agents, but their oxidative activity decreases with the oxidation state of the halogen element. This is explained by an increase in the strength of the acid anions, which in turn is related to the increased number of valence electrons in the central atom of the halogen element. Therefore, the strongest oxidizing agents are the weakest acids HClO, HBrO, and HIO.

Hypochlorous acid is a strong oxidizing agent and is reduced to chlorine:

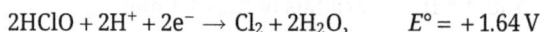

$$2HClO + 2H^+ + 2e^- \rightarrow Cl_2 + 2H_2O, \qquad E° = +1.64\,V$$

However, the hypochlorite anion is a weaker oxidizing agent that is usually reduced to Cl⁻:

$$ClO^- + H_2O + 2e^- \rightarrow Cl^- + 2OH^-, \qquad E° = +0.89\,V$$

It is the oxidizing ability of the hypochlorite ion that explains the whitening and bactericidal action of these compounds.

Hypochlorous acid oxidizes many simple and complex substances to their maximum oxidation states:

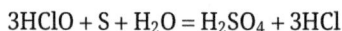

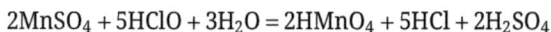

$$3HClO + S + H_2O = H_2SO_4 + 3HCl$$

$$2MnSO_4 + 5HClO + 3H_2O = 2HMnO_4 + 5HCl + 2H_2SO_4$$

Hypochlorites are of the greatest practical importance of the salts of hypohalogenous acids. They are obtained by electrolysis or disproportionation of chlorine in a basic medium. For example, when chlorine reacts with a solution of calcium hydroxide in the cold, a mixture of calcium chloride and hypochlorite is obtained:

$$2Cl_2 + 2Ca(OH)_2 = CaCl_2 + Ca(ClO)_2 + 2H_2O$$

This mixture of $CaCl_2$ and $Ca(ClO)_2$ is called chlorine lime. Its formula is written as CaCl(ClO) (a mixed salt of hydrochloric and hypochlorous acids). It is a strong oxidizing agent and is used for the disinfection of toilets, livestock premises, water bodies, etc.

Chlorous acid ($HClO_2$) is produced by reacting barium chlorite with dilute sulfuric acid:

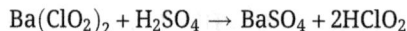

$$Ba(ClO_2)_2 + H_2SO_4 \rightarrow BaSO_4 + 2HClO_2$$

The acid exists only in solution and quickly decomposes. Theoretically, its anhydride (Cl_2O_3) does not exist. Chlorous acid is stronger than hypochlorous acid, but is a weaker oxidizing agent than hypochlorous acid. Its salts are called chlorites. Chlorites are generally more stable in aqueous solution, but they are still very perishable and decompose into perchlorates and other products. Chlorites are good oxidizing agents and are used to bleach paper and fabrics.

Chloric, bromic, and iodic acids are produced by reacting their barium salts with dilute sulfuric acid:

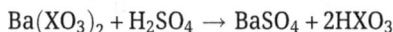

$$Ba(XO_3)_2 + H_2SO_4 \rightarrow BaSO_4 + 2HXO_3$$

Chloric and bromic acids exist only in solution, and iodic acid is isolated as a colorless crystalline substance.

Unlike chlorites, which are not of great interest, chlorates find wide practical applications. Sodium chlorate can be obtained from Cl_2 in a hot sodium hydroxide solution:

$$3Cl_2 + 6NaOH \rightarrow NaClO_3 + 5NaCl + 3H_2O$$

Of the salts of chloric acid ($HClO_3$), the most used is potassium chlorate ($KClO_3$), the so-called Berthollet salt. Potassium chlorate ($KClO_3$) is obtained by reacting chlorine with hot solutions of potassium hydroxide or carbonate:

$$3Cl_2 + 6KOH = KClO_3 + 5KCl + 3H_2O$$

$$3Cl_2 + 3K_2CO_3 = KClO_3 + 5KCl + 3CO_2$$

Potassium chlorate is stable when heated to 400 °C. At 400–550 °C, Berthollet salt decomposes by disproportionation:

$$4KClO_3 = 3KClO_4 + KCl$$

At temperatures of 550–600 °C, this salt decomposes by an intramolecular redox mechanism:

$$2KClO_3 = 2KCl + 3O_2$$

Unlike potassium nitrate, potassium chlorate can give up all the oxygen it contains. In addition, the decomposition process is exothermic, and for this reason, Berthollet salt surpasses many other compounds in oxidizing properties and its mixtures with combustible materials are easier to ignite. The decomposition of Berthollet salt is catalyzed by manganese(IV) and iron(III) oxides as well as by other catalysts. The process is accompanied by a decrease in the decomposition temperature to 200 °C. The mechanism of the process involving the catalyst is a good illustration of a classical catalytic reaction:

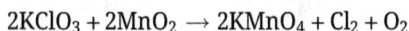

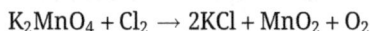

$$2KClO_3 + 2MnO_2 \rightarrow 2KMnO_4 + Cl_2 + O_2$$

$$2KMnO_4 \rightarrow K_2MnO_4 + MnO_2 + O_2$$

$$K_2MnO_4 + Cl_2 \rightarrow 2KCl + MnO_2 + O_2$$

Thus, oxygen is oxidized from −2 to 0 oxidation state, manganese goes from +4 through +7 and +6 and back to +4, and chlorine is reduced from +5 to 0 and to −1 oxidation state.

Berthollet salt is a strong explosive. When it comes into contact with phosphorus, sulfur, carbon, and especially organic matter, Berthollet salt reacts with an explosion upon impact or mechanical friction. Berthollet salt is a strong oxidizing agent, especially in the molten state and in a mixture with bases. Under these conditions, it oxidizes many inert compounds:

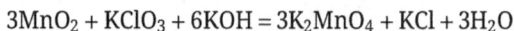

$$3MnO_2 + KClO_3 + 6KOH = 3K_2MnO_4 + KCl + 3H_2O$$

The main consumers of Berthollet salt are the match industry, pyrotechnics, and the production of explosives. In the chemical industry and in chemical analyses, it is used for oxidation and bringing the most inert compounds into a soluble form. It can be used to produce oxygen in the laboratory, most often in a mixture with manganese dioxide MnO_2, which lowers the decomposition temperature of chlorate.

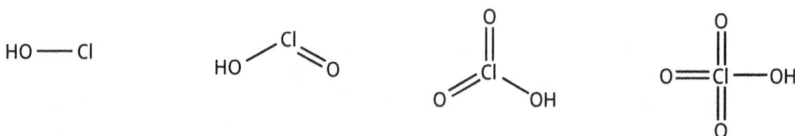

Figure 22.1: Structures of hypochlorous, chlorous, chloric, and perchloric acids.

The strongest of all known acids is *perchloric acid*. Pure acid is a colorless liquid that can easily be exploded. Due to its oxidizing properties and high oxygen content, its contact with organic materials, such as wood or paper, results in an explosive reaction. Concentrated perchloric acid, usually a 60% aqueous solution, is rarely used as an acid. Much more often it is used as a powerful oxidizing agent in the oxidation of metal alloys. Diluted solutions of perchloric acid in the cold are relatively safe.

Perchloric acid is produced by the reaction:

$$2KClO_4 + H_2SO_4 \rightarrow K_2SO_4 + 2HClO_4$$

The acid decomposes as follows:

$$4HClO_4 \rightarrow 2Cl_2 + 7O_2 + 2H_2O$$

$HClO_4$ salts, the so-called *perchlorates*, are obtained from the corresponding chlorates when heated in the absence of catalysts, whereby disproportionation occurs:

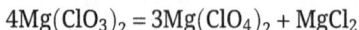

$$4Mg(ClO_3)_2 = 3Mg(ClO_4)_2 + MgCl_2$$

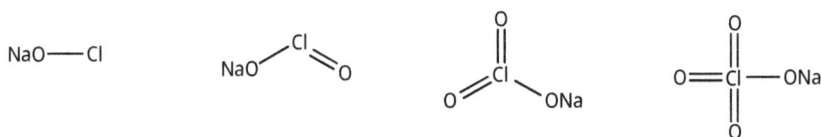

Figure 22.2: Structures of hypochlorites, chlorites, chlorates, and perchlorates.

Ammonium perchlorate is obtained by neutralizing perchloric acid with NH_3:

$$NH_3 + HClO_4 \rightarrow NH_4ClO_4$$

Potassium perchlorate is produced industrially by an exchange reaction between solutions of sodium perchlorate ($NaClO_4$) and potassium chloride (KCl). When these solutions are mixed, potassium perchlorate ($KClO_4$) is obtained in the form of a fine white precipitate. This ion exchange reaction is based on the lower solubility of potassium perchlorate compared to that of $NaClO_4$. It has been found that with an increase in the size of the cation, the solubility of the salts of a given group decreases. Usually, the large size (i.e., low electron density) of the ion means a small hydration energy. Thus, potassium perchlorate has a low solubility (20 g/L of water). On the other hand, silver perchlorate is quite soluble (up to 5 kg/L of water). The high solubility of silver perchlorate in weakly polar organic solvents, as well as in water, suggests that its bonding in the solid phase is essentially covalent rather than ionic. In this case, in order to dissolve the compound, only the dipole interactions must be overcome, in contrast to the much stronger electrostatic interactions in an ionic crystal lattice, which are overcome by strongly polar solvents.

Potassium perchlorate is a strong oxidizing agent – it vigorously gives its oxygen to oxidizing materials. Mixed with sugar, $KClO_4$ forms a mixture possessing the properties of a weak explosive. For example, it interacts with glucose in the following way:

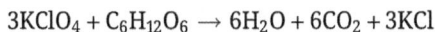

$$3KClO_4 + C_6H_{12}O_6 \rightarrow 6H_2O + 6CO_2 + 3KCl$$

Perchlorates are used as oxidizing agents and explosives. Magnesium perchlorate (anhydron) is used as a gas desiccant. Potassium perchlorate is used in the manufacture of fireworks and flares, as well as in the manufacture of ammonium perchlorate, which is widely used in rocket fuels. Other uses of perchlorates include temporary adhesives, electrolysis baths, batteries, air bags, etching agents, drying agents, cleaning agents and bleach, and oxygen generating systems. Many years ago, perchlorates were used as a medication to treat overactive thyroid glands, and they still have some medical uses.

Chlorine makes a major contribution to combating epidemics of cholera and typhoid. It is an effective water disinfectant against parasites, bacteria, and viruses. It provides a durable protection, is easy, and inexpensive to use. It can be used as Cl_2, $NaClO$, and $Ca(ClO)_2$. In water all of them yield the strong oxidizer – hypochlorite anion (ClO^-).

Chlorine and bromine in the body are typically found as hydrated ions Br^- and Cl^-. Fluorine and iodine have different behavior under the same conditions. Iodine characteristically forms C–I bonds. Fluorine usually binds to metals, for example, Ca, Mg, and Fe. The fluoride ion differs substantially from the other analogues. Cl^-, Br^-, and I^- anions are similar in their chemical properties and can easily replace each other in vivo, while showing synergy and antagonism.

Chapter 23
Eighth main group of the periodic table

23.1 General characteristics of the elements of VIIIA group

The noble gases helium (He), neon (Ne), argon (Ar), krypton (Kr), xenon (Xe), and radon (Rn) are located in the VIIIA group. Their atoms have complete configurations of their outer electron shell with two (He) or eight (from Ne to Rn) electrons, notable with their particularly high stability. The main characteristics and physical properties of the elements are given in Table 23.1.

Table 23.1: Characteristics and physical properties of elements of VIIIA group.

Element	He	Ne	Ar	Kr	Xe	Rn
Valence electrons	$1s^2$	$2s^2 2p^6$	$3s^2 3p^6$	$4s^2 4p^6$	$5s^2 5p^6$	$6s^2 6p^6$
Radius of the atom (nm)	0.122	0.160	0.192	0.198	0.218	0.220
Ionization potential (eV)	24.6	21.6	15.8	14.0	12.1	10.8
Boiling temperature (°C)	−269	−246	−186	−153	−108	−62
Melting temperature (°C)	−272	−249	−189	−157	−112	−71

In noble gas atoms, all electrons are paired. They have very high ionization potentials, so they exist in an atomic state with low melting and boiling points. However, as the sequence number of the element increases, their ionization potentials decrease, and in krypton, xenon, and radon they become equal to the ionization potentials of oxygen (13.6 eV), nitrogen (14.5 eV), and chlorine (13.0 eV). Of all the gases, they have relatively the highest electrical conductivity, which is essential for lighting technology, where they have an important application. The solubility of noble gases in water is high, and with an increase in the serial number in the group, it increases.

All elements in the group are colorless, monatomic gases at room temperature. They neither burn nor keep burning. Their very low melting and boiling points indicate that the dispersion forces possessed by atoms in the solid and liquid phases are very weak.

These are the chemically least active p-elements, which is why they are called inert or noble gases. Only Kr, Xe, and Rn participate in chemical interactions. Inert gases form clathrates with water in the solid state: $Kr \cdot 6H_2O$, $Xe \cdot 6H_2O$, and $Ar \cdot 6H_2O$. Until 1962, it was believed that these elements could not participate in chemical reactions and exist in the form of compounds. These inert gases are located in a separate (zero) group of the periodic table. In 1962, the first compounds of xenon were obtained, and later of krypton and radon. Currently, more than 300 compounds of these elements are known. The question of argon compounds is controversial, and helium and neon compounds are practically nonexistent. The arrangement of noble gases in group VIIIA is consistent with the number of their valence electrons.

https://doi.org/10.1515/9783111712246-023

23.2 Occurrence, properties, and application

Inert gases are formed in nature as a result of various nuclear processes. Air contains 55.24×10^{-4}% He, 1.82×10^{-3}% Ne, 0.934% Ar, 1.14×10^{-4}% Kr, 8.6×10^{-6}% Xe, and 6×10^{-20}% Rn (vol.%). On the Earth, argon is more abundant than other noble gases. Argon is obtained from air. The high concentration of argon in the atmosphere is the result of the radioactive decay of potassium-40, a naturally occurring radioactive isotope of potassium. Space is richer in noble gases than the Earth's atmosphere. Helium is widely distributed in space, as it is formed as a result of thermonuclear processes taking place in the bowels of stars. It was first discovered on the Sun when studying its spectrum, and only then on Earth. Noble gases are also found in natural gases and in some minerals. Helium is found in high concentrations in some underground gas deposits, where it is accumulated as a result of the decay of some radioactive elements in the earth's crust.

Compared to other elements, the *helium* atom has the most stable electronic structure. Due to the negligible polarizability of its atoms, its dispersion interaction is very weak, and its boiling and melting temperatures are the lowest. In the liquid state, helium exhibits superconductivity and superfluidity. Helium is the coldest liquid. When liquid helium evaporated, a temperature was obtained that was in close proximity to absolute zero.

Neon is close in properties to helium, but it is characterized by high polarizability of its atoms, so its boiling and melting temperatures are higher than those of helium.

In *argon*, compared to helium and neon, the size of the atom increases noticeably, and thus the ionization potential decreases. The polarizability of the atoms and the energy of the intermolecular dispersion interaction increase, which leads to an increase in the boiling and melting temperatures. There is a compound of argon with water, in which the argon atoms are located in the cavities of the crystal lattice of solid water because of intermolecular interaction. Such compounds are called *clathrates*. There are contradictory reports on the existence of valence compounds of argon. There are scientific reports about the production of argonids – compounds of argon with metals, in which argon is found in negative oxidation states. These data have not been confirmed so far, but have not been refuted. *Krypton, xenon*, and *radon* form chemical compounds with fluorine, oxygen, and other elements. Xenon compounds are of interest to chemistry.

Helium is produced from natural gases by deep cooling. It is used to create ultra-low temperatures and an inert environment in welding metals. It is also used in light diving equipment to form oxygen-helium gas mixtures, which are used in descent to great depths. It is often used as an inert medium in technical processes taking place at high temperature and as a substitute for hydrogen, which is lighter, but highly explosive. Neon is used to fill vacuum utensils and various lamps (headlights, signaling devices, advertising devices) with characteristic red lighting (neon lights). Argon is used in welding Al, Mg, and their alloys and in other areas of technology. Noble gases are mainly used in lighting technology for the preparation of neon, xenon lamps, etc.

23.3 Chemical compounds of xenon

The chemistry of inert gases is mainly represented by the chemistry of xenon.

23.3.1 Fluorides

The chemistry of xenon is mainly related to fluorine because only with it xenon interacts directly, forming fluorides XeF_2, XeF_4, XeF_6, and XeF_8. In 1962, Bartlett discovered that at 400 °C, xenon combines with F_2 to form XeF_4. In addition to it, other noble gas fluorides were obtained: XeF_2, XeF_6, and KrF_4. All of them are white crystalline substances soluble in an aqueous solution of KI. In water, the fluorides of the inert gases hydrolyze and form the corresponding hydroxides, such as $Xe(OH)_4$ and $Xe(OH)_6$, which exhibit weak acidic properties. Xenon reacts with fluorine, and depending on the conditions, different products are obtained:

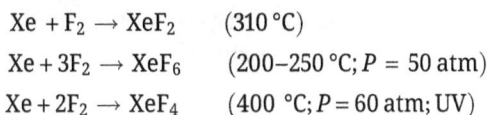

$$Xe + F_2 \rightarrow XeF_2 \quad (310\,°C)$$
$$Xe + 3F_2 \rightarrow XeF_6 \quad (200\text{–}250\,°C; P = 50\,atm)$$
$$Xe + 2F_2 \rightarrow XeF_4 \quad (400\,°C; P = 60\,atm; UV)$$

The product depends on the ratio of the starting reactants, on temperature and pressure. For the formation of xenon hexafluoride, a very high partial pressure of F_2 is required. All xenon fluorides are stable in terms of their dissociation to elements at ordinary temperature, that is, they have negative values of the free energy of formation at 25 °C. Xenon hexafluoride (with six bonding electron pairs) could take the form of a pentagonal bipyramid, a trigonal prism, or an octahedron. Structural studies of xenon hexafluoride in the gas phase show that it has an octahedral structure.

With an increase in the oxidation state of xenon, an increase in acidic properties is observed in fluorides. Xenon hexafluoride interacts with basic fluorides, forming fluorocomplexes:

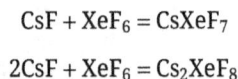

$$CsF + XeF_6 = CsXeF_7$$
$$2CsF + XeF_6 = Cs_2XeF_8$$

XeF_2 decomposes when interacting with water. At the hydrolysis, it is reduced to xenon:

$$2XeF_2 + 2H_2O \rightarrow 2Xe + O_2 + 4HF$$

XeF_6 is first hydrolyzed to oxidotetrafluoride $XeOF_4$, which in turn is hydrolyzed to XeO_3:

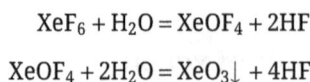

$$XeF_6 + H_2O = XeOF_4 + 2HF$$

$$XeOF_4 + 2H_2O = XeO_3\downarrow + 4HF$$

In some reactions, xenon compounds are oxidizing agents, and in others – reducing agents:

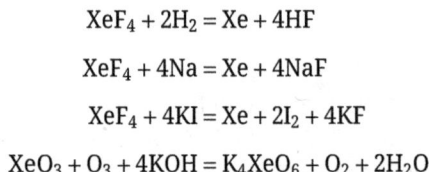

$$XeF_4 + 2H_2 = Xe + 4HF$$

$$XeF_4 + 4Na = Xe + 4NaF$$

$$XeF_4 + 4KI = Xe + 2I_2 + 4KF$$

$$XeO_3 + O_3 + 4KOH = K_4XeO_6 + O_2 + 2H_2O$$

Fluorides are strong fluorinating agents. For example, xenon difluoride is often used to fluoridate organic compounds. It is a very "pure" fluorinating agent since the inert xenon can be easily separated from the reaction product:

$$XeF_2 + CH_2 = CH_2 \rightarrow CH_2F - CH_2F + Xe$$

In addition, xenon fluorides are used as reagents in the fluoridation of elements to their highest oxidation state. Thus, XeF_4 oxidizes sulfur tetrafluoride to sulfur hexafluoride:

$$XeF_4 + 2SF_4 \rightarrow 2SF_6 + Xe$$

23.3.2 Oxides

Xenon forms two oxides: xenon trioxide and xenon tetroxide. *Xenon trioxide* is a colorless, solid, highly explosive substance that decomposes into xenon and oxygen during storage. The oxide is an extremely strong oxidizing agent, although its reactions are kinetically slower. Xenon trioxide is obtained by complete hydrolysis of XeF_4 and XeF_6:

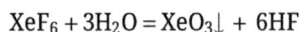

$$XeF_6 + 3H_2O = XeO_3\downarrow + 6HF$$

Due to the free electron pair, its molecule has the shape of a trigonal pyramid. The length of the bonds indicates a certain degree of multiplicity, similar to the compounds of silicon, phosphorus and sulfur.

Xenon trioxide dissolves in water to form the weak xenic acid H_2XeO_4. It reacts with dilute bases and forms xenates of the type Na_2XeO_4. In highly alkaline solutions, xenates are unstable and gradually disproportionate to the formation of xenon and perxenate ion XeO_6^{4-}, respectively:

$$XeO_3 + OH^- \rightarrow HXeO_4^-$$

$$2HXeO_4^- + 2OH^- \rightarrow XeO_6^{4-} + Xe + O_2 + 2H_2O$$

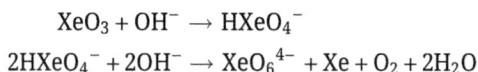

Perxenates are also obtained by oxidation of xenates with ozone:

$$3NaHXeO_4 + 9NaOH + O_3 = 3Na_4XeO_6 + 6H_2O$$

The alkaline and alkaline earth salts of the perxenate ion (isoelectronic with the periododate ion IO_6^{5-}) can be isolated in a crystal state. They are colorless, stable substances. In the perxenate ion, the xenon atom is surrounded by six O atoms in an octahedral structure. Perxenates are among the strongest known oxidizing agents, close in oxidation capacity to fluorine. For example, they rapidly oxidize Mn(II) ions to permanganate ions, reducing themselves to a hydrogenxenate ion:

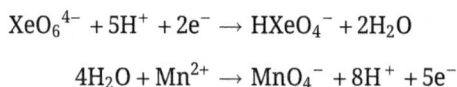

$$XeO_6^{4-} + 5H^+ + 2e^- \rightarrow HXeO_4^- + 2H_2O$$

$$4H_2O + Mn^{2+} \rightarrow MnO_4^- + 8H^+ + 5e^-$$

When perxenates react with anhydrous H_2SO_4 at a low temperature (−5 °C), gaseous xenon tetroxide (XeO_4) is formed. It is a perishable compound that decomposes into XeO_3, Xe, and O_2.

Xenon tetroxide is usually obtained by adding concentrated H_2SO_4 to barium perxenate:

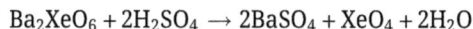

$$Ba_2XeO_6 + 2H_2SO_4 \rightarrow 2BaSO_4 + XeO_4 + 2H_2O$$

In this oxide, xenon is in +8 oxidation state. It is an explosive gas. It has been proven that its structure is tetrahedral. It is possible that xenon tetroxide exists deep in the earth. For decades, geochemists have been looking for an explanation for the low level of xenon in the atmosphere (the so-called xenon deficiency). It is assumed that xenon formed by radioactive decay in the earth does not spread on the surface since it is replaced by silicon in tetrahedral silicate structures at extreme temperatures and pressures in the earth's crust, which has been proven experimentally.

Section B.II: **Secondary groups of the periodic table**

Chapter 24
d-Elements

24.1 General characteristics of transition metals

The secondary B groups include elements only from the major periods. These are d- and f-elements because in them the electrons occupy the d- and f-states of the lower shells. The electron configuration of the penultimate $(n – 1)$ and last (n) shells of the d-elements is expressed in general terms as follows: $(n –1)s^2(n – 1)p^6(n – 1)d^{1-10}ns^{2(1)}$. In f-elements, electrons occupy the $(n – 2)$f-orbitals of the penultimate shell in sequence and their electron configuration of the outer shells is generally expressed by the formula: $(n – 2)s^2(n – 2)p^6(n – 2)d^{10}(n – 2)f^{1-14}ns^{2(1)}$. In the secondary groups of the periodic table the elements from scandium to zinc, from yttrium to cadmium, from lutetium to mercury, lanthanides and actinides are located. Many of these elements and their salts are paramagnetic. They have no electronic analogs in the small periods. According to their atomic number and the similarity of their electronic structure, they are arranged in eight secondary groups. The elements in a secondary group exhibit similar chemical properties because their atoms have an equal number of valence electrons arranged in the same type of subshells.

From the fourth period onward, in the long-period version of the periodic table, only d-elements appear in the IV and V periods, and then d- and f-elements (in the VI and VII periods). These elements are called *transition* elements because they make a transition from s- to p-elements in long periods. In the short-period version of the periodic table, the f-elements are exported separately.

According to some authors, the interchangeable use of the terms d-elements and transition metals is not entirely correct. Transition metals are generally characterized by a variable oxidation state and a tendency to form complexes. According to these concepts, scandium (and the other members of group IIIB) is excluded from the group of transition metals. Scandium is similar in properties to the elements of group IIIA, in particular, the metal aluminum. The other elements of group IIIB are similar to the 4f-elements in terms of their chemical properties and are discussed together with them. At the other end of the d-block are the elements with a filled d-subshell (zinc, cadmium, and mercury), which determines some differences between group IIB and the others. The elements from rutherfordium (element 104) to roentgenium (element 111) are also transition metals, although they are radioactive elements, and are often discussed together with actinides.

The electron configuration of the d-subshell of the penultimate $(n – 1)$ shell and of the s-subshell of the last shell for the elements of the B-groups is represented as follows: IB $(n – 1)d^{10}ns^1$; IIB $(n – 1)d^{10}ns^2$; IIIB $(n – 1)d^1ns^2$; IVB $(n – 1)d^2ns^2$; VB $(n –1)d^3ns^2$; VIB $(n –1)d^5ns^1$; VIIB $(n – 1)d^5ns^2$; VIIIB $(n – 1)d^6ns^2$; $(n – 1)d^7ns^2$; $(n – 1)d^8ns^2$. With a few exceptions, the configuration of the outer electron shells of the atoms of the

https://doi.org/10.1515/9783111712246-024

d-elements is ns². Therefore, all d-elements are metals and the changes in their properties with the increase of Z are not as sharp as in the case of the s- and p-elements. In their highest oxidation states, the d-elements exhibit a certain resemblance to the p-elements of the corresponding groups of the periodic table.

Here are some general regularities characterizing the properties of d-elements. A distinctive feature of d-elements is that in their atoms, two (sometimes one) valence electrons are located in the outer ns-subshell, and the rest are located in the penultimate $(n-1)$d-subshell. This subshell occurs after the filled 4s-subshell and can accommodate from 1 to 10 electrons. That is why the first d-element (Sc) does not appear until the fourth period after calcium (4s²). In each subsequent period, there are 10 (decade) d-elements. At the same time, in the first five elements of each decade, electrons fill the orbitals singly and the number of unpaired electrons at the d-subshell increases from one to five. In subsequent elements, the electrons complement each other and the number of unpaired electrons decreases. Thus, in the first elements of the respective decades (Sc, Y, and La) there is one unpaired electron, in the fifth (Mn, Tc, and Re) – five, and in the last (Zn, Cd, and Hg) there are no unpaired electrons. This is also the reason why zinc, cadmium, and mercury differ from other d-elements.

The highest valence of the d-elements is equal to the group number (excluding IB and VIIIB), and the lower valence is equal to the number of electrons in the outer electron shell.

The d-elements are metals with metallic chemical bonds and a metallic crystal lattice. Their common physical properties are solid state of matter (except Hg), metallic luster, thermal and electrical conductivity, different ductility, and elasticity. Some are light with a density <7 g/cm³, others heavy with a density >7 g/cm³ and have different melting and boiling temperatures. Their electrical conductivity varies widely. The electrical conductivity of Hg is taken as a reference, which is 1. The energy of the interatomic bonds in the crystal lattices of these metals is determined not only by the delocalized s-electrons but also by additional covalent bonds between the unpaired electrons of the d-orbitals. Therefore, the d-elements (without Zn, Cd, and Hg) are stable and heat-resistant, and this applies to those located in the middle part of the decades (Cr, Mo, W, Mn, and Tc).

The elements of the B groups exhibit common chemical properties characteristic of metals. They interact with oxygen, various nonmetals, acids, and salts. Due to their less pronounced metallic character, they do not react with water (unlike the elements of groups IA and IIA).

Between the d-elements, there are both very strong reducers (scandium, yttrium, lanthanum), and very weak reducers (precious metals). The reduction capacity of the metal depends on many factors such as the structure of the isolated atom, the number of electrons given up during oxidation, the energy spent on the separation of each electron (ionization energy), the energy spent on breaking the interatomic bonds (crystal lattice energy), and the energy released during the hydration.

In the formation of their compounds, s-electrons and all or part of the d-electrons are involved in chemical bonds. Therefore, d-elements are characterized by variable valence, as well as the variety and wide intervals of change of the acid-basic and oxidation-reduction properties of their compounds. In contrast to the main groups, the stability of compounds in higher oxidation states in the secondary groups increases. For example, in chromium, the most stable compounds are in an oxidation state of +3, in molybdenum – in an oxidation state of +4, and in tungsten – in an oxidation state of +6. For the same element, higher oxidation states are observed in compounds with the most electronegative and small-sized atoms of the elements (oxygen and fluorine). For example, V_2O_5, Re_2O_7, and ReF_7 are stable, but elements such as VCl_5 and $ReCl_7$ do not exist.

As the oxidation state of the element increases, the nature of the bond changes from ionic to more and more covalent in its homogeneous compounds. So, the lower oxides and hydroxides are basic, and the higher ones are acidic, the lower halides are ionic soluble salts, and the higher halides are molecular, volatile, hydrolysable substances. As the oxidation state increases, the oxidation rate increases and the reduction activity of the corresponding compounds decreases.

The d-elements are characterized by the formation of compounds with a variable composition. These include most of their oxides. Thus, if calcium or aluminum forms only one compound CaO and Al_2O_3 with oxygen, titanium forms multiple "oxides": Ti_2O, Ti_5O_4, TiO, Ti_2O_3, Ti_5O_9, $Ti_{10}O_{19}$, and TiO_2. As the oxygen content increases, the bonds change from metallic to covalent-polar or almost covalent. Compounds with variable composition also include hydrides, sulfides, carbides, and other binary compounds of d-elements.

Quite characteristic of d-elements is the formation of numerous and stable complex compounds. In complex compounds, d-elements form bonds with ligands by a donor-acceptor mechanism. Forming a variety of complexes with the numerous bioligands of living organisms, transition biometals essentially have the behavior of "organizers of life processes." Some d-elements, in addition to complex compounds, also form more complex cluster compounds. These compounds are made up of complex ions that contain covalent bonds of type E–E. For example, the compound Cr_2Cl_{12} contains a cluster of $(Cr_2Cl_9)^{3-}$ and its formula is $(Cr_2Cl_9)Cl_3$. The role of d-elements and their compounds in catalytic processes is great. d-Elements, such as platinum, palladium, nickel, and iron, accelerate numerous reactions involving inorganic and organic substances. The high catalytic activity of d-elements is also associated with their other features.

24.2 Occurrence and synthesis

Some d-elements are found in a free (native) state in nature (gold, platinum, silver, and mercury), but most are in the composition of compounds, that is, in an oxidized state. Metals are most often found in ores in the form of oxides, sulfides, carbonates, silicates, and phosphates. Ores with a high content of elements are almost depleted in nature, and at present, metals are often extracted from poor ore deposits containing many impurities (sand, clay, etc.). Therefore, the ores are pre-enriched, that is, the minerals are separated from the impurities. The enrichment of ores is most often carried out by nonchemical methods (magnetic separation, flotation, gravity enrichment), using for this purpose the differences in the physical properties of the mineral and impurities. Iron, manganese, and other ores containing metal oxides are reduced immediately after enrichment. For further processing of other minerals after enrichment, chemical processes are carried out, with the help of which the extracted element is brought into a soluble or other form convenient for further processing. For chemically inert minerals, various processes are carried out such as "baking" with bases in the presence of oxidizing agents, treatment with acids and complexing substances, chlorination or fluoridation, frying in an oxygen environment, etc. After these operations, the metals enter the form of oxides, chlorides, or fluoride for reduction.

For the reduction of metals, carbon, CO, hydrogen, hydrogen-containing gases, and active metals (aluminum, magnesium, calcium, and sodium) are used. The choice of a reducer is determined by the thermodynamics of the process and its economic feasibility. For example, carbon can be a potential reducer of practically all oxides, but in the case of active metal oxides, the reduction temperature is so high that the process is technically difficult and economically unjustified. Carbon and CO are used in the production of iron, manganese, nickel, cobalt, copper, zinc, cadmium, bismuth, lead, tin, and some other metals. Hydrogen is usually used in the reduction of tungsten, molybdenum, and rhenium. In the production of titanium, zirconium, hafnium, chromium, vanadium, scandium, yttrium, lanthanum, etc., the most commonly used are metallothermic processes.

All the mentioned methods for obtaining metals are called *pyrometallurgical* because the reduction is carried out at high temperature. These methods have many disadvantages including harmful working conditions and special requirements regarding the thermal resistance of the equipment. Therefore, *hydrometallurgical methods* are becoming more and more widespread methods in which all stages of extraction of metals from ores are carried out using solutions of acids, bases, and salts.

Modern technology requires metals of high purity. Special methods have been developed for their production such as vacuum distillation, thermal decomposition, and electrolytic refining. The *vacuum distillation* method is based on the difference in the volatility of the purified metal and the impurities contained in it. The metal melts in a vacuum when the air is continuously removed. Impurities, if they are more volatile, are separated from the metal and deposited in the cool part of the equipment. In case the basic metal is volatile, it is distilled in the cool part of the equipment, and the impurities remain. The method *of thermal decomposition of volatile compounds* is based on the formation of gaseous compounds of metals with subsequent decomposition of these compounds at high temperature. For example, when titanium reacts with iodine, the gaseous compound TiI_4 is formed in one part of the reactor at 200 °C and decomposes in the other part of the reactor at 1,300 °C. Purification is possible because iodine does not interact with impurities. This method is called the method of *transport reactions* because in the process of purifying the metal, it is transferred from one part of the reactor to another. *Electrolytic refining* is a method based on the fact that under certain conditions during electrolysis it is possible to have anodic dissolution (oxidation) of the metal. For example, during the electrolysis of nickel sulfate in a solution with a nickel anode, the metal of the anode is oxidized because the potential of nickel (−0.28 V) is much lower than the oxidation potentials of water (1.23 V) and sulfate ions (2.01 V). At the cathode, the opposite process takes place – nickel reduction. In this way, the metal is transferred from the anode to the cathode. The anode is made of technical nickel to be purified. The pure metal is separated at the cathode. Impurities of more active metals (Zn, Fe, Mg) are also oxidized at the anode, but are not released at the cathode, they accumulate in the electrolyte. Impurities of less active metals are not dissolved at the anode, but precipitated to the bottom of the electrolysis bath and then processed in order to extract the more valuable metals.

24.3 Biological role

Many of the transition metals are significant to the biochemistry of living systems, the most noticeable examples being eight biogenic metals (Cu, Zn, Fe, Mn, Co, Mo, Ni, and Cr), which have been broadly studied with reference to their biofunctions. The cations of these biometals are components of a large number of enzymes, proteins, and vitamins. These metals exist generally as coordination complexes in the living body. Examples of biologically active complexes are Fe in hemoglobin, cytochromes, and catalases; Mg in chlorophyll, Mg complex with ATP in the enzyme kinase; Co in vitamin B_{12} (cobalamine), cyanocobalamine, etc.

The typical biochemical properties of d-elements and their compounds are redox processes and complex formation reactions. Because of the unfilled d-orbitals, their oxidation states are variable, hence their affinity to participate in various oxidation-reduction bioreactions. On the other side, d-elements demonstrate a high ability to form complex compounds due to the free s- and p-orbitals and partly vacant d-orbitals which enables them to form donor-acceptor bonds in the complexes acting as perfect complexing agents in a number of bioreactions with various endogenous and exogenous (vitamins, drugs, toxic compounds, etc.) ligands.

The biological functions of transition metals have dual roles. In normal levels, they can be vital for the stabilization of the cell structures. In high levels they are toxic. The toxic effects of transition metals are mainly connected with their ability to form coordination complexes. On the other side, in case of deficiency, alternative pathways may be activated causing some diseases.

Many coordination complexes of d-elements are also used as therapeutic agents. The biological role and the functions of the transition metals are discussed in the next chapters.

Chapter 25
First B group

25.1 General characteristics

The first B group includes the following elements: copper (Cu), silver (Ag), and gold (Au). The electronic structure of their atoms is $(n - 1)d^{10}ns^1$, that is, they have a filled $(n - 1)d$-subshell. They have one electron in each of their outer electron shells, similar to the alkali elements of group IA, but have 18 electrons in their penultimate electron shell (Table 25.1). The ionization energy I_2 for the three elements is close and relatively low. Due to their peculiar electronic structure, they can exhibit in addition to +I, also +II and +III oxidation states. In chemical interactions, these elements donate electrons from their last electron shell, and copper and gold donate electrons from their penultimate electron shell. The oxidation state characteristic of all elements of the group is +1, and Cu and Au are characterized by higher oxidation states – for copper +2 and for gold +3. The atoms of the elements of group IB retain their valence electrons more firmly, which is why they have weaker chemical activity than the elements of the IA group. In the Cu–Ag–Au series, the chemical activity decreases. They are typical metals – highly fusible, soft, and plastic.

Table 25.1: Characteristics of atoms of elements of IB group.

Element	Cu	Ag	Au
Valence electrons	$3d^{10}4s^1$	$4d^{10}5s^1$	$5d^{10}6s^1$
Radius of the atom (nm)	0.128	0.144	0.144
Ionization potential (eV)	7.73	7.58	9.23
Density (g/cm^3)	8.96	10.5	19.3
T_m/T_b (°C)	1,083/2,543	960/2,167	1,063/2,880
Electrode potential (V)	0.34	0.80	1.50

25.2 Occurrence and synthesis

Copper is obtained mainly from sulfide ores (chalcopyrite $CuFeS_2$, chalcosine Cu_2S, covelline CuS, cuprite CuO) by pyrometallurgical and hydrometallurgical methods. In the pyrometallurgical method, the ore is burned in an oxygen medium "sulfide frying" with subsequent reduction, which forms sulfur dioxide and technical copper, which is purified from impurities by electrolysis:

https://doi.org/10.1515/9783111712246-025

$$2Cu_2S + 3O_2 \rightarrow 2Cu_2O + 2SO_2$$

$$2Cu_2O + Cu_2S \rightarrow 6Cu + SO_2$$

In the hydrometallurgical method, the ore is treated with various solutions that convert the copper in the form of Cu^{2+} cations into a solution:

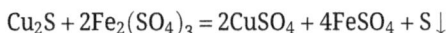

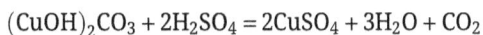

$$Cu_2S + 2Fe_2(SO_4)_3 = 2CuSO_4 + 4FeSO_4 + S\downarrow$$

$$(CuOH)_2CO_3 + 2H_2SO_4 = 2CuSO_4 + 3H_2O + CO_2$$

Copper is separated from the solutions with powdered iron or by electrolysis.

Silver (in the form of argentite, Ag_2S, etc.) is contained in the minerals of copper and nickel and is obtained from the sludges by electrolytic refining of metals.

Due to its very high reduction potential, gold is found in nature usually as a free metal in its native state in gold-bearing sand, from which it is separated by washing or by chemical methods. Since gold is a "soft acid," well-known gold minerals, such as calaverite $AuTe_2$ and sylvanite $AuAgTe_4$, include the "soft base" in their composition tellurium. The same processes as for silver are used to extract metallic gold:

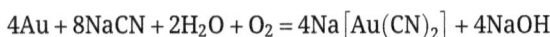

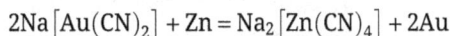

$$4Au + 8NaCN + 2H_2O + O_2 = 4Na\left[Au(CN)_2\right] + 4NaOH$$

$$2Na\left[Au(CN)_2\right] + Zn = Na_2\left[Zn(CN)_4\right] + 2Au$$

25.3 Properties, usage, and biological role

Copper, silver, and gold were once called coin metals because they were used for this purpose in the past. The reasons for this are, on the one hand, that they can be found in a metallic state, and on the other hand, they are soft, chemically inactive, and relatively rare metals. Copper, silver, and gold are metals of red, white, and yellow color, respectively. Their melting and boiling temperatures are significantly higher than those of the alkali metals of the IA group because not only ns- but also $(n - 1)$d-electrons are involved in the formation of the bond. These are the metals with the highest electrical conductivity, and they are also easy to process because they are soft, malleable, and pullable. Because of these qualities and their relative chemical resistance, they are used in electrical engineering, alloy production, jewelry industry, dental technology, etc.

The elements of group IB have low chemical activity. In the air, gold retains its luster, silver gradually darkens due to the formation of Ag_2S, and copper is covered with a green layer of basic copper carbonate $Cu(OH)_2 \cdot CuCO_3$ (called "patina"), which protects it from further corrosion. The formation of sulfide Ag_2S in silver is probably due, on the one hand, to the minimal tendency of silver to dissolve in water, and on the other hand, to the presence of sulfide ions (or traces of hydrogen sulfide H_2S) in the atmosphere, a process that takes place very slowly over time.

In the row of relative activity, these metals are found after hydrogen. Their reduction activity in the Cu–Ag–Au series decreases, with gold being the weakest reducing agent. These metals do not interact with acids that do not exhibit oxidizing action. Copper and silver are oxidized into concentrated sulfuric and nitric acids and gold in aqua regia:

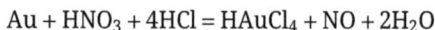

$$Cu + 2H_2SO_4 = CuSO_4 + SO_2 + 2H_2O$$

$$3Ag + 4HNO_3 = 3AgNO_3 + NO + 2H_2O$$

$$Au + HNO_3 + 4HCl = HAuCl_4 + NO + 2H_2O$$

In solutions and melts of bases, silver and gold are resistant, and copper interacts with base melts in the presence of oxidizing agents:

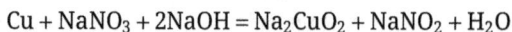

$$Cu + NaNO_3 + 2NaOH = Na_2CuO_2 + NaNO_2 + H_2O$$

Only copper reacts with oxygen when heated to form CuO ($T < 400$ °C) and Cu_2O ($T > 400$ °C). With hydrogen, nitrogen, and carbon, these metals do not form compounds, but are actively oxidized by the halogen elements. The three metals interact directly with the halogen elements, resulting in the corresponding halides – CuX_2, AgX, and AuX_3.

These metals are unstable in solutions of cyanides and ammonia because in this case their reduction activity increases due to the formation of complex compounds. Under these conditions, they are oxidized by oxygen and even by water:

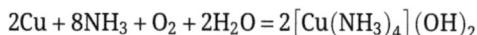

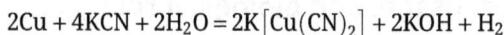

$$2Cu + 8NH_3 + O_2 + 2H_2O = 2\left[Cu(NH_3)_4\right](OH)_2$$

$$2Cu + 4KCN + 2H_2O = 2K\left[Cu(CN)_2\right] + 2KOH + H_2$$

The complexation reactions are very characteristic of all three metals. Typical for them are the ammonia complexes $[Cu(NH_3)_4](OH)_2$, $[Ag(NH_3)_2](OH)$. Similar complexes, but with even greater stability, they form with cyanide ions. These complexes are colorless: $[Cu(CN)_4]^{2-}$ and $[Ag(CN)_2]^-$. The complex-forming properties of Ag^+ are the basis of photographic processes. A photosensitive layer is applied to the photographic plate – finely dispersed AgX (most often AgBr) and gelatin. When light hits the plaque, in some places, AgX is reduced to Ag and the so-called hidden image appears. After the image is "developed," it is "fixed" – this means the removal of the AgX unaffected by the light. This process is basically a process of complexation, using a solution of sodium thiosulfate $Na_2S_2O_3$ – the so-called "fixation," which forms $Na_3[Ag(S_2O_3)_2]$ and the corresponding halogenide. The resulting substances are soluble in water and are easily removed.

In silver chemistry, Ag^+ ions (d^{10}) are dominant, as a result of which silver often behaves like most metals from the main group (IA). Gold, on the other hand, resembles platinum metals. For example, like platinum and other 5d transition metals, gold forms a hexafluoroaurate(V) ion whose salts can be isolated, for example, $(O_2)^+[AuF_6]^-$.

Copper is used to make electrical wires, electrodes, and various types of alloys such as brass (from Cu and Zn), bronze (from Cu and Sn), and cupronickel (from Cu and Ni). Copper sulfate "blue stone" is used to control pests of agricultural crops. Silver was one of the first metals known to humans. Due to its beauty and intense brilliance, Ag has been associated with the Moon since ancient times and called "moonstone." The main users of silver are electrical engineering and radio electronics, and in the form of silver bromide it is used in photography. In aviation and space technology, silver-zinc batteries are used, which are notable by their high energy consumption. Silver is used for the manufacture of mirrors, chemical equipment, medical instruments, wires, jewelry, and household goods. Gold is the equivalent of money, and much of it is stored in banks. Its application is mainly in the jewelry industry and in coinage. However, not a small part is used in radio-electronic devices, in the manufacture of dentures and in modern microelectronics in the preparation of corrosion-resistant contacts. The gold content in its alloys is often expressed by the so-called "carats." Pure gold is 24 carat. Carats indicate how many twenty-fourths of an alloy is gold. For example, an alloy of 8 carats contains one-third of gold, and an alloy of 12 carats – 50%. The rest of these alloys contain copper or silver.

Copper belongs to the group of the most important microbiological elements. In small quantities, copper catalyzes a number of processes in living organisms. Deficiency of copper ions (Cu^{2+}) causes anemia in humans and makes it difficult for plants to absorb nitrogen. In larger quantities, these ions have a toxic effect since all copper compounds are poisonous. In the human body, copper is found mainly in the liver and in some parts of the brain. In plant and animal organisms, copper is found in the form of coordination compounds, mainly copper-containing proteins, including enzymes such as hemocyanin (for oxygen transport) and cytochrome oxidase (for electron transfer). The enzymes containing copper in the active site are more than 20, most of which are oxidases. Their biological role is related to the processes of hydroxylation, oxidative catalysis, and oxygen transfer. The role of copper in the enzyme cytochrome oxidase, which controls reactions of the type $O_2 \rightarrow H_2O$, $O_2 \rightarrow H_2O_2$, as well as the important for the body reaction of disproportionation $O_2^- \rightarrow O^{2-} + O^0$, which occurs with the participation of the enzyme superoxide dismutase, has been studied in detail. Hemocyanin is an oxygen carrier, which (unlike hemoglobin) is extracellular, which is typical for Cu-containing proteins in general. The daily norm of Cu in humans is about 2 mg. A person ingests 2–3 mg of Cu with food per day, of which a total of 30% is absorbed. Copper provides better iron resorption and takes part in the synthesis of hemoglobin. In addition to humans, copper plays an essential role in animals, plants, and microorganisms. Some copper compounds are used in agriculture and for the preparation of pharmaceuticals. Deviations in the copper content from the norms lead to severe and often irreversible diseases. The accumulation of copper in the liver or brain leads to rheumatoid arthritis and Wilson's disease. Copper complexes with thiosemicarbazones and Schiff bases are used as bactericidal agents.

Silver ions have a pronounced bactericidal effect. Water placed in a silver container is sterilized despite the negligible concentration of silver ions in it. This water, called silver water, is used in surgery, ophthalmology, dermatology, etc. The preparations protargol, collargol, adsorgan, etc., are known, which contain colloidal silver stabilized with proteins.

Treatment with gold preparations, called chrysotherapy, was known as early as 2500 BC in China. In the form of official pharmaceuticals, gold compounds were used in the 1920sas a means of combating tuberculosis, arthritis, etc. The action of Au preparations is related to the fact that the intravenous complexes are dissociated in the blood plasma and the free Au^+ ions bind to the thiol (–SH) groups of the proteins of the blood, thus quickly spreading throughout the body. It is believed that Au^+ ions block excess sulfhydryl groups, but they can also act in other ways, for example, to inhibit the active forms of OH and O_2^- radicals. The main disadvantage of Cu, Ag, and Au preparations is their poor tolerance in the stomach.

The compounds of Cu, Ag, and Au differ significantly in properties, given the difference in their typical oxidation states. Therefore, they will be discussed separately for each of the elements.

25.4 Chemical compounds of copper

25.4.1 Cu(II) compounds

In copper, +2 oxidation state is more characteristic. Although copper forms compounds in both oxidation states +1 and +2, +2 is still dominant in aqueous solutions. In aqueous solution, almost all Cu(II) salts are colored blue, the color being due to the presence of hexaaquacopper(II) ions $[Cu(H_2O)_6]^{2+}$. The exception is Cu(II) chloride. The concentrated aqueous solution of this compound is green in color due to the presence of complex ions, for example, the plane tetrachlorocuprate(II) ions $[CuCl_4]^{2-}$. Upon dilution, the color of the solution changes to blue due to the substitution of Cl^- ions in the complex with water:

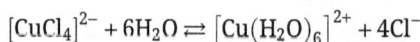

$$[CuCl_4]^{2-} + 6H_2O \rightleftharpoons [Cu(H_2O)_6]^{2+} + 4Cl^-$$

When copper is heated to 400 °C in the presence of oxygen, as well as when the hydroxide and copper(II) salts decompose, the *oxide* CuO is formed. It reacts with acids to form salts, most of which are soluble in water. When CuO is melted with bases, cuprates(II) of the type K_2CuO_2 are formed. Copper(II) oxide easily dissolves in ammonia solution to form a complex compound:

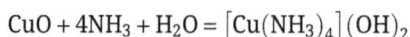

$$CuO + 4NH_3 + H_2O = [Cu(NH_3)_4](OH)_2$$

Copper(II) hydroxide ($Cu(OH)_2$) is a pale blue puffy precipitate that is produced by the interaction of soluble salts with bases: $Cu^{2+} + 2OH^- \rightarrow Cu(OH)_2$.

Under low heating (60–70 °C), hydroxide decomposes into black CuO and H_2O, and in concentrated solutions of bases, when heated, it forms dark blue hydroxocuprates $Me_2[Cu(OH)_4]$:

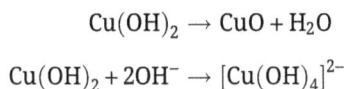

$$Cu(OH)_2 \rightarrow CuO + H_2O$$

$$Cu(OH)_2 + 2OH^- \rightarrow [Cu(OH)_4]^{2-}$$

The resulting complexes are stable only in concentrated alkaline solutions due to the weakly expressed acidic properties of $Cu(OH)_2$. Therefore, $Cu(OH)_2$ has an amphoteric character with more pronounced basic ratios. Copper dihydroxide is dissolved in ammonia, resulting in a solution of bright blue color $[Cu(NH_3)_4](OH)_2$ (Schweitzer reagent). It is used to prove certain organic substances and in the processing of cellulose:

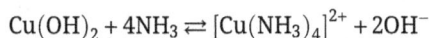

$$Cu(OH)_2 + 4NH_3 \rightleftharpoons [Cu(NH_3)_4]^{2+} + 2OH^-$$

All Cu(II) *halides* are known. Their stability decreases from fluoride to iodide, and CuI_2 decomposes at room temperature to CuI.

Of the oxysalts of Cu(II), the most widely used in practice is the *sulfate* $CuSO_4 \cdot 5H_2O$. It is a solid crystalline substance with a blue color, which is due to hydrated copper ions. It dissolves very well in water. As a result of hydrolysis, the aqueous solution of $CuSO_4$ has an acidic character (salt of a weak base and a strong acid). It reacts with bases to produce $Cu(OH)_2\downarrow$. The reaction is used to detect Cu(II). It is used in agriculture. In the form of Bordeaux solution (a solution of $CuSO_4$ with a mass fraction of 1%, mixed with milk of lime to a neutral reaction) it is used against mildew and other mold and fungal diseases of vines and vegetables. Anhydrous copper sulfate is a white crystalline substance and is used to detect water in alcohol, aviation gasoline, and other organic liquids. Its aqueous solutions are blue-colored. Crystal hydrates of $CuSO_4$ are also known, which are obtained by successive dehydration of $CuSO_4 \cdot 5H_2O$, for example, $CuSO_4 \cdot 3H_2O$ (at 90 °C), $CuSO_4 \cdot H_2O$ (at 100 °C), and $CuSO_4$ (at 250 °C), the latter decomposing to Cu, SO_2, and O_2 at 650 °C.

In most complex compounds, Cu(II) is the thermodynamically more stable oxidation state, although reducing ligands, such as iodide ions, reduce Cu(II) ions to Cu(I). When potassium iodide is added to a solution of $CuSO_4$, a redox reaction occurs:

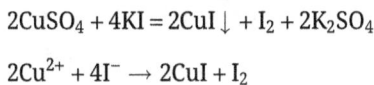

$$2CuSO_4 + 4KI = 2CuI \downarrow + I_2 + 2K_2SO_4$$

$$2Cu^{2+} + 4I^- \rightarrow 2CuI + I_2$$

The comparison of the redox potentials of the two half-reactions:

$$Cu^{2+} + e^- = Cu^+ \qquad (J° = 0.15 \, V)$$
$$Cu^{2+} + I^- + e^- = CuI \downarrow \quad (J° = 0.86 \, V)$$

indicates that Cu^{2+} cations are weak oxidizing agents, but their oxidizing ability increases with the formation of insoluble CuI.

25.4.2 Cu(I) compounds

There is a *hydride* of Cu(I), obtained indirectly – CuH. This is the only hydride in the group because silver and gold do not interact with hydrogen. In +1 oxidation state, there is oxide Cu_2O, which is obtained by heating CuO at 1,100 °C or by reducing copper(II) salts:

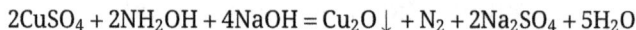

$$4CuO \rightarrow 2Cu_2O + O_2$$

$$2CuSO_4 + 2NH_2OH + 4NaOH = Cu_2O \downarrow + N_2 + 2Na_2SO_4 + 5H_2O$$

Cu_2O is a red, water-insoluble substance. In nature, it is found as a red copper ore – cuprite. Cu_2O is used as a pigment in paints for painting ships. Due to its toxic effect, marine animals do not stick to the bottoms. It is the most durable compound of monovalent copper with application.

This oxide does not react with water, reacts with acids to form Cu(I) salts, and when melted with alkali metal oxides, it turns into cuprate(I) of the NaCuO type. Cu_2O when heated with dilute H_2SO_4 disproportionates to the formation of metallic Cu, and concentrated H_2SO_4 oxidizes it:

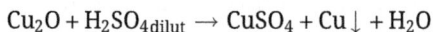

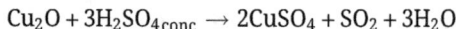

$$Cu_2O + H_2SO_{4\,dilut} \rightarrow CuSO_4 + Cu \downarrow + H_2O$$

$$Cu_2O + 3H_2SO_{4\,conc} \rightarrow 2CuSO_4 + SO_2 + 3H_2O$$

Cu_2O also dissolves in ammonia solution:

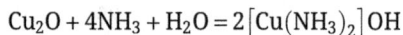

$$Cu_2O + 4NH_3 + H_2O = 2\left[Cu(NH_3)_2\right]OH$$

Cu(I) hydroxide (CuOH) does not exist because it decomposes into Cu_2O and water at the moment of its formation. In aqueous solution, hydrated Cu(I) ions are unstable and disproportionate to Cu(II) ions and elemental copper:

$$2Cu^+ \rightleftarrows Cu^{2+} + Cu$$

In solutions, Cu(I) salts are unstable. They are oxidized by oxygen or disproportionate:

$$4CuCl + O_2 + 2H_2O = 4(CuOH)Cl$$

$$Cu_2SO_4 = Cu + CuSO_4$$

However, this does not happen with copper(I) complex compounds, such as $K[CuCl_2]$ or $[Cu(NH_3)_2]_2SO_4$, because complexation stabilizes the Cu(I) state.

Usually, in the solid phase, Cu(I) ions are stabilized by low-charge anions. Chloride, bromide, iodide, and cyanide of Cu(I) have been synthesized. The approach to the formation of stable compounds of Cu(I) can be illustrated by the reaction of metallic copper with boiling hydrochloric acid. This reaction occurs unexpectedly since hydrochloric acid is not a strong oxidizing agent. The Cu^+ ion formed during oxidation rapidly forms a colorless dichlorocuprate(I) complex ion $[CuCl_2]^-$, which directs the equilibrium to the right:

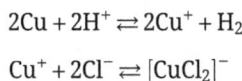

$$2Cu + 2H^+ \rightleftharpoons 2Cu^+ + H_2$$

$$Cu^+ + 2Cl^- \rightleftharpoons [CuCl_2]^-$$

The complex compound is unstable and decomposes into a white precipitate of CuCl:

$$[CuCl_2]^- \rightarrow CuCl + Cl^-$$

The combination of air and moisture oxidizes the compounds of Cu(I) to Cu(II). In organic chemistry, $[CuCl_2]^-$ is used to convert benzenediazonium chloride into chlorobenzene:

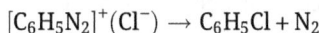

$$[C_6H_5N_2]^+(Cl^-) \rightarrow C_6H_5Cl + N_2$$

In general, Cu(I) compounds are colorless or white because the electron configuration of the ion is d^{10}. Filled d-orbitals do not allow electronic transitions that cause absorption of light in the visible region of the spectrum. An oxidation state of +3 for copper is not characteristic, although a small number of compounds are known – Cu_2O_3, $KCuO_2$, $Ba(CuO_2)_2$, and $K[Cu(OH)_4]$, which exhibit strong oxidative activity.

25.5 Chemical compounds of silver

In silver, the most characteristic is +1 oxidation state. Of its numerous salts, the most important is the soluble silver nitrate $AgNO_3$, which is obtained by reacting silver with HNO_3 and which is the starting substance for the production of other compounds. When a solution of $AgNO_3$ reacts with bases, the *oxide* Ag_2O is released:

$$2AgNO_3 + 2KOH = Ag_2O \downarrow + 2KNO_3 + H_2O$$

The formation of the oxide is due to the fact that AgOH is unstable and does not exist at room temperature. Silver(I) oxide is dissolved in ammonia solution:

$$Ag_2O + 4NH_3 + H_2O = 2[Ag(NH_3)_2]OH$$

When melted with alkali metal oxides, it forms argentates(I) of the type KAgO and interacts with acids to form simple salts (AgNO₃, Ag₂SO₄, and AgCl). At temperatures higher than 200 °C, Ag₂O decomposes into silver and oxygen, and at 40 °C it is reduced by hydrogen.

Silver *halides* are of great importance, namely the soluble AgF and the insoluble AgCl (white), AgBr (pale yellow), and AgI (yellow). The insolubility of silver chloride, bromide, and iodide is related to the partially covalent nature of the bonds. Silver fluoride (AgF) is a white, soluble solid and is considered as ionic in both solid and aqueous solution. Silver halides are dissolved in ammonia and in a solution of sodium thiosulfate to complex compounds:

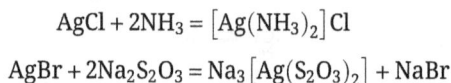

$$AgCl + 2NH_3 = \left[Ag(NH_3)_2\right]Cl$$

$$AgBr + 2Na_2S_2O_3 = Na_3\left[Ag(S_2O_3)_2\right] + NaBr$$

Silver chloride, bromide, and iodide decompose into simple substances under the action of light, and fluoride – under the action of ultraviolet irradiation. Silver chloride, bromide, and iodide are sensitive to light, due to the reduction of silver ions, which is why silver compounds and their solutions are stored in bottles of dark color because of the reaction $Ag^+ + e^- \rightarrow Ag$.

When interacting with reducing agents, silver(I) compounds are easily reduced to a metal. In reduction with organic compounds, silver is separated as a shiny precipitate adhering to the surface of the reaction vessel ("silver mirror"):

$$2[Ag(NH_3)_2]Cl + CH_2O + H_2O = 2Ag\downarrow + HCOONH_4 + 2NH_4Cl + NH_3$$

Silver compounds in oxidation state +2 (AgO, AgF₂, etc.) are very strong oxidizing agents, but have no practical application. In silver, no compounds in the oxidation state of +3 are known.

25.6 Chemical compounds of gold

The characteristic oxidation state of gold is +3, and its most famous compounds are tetrachlorogold(III) acid (HAuCl₄, a product of the interaction of gold with aqua regia) and AuCl₃ (formed by the reaction of gold with chlorine). When bases react with solutions of these compounds, a precipitate of the *hydroxide* Au(OH)₃ is formed, which turns into AuO(OH) when dried, and into Au₂O₃ oxide when heated to 140 °C. Gold(III) oxide is unstable and at 160 °C it decomposes into gold and oxygen. Au₂O₃ and Au(OH)₃ are amphoteric compounds, when they interact with bases, hydroxaurates K[Au(OH)₄], Ba[Au(OH)₄]₂, and others are formed. Numerous complex compounds of gold(III) are also known.

Gold(III) *chloride*, which has a bridge structure similar to aluminum chloride, is most often represented by the formula Au₂Cl₆. It can be obtained directly:

$$2Au + 3Cl_2 \rightarrow Au_2Cl_6$$

The dissolution of Au(III) chloride in concentrated HCl leads to the formation of tetrachloroaurate(III) ions $[AuCl_4]^-$, one of the components in "liquid gold" (a solution of gold with thiol ligands, which, when heated, forms a layer of metallic gold).

A series of gold(I) compounds have also been obtained, such as Au_2O, $AuCl$, $AuBr$, AuI, and Au_2S, which are unstable when dissolved. Gold(I) oxide Au_2O is one of the few stable gold compounds in the +1 oxidation state. As with copper, this oxidation state is stable only in the solid state since the aqueous solutions of all Au(I) salts are disproportionate to metal and Au(III) ions:

$$3Au^+ \rightarrow 2Au + Au^{3+}$$

All gold compounds are strong oxidizing agents, except for the $[Au(CN)_4]^-$ and $[Au(CN)_2]^-$ complexes, in which the Au(III) state is stabilized by ligands with a strong ligand field.

Chapter 26
Second B group

26.1 General characteristics

In IIB group of the periodic table the elements zinc Zn, cadmium Cd and mercury Hg are located. A feature of the electronic structure of the atoms of these elements is the fully formed penultimate d^{10}-subshell and the penetration of the outer ns^2-electrons under the shield of $(n–1)d^{10}$-electrons. That is why the atoms of these elements have higher ionization potentials, and the metals have low melting and boiling temperatures (Table 26.1).

Table 26.1: Characteristics of atoms of elements of IIB group.

Element	Zn	Cd	Hg
Valence electrons	$3d^{10}4s^2$	$4d^{10}5s^2$	$5d^{10}6s^2$
Radius of the atom (nm)	0.139	0.156	0.160
Ionization potential (eV)	9.39	8.99	10.43
Density (g/cm^3)	7.1	8.7	13.6
T_m/T_b (°C)	419/907	321/767	−39/357
Electrode potential (V)	−0.76	−0.40	+0.85

The second B group, although located at the end of the transition metal series, has a rather similar behavior to that of the main IIA group. Due to the filled d-subshell, most of the compounds of the metals of group IIB are white, except for those in which the anion is colored. But in reality, the chemistry of these elements is different. The melting points of zinc and cadmium are 419 °C and 321 °C, respectively, much lower than the typical values for transition metals close to 1,000 °C.

Zinc and cadmium are very similar in terms of their chemical properties due to their constant +2 oxidation state in all their compounds. They are soft metals with pronounced chemical activity. Zinc reacts with dilute acids and, when slightly heated, burns in a chlorine atmosphere:

$$Zn + 2H^+ \rightarrow Zn^{2+} + H_2$$

$$Zn + Cl_2 \rightarrow ZnCl_2$$

The only real similarity between group IIB and the other transition metals is complexation, especially with ligands such as ammonia, cyanide and halide ions. All metals in the group (especially Hg) tend to form covalent instead of ionic compounds. Strong similarities in the chemical behavior of magnesium and zinc have been demonstrated.

https://doi.org/10.1515/9783111712246-026

On the other hand, there is a relationship between Zn(II) and Sn(II) and between Cd(II) and Pb(II).

Mercury is quite different from zinc and cadmium. It is the only liquid metal under ordinary conditions. It has a positive electrode potential, forms compounds in oxidation states of +1 and +2, although Hg^+ ions do not exist independently (they are in the form of Hg_2^{2+} ions). Many of the mercury compounds are unstable. The differences between mercury from zinc and cadmium are explained by the high stability of the $6s^2$-electron pair.

26.2 Occurrence and synthesis

Zinc occupies the 23rd place in terms of distribution in nature (2×10^{-2}%). Cadmium (1.3×10^{-5}%) and mercury (8.3×10^{-6}%) are presented in nature more modestly. Their main minerals are sphalerite (ZnS), smithsonite ($ZnCO_3$), greenockite CdS, cinnabar HgS, and mercury is also found in its native form. Zinc in the form of alloys has been known since antiquity. It was obtained in its pure form only at the end of the eighteenth century.

To obtain zinc by the pyrometallurgical method, sphalerite is subjected to oxidative frying (at 800 °C), and then ZnO – to carbon reduction:

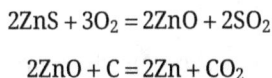

$$2ZnS + 3O_2 = 2ZnO + 2SO_2$$

$$2ZnO + C = 2Zn + CO_2$$

In the hydrometallurgical method, zinc oxide is converted to sulfate by reacting with sulfuric acid, and then zinc is reduced from sulfate by electrolysis.

Cadmium, usually accompanying zinc, is hydrometallurgically separated from solutions using sputtered zinc:

$$CdSO_4 + Zn = ZnSO_4 + Cd\downarrow$$

Mercury is obtained directly from cinnabar (HgS), because its oxide at the frying temperature (500–600 °C) is unstable:

$$HgS + O_2 = Hg + SO_2$$

Mercury can also be produced by heating HgS with iron or calcium oxide:

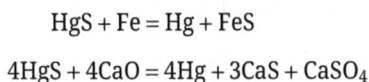

$$HgS + Fe = Hg + FeS$$

$$4HgS + 4CaO = 4Hg + 3CaS + CaSO_4$$

26.3 Properties

26.3.1 Properties of zinc and cadmium

Zinc is a metal with a gray-white color and a metallic luster. In air, its surface becomes covered with an oxide layer and becomes cloudy. In cool conditions, zinc is quite brittle, but at temperatures of 100–150 °C it is easy to process. In terms of chemical and physical properties, cadmium is very similar to zinc. It is a silvery-white, soft metal, easy to forge and roll. Zinc and cadmium are resistant to air due to the protective oxide layer covering them. Mercury does not interact with oxygen due to its chemical inertness.

Zinc and cadmium actively displace hydrogen from acids with non-oxidative action:

$$Zn + 2HCl = ZnCl_2 + H_2$$

$$Cd + H_2SO_4 = CdSO_4 + H_2$$

Acids with oxidative action dissolve both metals. This way, Zn reduces H_2SO_4 to sulfur or H_2S, and Cd to SO_2:

$$4Zn + 5H_2SO_{4conc} \rightarrow 4ZnSO_4 + H_2S\uparrow + 4H_2O$$

$$Cd + 2H_2SO_{4conc} \rightarrow CdSO_4 + SO_2\uparrow + 2H_2O$$

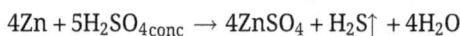

When interacting with nitric acid, various products are formed:

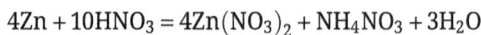

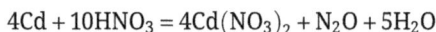

$$3E + 8HNO_{3conc} = 3E(NO_3)_2 + 2NO + 4H_2O$$

$$4Zn + 10HNO_3 = 4Zn(NO_3)_2 + NH_4NO_3 + 3H_2O$$

$$4Cd + 10HNO_3 = 4Cd(NO_3)_2 + N_2O + 5H_2O$$

Only amphoteric zinc actively reacts with bases:

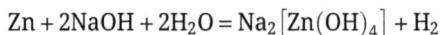

$$Zn + 2NaOH + 2H_2O = Na_2\left[Zn(OH)_4\right] + H_2$$

Zinc dissolves in an aqueous solution of ammonia, as it forms a well-soluble complex:

$$Zn + 4NH_4OH = \left[Zn(NH_3)_4\right](OH)_2 + H_2 + 2H_2O$$

Zinc displaces less active metals from solutions of their salts:

$$CuSO_4 + Zn = ZnSO_4 + Cu$$

$$CdSO_4 + Zn = ZnSO_4 + Cd$$

When heated, all three metals interact with oxygen, sulfur (mercury – even at room temperature), as well as with halogen elements:

$$2Zn + O_2 = 2ZnO$$
$$Zn + Cl_2 = ZnCl_2$$
$$Zn + S = ZnS$$

When heated, zinc reacts with NH_3 to form nitride, with water and with hydrogen sulfide:

$$3Zn + 2NH_3 = Zn_3N_2 + 3H_2$$
$$Zn + H_2O = ZnO + H_2$$
$$Zn + H_2S = ZnS + H_2$$

The sulfide formed on the surface of zinc protects it from further interaction H_2S.

26.3.2 Properties of mercury

Mercury is a silvery-white liquid metal, that in its solid state possesses good malleability and elasticity. Itsy vapor is very toxic. Hg accumulates and is not excreted from the human body.

Mercury is a weakly active metal. It interacts with oxygen only when heated:

$$2Hg + O_2 = 2HgO$$

Mercury interacts with chlorine in the cold, forming $HgCl_2$:

$$Hg + Cl_2 = HgCl_2$$

It easily interacts with powdered sulfur, forming a stable sulfide:

$$Hg + S = HgS$$

This reaction is used to collect dispersed mercury by sprinkling the area with sulfur powder.

Mercury does not dissolve in water and alkalis, but only in acids with oxidizing action. Mercury, located after hydrogen in the row of relative activity, does not interact with hydrochloric and dilute sulfuric acid, but is oxidized by concentrated sulfuric and nitric acids:

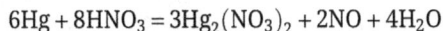

$$Hg + 2H_2SO_4 = HgSO_4 + SO_2 + 2H_2O$$
$$Hg + 4HNO_3 = Hg(NO_3)_2 + 2NO_2 + 2H_2O$$
$$6Hg + 8HNO_3 = 3Hg_2(NO_3)_2 + 2NO + 4H_2O$$

The processes take place under different conditions: in concentrated sulfuric acid – when heated, and in nitric acid – in cold. Depending on the quantitative ratio of the starting acids and IIg, mercury salts are formed in +1 or +2 oxidation states.

Hg(II) chloride ($HgCl_2$) is reduced by metallic mercury to Hg(I), forming Hg_2Cl_2 (calomel):

$$HgCl_2 + Hg = Hg_2Cl_2$$

Mercury dissolves many metals and forms amalgams. In them, the metals behave as if in a free state, but become less active. The formation of amalgam reduces the activity, which is analogous to the dilution in solutions.

26.4 Use and biological role

Zinc is a part of a number of important alloys. It is used in large quantities as an anti-corrosion coating of iron. This electrochemical process is called galvanizing. A protective layer is formed in moist air, which is initially an oxide, but over time it turns into basic carbonate $Zn_2(OH)_2CO_3$. The advantage of galvanizing is that zinc oxidizes before iron. This is due to the more negative potential of zinc than that of iron. At these conditions Zn acts as an anode:

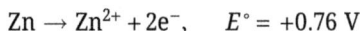

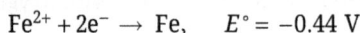

$$Zn \rightarrow Zn^{2+} + 2e^-, \quad E^\circ = +0.76 \text{ V}$$
$$Fe^{2+} + 2e^- \rightarrow Fe, \quad E^\circ = -0.44 \text{ V}$$

ZnS has luminescence properties, which is why it is widely used for luminescent lamps, cathode ray tubes for TV screens, etc.

Cadmium is used for anticorrosion coatings, in the production of control devices for nuclear reactors. It is part of some low-melting alloys, and is used in the manufacture of batteries, plasticizers and pigments. Some cadmium compounds are brightly colored, for example, CdS produces yellow paints and varnish coatings of various shades. Cadmium sulfide, selenide, and telluride, in addition to being pigments, are used as phosphors and semiconductors.

Mercury is used in the manufacture of thermometers, barometers and electrodes to obtain sodium base and chlorine in the electrolysis of a solution of NaCl. It is also used in amalgam metallurgy, in mercury lamps, quartz lamps, diffusion vacuum pumps, etc. Solutions of other metals in mercury are called amalgams. Sodium and zinc amalgams are used as laboratory reducing agents. The most common of all amalgams is the dental amalgam (containing mercury mixed with other metals, such as silver, tin and copper), which is used to make fillings. Mercury compounds are used in agriculture and horticulture as fungicides and as wood preservatives.

Mercury salts are primarily used as catalysts for many chemical processes. For example, $HgCl_2$ catalyzes the hydrochlorination reaction of acetylene:

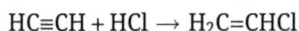

$$HC{\equiv}CH + HCl \rightarrow H_2C{=}CHCl$$

$HgSO_4$ is used as a catalyst for the hydration of acetylene by the Kucherov reaction:

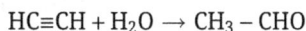

$$HC\equiv CH + H_2O \rightarrow CH_3 - CHO$$

The hard-to-dissolve calomel Hg_2Cl_2 is used in the preparation of standard electrometers for electrometric instruments.

Zinc is an essential trace element necessary for the normal functioning of humans, animals and plants. It is nontoxic. The human body contains about 2 g of zinc. It is concentrated in the liver, bones, teeth, brain, heart, and in the glands – thyroid, pancreas, sexual, and milk. Deviations from the zinc content from the norm lead to serious diseases. Zinc is necessary for the proper functioning of cellular systems. Zn^{2+} forms complexes with ligands with O and N donor atoms. Zinc is part of the active centers of many important enzymes, catalyzing mainly the hydrolysis reactions of peptides, collagen, phospholipids, etc. Zinc activates the enzyme carbonic anhydrase, responsible for the hydration of CO_2 in biofluids and the transfer of H^+ ions to CO_3^{2-}, which regulates one of the body's most important buffer systems. Zinc is a part of the hormone insulin, which regulates blood sugar, normalizes sugar metabolism and enhances the action of pituitary gland hormones. Zn-containing preparations are designed for the proper functioning of Zn-dependent enzymes, which is important, for example, in the treatment of diabetes mellitus. Soluble zinc salts are applied to induce vomiting. Simple noncomplex zinc compounds are used as antiseptic and astringent agents (ZnS, ZnO, etc.). Insoluble zinc compounds are applied externally as adsorbing and drying agents.

Cadmium has no beneficial biological role. Metallic cadmium and its soluble salts are toxic. Cadmium is a metal that occurs naturally along with zinc. For this reason, the most common source of air pollution is the production of zinc. Other metallurgical industries can also be a source of cadmium contamination. It is released into the atmosphere when waste is burned. It is also found in tobacco smoke. Cadmium enters natural waters as a result of the decomposition of aquatic organisms that have the ability to accumulate it. The deposition of heavy metals, such as cadmium, in water causes characteristic diseases, for example, cadmium disease. The maximum permissible concentration of cadmium in drinking water is 0.005 mg/L. Lowering the concentration of dissolved Cd compounds is achieved by precipitating cadmium as hydroxide and carbonate or by using organisms that accumulate it. Cadmium intoxication is risky for those working in an environment related to its extraction and electrolysis, electroplating technicians and others. The penetration of cadmium occurs through the respiratory tract and through injured skin. It is deposited in the liver from there it slowly passes to the kidneys, where its highest concentrations are established. The intake of an increased amount of cadmium in the human body leads to anemia, liver and kidney damage, cardiopathies (a term used for heart diseases with an unclear cause), emphysema of the lungs (expansion of the normal size of the pulmonary alveoli), osteoporosis, skeletal deformity, development of hypertension, etc.

Mercury is extremely toxic. It enters the body usually through the respiratory, digestive system, and through the skin. Signs of acute mercury poisoning are severe intestinal upset, vomiting, swelling of the gums, weakened memory, chills, nervous disturbances, etc. Mercury(II) chloride ($HgCl_2$), known as sublimate, is used as a disinfectant and antiseptic agent in surgery, gynecology, ophthalmology, etc. Hg(I) chloride (Hg_2Cl_2), known as calomel, is used as a laxative and disinfectant in intestinal diseases, as well as a diuretic. Today, organic mercury compounds are used as diuretics. Simple noncomplex mercury compounds have been used as antiseptics in the past, for example, $HgNH_2Cl$ and $HgCl_2$. Previously, $HgCl_2$ was used as an antiseptic in leprosy patients. In the treatment of skin diseases, mercury-sulfur ointments have been applied. Currently, mercury compounds and mercury-organic preparations are avoided due to their high toxicity.

26.5 Chemical compounds

26.5.1 Chemical compounds of zinc and cadmium

In their compounds, zinc and cadmium exhibit a constant +2 oxidation state. They form simple, double and complex compounds with coordination numbers 4 and 6.

ZnO and CdO *oxides* are formed by the interaction of metals with oxygen, as well as by the decomposition of hydroxides and oxygen-containing salts:

$$E(OH)_2 = EO + H_2O\uparrow$$

$$ECO_3 = EO + CO_2\uparrow$$

$$2E(NO_3)_2 = 2EO + 4NO_2\uparrow + O_2\uparrow$$

Zinc oxide is a white powder that melts at a temperature of about 2,000 °C and it does not easily dissolve in water. Due to its amphoteric properties, it dissolves easily in acids and bases, forming tetrahydroxozincates, and zincates:

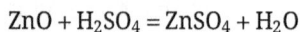

$$ZnO + H_2SO_4 = ZnSO_4 + H_2O$$

$$ZnO + 2KOH + 2H_2O = K_2\left[Zn(OH)_4\right]$$

In melts, it interacts with basic and acidic oxides:

$$ZnO + CaO = CaZnO_2$$

$$ZnO + SiO_2 = ZnSiO_3$$

Zinc oxide is used as a catalyst in many chemical processes.

The hydroxides $Zn(OH)_2$ and $Cd(OH)_2$ precipitate by reacting bases with soluble salts:

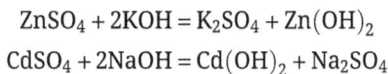

$$ZnSO_4 + 2KOH = K_2SO_4 + Zn(OH)_2$$
$$CdSO_4 + 2NaOH = Cd(OH)_2 + Na_2SO_4$$

Zinc hydroxide, due to its amphoteric properties, is easily soluble in acids, bases, and NH_3:

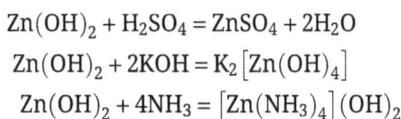

$$Zn(OH)_2 + H_2SO_4 = ZnSO_4 + 2H_2O$$
$$Zn(OH)_2 + 2KOH = K_2[Zn(OH)_4]$$
$$Zn(OH)_2 + 4NH_3 = [Zn(NH_3)_4](OH)_2$$

The amphoteric properties of cadmium hydroxide $Cd(OH)_2$ are very weak. In concentrated solutions of alkali and alkaline earth hydroxides, the precipitate can be dissolved, as complex compounds with coordination numbers 4 and 6 are formed. Tetrahydroxocadmate(II) – $[Cd(OH)_4]^{2-}$ or hexahydroxocadmate(II) – $[Cd(OH)_6]^{4-}$ are obtained. $Cd(OH)_2$ reacts with ammonia and forms a soluble complex – tetraaminecadmium(II) hydroxide:

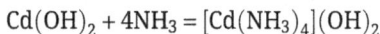

$$Cd(OH)_2 + 4NH_3 = [Cd(NH_3)_4](OH)_2$$

Most zinc salts are soluble in water, and these solutions contain the colorless hexaaquazinc(II) ion, $[Zn(H_2O)_6]^{2+}$. Solid salts are often hydrated, for example, nitrate is hexahydrate and sulfate is heptahydrate, such as magnesium(II) and cobalt(II). The structure of sulfate heptahydrate is $[Zn(H_2O)_6]^{2+}[SO_4.H_2O]^{2-}$. The zinc ion has a d^{10} electron configuration, so the crystal field stabilization energy is 0. Therefore, the size and charge of the anion determine the stereochemistry of Zn^{2+}, which is octahedral or tetrahedral, respectively. Solutions of zinc salts are acidic as a result of hydrolysis, similar to that of aluminum(III) or iron(III) salts:

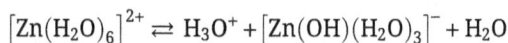

$$[Zn(H_2O)_6]^{2+} \rightleftharpoons H_3O^+ + [Zn(OH)(H_2O)_3]^- + H_2O$$

The addition of OH^- ions to the solution causes the white zinc hydroxide to precipitate:

$$Zn^{2+} + 2OH^- \rightarrow Zn(OH)_2$$

An excess of OH^- ions produces a soluble tetrahydroxozincate(II) anion, $[Zn(OH)_4]^{2-}$:

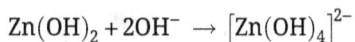

$$Zn(OH)_2 + 2OH^- \rightarrow [Zn(OH)_4]^{2-}$$

The most commonly used zinc salt is zinc chloride. It is obtained as the dihydrate $Zn(H_2O)_2Cl_2$ and as anhydrous zinc chloride. The latter is easily liquefied and is extremely soluble in water. They also dissolve in organic solvents such as ethanol and acetone, and this property proves the covalent nature of the bonds. Zinc chloride $ZnCl_2$ is obtained by dissolving zinc or its oxide in hydrochloric acid. $ZnCl_2$ forms with hydrochloric acid the complex acid H_2ZnCl_4, which dissolves metal oxides, but not metals. A solution of $ZnCl_2$ in hydrochloric acid is used to treat metal surfaces. Zinc chloride ($ZnCl_2$) and zinc sulfate ($ZnSO_4$) are used in medicine as antiseptics.

When zinc soluble salt reacts with a solution of Na_2S, a white precipitate of ZnS is obtained:

$$ZnSO_4 + Na_2S = ZnS + Na_2SO_4$$

Zinc sulfide is insoluble in water. It melts at 1,800–1,900 °C under pressure, and at 1,180 °C it sublimates. It dissolves easily in acids:

$$ZnS + 2HCl = ZnCl_2 + H_2S$$

It is part of the white pigment lithopone, obtained by mixing BaS with zinc sulfate:

$$BaS + ZnSO_4 = BaSO_4 + ZnS$$

Lithopone is significantly cheaper than lead white pigments, but it is less resistant to light. Under the influence of ultraviolet and radioactive rays, zinc sulfide glows. That is why it is used as a phosphor in electron-beam tubes.

When a soluble cadmium salt reacts with a solution of Na_2S, a yellow or orange precipitate of CdS is produced (depending on the pH of the medium):

$$CdSO_4 + Na_2S = CdS + Na_2SO_4$$

The CdS precipitate is dissolved in a concentrated HCl solution:

$$CdS + 2HCl = CdCl_2 + H_2S$$

Cadmium sulfide is used as a pigment. Although cadmium compounds are highly toxic, sulfide is so poorly soluble that it does not pose much risk.

26.5.2 Chemical compounds of mercury

In its compounds state, mercury exhibits +1 and +2 oxidation states. Mercury compounds in +1 oxidation state contain the group $-Hg-Hg-$ or (Hg_2^{2+}), for example, Hg_2Cl_2. Mercury forms simple and complex compounds with a coordination number four.

Mercury(II) compounds are very different from similar compounds of zinc and cadmium. For example, the formation energy of HgO is equal to −58.6 kJ/mol and it decomposes into simple substances at 400 °C, while in ZnO these characteristics are equal to − 318 kJ/mol and 1,950 °C, respectively. Once produced, it decomposes into oxide and water:

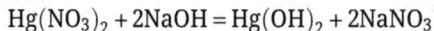

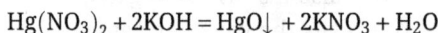

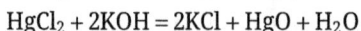

$$Hg(NO_3)_2 + 2NaOH = Hg(OH)_2 + 2NaNO_3$$
$$Hg(OH)_2 = HgO + H_2O$$
$$Hg(NO_3)_2 + 2KOH = HgO\downarrow + 2KNO_3 + H_2O$$
$$HgCl_2 + 2KOH = 2KCl + HgO + H_2O$$

HgO is a substance of yellow or red color, insoluble in water. When heated, it easily decomposes into oxygen and mercury. It exhibits basic properties. HgO dissolves in acids with which mercury forms soluble salts:

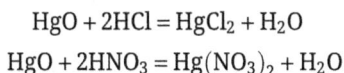

$$HgO + 2HCl = HgCl_2 + H_2O$$
$$HgO + 2HNO_3 = Hg(NO_3)_2 + H_2O$$

Mercury(II) salts are easily reduced by strong reducers, such as $SnCl_2$. When $Hg(NO_3)_2$ reacts with $SnCl_2$, a white precipitate of Hg_2Cl_2 is obtained, which in excess of $SnCl_2$ turns black due to the release of metallic mercury in a dispersed state:

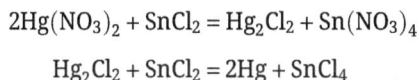

$$2Hg(NO_3)_2 + SnCl_2 = Hg_2Cl_2 + Sn(NO_3)_4$$
$$Hg_2Cl_2 + SnCl_2 = 2Hg + SnCl_4$$

When a soluble Hg(II) salt reacts with a soluble sulfide, a black HgS is obtained:

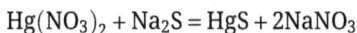

$$Hg(NO_3)_2 + Na_2S = HgS + 2NaNO_3$$

Mercury(II) sulfide dissolves in aqua regia when heated:

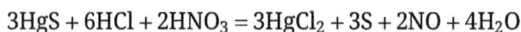

$$3HgS + 6HCl + 2HNO_3 = 3HgCl_2 + 3S + 2NO + 4H_2O$$

Mercury easily forms complex compounds in solutions. If a solution of KI is added to a solution of $Hg(NO_3)_2$, an orange-yellow precipitate of HgI_2 is obtained. In excess of KI, the precipitate dissolves, a soluble complex is obtained – potassium tetraiodomercurate(II) ($K_2[HgI_4]$):

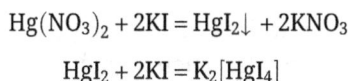

$$Hg(NO_3)_2 + 2KI = HgI_2\downarrow + 2KNO_3$$
$$HgI_2 + 2KI = K_2[HgI_4]$$

Mercury complexes are more resistant than the similar compounds of zinc and cadmium. Interestingly, mercury, located in the row of relative activity after hydrogen, displaces it from HI due to the formation of a very stable complex:

$$Hg + 4HI = H_2[HgI_4] + H_2$$

Mercury forms a small number of compounds in which it is found in the form of the Hg_2^{2+} cation, namely Hg_2O oxide, halides Hg_2X_2 and some other salts. In mercury compounds with a +1 oxidation state, the two mercury atoms are bonded together by a covalent bond. In these compounds, mercury is divalent, but one valence is used for a bond between mercury atoms (analogy with oxygen in H_2O_2), Therefore, the formal oxidation state of mercury in these compounds is +1.

A common property of mercury(I) compounds is their disproportionation to form metallic mercury and mercury(II) compounds. For example, Hg_2Cl_2 decomposes under low heating:

$$Hg_2Cl_2 = Hg + HgCl_2$$

In some cases, the disproportionation proceeds so rapidly that the Hg_2^{2+} compounds decompose at the moment of formation:

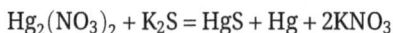

$$Hg_2(NO_3)_2 + K_2S = HgS + Hg + 2KNO_3$$

When alkalis react with mercury(I) salts, metallic mercury and Hg(II) oxide are released:

$$Hg_2(NO_3)_2 + 2NaOH = HgO + Hg + 2NaNO_3 + H_2O$$
$$Hg(NO_3)_2 + 2KOH = 2KNO_3 + HgO + Hg + H_2O$$

If a solution of NaOH is added to a solution of $Hg_2(NO_3)_2$, $Hg_2(OH)_2$ is obtained:

$$Hg_2(NO_3)_2 + 2NaOH = Hg_2(OH)_2 + 2NaNO_3$$

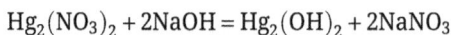

The hydroxide $Hg_2(OH)_2$ is very unstable and dehydrates at ordinary temperature:

$$Hg_2(OH)_2 = Hg_2O + H_2O$$

The resulting Hg_2O is also unstable and breaks down into HgO and Hg:

$$Hg_2O = HgO + Hg$$

The precipitate has a black color, which is due to the Hg obtained in the dispersed state. When nitric acid is added, an interaction occurs until complete dissolution:

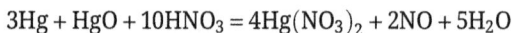

$$3Hg + HgO + 10HNO_3 = 4Hg(NO_3)_2 + 2NO + 5H_2O$$

Although chloride and nitrate of Hg(I) exist with a defined composition of Hg_2Cl_2, $Hg_2(NO_3)_2$, compounds with other common anions, such as sulfides, etc., have never been synthesized. This is due to the fact that under normal conditions there is a tendency for Hg(I) ions to disproportionate to Hg(II) and metallic mercury:

$$Hg_2^{2+} + S^{2-} \rightarrow Hg + HgS$$

The sludge containing HgS and Hg dissolves in aqua regia ($HCl:HNO_3 = 3:1$):

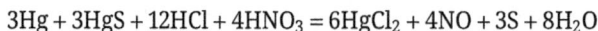

$$3Hg + 3HgS + 12HCl + 4HNO_3 = 6HgCl_2 + 4NO + 3S + 8H_2O$$

When soluble salts of mercury(I) react with soluble chlorides, a white slightly soluble Hg(I) chloride (Hg_2Cl_2) is obtained:

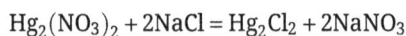

$$Hg_2(NO_3)_2 + 2NaCl = Hg_2Cl_2 + 2NaNO_3$$

The resulting white precipitate of Hg_2Cl_2 does not dissolve in acids and bases. When ammonia is added, the precipitate turns black due to the release of dispersed mercury:

$$Hg_2Cl_2 + 2NH_3 = HgNH_2Cl + Hg + NH_4Cl$$

White mercury(II) amidochloride and mercury are obtained. The released mercury is in a finely dispersed state in the precipitate and determines its black color.

In compounds of mercury with nonmetals, chemical bonds are predominantly covalent, so such compounds (for example, halides) easily sublimate, and their solutions are weak electrolytes.

The compounds of mercury(I) and mercury(II) are easily reduced. When Hg^{2+} is reduced, Hg_2^{2+} is formed first, and then metallic mercury:

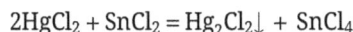

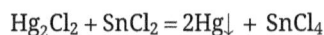

$$2HgCl_2 + SnCl_2 = Hg_2Cl_2\downarrow + SnCl_4$$

$$Hg_2Cl_2 + SnCl_2 = 2Hg\downarrow + SnCl_4$$

The degree of the poisonous effects is dependent on the forms of mercury compounds, the exposure degree, the period and the pathway of exposure. All mercury compounds induce toxic effects, especially the soluble ones. Inorganic mercury compounds are not a big problem because they are mostly insoluble. The most toxic and hazardous form of mercury is represented by organomercury compounds which commonly arise from biological sources, primarily freshwater or saltwater fish. These compounds can readily be dispersed and accumulated in the kidney, brain and liver. Organomercury compounds easily cross the placenta leading to neurological and teratogenic effects. All Hg forms are toxic to the fetus, the most dangerous being $CH_3-Hg^+X^-$. Organic mercury compounds are mostly of the aryl- and alkyl-classes, which have been broadly used as fungicides. For the treatment of mercury intoxication thiol-based chelation agents, such as DMSA, DMPS, dimercaprol, and penicillamine are usually administrated.

Chapter 27
Third B group

27.1 General characteristics

The secondary III B group of the periodic table is made up of the elements scandium (Sc), yttrium (Y), and lanthanum (La), which is generally considered with the lanthanoids, and actinium Ac, which is classified with the actinoids. The group is also called the scandium group or scandium family after its lightest member.

In terms of reduction ability, which can be adjudicated by the values of ionization and electrode potentials, these elements are superior to the elements of the IIIA group and are closer to Mg. The main characteristics of atoms of the elements are presented in Table 27.1.

Table 27.1: Characteristics of atoms of elements of IIIB group.

Element	Sc	Y	La
Valence electrons	$3d^14s^2$	$4d^15s^2$	$5d^16s^2$
Atomic radius (nm)	0.164	0.181	0.187
Ionization potential (eV)	6.56	6.22	5.58
Electrode potential (V)	−2.08	−2.37	−2.52

Scandium has quite a few similarities to aluminum, referring to their close melting points and electrode potentials as well as their chemical relationships. In solution, the salts of Sc^{3+} hydrolyze significantly, and their solutions are acidic, similar to those of Al^{3+}. The addition of hydroxide ions to the corresponding solutions results in the formation of gel-like precipitates from the hydroxides of aluminum and scandium. The latter dissolve in excess of the base to hydroxyaminos. Both metals form isomorphic compounds of the type Na_3MF_6, where M = Al or Sc. Al not only resembles Sc but also differs quite differently in activity from its counterpart Ga.

The chemistry of the elements in the group is more similar to that of the lanthanides than that of the transition metals. That is why these elements are usually discussed together with the lanthanides, which have taken the common name of rare earth elements (REEs).

In the outer electron shell of the elements of Sc subgroup there are three electrons, as well as in the main group. However, these elements refer to d-elements whose electron configuration of the valence shell is d^1s^2. In all three elements, valence electrons are given quite easily.

In contrast, the lanthanides have a different configuration of their outer electron shell: they build the 4f-subshell instead of the d-subshell. Starting from cerium, all these elements, excluding gadolinium and lutetium, have an electronic configuration

https://doi.org/10.1515/9783111712246-027

of their outer electron shell $4f^n6s^2$ (gadolinium and lutetium have $5d^1$-electrons). The number n changes from 2 to 14. Therefore, s- and f-electrons take part in the formation of their valence bonds. The most common oxidation state for lanthanides is +3 and less often +4. The electronic structure of the valence shell of the actinides is very similar to the electronic structure of the valence shell of the lanthanides. Lanthanides and actinides are typical metals. Their properties are discussed in a separate chapter.

27.2 Occurrence

The compounds of scandium, yttrium, lanthanum, and lanthanides were known as early as the beginning of the nineteenth century. The content of scandium, yttrium, and lanthanum in nature is negligible. These are scattered elements found in the form of impurities in about 250 minerals. The source for their production is the mineral monazite, which is a mixture of phosphates of these elements. Pure scandium was obtained by Nilsson in 1879. Yttrium was discovered by Gadolin in 1794 and its content in the earth's crust is 2.9×10^{-3}%. The content of lanthanum, discovered by Mosander in 1839, is 4.9×10^{-3}%. Actinium is an unstable element, the content of which in nature is negligible. Actinides are the heaviest elements following actinium in the periodic table.

27.3 Synthesis

The extraction of scandium, yttrium, and lanthanum from ores is carried out using a complex technology, including multistage enrichment and separation. The final products are the relevant oxides, chlorides, or fluorides. The metals are obtained by reducing oxides or fluorides with aluminum, magnesium, or calcium:

$$La_2O_3 + 2Al = Al_2O_3 + 2La$$
$$La_2O_3 + 3Ca = 3CaO + 2La$$

27.4 Properties and compounds

Scandium and yttrium in air are covered with a thin layer of oxides and retain their silvery-white color, while the most active lanthanum darkens, becoming covered with a layer of hydroxide. Lanthanum is pyrophoric, that is, in powder state it ignites spontaneously in air. All three metals under normal conditions actively interact with acids, and when heated – with nonmetals. Y and La are in oxidation states of +3, which corresponds to the simultaneous participation of the three valence electrons in chemical bonds.

Their hard-to-melt refractory *oxides* E_2O_3 are formed by the interaction of metals with oxygen, during the decomposition of hydroxides and some of their salts. In the order Sc_2O_3–Y_2O_3–La_2O_3, an increase in the basic properties of oxides is observed, which can be explained by the change in sign and the numerical significance of the Gibbs energy for their reaction with water:

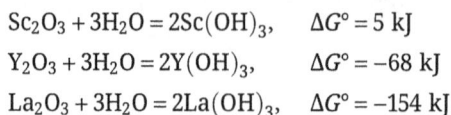

$$Sc_2O_3 + 3H_2O = 2Sc(OH)_3, \quad \Delta G° = 5 \text{ kJ}$$
$$Y_2O_3 + 3H_2O = 2Y(OH)_3, \quad \Delta G° = -68 \text{ kJ}$$
$$La_2O_3 + 3H_2O = 2La(OH)_3, \quad \Delta G° = -154 \text{ kJ}$$

The amphoteric properties of Sc_2O_3 are manifested when interacting with bases in melts:

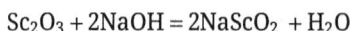

$$Sc_2O_3 + 2NaOH = 2NaScO_2 + H_2O$$

Yttrium and lanthanum oxides are practically nonamphoteric.

Slightly soluble *hydroxides* $E(OH)_3$ are released as precipitates when soluble salts react with solutions of alkali bases and NH_4OH. Hydroxides $Y(OH)_3$ and $La(OH)_3$ are nonamphoteric and do not dissolve in excess of bases, while scandium hydroxide reacts with base solutions when heated, similar to aluminum hydroxide:

$$Sc(OH)_3 + 3NaOH = Na_3[Sc(OH)_6]$$

Scandium, yttrium, and lanthanum form many *salts* – nitrates, sulfates, carbonates, salts of hydrohalic acids, as well as double salts. Complex compounds with coordination numbers of 6 (in Sc) and 8 (in Y and La) are also known.

27.5 Application

The use of scandium, yttrium, and lanthanum is difficult due to the large losses in their production. However, there are some important areas of their application where they are irreplaceable. These metals are mainly used to produce special alloys that exhibit specific electrical and magnetic properties. Scandium is used to make structural alloys for rocketry and aviation. Yttrium improves the properties of alloys used at high temperatures. A small addition of yttrium (about 0.4%) to nichrome increases its shelf-life by about 10 times.

Oxides and other compounds of the elements of the group are used as catalysts and activators of catalysts, as phosphors, components of refractory ceramics, and as magnetic materials.

Although scandium is a comparatively widespread element in the earth crust, it has unknown biological role. It is a supposed carcinogen.

Scandium radionuclides have shown a great potential in nuclear medicine. Among the scandium radioisotopes, [43]Sc and [44]Sc are positron emitter radionuclides for PET imaging, while ^{47}Sc is of interest for radiotherapy. Concerning toxicity, elemental scandium is considered nontoxic, although some of its compounds might be cancerogenic. The main risks of scandium exposure are aerosols and gasses within a working environment because of some Sc-based damps and gasses in the air. These emissions can cause lung embolisms, particularly in prolonged exposure. Scandium is mostly dangerous for the liver at body accumulation. Nevertheless, scarce in vivo testing of scandium compounds has been done.

Scandium and yttrium are found in identical ores as the other rare earth metals and show analogous properties. There are no known significant functions of yttrium in living systems. However, it occurs in organisms and has a tendency to concentrate and accumulate in many human organs. Elemental Sc is considered nontoxic, although some of its compounds might be cancerogenic.

Yttrium is usually dangerous in workplaces because of the humidity and gases released into the air. This causes pulmonary embolism, particularly at longstanding exposure. Yttrium can also increase the risk of lung cancer when it is inhaled.

The biological role of lanthanum is discussed in the next chapters, as the lanthanide series contains 15f-block elements, including lanthanum. Lanthanides along with Sc and Y belong to the group of REEs.

Chapter 28
Fourth B group

28.1 General characteristics

The secondary IVB group of periodic table consists of titanium (Ti), zirconium (Zr), hafnium (Hf), and rutherfordium (Rf), which is a synthetic, radioactive element. Their physical characteristics are listed in Table 28.1.

Table 28.1: Characteristics of atoms and physical features of elements of IVB group.

Element	Ti	Zr	Hf
Valence electrons	$3d^2 4s^2$	$4d^2 5s^2$	$5d^2 6s^2$
Radius of the atom (nm)	0.146	0.160	0.159
Ionization potential (eV)	6.82	6.84	7.50
Electrode potential (V)	−1.63	−1.53	−1.57
Density (g/cm^3)	4.51	6.50	13.30
T_m/T_b (°C)	1,668/3,260	1,855/4,330	2,222/5,690

It can be seen that the elements are complete electron analogues. The presence of four valence electrons explains their location in the fourth B group. The elements are metals, and their metallic properties are more pronounced than in the metals of the main group – lead and tin. During the transition from titanium to zirconium, the radius of the atom increases. In zirconium and hafnium, the atomic (as well as ionic) radii are practically the same, which is explained by the lanthanide contraction, namely the location between zirconium and hafnium of f-elements (lanthanides), in which the size of the atoms monotonically decreases (contraction). That is why the properties of zirconium and hafnium are very close, and their technological separation is one of the most complex problems in inorganic chemistry.

In the row of relative activity of the metals, titanium, zirconium and hafnium are found next to aluminum, that is, this defines them as active metals. At ordinary temperatures, they are chemically resistant to air, as well as in very aggressive environments, which is explained by the formation of a protective layer of oxides on their surface.

28.2 Occurrence

Titanium is the tenth most-abundant element with a content of 0.41% in the Earth's crust. The most important minerals of titanium are rutile TiO_2, ilmenite $FeTiO_3$, perovskite $CaTiO_3$ and titanomagnetite $FeTiO_3 . Fe_3 O_4$. The zirconium content is $2.1 \times 10^{-2}\%$. Zirconium is the fourth most abundant transition metal in the earth's crust (after

https://doi.org/10.1515/9783111712246-028

iron, titanium and manganese). Its most important minerals are baddeleyite ZrO_2 and zirconyl $ZrSiO_4$. Hafnium does not form its own minerals, but is present as an impurity in all zirconium minerals, with a content of $4.2 \times 10^{-4}\%$. Some ores, such as alvite $(Hf,Zr)SiO_4.xH_2O$, can sometimes contain more hafnium than zirconium. Rutherfordium does not exist in nature, it is produced by nuclear fusion.

Natural titanium has the following isotopes: 46 (8.0%), 47 (7.3%), 48 (73.9%), 49 (5.5%), and 50 (5.3%). The isotopes of zirconium are: 90 (51.5%), 91 (11.2%), 92 (17.1%), 94 (17.4%), and 96 (2.8%). The isotopes of hafnium are: 174 (0.2%), 176 (5.2%), 177 (18.6%), 178 (27.1%), 179 (13.7%), and 180 (35.2%).

Since titanium, zirconium, and hafnium exhibit high chemical activity at high temperatures, their separation in their pure form is a rather complex task.

28.3 Synthesis

Titanium is difficult to be obtained from its most common ore, rutile TiO_2. The reduction of TiO_2 with carbon results in a metallic carbide rather than a metal. The only practical way involves the initial conversion of TiO_2 to Ti(IV) chloride, which takes place when the oxide is heated with carbon and chlorine:

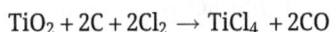

$$TiO_2 + 2C + 2Cl_2 \rightarrow TiCl_4 + 2CO$$

In general, the processing of ores of these metals most often leads to the formation of chlorides ($TiCl_4$ and $ZrCl_4$) or fluorocomplex compounds (K_2ZrF_6 and K_2HfF_6). From them, the corresponding metals of technical purity are obtained by magnesium thermal or sodium-thermal reduction in an atmosphere of argon or helium:

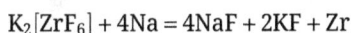

$$TiCl_4 + 2Mg = 2MgCl_2 + Ti$$
$$K_2[ZrF_6] + 4Na = 4NaF + 2KF + Zr$$

High-purity metals are usually obtained by thermolysis of their volatile iodides.

28.4 Physical and chemical properties

Titanium is a hard silvery-white metal that has the lowest density (4.5 g/cm^3) among transition metals. This combination of high strength and low density makes it a preferred metal in the production of aircraft and nuclear submarines. Of all the 4d and 5d pairs of transition metals, zirconium and hafnium are the closest in properties. They have almost identical ionic radii. Zirconium and hafnium are silvery metals with very high melting and boiling points. The only big difference is in their density. As with other 4d–5d pairs, hafnium is much denser than zirconium.

Titanium, zirconium, and hafnium do not interact with nonoxidizing acids at room temperature, but passive in nitric acid. They interact with acids only in those cases when their oxidation is accompanied by the formation of anionic complexes, that is, with HF, aqua regia, a mixture of HNO_3, and HF:

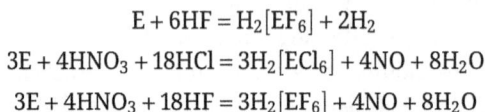

$$E + 6HF = H_2[EF_6] + 2H_2$$
$$3E + 4HNO_3 + 18HCl = 3H_2[ECl_6] + 4NO + 8H_2O$$
$$3E + 4HNO_3 + 18HF = 3H_2[EF_6] + 4NO + 8H_2O$$

When interacting with concentrated H_2SO_4, zirconium and hafnium form complex acids:

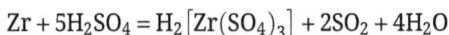

$$Zr + 5H_2SO_4 = H_2\left[Zr(SO_4)_3\right] + 2SO_2 + 4H_2O$$

Unlike zirconium and hafnium, titanium reacts when heated with hydrochloric and dilute sulfuric acid with the release of hydrogen and the formation of violet solutions from titanium(III) salts. These are respectively chloride $TiCl_3$ and sulfate $Ti_2(SO_4)_3$, which in solution are in the form of aquacomplexes. Concentrated sulfuric acid oxidizes titanium, which forms oxysalts:

$$Ti + 3H_2SO_4 = TiOSO_4 + 2SO_2 + 3H_2O$$

Zirconium and hafnium do not interact with solutions and base melts even in the presence of oxidizing agents, while titanium in these cases reacts:

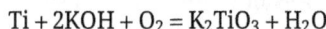

$$Ti + 2KOH + O_2 = K_2TiO_3 + H_2O$$

All three metals are pyrophoric. At temperatures higher than 600 °C, pure metals burn in an oxygen environment to form dioxides EO_2. They interact with halogen elements, and at temperatures higher than 800 °C they react with practically all nonmetals, forming nitrides, carbides, sulfides, phosphides, borides, etc.

28.5 Chemical compounds

Titanium, zirconium and hafnium form compounds in oxidation states +2, +3, and +4. In the group from the top to the bottom, the common property of the d-elements is manifested, namely the stabilization of the higher oxidation state +4 and the reduction of the stability of the lower oxidation states. The excitation of the tetravalent state in them is much easier than in the elements of IVA group.

Compounds in the oxidation state of +2 are formed mainly by titanium. They are few in number (oxide, hydroxide, sulfide, halides), as well as TiSe, TiTe, and Ti_2Si. All of them are unstable, because they are strong reducers:

$$2Ti(OH)_2 + 2H_2O = 2Ti(OH)_3 + H_2$$

The direct formation of *hydrides* EH_2 occurs with the release of heat. Under normal conditions, they are resistant to air, but when ignited, they burn. They are quite inert to most substances. The absorption of H_2 by the metal varies depending on pressure and temperature. Titanium hydride is a catalyst in some hydrogenation reactions of organic compounds. Zirconium hydride is of interest to nuclear energy. By thermal decomposition of EH_2, the corresponding metals of high purity are obtained, which are used in a number of fields of technology.

All *halogenides* EX_2 are formed by heating and decomposition of the corresponding EX_3 halogenides in the absence of air. At higher temperatures, the EX_2 themselves decompose:

$$2EX_3 = EX_4 + EX_2$$
$$2EX_2 = EX_4 + E$$

A lot of titanium compounds in the oxidation state of +3 are known: Ti_2O_3, $Ti(OH)_3$, halides TiX_3, sulfide Ti_2S_3, sulfate $Ti_2(SO_4)_3$, $CsTi(SO_4)_2 \cdot 12H_2O$, and complex compounds such as K_3TiCl_6 and $K_3[Ti(CN)_6]$. These compounds are easily oxidized:

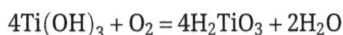

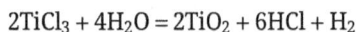

$$4Ti(OH)_3 + O_2 = 4H_2TiO_3 + 2H_2O$$
$$2TiCl_3 + 4H_2O = 2TiO_2 + 6HCl + H_2$$

In zirconium and hafnium, compounds in oxidation state +3 are few in number and are more unstable than Ti(III) compounds.

The most stable and of the greatest practical importance are the compounds of these elements in the oxidation state of +4. When metals interact with oxygen, during thermal decomposition of hydroxides and some salts, thermodynamically stable hard-to-melt *oxides* with a composition of EO_2 are formed. Titanium dioxide is an amphoteric oxide with weak acidic character. It interacts with concentrated sulfuric acid and with melts of bases:

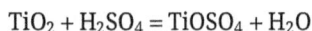

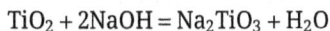

$$TiO_2 + H_2SO_4 = TiOSO_4 + H_2O$$
$$TiO_2 + 2NaOH = Na_2TiO_3 + H_2O$$

Although there are some similarities between Ti(IV) and Si(IV), there is a much greater similarity between Ti(IV) and Sn(IV). In fact, this pair is among the most similar elements found in different groups of the periodic table. For example, the white Ti(IV) and Sn(IV) oxides are isostructural and turn yellow when heated, the so called thermochromism.

In the TiO_2–ZrO_2–HfO_2 series, the basic properties of the oxides are enhanced, but both ZrO_2 and HfO_2 exhibit some weak acidic character. In melts of bases and carbonates of alkali metals, they form zirconates M_2ZrO_3 and hafnates M_2HfO_3.

Titanium(IV) *hydroxide* is a slightly soluble amphoteric compound with predominantly acidic properties. A distinction is made between orthotitanic H_4TiO_4 and metatitanic H_2TiO_3 acids. Their formulas are conditional, because in reality it is a hydrated precipitate of the oxide $TiO_2 \cdot xH_2O$. Orthotitanic acid is a freshly precipitated amorphous precipitate, and metatitanic acid is formed by heating or long-term storage of orthotitanic acid. Both acids are very weak. In melts of bases, these acids form meta-, ortho-, or polytitanates (depending on the amount of base).

Zirconium hydroxide is an amphoteric compound, the composition of which is expressed by different formulas: $Zr(OH)_4$, $ZrO_2 \cdot xH_2O$, $ZrO(OH)_2$, etc. In melts of bases, $Zr(OH)_4$ forms meta-, ortho-, and polyzirconates. Close, but more pronounced basic properties are exhibited by hafnium hydroxide $Hf(OH)_4$.

By methods of anhydrous synthesis, *halogens*, *sulfides*, and other binary compounds are obtained, which are hydrolyzed to form oxysalts:

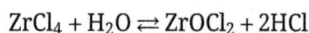

$$ZrCl_4 + H_2O \rightleftharpoons ZrOCl_2 + 2HCl$$

Due to the similarity in the properties of Ti(IV) and Sn(IV) compounds, chlorides also have very close similarities in terms of melting and boiling points: Ti(IV) chloride (m.p. -24 °C, b.p. 136 °C) and Sn(IV) chloride (m.p. -33 °C, b.p. 114 °C). The two chlorides behave as Lewis acids and are hydrolyzed in water:

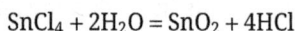

$$TiCl_4 + 2H_2O = TiO_2 + 4HCl$$

$$SnCl_4 + 2H_2O = SnO_2 + 4HCl$$

When tetrafluorides and tetrachlorides interact with concentrated solutions of HF, HCl and their salts, complex compounds are formed, for example, M_2EX_6, H_2TiCl_6, H_4ZrF_8, Na_2TiF_6, K_4HfF_8, etc. M_2EX_6 are hydrolyzed much less than the initial halides EX_4, which is evidence of the stability of EX_6^{2-} in solution. The stability of complex compounds increases from titanium to hafnium and decreases in the order F–Cl–Br–I. The complex salts of Zr and Hf are colorless, while the color of the titanium derivatives depends on the nature of the halogen element, for example, $H_2[TiF_6]$ is colorless, $H_2[TiCl_6]$ is yellow, $H_2[TiBr_6]$ is red, and $H_2[TiI_6]$ is dark red.

Oxidizing properties for Ti, Zr, and Hf compounds in oxidation state +4 are not typical.

28.6 Use and biological role

Titanium and zirconium are used as alloying additives to steels for increasing their strength, corrosion resistance, elasticity, and hardness. Light titanium and its alloys are used in rocketry, aerospace, and supersonic aircraft. The unique corrosion resistance of titanium in seawater is widely used. Titanium is inert in terms of blood and substances produced by the human body, so it is used in surgery (prosthetic bones).

Zirconium practically does not absorb neutrons, so it is used in nuclear reactors. Hafnium, accompanying zirconium, on the contrary, is quite active in absorbing neutrons, which necessitates the serious purification of zirconium from hafnium. This difficult task is solved by the methods of recrystallization of complex fluorinated salts, rectification of volatile halides, ion exchange, and selective extraction.

Some compounds of the elements of the group are widely used. Titanium dioxide is used as a white pigment resistant to weather conditions and also as a catalyst in organic synthesis. The highly durable and very hard carbides, nitrides, and borides of these metals are used in the processing and grinding of hard alloys, glass products and precious stones. A technology has been developed for obtaining large transparent crystals of zirconium dioxide – cubic zirconia. Zirconium dioxide is used in the production of optical products and for the preparation of jewelry.

Very little is reported on the functions of titanium in the life processes. In general, titanium and its analogues are nontoxic. Their chemical compounds are in most cases insoluble. The element does not participate in the biochemical processes of living organisms. It is one of the few d-elements with an unclear or not identified biological role in the human body, though it has been proven that it can act as a stimulant. Negligible amounts of titanium are constantly contained in the organisms of plants and animals. Titanium is useful for the energy production of certain plants. Ti is considered as one of the most biocompatible metals. It is not damaging or poisonous to living tissues because of its resistance to body fluids.

Though titanium, zirconium, and hafnium with little harmfulness are not regarded as toxic elements, nevertheless as heavy metals, they have harmful health properties in working conditions, disturbing lung functions and producing pulmonary diseases. Highly dispersed zirconium causes irritation if it comes in contact with the skin and eyes. The usage of zirconium in medicine and dentistry has rapidly expanded. Hafnium is an unreactive metal, and cannot be affected by air, water, acids, and bases. Its toxicological and biological properties of hafnium are practically unknown. The radioactivity of rutherfordium makes it toxic to living organisms.

Chapter 29
Fifth B group

29.1 General characteristics

In the secondary VB group of the periodic table, the d-elements vanadium (V), niobium (Nb), tantalum (Ta), and dubnium (Db) are located. Their physical characteristics are listed in Table 29.1.

Table 29.1: Characteristics of atoms and physical features of elements of VB group.

Element	V	Nb	Ta
Valence electrons	$3d^3 4s^2$	$4d^4 5s^1$	$5d^3 6s^2$
Radius of the atom (nm)	0.134	0.146	0.146
Ionization potential (eV)	6.74	6.88	7.89
T_m/T_b (°C)	1,900/3,400	2,470/4,760	3,015/5,500
Density (g/cm^3)	5.96	8.57	16.6
Electrode potential (V)	−1.18	−1.10	−0.81

From the data given, it can be seen that the atoms of all elements of the group contain five valence electrons each. The atomic radius from vanadium to niobium increases, with the radii of niobium and tantalum being the same, which is explained by the lanthanide contraction. The metals of the group have high T_m. In terms of reducing activity, all these elements are superior to zinc.

29.2 Occurrence and synthesis

The content of vanadium in the earth's crust is 1.5×10^{-2}%, which is more than that of copper, lead, and zinc, but it belongs to the scattered elements, because there are no large deposits of its own minerals. The main part of vanadium is obtained from iron ores, the deposits of which contain about 1% vanadium. Its most important minerals are sulvanite Cu_3VS_4, alaite $V_2O_3 \cdot H_2O$, vanadinite $Pb_5(VO_4)_3Cl$, etc. In most cases, the product of their processing is V_2O_5, which is reduced with Al, or VCl_3, for the reduction of which Mg is used.

The distribution of niobium (1×10^{-3}%) and tantalum (2×10^{-4}%) is less than that of vanadium. These are rare scattered elements present in iron, uranium and lead ores. Both zirconium and hafnium, as well as niobium and tantalum, are closely related and are usually found together in their minerals of the type ME_2O_6 (M = Fe, Mn). The ore has a common composition $(Fe^{2+}, Mn^{2+})(Nb^{5+}, Ta^{5+})_2(O^{2-})_6$. If the mineral is

https://doi.org/10.1515/9783111712246-029

richer in niobium, it is called columbite $Fe(NbO_3)_2$, and if it is richer in tantalum, it is called tantalite $Fe(TaO_3)_2$. After complex processing from the ores of niobium and tantalum, fluorocomplex compounds K_2NbF_7 and K_2TaF_7 are obtained, from which with the help of Na or K, the corresponding metals with technical purity are reduced. Due to the similarity in their chemical properties, the separation of niobium and tantalum from their ores presents a serious difficulty. Vanadium, niobium, and tantalum of high purity are produced by the thermal decomposition of their iodides.

Db is a radioactive element artificially obtained by nuclear reactions. Several isotopes of this element are known, of which the most stable is ^{262}Db with a half-life of 34 s.

29.3 Properties

Vanadium, niobium, and tantalum are light gray metals with a cubic volumetric-centered crystal lattice and metallic luster. Metals are easy to process. Their physical characteristics depend significantly on the purity of the metals. Impurities such as O, H, N, and C sharply reduce their ductility and increase their hardness.

Under normal conditions, V, Nb, and Ta are protected from aggressive media by their surface oxide layers, which are particularly stable for Nb and Ta. These two metals do not interact with any acid, are stable in aqua regia, and react only with a mixture of HNO_3 and HF:

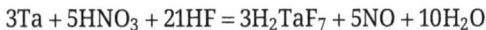

$$3Ta + 5HNO_3 + 21HF = 3H_2TaF_7 + 5NO + 10H_2O$$

Vanadium, which is highly reactive, interacts with HF (up to VF_3), with concentrated H_2SO_4 (up to $VOSO_4$) and HNO_3 (up to VO_2NO_3). All three metals interact with melts of bases in the presence of oxidizing agents:

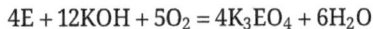

$$4E + 12KOH + 5O_2 = 4K_3EO_4 + 6H_2O$$

At high temperatures, vanadium, niobium, and tantalum interact with oxygen, halogen elements, sulfur, nitrogen, carbon, hydrogen, and water vapor. Their hydrides EH are metallic powders of gray or black color and variable composition. They exhibit chemical resistance to water and dilute acids. Most binary compounds of d-elements have a variable composition, which depends on how they are obtained. Nitrides, carbides, borides, etc. compounds of group VB elements with low-active nonmetals exhibit high chemical stability. Their oxides have a diverse composition and their physicochemical properties change in wide limits. With an increase in oxygen content, the metallic bond decreases and the degree of covalent bond increases.

Vanadium, niobium, and tantalum, both with each other and with metals closely located to them (from the groups of iron, titanium, and chromium), form solid metal solutions.

The most interesting feature of niobium and tantalum is the formation of metal cluster halides. A typical example is the series with the general formula M_6X_{16}, where

M is niobium or tantalum and X is the halogen element. In fact, compounds have a formula with a composition $[M_6X_{12}]^{2+} \cdot 2X^-$. The cluster ion $[M_6X_{12}]^{2+}$ has an octahedral structure, made up of metal and halogen atoms with bridge bonds between metal atoms. These ions are extremely stable and can form $[M_6X_{18}]^{2-}$ ions during redox processes.

29.4 Chemical compounds

Vanadium, niobium, and tantalum form compounds in oxidation states of +5, +4, +3, and +2, which in vanadium correspond to d^0, d^1, d^2, and d^3 electron configurations. The higher oxidation state of +5 is more characteristic for niobium and tantalum, while vanadium primarily forms compounds in the +4 oxidation state.

In the +2 oxidation state, the mainly known vanadium compounds are: VO oxide, $V(OH)_2$ hydroxide, halides VF_2 and VCl_2, sulfate $VSO_4 \cdot 6H_2O$ and double salts of the type $K_2SO_4 \cdot VSO_4 \cdot 6H_2O$. All of them are reducers and are quite unstable. Violet solutions of $[V(H_2O)_6]^{2+}$ are oxidized even in air to $[V(H_2O)_6]^{3+}$ with a green color. Without oxidizing agents, solutions of V(II) compounds gradually decompose water with the release of hydrogen. For example, vanadium dichloride in solution is oxidized by water:

$$2VCl_2 + 2H_2O = 2VOCl + 2HCl + H_2$$

Vanadium(II) oxide (VO) is gray in color and metallic. It is characterized by high electrical conductivity. Its crystal lattice is of the NaCl type. It is obtained by reducing V_2O_5 with hydrogen. It does not interact with water. It is easily dissolved in dilute acids and exhibits basic properties.

The Nb(II) and Ta(II) derivatives are cluster compounds that contain metal-metal bonds.

Compounds with an oxidation state of +3 are also characteristic mainly in vanadium. Amphoteric compounds with predominantly basic signs are oxide V_2O_3 and hydroxide $V(OH)_3$, halides VF_3, VCl_3, VBr_3, and VI_3, sulfates $V_2(SO_4)_3$, $K_2SO_4 \cdot V_2(SO_4)_3 \cdot 12H_2O$, $(NH_4)_2SO_4 \cdot V_2(SO_4)_3 \cdot 12H_2O$, etc. Stable in the solid state, in aqueous solutions these compounds are oxidized from dissolved oxygen to vanadium(IV) compounds:

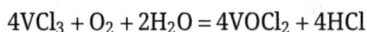

$$4VCl_3 + O_2 + 2H_2O = 4VOCl_2 + 4HCl$$

The vanadium(III) oxide (V_2O_3) is black in color, with a crystal lattice of the α-Al_2O_3 corundum type. The corresponding hydroxide $V(OH)_3$ is green in color with a variable composition ($V_2O_3 \cdot nH_2O$). It is obtained by precipitation from vanadium(III) solutions. V_2O_3 and $V(OH)_3$ dissolve in acids. Melts of oxide and hydroxide with bases form vanadade MVO_2, where M is an alkali metal.

Vanadium(III) halides are crystalline substances. When VF_3 and VCl_3 react with fluorides and chlorides of alkali metals, complex compounds of halogenvanate are formed $K_3[VF_6]$, $Na[VCl_4]$, $K_3[VCl_6]$, $K_3[V_2Cl_9]$, etc.:

$$3KF + VF_3 = K_3[VF_6]$$

$$3KCl + 2VCl_3 = K_3[V_2Cl_9]$$

Vanadium(III) halides, when heated, decompose by disproportionation:

$$2VCl_3 = VCl_2 + VCl_4$$

For the technique, the yellow-bronze nitride VN ($T_m = 2,050$ °C), resistant to water and acids, as well as the carbide VC ($T_m = 2,800$ °C), which has high hardness, are important.

In Nb and Ta, only halides are obtained in the oxidation state of +3 as cluster compounds.

For vanadium, the most characteristic oxidation state is +4. V(III) compounds are readily oxidized to V(IV), and V(V) compounds are easily reduced to V(IV) derivatives.

Vanadium dioxide (VO_2), vanadium tetrafluoride (VF_4), and vanadium tetrachloride (VCl_4) are known, as well as numerous compounds containing vanadyl cation VO^{2+}: $VO(OH)_2$, $VOCl_2$, $VOBr_2$, VOI_2, $VOSO_4$, etc. These compounds are formed in solutions as a result of hydrolysis:

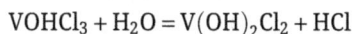

$$VCl_4 + H_2O = VOHCl_3 + HCl$$

$$VOHCl_3 + H_2O = V(OH)_2Cl_2 + HCl$$

$$V(OH)_2Cl_2 = VOCl_2 + H_2O$$

VO_2 is an amphoteric oxide. When it interacts with acids, vanadyl salts are formed:

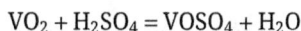

$$VO_2 + H_2SO_4 = VOSO_4 + H_2O$$

When interacting with melts of bases, VO_2 forms metavanadates(IV):

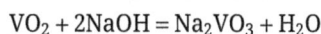

$$VO_2 + 2NaOH = Na_2VO_3 + H_2O$$

When reacting with basic solutions, the oxide forms tetravanadates(IV):

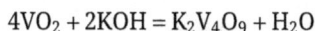

$$4VO_2 + 2KOH = K_2V_4O_9 + H_2O$$

Of the vanadium(IV) halides, brown VF_4 and dark red VCl_4 are known, as well as oxo-halogenides of the type VOX_2. Vanadium tetrahalides are easily hydrolyzed:

$$VCl_4 + H_2O = VOCl_2 + 2HCl$$

For niobium and tantalum, the dioxides NbO_2 and TaO_2, halides and oxohalogenides, which are characterized by reducing properties, are known. For example, when $TaCl_4$ reacts with bases, a redox process takes place:

$$2TaCl_4 + 14NaOH = 2Na_3TaO_4 + 8NaCl + 6H_2O + H_2$$

These compounds belong to the group of cluster compounds.

In the oxidation state +5, the stability of the same type of compounds in the group increases in the order V(V)–Nb(V)–Ta(V), which is seen when comparing the values of the Gibbs formation energies for the oxides: V_2O_5 (−1,427 kJ/mol), Nb_2O_5 (−1,776 kJ/mol), Ta_2O_5 (−1,908 kJ/mol).

Only the oxide V_2O_5 and fluoride are known for vanadium(V), while for niobium (V) and tantalum(V) all other halides EX_5 are also known. The E(V) state is also characterized by oxyhalides of the type EOX_3.

The oxides of E(V) are refractory crystalline substances (V_2O_5 – red, N_2O_5 and Ta_2O_5 – white), exhibiting acidic properties.

Vanadium(V) oxide is obtained by thermal decomposition of ammonium vanadate. It has a low solubility in water (forms an acidic solution of light-yellow color). It interacts with bases, and with acids – only during prolonged heating, which is why its acidic properties predominate. In a melt of V_2O_5 with bases, metavanadates $NaVO_3$ are formed. When interacting with basic solutions, depending on the pH and concentration, orthovanadates K_3VO_4, divanadates $K_4V_2O_7$, trivanadates $K_3V_3O_9$, as well as more complex polyvanadates are formed. When these solutions are neutralized, a precipitate of hydrated oxide $V_2O_5 \cdot xH_2O$ is formed, which is a weak acid, most often expressed by the formulas HVO_3 and H_3VO_4. Vanadium(V) oxide interacts with concentrated acids, exhibiting weakly expressed basic properties. In this interaction, salts of the VO_2^+ cation are formed, such as chloride VO_2Cl, nitrate VO_2NO_3, etc. Vanadium(V) compounds are oxidizing agents of medium activity:

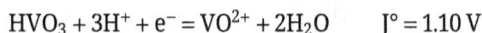

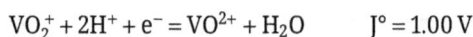

$$HVO_3 + 3H^+ + e^- = VO^{2+} + 2H_2O \qquad J° = 1.10 \text{ V}$$

$$VO_2^+ + 2H^+ + e^- = VO^{2+} + H_2O \qquad J° = 1.00 \text{ V}$$

Vanadium(V) compounds in an acidic solution oxidize concentrated hydrochloric acid:

$$V_2O_5 + 6HCl = 2VOCl_2 + Cl_2 + 3H_2O$$

The oxides Nb_2O_5 and Ta_2O_5 are chemically inactive. These oxides do not react with water and acids, but interact with hydroxides and carbonates of alkali metals only in melts:

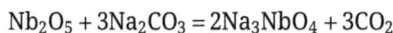

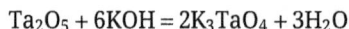

$$Nb_2O_5 + 3Na_2CO_3 = 2Na_3NbO_4 + 3CO_2$$

$$Ta_2O_5 + 6KOH = 2K_3TaO_4 + 3H_2O$$

The oxides Nb_2O_5 and Ta_2O_5 interact with aqueous solutions of KF and HF:

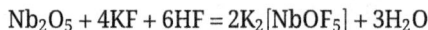

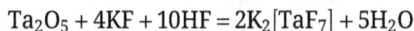

$$Nb_2O_5 + 4KF + 6HF = 2K_2[NbOF_5] + 3H_2O$$

$$Ta_2O_5 + 4KF + 10HF = 2K_2[TaF_7] + 5H_2O$$

The hydrated oxides $Nb_2O_5 \cdot xH_2O$ and $Ta_2O_5 \cdot xH_2O$ are formed by reacting acids with tantalates and niobates and by hydrolysis of pentahalogenides. In their chemical relations, they are weak acids. They interact with solutions of bases to form polymeric niobates and tantalates, for example, $K_3Nb_6O_{19}$, $Na_7Ta_5O_{16}$, $K_2Nb_2O_7$, etc. Oxidizing properties for Nb(V) and Ta(V) compounds are not characteristic.

Oxovanadates(V), oxoniobathes(V), and oxotanthanthates(V) are crystalline substances with complex composition and structure, most of which are polymeric compounds. Only the derivatives of the s-elements of group IA and NH_4^+ of the type KEO_3 and K_3EO_4 are soluble.

The pentahalogenides of vanadium, niobium, and tantalum are refractory, volatile, chemically active compounds. They dissolve in organic solvents. In water, they are easily hydrolyzed, forming amorphous precipitates of hydrated oxides $E_2O_5 \cdot nH_2O$:

$$2EX_5 + 5H_2O = E_2O_5 + 10HX$$

E(V) halides, exhibiting acidic ratios, react with basic halides to form anionic complexes:

$$KF + VF_5 = K[VF_6]$$

$$2KF + TaF_5 = K_2[TaF_7]$$

Oxohalogenides react in a similar way. They are easily hydrolyzed in water to form hydrated oxides $E_2O_5 \cdot nH_2O$, and with basic halides they form anionic complexes:

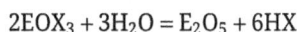

$$2EOX_3 + 3H_2O = E_2O_5 + 6HX$$

$$2KF + VOF_3 = K_2[VOF_5]$$

29.5 Use and biological role

More than 90% of the vanadium produced is used as an alloying additive to steels, which gives them resistance to cold, i.e. strength at negative temperatures. Vanadium is added to steel in the form of an alloy with iron (ferrovanadium), which is cheaper than pure vanadium. Vanadium compounds are used as catalysts in the production of sulfuric acid, as well as in organic synthesis.

Niobium and tantalum are widely used due to their extremely valuable properties: corrosion resistance, high melting point, and low coefficient of thermal expansion. They are used to make cutting tools and corrosion-resistant steels. Niobium is used in radars and X-ray equipment. Tantalum is used as a substitute for gold, silver, and platinum in the preparation of special chemical equipment. In surgery, in the form of thin wires, it is used to connect blood vessels and nerves. Alloys of niobium with tin possess superconductivity (at 20 K) and are used in the preparation of superconducting magnets. Alloys of niobium and tantalum, as well as tantalum and tungsten, withstanding up to 2,300 °C, are used in jet and space engineering. Niobium and tantalum carbides are used as extremely hard and heat-resistant materials.

Of the elements of the fifth B group, only vanadium is identified with a certain role in the biochemistry of living systems. It is conditionally accepted as a trace element with a proven and recognized biological function, although no problems with vanadium deficiency have been observed. Vanadium is widely distributed in water, air, soil, plant, and animal tissues. The body of an adult contains 110–125 µg of vanadium (mainly in teeth and bone tissue, as well as in the lungs, hair, liver, and kidneys), which is why in terms of content in the body it is classified as ultratrace elements. The vanadium content decreases significantly with age. The daily intake of vanadium with food is small. It is found in the largest quantities in seafood, mushrooms, soybeans, parsley, and black pepper. It is interesting to note that the poorest in vanadium are fresh fruits and vegetables. The biological role of vanadium in animal organisms is not yet fully understood. It is known that the element is a powerful inhibitor and regulator of membrane-bound Na^+–K^+–ATPases, ribonucleases and other enzymes, and is supposed to be involved in the regulation of glucose and cholesterol metabolism. It has been found that it participates in the regulation of cardiovascular activity, in the metabolism of bone and dental tissues, in the binding of hemoglobin with oxygen. It catalyzes the oxidation of phospholipids, affects some functions of the eyes, kidneys, myocardium, and nervous system. Vanadium is also a potential anticarcinogenic agent. Convincing evidence has been accumulated for the biological role of vanadium in microorganisms as well as in plants. Although not widespread in nature, vanadium seems to be vital for the simplest groups of marine organisms, the membranous organisms (located between invertebrates and vertebrates). They use very high levels of vanadium to transport oxygen in their blood plasma.

Niobium is not an essential element with unknown biological role. Inhaled niobium is kept mostly in lungs and bones. It interferes with Ca(II) as an enzyme activator. Niobium is one of the most inert metals, making it an ideal safe and hypoallergenic body implant material.

The biological role of tantalum is unknown. The metal hardness, chemical inertness, and resistance to foreign agents' attacks make it compatible with human body tissues. As a result, the metal can be used in many biomedical applications, analogous to Ti and Nb. This metal is supposed to be the only material that is fully compatible with the human body. It is usually used in surgical implants and for repairing of bone defects due to its exceptional corrosion resistance.

Chapter 30
Sixth B group

30.1 General characteristics

The sixth secondary group of the periodic table includes the d-elements: chromium (Cr), molybdenum (Mo), tungsten (W), and element with no. 106. Their physical characteristics are presented in Table 30.1.

Table 30.1: Characteristics of atoms and physical features of elements of VIB group.

Element	Cr	Mo	W
Valence electrons	$3d^54s^1$	$4d^55s^1$	$5d^46s^2$
Radius of the atom (nm)	0.127	0.137	0.140
Ionization potential (eV)	6.77	7.10	7.98
T_m/T_b (°C)	1,890/3,390	2,620/4,800	3,380/5,900
Density (g/cm^3)	7.2	10.2	19.3
Electrode potential (V)	−0.74	−0.20	−0.15

From the above data, it can be seen that in the ground state of chromium and molybdenum atoms, one electron from the outer s-subshell passes to the d-subshell, at which the latter becomes semi-filled (d^5), corresponding to a state with increased energy stability. In tungsten, this is not observed, but it easily passes into an excited state of $5d^56s^1$, analogous to the ground state of chromium and molybdenum atoms. The atomic radii of Mo and W are close due to lanthanide contraction. Compared to the d-elements of the fifth B group, chromium, molybdenum and tungsten have a decrease in atomic radii and an increase in ionization potentials, density, melting and boiling temperatures. All this leads to a decrease in their reduction capacity, especially in molybdenum and tungsten, which is evident from the values of the electrode potentials.

30.2 Occurrence and synthesis

Chromium occupies the 21st place in terms of distribution in the earth's crust, about 0.02%. Its main mineral is chromite $Fe(CrO_2)_2$ or $FeO.Cr_2O_3$. Molybdenum and tungsten are less common and have almost the same content (10^{-4}%), their main minerals are molybdenite MoS_2, *powellite* $CaMoO_4$, scheelite $CaWO_4$, and ferberite $FeWO_4$.

Chromium is obtained by reducing enriched chromite with carbon. It is obtained in the form of an alloy with iron (ferrochrome), which is much more widely used than pure chromium:

https://doi.org/10.1515/9783111712246-030

$$FeO.Cr_2O_3 + 4C = FeCr_2(ferrochrome) + 4CO$$

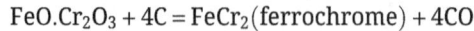

To obtain pure chromium, chromite is chemically degraded at first:

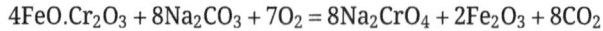

$$4FeO.Cr_2O_3 + 8Na_2CO_3 + 7O_2 = 8Na_2CrO_4 + 2Fe_2O_3 + 8CO_2$$

The resulting sodium chromate, when dissolved in water, is separated from the iron oxide, and during subsequent treatment with sulfuric acid, it turns into dichromate:

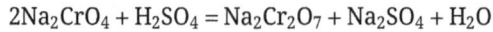

$$2Na_2CrO_4 + H_2SO_4 = Na_2Cr_2O_7 + Na_2SO_4 + H_2O$$

When concentrating the solution, the less soluble $Na_2Cr_2O_7$ crystallizes first, which is reduced to Cr_2O_3 with carbon:

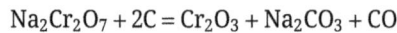

$$Na_2Cr_2O_7 + 2C = Cr_2O_3 + Na_2CO_3 + CO$$

Metallic chromium from Cr_2O_3 cannot be reduced with the use of carbon, because chromium carbides are formed, so CO, aluminum, or silicon are used as reducers:

$$Cr_2O_3 + 2Al = Al_2O_3 + 2Cr$$

$$2Cr_2O_3 + 3Si + 3CaO = 3CaSiO_3 + 4Cr$$

Chromium is also obtained by electrolysis of aqueous solutions of chromium compounds.

Molybdenite is used to obtain molybdenum. Molybdenite is first oxidized, after which MoO_3 is reduced with hydrogen:

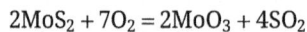

$$2MoS_2 + 7O_2 = 2MoO_3 + 4SO_2$$

$$MoO_3 + 3H_2 = Mo + 3H_2O$$

Scheelite and wolframite are processed using a complex technology, tungsten oxide (WO_3) is obtained, which is also reduced with hydrogen.

30.3 Physical and chemical properties

Chromium, molybdenum, and tungsten are grayish metals with a characteristic metallic luster. As the serial number increases, their melting and boiling temperatures increase. At room temperature, metals are inert due to the formation of thin but very strong oxide layers on their surface.

In the Cr–Mo–W series, chemical activity decreases. With an increase in the oxidation state of the elements, the basic properties weaken and the acidic properties of their oxides and hydroxides increase. The higher oxides of EO_3 correspond to the acids H_2EO_4. In the same direction, the oxidizing properties of compounds are enhanced.

Under normal conditions, chromium reacts only with fluorine. At high temperatures (above 600 °C), chromium and other metals of the group interact with oxygen and other nonmetals:

$$4Cr + 3O_2 \rightarrow 2Cr_2O_3$$
$$2Cr + 3Cl_2 \rightarrow 2CrCl_3$$
$$2Cr + N_2 \rightarrow 2CrN$$
$$2Cr + 3S \rightarrow Cr_2S_3$$

In the molten state, chromium reacts with water vapor:

$$2Cr + 3H_2O \rightleftharpoons Cr_2O_3 + 3H_2$$

The metals of the group interact with melts of bases and carbonates of alkali metals in the presence of oxidizing agents:

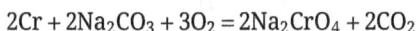

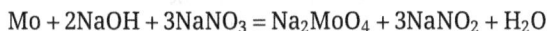

$$2Cr + 2Na_2CO_3 + 3O_2 = 2Na_2CrO_4 + 2CO_2$$
$$Mo + 2NaOH + 3NaNO_3 = Na_2MoO_4 + 3NaNO_2 + H_2O$$

In aqua regia, in concentrated sulfuric and nitric acid in different concentrations, due to the formation of an oxide layer, chromium is passivated. It displaces hydrogen from acids with nonoxidative action (dilute HCl, H_2SO_4) with the formation of $CrCl_2$ and $CrSO_4$. In the absence of air, salts of Cr^{2+} are formed, while in the presence of O_2, salts of Cr^{3+}:

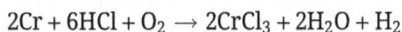

$$Cr + 2HCl \rightarrow CrCl_2 + H_2$$
$$2Cr + 6HCl + O_2 \rightarrow 2CrCl_3 + 2H_2O + H_2$$

With significant heating, chromium also interacts with concentrated HNO_3 and H_2SO_4, oxidizing to a trivalent state:

$$Cr + 6HNO_3 = Cr(NO_3)_3 + 3NO_2 + 3H_2O$$

Tungsten does not react with any acids. It is also stable in aqua regia and is oxidized only by a mixture of HNO_3 and HF:

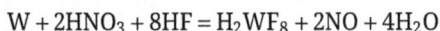

$$W + 2HNO_3 + 8HF = H_2WF_8 + 2NO + 4H_2O$$

Molybdenum also reacts with a mixture of HNO_3 with HF, as well as with aqua regia and with concentrated nitric acid (when heated):

$$Mo + 6HNO_3 = H_2MoO_4 + 6NO_2 + 2H_2O$$

30.4 Chemical compounds

Chromium, molybdenum, and tungsten form compounds in oxidation states of +2, +3, +4, +5, and +6. The most characteristic oxidation state for chromium is +3, while for molybdenum and tungsten it is +6. Therefore, there is a common pattern for all d-elements, namely from the top to the bottom in the group, the stability of the highest oxidation state increases.

In the *oxidation state of +2*, chromium is known to have its oxide CrO, hydroxide $Cr(OH)_2$, halides CrX_2, sulfate $CrSO_4$, acetate $Cr(CH_3COO)_2$, and a small number of complex compounds, such as $[Cr(NH_3)_6]X_2$ and $M_2[CrX_4]$, where X is a halogen element and M is an alkali metal. Chromium(II) oxide and hydroxide are basic:

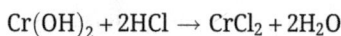

$$Cr(OH)_2 + 2HCl \rightarrow CrCl_2 + 2H_2O$$

Soluble chromium(II) salts are obtained by reacting metallic chromium with non-oxidative acids or by reducing chromium(III) compounds with zinc:

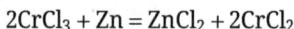

$$Cr + 2H^+ = Cr^{2+} + H_2$$

$$2CrCl_3 + Zn = ZnCl_2 + 2CrCl_2$$

Chromium(II) salts are blue in color, which is due to the aquacomplex $[Cr(H_2O)_6]^{2+}$.

In molybdenum and tungsten, only the halides are obtained in the oxidation state of +2. All compounds of M(II) are strong reducers. They are oxidized to M(III) compounds by the action of oxygen from the air and reduce hydrogen from water, so they are unstable in aqueous solutions:

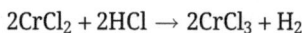

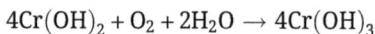

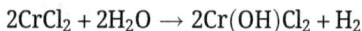

$$2CrCl_2 + 2HCl \rightarrow 2CrCl_3 + H_2$$

$$4Cr(OH)_2 + O_2 + 2H_2O \rightarrow 4Cr(OH)_3$$

$$2CrCl_2 + 2H_2O \rightarrow 2Cr(OH)Cl_2 + H_2$$

The state corresponding to *an oxidation state of +3* is most stable in chromium. In this oxidation state, chromium has many simple and complex compounds, crystal hydrates and alum. In composition and properties, many of Cr(III) compounds are close to the compounds of Al.

Chromium(III) oxide (Cr_2O_3) is a green-colored, water-insoluble powdery substance. It is obtained by heating chromium(III) hydroxide or potassium and ammonium dichromate:

$$2Cr(OH)_3 \rightarrow Cr_2O_3 + 3H_2O$$

$$4K_2Cr_2O_7 \rightarrow 2Cr_2O_3 + 4K_2CrO_4 + 3O_2$$

$$(NH_4)_2Cr_2O_7 \rightarrow Cr_2O_3 + N_2 + 4H_2O$$

Amphoteric chromium(III) oxide is highly soluble and chemically inert, but in a highly dispersed state it dissolves in acids and basic solutions to form aqua- and hydroxo-complexes:

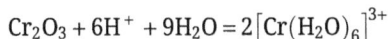

$$Cr_2O_3 + 6H^+ + 9H_2O = 2\left[Cr(H_2O)_6\right]^{3+}$$

$$Cr_2O_3 + 6OH^- + 3H_2O = 2\left[Cr(OH)_6\right]^{3-}$$

Metachromites are formed in melts of bases, oxides and carbonates of alkali metals. Chromates in the oxidation state of +6 are formed in the presence of oxidizing agents:

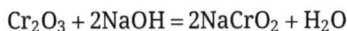

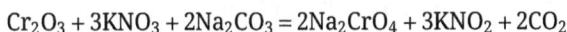

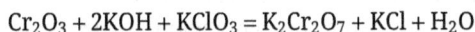

$$Cr_2O_3 + 2NaOH = 2NaCrO_2 + H_2O$$

$$Cr_2O_3 + 3KNO_3 + 2Na_2CO_3 = 2Na_2CrO_4 + 3KNO_2 + 2CO_2$$

$$Cr_2O_3 + 2KOH + KClO_3 = K_2Cr_2O_7 + KCl + H_2O$$

The hydroxide $Cr(OH)_3$ precipitates by reacting bases or ammonium hydroxide with solutions of Cr(III) salts, and by mixing solutions of salts mutually enhancing their hydrolysis:

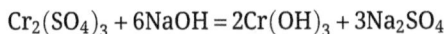

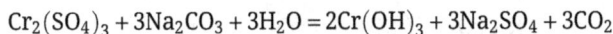

$$Cr_2(SO_4)_3 + 6NaOH = 2Cr(OH)_3 + 3Na_2SO_4$$

$$Cr_2(SO_4)_3 + 3Na_2CO_3 + 3H_2O = 2Cr(OH)_3 + 3Na_2SO_4 + 3CO_2$$

The green amorphous chromium trihydroxide precipitate "ages" or crystallizes over time. When heated, $Cr(OH)_3$ first passes into $CrO(OH)$ and then into Cr_2O_3. Chromium trihydroxide reacts with acids and bases to form the same products like chromium(III) oxide:

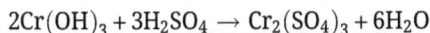

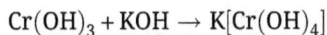

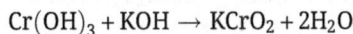

$$2Cr(OH)_3 + 3H_2SO_4 \rightarrow Cr_2(SO_4)_3 + 6H_2O$$

$$Cr(OH)_3 + KOH \rightarrow K[Cr(OH)_4]$$

$$Cr(OH)_3 + KOH \rightarrow KCrO_2 + 2H_2O$$

Chromium(III) salts are violet or green in color, chemically similar to the colorless salts of aluminum. When $Cr(OH)_3$ is dissolved in hydrochloric acid, salts of different colors are formed, which is due to hydrate isomerism, that is, the number of coordinated water molecules: $[Cr(H_2O)_6]Cl_3$ – violet, $[Cr(H_2O)_5Cl]Cl_2$ – dark green, and $[Cr(H_2O)_5Cl_2]Cl$ – light green aquacomplex. Hydroxycomplexes of chromium $Na[Cr(OH)_4(H_2O)_2]$ and $Na_3[Cr(OH)_6]$ are bright green in color.

Cr(III) compounds can exhibit both oxidative and reducing properties:

$$Zn + 2CrCl_3 \rightarrow 2CrCl_2 + ZnCl_2$$

$$2CrCl_3 + 16NaOH + 3Br_2 \rightarrow 6NaBr + 6NaCl + 8H_2O + 2Na_2CrO_4$$

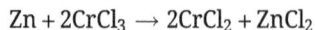

A comparison of the lower two half-reactions shows that chromium(III) compounds can be relatively easily oxidized to chromium(VI) in an alkaline environment as well as in an acidic environment in the presence of very strong oxidizing agents:

$$Cr(OH)_3 + 5OH^- - 3e^- = CrO_4^{2-} + 4H_2O, \quad J° = -0.13 \text{ V}$$
$$2Cr^{3+} + 7H_2O - 6e^- = Cr_2O_7^{2-} + 14H^+, \quad J° = 1.33 \text{ V}$$

In *the oxidation state +4*, the oxides of these metals MO_2, as well as other binary compounds, were obtained. Molybdenum(IV) sulfide, which is called disulfide, is one of the molybdenum(IV) compounds of industrial importance. The purified black Mo(IV) sulfide MoS_2 has a layered structure of tetrahedrally coordinated molybdenum atoms bonded by sulfide bridges, similar to graphite.

In *the oxidation state of +5*, Mo(V) oxide Mo_2O_5 and Mo(V) metahydroxide $MoO(OH)_3$ were obtained. The compounds of the elements of the group in oxidation states +4 and +5 are unstable and of no practical interest.

Numerous compounds of chromium, molybdenum and tungsten are known in *the oxidation state of +6*. These are oxides, acids, halides, chromates, molybdates, tungstates, etc. Cr(VI) trioxide CrO_3 is a bright red crystalline substance with good solubility in water. It is obtained by reacting potassium chromate (or dichromate) with a concentrated H_2SO_4:

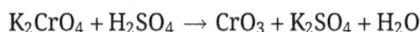

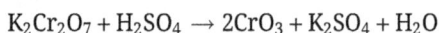

$$K_2CrO_4 + H_2SO_4 \rightarrow CrO_3 + K_2SO_4 + H_2O$$

$$K_2Cr_2O_7 + H_2SO_4 \rightarrow 2CrO_3 + K_2SO_4 + H_2O$$

CrO_3 oxide belongs to the acidic oxides. With bases, it forms yellow chromates CrO_4^{2-}:

$$CrO_3 + 2KOH \rightarrow K_2CrO_4 + H_2O$$

In an acidic environment, chromates turn into orange-colored dichromats $Cr_2O_7^{2-}$:

$$2K_2CrO_4 + H_2SO_4 \rightarrow K_2Cr_2O_7 + K_2SO_4 + H_2O$$

In a basic environment, this reaction proceeds in the opposite direction:

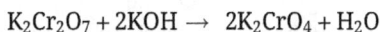

$$K_2Cr_2O_7 + 2KOH \rightarrow 2K_2CrO_4 + H_2O$$

When CrO_3 reacts with water, chromic acids are formed according to the scheme:

$$H_2O + CrO_3 \rightarrow H_2CrO_4 + CrO_3 \rightarrow H_2Cr_2O_7 + CrO_3 \rightarrow H_2Cr_3O_{10} + CrO_3 \rightarrow H_2Cr_4O_{13}$$

Of the acids formed, chromic H_2CrO_4 and dichromic $H_2Cr_2O_7$ acids, as well as their chromate and dichromate salts, are of the greatest importance. Chromates are obtained by oxidation of chromium(III) compounds in an alkaline medium or by their melting in oxidation-alkaline mixtures:

$$2K_3\left[Cr(OH)_6\right] + 3Cl_2 + 4KOH = 2K_2CrO_4 + 6KCl + 8H_2O$$
$$Cr_2(SO_4)_3 + 3KNO_3 + 5K_2CO_3 = 2K_2CrO_4 + 3K_2SO_4 + 3KNO_2 + 5CO_2$$

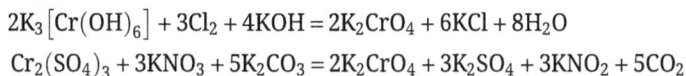

As noted, chromates are stable in alkaline solutions, and when the solutions are acidified, they pass into dichromates, that is, between chromate and dichromate ions, the following chemical equilibrium in solution occurs:

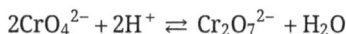

$$2CrO_4{}^{2-} + 2H^+ \rightleftarrows Cr_2O_7{}^{2-} + H_2O$$

Chemical equilibrium is pH-dependent and can shift in one direction or another when an acid or base is added. In solutions of chromic acid salts, the chromate $\rightleftarrows$ dichromate equilibrium shifts towards chromate not only with a change in the pH of the medium, but also with the addition of Ba^{2+}, Pb^{2+}, Ag^+, etc. ions to the corresponding dichromats. These cations form monochromates:

$$2K_2Cr_2O_7 + 4AgNO_3 + H_2O = 2Ag_2CrO_4\downarrow + 4KNO_3 + H_2Cr_2O_7$$

Cr(VI) compounds exhibit strong oxidizing properties, reducing themselves to Cr^{3+}. With such reducers as hydrogen sulfide, sulfides, sulfites, nitrites, etc., chromium(VI) reacts both in acidic and basic environments. Of all chromium(VI) compounds, dichromates in acidic environments have the highest oxidizing activity:

$$Cr_2O_7{}^{2-} + 14H^+ + 6e^- = 2Cr^{3+} + 7H_2O, \quad J^\circ = 1.33 \text{ V}$$

Compared to them, chromates in a basic medium are weaker oxidizing agents:

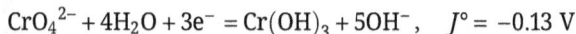

$$CrO_4{}^{2-} + 4H_2O + 3e^- = Cr(OH)_3 + 5OH^-, \quad J^\circ = -0.13 \text{ V}$$

In molybdenum and tungsten, +6 is the most characteristic oxidation state. In the order CrO_3–MoO_3–WO_3, the acidic properties of the oxides weaken. Molybdenum and tungsten trioxides interact with solutions and melts of bases to form molybdates (Na_2MoO_4) and tungstates (K_2WO_4). When acidification, solutions of these compounds form precipitates from the weak, slightly soluble molybdic H_2MoO_4 and tungstic H_2WO_4 acids:

$$Na_2MoO_4 + 2HCl = H_2MoO_4\downarrow + 2NaCl$$
$$Na_2WO_4 + 2HCl + H_2O = H_2WO_4.H_2O\downarrow + 2NaCl$$

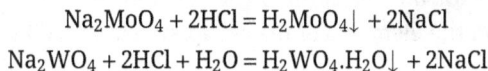

Molybdic and tungstic acids are soluble in acids and bases.

Molybdenum and tungsten are characterized by the formation of heteropolycompounds, including many simple polyacids and polysalts $K_2Mo_2O_7$, $K_2Mo_3O_{10}$, $K_6Mo_6O_{21}$, $K_4Mo_8O_{26}$, as well as heteropolyacids, that is, polyacids containing in their anion, in addition to oxygen, molybdenum (tungsten) and another element (P, Si, B, S, Se, and Te). Some of the salts of these acids are also known, for example: $Na_4[SiW_{12}O_{40}]$, $(NH_4)_3[PMo_{12}O_{40}]$. Just as silicates form groups of SiO_4 tetrahedra and polysilicates, molybdates ($MoO_4{}^{2-}$) and tungstates ($WO_4{}^{2-}$) form octahedral clusters of MO_6 units.

This type of clusters can include heteroions, such as phosphate (PO_4^{3-}) and silicate (SiO_4^{4-}). For example, when molybdate and phosphate ions are mixed and acidified, a phosphomolybdate ion is obtained, $[PMo_{12}O_{40}]^{3-}$:

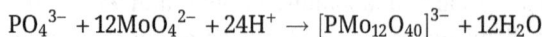

$$PO_4^{3-} + 12MoO_4^{2-} + 24H^+ \rightarrow [PMo_{12}O_{40}]^{3-} + 12H_2O$$

Tungstensilicic acid is obtained at low pH values:

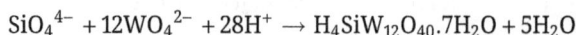

$$SiO_4^{4-} + 12WO_4^{2-} + 28H^+ \rightarrow H_4SiW_{12}O_{40}.7H_2O + 5H_2O$$

Salts of these acids with small-sized cations are very soluble in water, while salts of larger cations, such as cesium and barium, are insoluble.

The compounds of Mo(VI) and W(VI) are quite stable and exhibit their oxidizing properties only when they interact with very strong reducers, for example, with zinc in an acidic environment. In this case, molybdenum is reduced to molybdenum blue $MoO_{3-n}(OH)_n$ ($0 \le n \le 2$), and tungsten to tungsten blue $WO_{3n}(OH)_n$ ($0.1 \le n \le 0.5$). In these compounds, molybdenum is in oxidation states +4, +5, and +6, and tungsten – in oxidation states +5 and +6.

Halogenides EX_6 and oxohalogenides EOX_4 and EO_2X_2 are molecular compounds with covalent bonds. They are low-melting, volatile substances. Chromyl chloride CrO_2Cl_2 is a red oily liquid with a tetrahedral structure. When concentrated H_2SO_4 is added to a mixture of potassium dichromate and NaCl, the mixture darkens as a result of the formation of chromyl chloride:

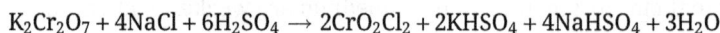

$$K_2Cr_2O_7 + 4NaCl + 6H_2SO_4 \rightarrow 2CrO_2Cl_2 + 2KHSO_4 + 4NaHSO_4 + 3H_2O$$

With light and very careful heating, chromium chloride evaporates to a dark red toxic gas. When this gas is liquefied and a base is added, it hydrolyzes to obtain yellow chromate:

$$CrO_2Cl_2 + 4OH^- \rightarrow CrO_4^{2-} + 2Cl^- + 2H_2O$$

With bromine and iodine, no analogous chromyl halides are formed, so the above reaction is a specific test for the detection of chloride ions.

The sulfides ES_3 of the elements of the group are classified as sulfoanhydrides.

30.5 Usage and biological role

Chromium, molybdenum, and tungsten, as well as their compounds, have wide practical applications. Chromium in the composition of an alloy with iron (ferrochrome) is used for alloying steels. Chromium as an alloying additive is found in instrument, structural and stainless steels and is the main component of heat-resistant alloys. Chrome-plating of various products is widely used in practice. Molybdenum and tungsten are part of steels that retain their strength at high temperatures. They are used in electrovacuum devices, in lamps, in high-temperature furnaces, etc. Silicon and tungsten carbides are highly hard. They are part of alloys from which machining instruments are made. Potassium dichromate is used as an oxidizing agent in inorganic and organic synthesis. Potassium-chromium alum is used in leather processing. The compounds Cr_2O_3, $PbCrO_4$, and $ZnCrO_4$ are used as pigments in paints. Chromium oxide enters abrasive and polishing compositions. A mixture of $K_2Cr_2O_7$ with concentrated sulfuric acid (the so-called chromium mixture) is used to clean laboratory dishes. Molybdenum disulfide (MoS_2) is an important semiconductor. It is also used as a solid lubricant in removable electrical outlets. Numerous compounds of chromium, molybdenum, and tungsten are used as catalysts.

The biological role of chromium is associated with the regulation of carbohydrate metabolism and blood glucose levels. Although chromium(VI) is toxic and carcinogenic, chromium(III) in small quantities is necessary for the human body. Cr(III) ions bind to insulin and help transport glucose (blood sugar) across cell membranes and inside cells, where it can be burned and converted into energy. Deficiency or inability to absorb Cr(III) ions can lead to some forms of diabetes. Insufficient levels of chromium(III) can lead to weight loss, growth disorders, as well as some effects on the nervous system. The chromium content in the body of an adult is in the range of 6–12 mg, much of which is concentrated in the skin, bones and muscles. All chromium compounds are toxic. Taken orally (especially those with an oxidation state of +6) they cause severe damage to the kidneys and the nervous system. Chromium(VI) is genotoxic and is classified as a proven carcinogen. The acute toxicity of chromium(VI) is due to its strong oxidative properties. Once it enters the bloodstream, it harms the kidneys, liver, and blood cells through oxidative reactions. Inhalation of potassium dichromate powder causes atrophy of the nasal mucosa, followed by ulceration and perforation of the nasal septum. The lethal dose in humans is 0.25–0.35 g of $K_2Cr_2O_7$. Chromium compounds are also common in the soil and groundwater of abandoned industrial sites. Chromium salts (chromates) are the cause of allergic reactions in some people. Chromates are often used in the manufacture of leather goods, paints, cement, mortar and anticorrosion materials. Hexavalent chromium compounds can penetrate even through healthy skin. Differences in the absorption of chromium(III) and chromium(VI) are likely. The ingestion of hexavalent chromium can damage the liver, increases the risk of lung cancer and causes asthmatic bronchitis. This is due to the fact that Cr(VI) is much easier to absorb through body barriers (lung tissue, gastrointestinal tract and skin) than trivalent chromium. However, in the stomach, hexavalent chromium is largely reduced to chromium(III).

Molybdenum is the most important member of the group from a biological point of view. It is the heaviest of the elements with such a wide range of functions in living organisms. It is known that dozens of enzymes are associated with molybdenum, which is normally absorbed as molybdate ion, $[MoO_4]^{2-}$. The most important molybdenum enzyme, which also contains iron, is nitrogenase. This family of enzymes is found in bacteria that reduce "inert" atmospheric nitrogen to ammonia, which is used in the synthesis of proteins in plants. Some of these bacteria are in symbiosis with legumes. It is the presence of Mo that allows legumes to assimilate atmospheric nitrogen. In animal organisms, Mo is part of redox enzymes, including xanthine oxidase, which is involved in the exchange of purines and the transport of O_2. The antitumor capabilities of Mo are investigated due to its strong antioxidant effect. When there is an excess of Mo in the soil, it accumulates in the organisms, which contributes to the activation of xanthine oxidase and the synthesis of an additional amount of uric acid. As a result, calcium salts of uric acid – urates are formed, which are deposited in the joints, causing gout. Mo-containing enzymes usually catalyze the oxidation and reduction of small molecules, for example, the oxidation of SO_3^{2-}, AsO_2^-, xanthine, aldehydes, CO, or the reduction of NO_3^-, dimethyl sulfoxide, etc. Deficiency of these enzymes, for which there is usually a genetic cause, causes serious pathologies. An interesting question is – why such a rare metal as molybdenum is so important from a biological point of view? Obviously, there are several possible reasons. The molybdate ion has a high solubility near neutral pH values, which makes it easily transportable by biological liquids. The ion has a negative charge, which in different environments makes it more suitable than the cations of 3d metals. In fact, it is claimed that the molybdate ion is transported by the same mechanism as the sulfate ion, which is an example of the similarity of ions of the sixth main and secondary groups. The element has variable coordination numbers (4, 5, 6, and 8) and a wide range of oxidation states (+4, +5, and +6), who's reduction potentials overlap with those of biological systems. The accumulation of molybdenum in the body is harmful and can be due to industrial intoxication or its high content in drinking water. Molybdenum ranks 18th among metals in terms of its distribution in seawater.

The biological action of tungsten is due to its characteristic variable oxidation states (+4, +5, and +6). Tungsten-containing enzymes have been found in some hyperthermal bacteria. Since these bacteria exist at very high temperatures up to 110 °C, enzymes use more tungsten than molybdenum, as tungsten forms stronger metal-ligand bonds, allowing the enzyme to operate at high temperatures. Thus, the reaction rate of tungsten enzymes at about 110 °C is comparable to that of molybdenum enzymes at 37 °C.

Chapter 31
Seventh B group

31.1 General characteristics

The seventh B group of the periodic table contains the elements: manganese (Mn), technetium (Tc), rhenium (Re), and element borium (Bh) with no. 107. Their physical characteristics are listed in Table 31.1.

Table 31.1: Characteristics of atoms and physical features of elements of VIIB group.

Elements	Mn	Tc	Re
Valence electrons	$3d^54s^2$	$4d^55s^2$	$5d^56s^2$
Radius of the atom (nm)	0.130	0.136	0.137
Ionization potential (eV)	7.44	7.28	7.88
T_m/T_b (°C)	1,244/2,120	2,200/4,600	3,180/5,640
Density (g/cm^3)	7.4	11.5	21.0
Electrode potential (V)	−1.18	0.47	0.37

The atom of each element in the group contains seven valence electrons. The radii and ionization potentials of the atoms are close, but in terms of melting temperatures, highly meltable technetium and rhenium are very different from manganese, which indicates an increase in the strength of their crystal lattices. This can explain the sharp weakening of the reduction activity in the transition from manganese to technetium and rhenium. In the row of relative activity of metals, manganese is located before zinc, and Tc and Re – after hydrogen.

31.2 Occurrence and synthesis

Manganese was first discovered in 1774 and ranks 14th in terms of distribution on Earth (0.09% in the Earth's crust). Its main minerals are pyrolusite (MnO_2) and rhodochrosite ($MnCO_3$).

The existence of technetium and rhenium was predicted by Mendeleev in 1870. Technetium is the only transition element that does not have any stable natural isotopes, that is, it does not exist in nature. The name technetium comes from the Greek word technetos "artificial." For the first time, the element was obtained only in 1937 artificially by nuclear fusion. The artificially synthesized isotope technetium-99 has a long half-life of 2×10^5 years. The radiation of technetium-99 is low enough to make it possible to study the chemical properties of the isotope in a conventional laboratory.

https://doi.org/10.1515/9783111712246-031

So far, about 20 isotopes of technetium have been successfully obtained, some of which are used in scientific research. Technetium has important medical applications in radiotherapy. In terms of chemical properties, it is close to manganese and especially to rhenium.

Rhenium was discovered in 1928 and is a very rare ($\sim10^{-7}$%) and scattered element. Rhenium, whose stable isotopes exist, is found in concentrations of up to 0.2% in molybdenite (MoS_2). This kind of "diagonal connection" between the transition 4d and 5d metals of adjacent groups is the result of similar atomic radii (190 pm for molybdenum and 188 pm for rhenium). The inclusion of Re(IV) in Mo(IV) sulfide is also a reflection of the fact that, like manganese, rhenium readily forms compounds in the oxidation state of +4 (together with +6 and +7).

Borium was obtained by nuclear fusion in 1976.

The annual world production of manganese is several million tons. It is obtained from pyrolusite by carbothermy:

$$MnO_2 + C = Mn + CO_2$$

High-purity manganese is also obtained by electrolysis of a solution of $MnSO_4$.

Rhenium is separated from waste products during the processing of molybdenum and other ores in the form of potassium perrenate, which is reduced with hydrogen:

$$2KReO_4 + 7H_2 = 2Re + 2KOH + 6H_2O$$

Technetium is a waste product in the processing of uranium in nuclear reactors.

31.3 Physical and chemical properties

Manganese, technetium, and rhenium are silvery-gray, heavy, hard, and highly meltable metals. The reactivity of the elements of group VIIB decreases in the order Mn–Tc–Re. At ordinary temperatures, Mn, Tc, and Re in a powdery state oxidize in moist air and turn into MnO, TeO, and $HPeO_4$. Powdered manganese when heated interacts with water:

$$Mn + 2H_2O = Mn(OH)_2 + H_2$$

When heated, the metals of VIIB group react with O_2, $Г_2$, Cl_2, S, P, and Si. At high temperatures, manganese reacts with oxygen to form Mn_2O_3, and technetium and rhenium are oxidized to Tc_2O_7 and Re_2O_7. Manganese is highly active against oxygen, forming an oxide layer in air. Five manganese oxides are known in +2, +3, +4, +6, and +7 oxidation states: MnO, Mn_2O_3, MnO_2, MnO_3, and Mn_2O_7. As the oxidation state of manganese in oxides increases, their chemical character changes from basic for MnO and Mn_2O_3, through amphoteric (MnO_2), to acidic in MnO_3 and Mn_2O_7. Manganese oxides in the higher and intermediate oxidation states are strong oxidizers.

With sulfur, manganese forms sulfide MnS, and Tc and Re – disulfides TcS_2 and ReS_2. With halogen elements, manganese forms $MnCl_2$, $MnBr_2$, MnI_2 (with fluorine –

MnF_4), and technetium and rhenium give compounds with higher oxidation states ReF_7, TcF_6, $ReCl_5$, etc. These metals do not interact with hydrogen. With carbon and nitrogen at 1,200 °C, manganese forms carbides and nitrides with nonstoichiometric compositions. With metals, the elements of the group form alloys that are used in the production of steel.

When Mn reacts with dilute nonoxidative acids (especially when heated), H_2 is released:

$$Mn + 2HCl = MnCl_2 + H_2$$

$$Mn + H_2SO_4 = MnSO_4 + H_2$$

The Mn^{2+} ion in water exists as an aqua complex:

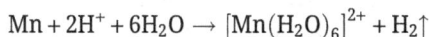

$$Mn + 2H^+ + 6H_2O \rightarrow [Mn(H_2O)_6]^{2+} + H_2\uparrow$$

Manganese, technetium and rhenium interact with acids with oxidative action only when heated, since passivation takes place at room temperature. In these interactions, manganese is oxidized to an oxidation state of +2, and Tc and Re – to their maximum oxidation states:

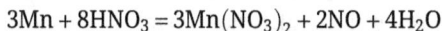

$$Mn + 2H_2SO_4 = MnSO_4 + SO_2 + 2H_2O$$

$$3Mn + 8HNO_3 = 3Mn(NO_3)_2 + 2NO + 4H_2O$$

$$3Tc(Re) + 7HNO_3 = 3HTcO_4(HReO_4) + 7NO + 2H_2O$$

None of the metals reacts with basic solutions, while in melts they all react, but in the presence of oxidizing agents. In these interactions, manganese is oxidized to an oxidation state of +6, and Tc and Re – to their maximum oxidation states:

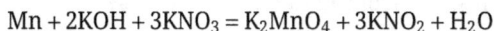

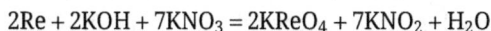

$$Mn + 2KOH + 3KNO_3 = K_2MnO_4 + 3KNO_2 + H_2O$$

$$2Re + 2KOH + 7KNO_3 = 2KReO_4 + 7KNO_2 + H_2O$$

At present, few technetium and rhenium compounds have been obtained and they have not been sufficiently studied. It is known that compounds in oxidation states +2 ($ReCl_2$, ReS), +3 (Re_2O_3), +4 (ReO_2, TcO_2, EX_4), and +6 (EO_2, EX_6, M_2EO_4) are easily oxidized to the most stable oxidation state for these elements +7.

Rhenium forms some compounds with valuable practical applications. It is well known that the search for new superhard materials has traditionally been directed towards the borides, carbides and nitrides of light elements. The compound is not only as hard as diamond, but is extremely easy to be synthesized – without the use of ultra-high pressures and temperatures:

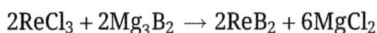

$$2ReCl_3 + 2Mg_3B_2 \rightarrow 2ReB_2 + 6MgCl_2$$

Rhenium diboride is a metal-like material with a high density close to that of rhenium itself, since the B atoms in the lattice are located between the Re atoms. The hardness is due to the very strong covalent bonds, which are almost nonpolar, since Re and boron have close electronegativity.

In oxidation state +7, technetium and rhenium oxides and their oxyanions are more stable than analogous compounds of manganese. The heats of formation of the compounds of M(VII) are compared in Table 31.2.

Table 31.2: The heats of formation of M(VII) compounds.

Compounds	Mn_2O_7	$Tc_2O_{7(s)}$	$Re_2O_{7(s)}$	MnO_4^- (l)	TcO_4^- (l)	ReO_4^- (l)
$\Delta G°_{form}$ (kJ/mol)	−543.4	−936.3	−1165.9	−426.4	−631.2	−698.1

The strength of acids in the $HMnO_4$–$HTcO_4$–$HReO_4$ order decreases very slightly, and their stability increases. Due to the increased resistance to acids and their acid residues, pertechnates, and perrenates are weaker oxidizing agents than permanganates.

31.4 Chemical compounds of manganese

Of the elements in the group, the properties and compounds of Mn are the most studied. Manganese exists in several modification forms, stable at different temperatures. The most characteristic oxidation state of Mn is +2. It forms compounds in which it exhibits other oxidation states +3, +4, +6, and +7, of which the most common are +4 and +7. The coordination number of manganese is six, and rarely four. Divalent manganese is a weak complexing agent.

The chemistry of manganese compounds is complex, because in its derivatives there are all oxidation states from +2 to +7.

In *the oxidation state of +2*, its oxide MnO exhibits basic properties. It occurs in nature in the form of small gray-green crystals, insoluble in water. It is obtained from $MnCO_3$:

$$MnCO_3 \rightarrow MnO + CO_2$$

Mn(II) oxide dissolves in acids by the reaction:

$$MnO + 2H^+ + 5H_2O \rightarrow [Mn(H_2O)_6]^{2+}$$

When MnO reacts with acids, numerous salts of Mn(II) are formed: halides, sulfate, nitrate, carbonate, oxalate, etc. From the solutions of these salts at pH = 8.7, the slightly soluble manganese(II) hydroxide ($Mn(OH)_2$) with weakly expressed basic properties precipitates:

$$[Mn(H_2O)_6]^{2+} + 2OH^- \rightarrow Mn(OH)_2\downarrow + 6H_2O$$

The compounds of divalent manganese are weak reducers. Their oxidation is possible only from such strong oxidizing agents as PbO_2 (in a sulfuric acid medium) and $KClO_3$ (in a KOH melt). The oxidation in an acidic medium reaches to oxidation state of +7 (Mn^{2+} is oxidized to MnO_4^-), and in an alkaline medium – up to +6 (Mn^{2+} is oxidized to MnO_4^{2-}):

$$2MnSO_4 + 5PbO_2 + 6HNO_3 = 2HMnO_4 + 3Pb(NO_3)_2 + 2PbSO_4 + 2H_2O$$

$$3MnSO_4 + 2KClO_3 + 12KOH = 3K_2MnO_4 + 2KCl + 3K_2SO_4 + 6H_2O$$

Freshly obtained MnO and $Mn(OH)_2$ at room temperature are oxidized by oxygen in the air and by other oxidizing agents to form *compounds of Mn(III)* – Mn_2O_3, H_2MnO_3, and manganites:

$$4MnO + O_2 = 2Mn_2O_3$$

$$4Mn(OH)_2 + O_2 = 4MnO(OH) + 2H_2O$$

$$Mn(OH)_2 + H_2O_2 \rightarrow H_2MnO_3\downarrow + H_2O$$

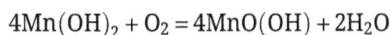

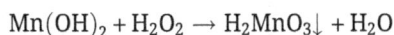

Dimanganese trioxide (Mn_2O_3) occurs in nature as the mineral bixbyite. It is practically insoluble in water and exhibits basic properties. Manganese(III) salts practically do not exist, because when Mn_2O_3 and MnO(OH) react with acids, oxidation-reduction processes take place:

$$Mn_2O_3 + H_2SO_4 = MnSO_4 + MnO_2\downarrow + H_2O$$

$$2Mn_2O_3 + 8HNO_3 = 4Mn(NO_3)_2 + O_2 + 4H_2O$$

$$Mn_2O_3 + 6HCl = 2MnCl_2 + Cl_2 + 3H_2O$$

Manganese(II) manganese(III) oxide (Mn_3O_4) is a red-brown solid of interest because of its formula. In the crystalline state, the Mn^{2+} and Mn^{3+} ions are present, thus the structure of the compound is represented as $(Mn^{2+})(Mn^{3+})_2(O^{2-})_4$. The structure of Fe_3O_4 is similar.

In the *oxidation state +4*, manganese exists in its natural compound manganese dioxide MnO_2 (pyrolusite). MnO_2 has a strong crystal lattice of the rutile (TiO_2) type. It is a dark brown substance with an amphoteric character and redox duality. MnO_2 does not interact with dilute acids and bases, while with concentrated acids there are reactions similar to those with Mn_2O_3:

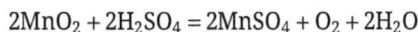

$$2MnO_2 + 2H_2SO_4 = 2MnSO_4 + O_2 + 2H_2O$$

$$2MnO_2 + 4HNO_3 = 2Mn(NO_3)_2 + O_2 + 2H_2O$$

$$MnO_2 + 4HCl = MnCl_2 + Cl_2 + 2H_2O$$

For this reason, salts of Mn^{4+} practically do not exist. The freshly precipitated hydrated manganese dioxide $MnO_2.xH_2O$ obtained in some reactions is considered to be manganese(IV) hydroxide with the conditional formula $Mn(OH)_4$. Manganese dioxide (MnO_2) and manganese(IV) hydroxide ($Mn(OH)_4$) are amphoteric compounds. In melts with bases and oxides of active metals (without access to air), manganites are formed (K_2MnO_3, $CaMnO_3$, K_4MnO_4, etc.), which are considered salts of H_2MnO_3 and H_4MnO_4 acids. Manganese dioxide can occur both as an oxidizing agent and as a reducer:

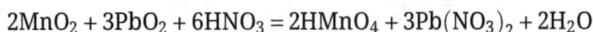

$$MnO_2 + 2KI + 2H_2SO_4 = MnSO_4 + I_2 + K_2SO_4 + 2H_2O$$

$$2MnO_2 + 3PbO_2 + 6HNO_3 = 2HMnO_4 + 3Pb(NO_3)_2 + 2H_2O$$

An oxidation state of +5 in manganese is not characteristic, and only a few compounds in this oxidation state are known, for example, M_3MnO_4, where M is an alkali metal.

Manganese compounds in *oxidation state +6* (manganates) are formed in manganese dioxide melts with bases in the presence of oxidizing agents:

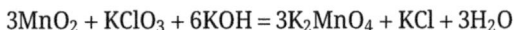

$$3MnO_2 + KClO_3 + 6KOH = 3K_2MnO_4 + KCl + 3H_2O$$

Mn(VI) compounds are too unstable. In solutions, they turn into compounds of Mn(II), Mn(IV), and Mn(VII). These are oxide MnO_3, weak manganic acid (H_2MnO_4), which exists only in aqueous solution, and its salts (manganates). Manganates are compounds of bright green color. They are stable only in a strongly alkaline environment, while in neutral and acidic solutions they disproportionate:

$$3K_2MnO_4 + 2H_2O = MnO_2 + 2KMnO_4 + 4KOH$$

For this reason, obtaining manganic acid (H_2MnO_4) is impossible. Manganates are easily decomposed as a result of hydrolysis and when heated.

The oxide MnO_3, to which manganic acid (H_2MnO_4) corresponds is an acidic oxide. At 50 °C, it decomposes:

$$2MnO_3 \rightarrow 2MnO_2 + O_2$$

When dissolved in water, MnO_3 is hydrolyzed to MnO_2 and $HMnO_4$:

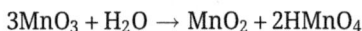

$$3MnO_3 + H_2O \rightarrow MnO_2 + 2HMnO_4$$

Manganates are very strong oxidizing agents in an acidic environment:

$$MnO_4^{2-} + 4H^+ + 2e^- = MnO_2 + 2H_2O, \quad J° = 2.26 \text{ V}$$

In a basic medium, they manifest themselves as reducers and oxidize to permanganates:

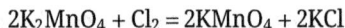

$$2K_2MnO_4 + Cl_2 = 2KMnO_4 + 2KCl$$

In the *oxidation state +7*, the entire genetic order of manganese compounds is known: oxide, acid, and salts. The higher manganese oxide Mn_2O_7 is a liquid substance with a molecular structure. The heptoxide is a green-brown oily liquid. It is unstable with very strong oxidizing properties, and when in contact with it, organic substances spontaneously ignite. When gradually heated, Mn_2O_7 decomposes at 55 °C. In case of mechanical shock, it decomposes with an explosion according to the reaction:

$$2Mn_2O_7 = 4MnO_2 + 3O_2$$

When this acidic oxide reacts with water, the very strong permanganic acid $HMnO_4$ is formed, which is stable in a solution of up to 20% content, because at a higher concentration it decomposes according to the equation:

$$4HMnO_4 = 4MnO_2 + 3O_2 + 2H_2O$$

The salts of this acid (permanganates) are thermally unstable compounds in the solid state. When permanganates are heated, intramolecular oxidation-reduction processes take place:

$$2KMnO_4 = K_2MnO_4 + MnO_2 + O_2$$

Permanganates in solution are violet in color. They form crystal hydrates of the type $EMnO_4 \cdot nH_2O$, where $n = 3$–6, E = Li, Na, Mg, Ca, Sr.

Manganese(VII) compounds have strong oxidizing properties. Permanganates as oxidizing agents are widely used in laboratory practice. Depending on the acidity of the medium, different processes of reduction of permanganate ions are possible. In an acidic medium, Mn^{2+} is obtained:

$$MnO_4^- + 8H^+ + 5e^- = Mn^{2+} + 4H_2O, \quad J^\circ = 1.51 \text{ V}$$

In a neutral or weakly basic medium, permanganate ions are reduced to Mn^{4+}:

$$MnO_4^- + 2H_2O + 3e^- = MnO_2 + 4OH^-, \quad J^\circ = 1.23 \text{ V}$$

In a highly basic medium, permanganate ions are reduced to Mn^{6+}:

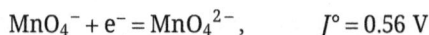

$$MnO_4^- + e^- = MnO_4^{2-}, \quad J^\circ = 0.56 \text{ V}$$

Therefore, the oxidizing properties of permanganates are most pronounced in acidic environments. The oxidative action of $KMnO_4$ is also manifested in its interaction with other manganese compounds in which the oxidation state is lower. In an aqueous solution of $MnSO_4$ and $KMnO_4$, a disproportionation reaction is possible. The product of this reaction is the brown precipitate of MnO_2:

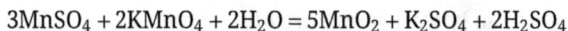

$$3MnSO_4 + 2KMnO_4 + 2H_2O = 5MnO_2 + K_2SO_4 + 2H_2SO_4$$

Potassium permanganate is such a strong oxidizing agent that it oxidizes even concentrated hydrochloric acid to Cl_2, and this is one of the methods of obtaining Cl_2 in laboratory conditions:

$$2KMnO_4 + 16HCl \rightarrow 2KCl + 2MnCl_2 + 8H_2O + 5Cl_2$$

31.5 Usage and biological role

Manganese in the form of an alloy with iron binds sulfur and oxygen in steel, which increases its strength and resistance to friction and corrosion. Manganese alloys with aluminum are used as strong magnets. Manganese alloy, which contains copper and nickel in addition to manganese, has a negligible temperature coefficient of electrical resistance and is widely used in electrical measuring instruments. Manganese dioxide is used in galvanic cells, and as a catalyst and starting substance for the production of multiple manganese compounds. Potassium permanganate is a convenient and widely used oxidizing agent in chemical syntheses and analyzes. With its help, the redox titration method is used to determine the content of various reducers in solutions. The convenience of using $KMnO_4$ lies in the fact that in an acidic environment the violet solution is discolored during reduction.

Rhenium is used in radio electronics (in the details of electrovacuum instruments) and as a catalyst, although it is a very expensive metal. Rhenium is also used as an additive to alloys for the preparation of electrodes in X-ray tubes and radio tubes. By precipitating metallic rhenium, mirrors with high reflectivity are obtained. Technetium, due to its radioactivity and expensive method of production, is only used for scientific research.

Manganese is one of the important biogenic trace elements. The Mn^{2+} ion, whose radius is close to that of Mg^{2+}, forms complexes with O- and N-donor bioligands. It is part of the active centers of a large number of enzymes – oxidoreductases, transferases, lyases, lecithin, etc. Manganese has an effect on blood formation, bone tissue formation, it affects the action of insulin and regulates carbohydrate metabolism. In the composition of the enzymes arginase, cholinesterase, and manganese participates in blood clotting. In the composition of peroxidase and aminophenol oxidase, it regulates oxygen exchange. In the composition of pyruvate carboxylase and phosphoglucose dismutase, it affects hydrocarbon metabolism. Manganese participates in the synthesis of vitamins B and affects the synthesis of hemoglobin. The specific role of manganese in the synthesis of mucopolysaccharides of cartilage tissue has been proven. Manganese is essential for the formation of thyroxine, the main thyroid hormone that affects the normal functioning of the central nervous system.

Manganese enters the body through breathing and, to a lesser extent, through food and drinking water. In the body, manganese is concentrated in the kidneys, liver and pancreas. Absorbed manganese is primarily excreted through bile. With a very high exposure that could be observed in a work environment, the toxic manifestations of manganese are predominantly neurological and provoke a disease similar to Parkinson's disease called manganism. Despite its undeniable biological role, manganese is toxic at high concentrations. Toxic effects are found on the central nervous system and lungs. Manganese deficiency is associated with nausea, vomiting, poor glucose tolerance (high blood sugar levels), skin rash, hair color loss, low cholesterol levels, dizziness, hearing loss, and impaired reproductive function. However, it is important to emphasize that manganese deficiency is very rare in humans and usually does not develop.

Dilute solutions of potassium permanganate $KMnO_4$ are used as a disinfectant in medical practice and as an antidote to phosphorus poisoning. It oxidizes organic matter in cells.

Technetium is the lightest radioactive element in the periodic table, having no stable isotopes (26 unstable and 11 metastable isotopes-isomers) and all of its isotopes are radioactive. It has no known biological role and is not normally found in the body. Of the elements of the VII secondary group, it is mostly used in medical radiation imaging as a tracer. The application of ^{99}Tc in radiopharmacy and imaging has been extensively studied. It is used as a complex coordination compound in diagnosis of heart diseases, as well as in diseases of bones, kidneys, and liver. The pertechnetate ion, which is close in size and charge to the iodide ion, has been used in imaging of the iodine-rich thyroid gland. The metastable radioisotope ^{99m}Tc, a strong γ-emitter with half-life of 6 h, is used in single-photon emission computed tomography (SPECT) tolerating rapid diagnosis. Technetium-99m is useful in the noninvasive examination and diagnosis of metabolic processes, brain, heart, kidneys, and thyroid, without serious damage to the patient. This radioactive isotope is also used in radiation treatment of cancers. The choice of ^{99m}Tc is acceptable because the half-life of this isotope is very low with a quick excretion from the body. Consequently, its adverse effects and radiation exposure for the patients are minimal. Moreover, technetium-99m is a γ-emitter, making it a very negligible risk of toxicity. Due to its unstable isotopes, metallic technetium does not occur naturally in the biosphere and therefore it never presents a risk. All of the technetium compounds are considered as highly toxic, mainly because of its radiotoxicity.

Rhenium is also a very rare microelement. That is why it gives no ecological problems in nature and seems to have no known biological functions in living systems. It is less reactive to air than manganese, which is located in the same group of the periodic table. There is no information about the toxicity of the metallic rhenium or of its compounds. Rhenium is readily soluble in water, and by this interaction the colorless perrhenic acid, $HReO_4$ is forming. Its salts ($MReO_4$) contain a tetrahedral ReO_4^- ion. Ammonium perrhenate NH_4ReO_4 is mostly important for the production of other Re compounds. Rhenium has one stable isotope, ^{185}Re, which however occurs in very small abundance in the nature. Complexes of its isotope ^{186}Re with potential antineoplastic activity have been studied and used in radiotherapy. Among other metals, rhenium merits special consideration, due to its varied range of oxidation states (from −1 to +7), giving the opportunity for excessive structural diversity of its coordination complexes and for the ability to modify the redox status of cancer cells.

Chapter 32
Eighth B group

The eighth B group contains nine d-elements, excluding the recently synthesized elements numbered 108 to 110. The eighth B group is a vertical row of three triplets of elements (triads) arranged horizontally in the periodic table. The first trio consists of iron (Fe), cobalt (Co), and nickel (Ni) (IV period), the second of ruthenium (Ru), rhodium (Rh) and palladium (Pd) (V period) and the third of osmium (Os), iridium (Ir), and platinum (Pt) (VI period). Since the eighth B group does not contain subgroups of elements that differ sharply in electronic structure and in the properties of atoms, there is a greater similarity in the properties of the elements in each triad than that between the individual triads. In the chemical literature, it is customary to consider iron (Fe), cobalt (Co), and nickel (Ni) separately, as they show horizontal similarity regardless of the different number of valence electrons. On the other hand, the similarity of most of the properties of the elements in the Ru–Rh–Pd and Os–Ir–Pt series makes it possible to unite them in a family of platinum metals, which are considered together.

The peculiarities in the properties of the elements of the triads of group VIIIB are explained by the fact that their d-subshells are not completely complete. Therefore, iron, cobalt, nickel, and platinum metals, as a rule, do not tend to form compounds in their higher oxidation states, with the exception of ruthenium and osmium in their oxides RuO_4 and OsO_4.

32.1 Iron subgroup

32.1.1 General characteristics

The iron subgroup refers to elements that are in some way related to iron in period 4 of the periodic table. Their physical characteristics are listed in Table 32.1.

Table 32.1: Characteristics of atoms and physical features of elements of iron subgroup.

Elements	Fe	Co	Ni
Valence electrons	$3d^64s^2$	$3d^74s^2$	$3d^84s^2$
Radius of the atom (nm)	0.126	0.134	0.135
Ionization potential (eV)	7.89	7.34	8.50
Electronegativity	1.64	1.70	1.75
Density (g/cm^3)	7.91	8.90	8.90
T_m/T_b (°C)	1,539/2,872	1,494/2,957	1,455/2,897
Electrode potential (V)	−0.44	−0.28	−0.25

https://doi.org/10.1515/9783111712246-032

The similarity in the properties of the elements is also reflected in their compounds. For iron, cobalt and nickel, the most characteristic oxidation state in the compounds is +2. Their compounds of the same type are similar in electronegativity, enthalpy of formation, solubility, etc. The reason for this similarity is the close sizes of their atoms. All three elements exhibit +2 and +3 oxidation states, with +3 being more characteristic for iron and +2 for cobalt and nickel. The related features of these metals are manifested in their characteristic ferromagnetism, catalytic activity, ability to form colored ions and complexation. Despite similar properties, iron is notable in the triad by its magnetic properties. The reduction activity of iron is significantly more pronounced than that of nickel and cobalt.

32.1.2 Occurrence and synthesis

Iron is the fourth most abundant element after oxygen, silicon, and aluminum (4.7%). Its main mineral ores are hematite (Fe_2O_3), goethite ($FeO(OH)$), limonite ($2Fe_2O_3.3H_2O$), magnetite (Fe_3O_4), pyrite (FeS_2), and siderite ($FeCO_3$). The iron is obtained in blast furnaces, where at high temperature the oxides are reduced with carbon or CO:

$$Fe_2O_3 + 3C \rightarrow 2Fe + 3CO$$

$$Fe_2O_3 + 3CO \rightarrow 2Fe + 3CO_2$$

At high temperatures, iron absorbs part of the carbon, resulting in pig iron (containing 3–4% carbon) or steel (about 1% carbon) through so-called temperature regulation of the carbon content. Pig iron is formed by the reduction of iron ore in blast furnaces at 1,300–1,600 °C in the molten state. Carbon in the form of coke obtained by thermolysis of coal is used as a reducer.

Steel is obtained after additional processing of pig iron to separate excess carbon. The scarcity and high cost of coke, due to reduced natural reserves of coking coal, stimulate the demand for new methods of steel production. In more modern methods, iron ore after enrichment is reduced with hydrogen at 1,000–1,100 °C or with a mixture of H_2 and CO to form porous hollow iron, which is then melted in electric furnaces to form steel. High purity iron is obtained by electrolysis or thermal decomposition of carbonyl ($Fe(CO)_5$).

The content of cobalt and nickel in the earth's crust are 4×10^{-3}% and 8×10^{-3}%, respectively. The main minerals of cobalt and nickel are cobaltite (CoAsS), gersdorffite (NiAsS), and pentlandite (FeNiS). Production of metals is a complex technological process, due to the low content of Co and Ni in the ore, the need to separate them from iron and copper, accompanying these minerals, from the similar properties of cobalt and nickel themselves. In the final stages of processing, oxides are obtained which are then reduced with carbon. Purification of cobalt and nickel from impurities is carried out by electrolysis and carbonyl methods.

32.1.3 Physical and chemical properties

Iron and nickel are silvery-white in color, and cobalt is bluish-white. Iron has several allotropic modifications –α, β, γ, and δ – iron. At room temperature the α-modification is stable, that above 769 °C turns into β-iron, above 910 °C – into γ-iron and above 1,390 °C – into δ-iron. α- and β-Iron have a cubic crystal lattice (a cube, at the vertices and in the center of which the iron atoms are located). The two modifications differ in magnetic properties: α-iron is a magnet, and β-iron has no magnetic properties, i.e. at temperatures higher than 769 °C, the magnetic properties of iron are lost. γ-Iron has a wall-centered cubic lattice (a cube in which there are iron atoms at the vertices and centers of each wall). Iron, cobalt, and nickel are highly meltable metals, they are easily subjected to mechanical processing – forging, drawing, and rolling.

Cobalt and nickel are resistant to air due to the formation of a protective oxide layer on their surface, while iron is slowly oxidized by atmospheric oxygen and moisture:

$$4Fe + 6H_2O + 3O_2 = 4Fe(OH)_3$$

This process is called atmospheric corrosion, due to which about 10% of the iron produced is lost annually. Pure iron, containing less than 0.01% of impurities of carbon, sulfur, and phosphorus, is considered as corrosion-resistant. Additional resistance to corrosion processes is obtained by heat and surface treatment (nitriding, polishing, etc.) with chromium, nickel, titanium and other alloying additives.

Cobalt is placed before hydrogen in the row of relative activity of metals, and the other chemical elements of the subgroup have positive standard electrode potentials. In this regard, there is a significant difference in their activity.

Iron at ordinary temperature, and cobalt and nickel when heated, interact with nonoxidative acting acids to form salts ($MeCl_2$ and $MeSO_4$) in oxidation state of +2, which reduces H_2:

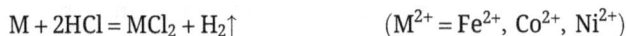

$$M + 2HCl = MCl_2 + H_2\uparrow \qquad (M^{2+} = Fe^{2+}, Co^{2+}, Ni^{2+})$$

In cold concentrated H_2SO_4 and HNO_3, the metals are passivated, which allows concentrated sulfuric acid to be transported in iron tanks. In dilute nitric acid, iron is oxidized to an oxidation state of +3, and cobalt and nickel – usually to a divalent state:

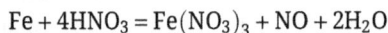

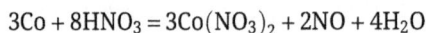

$$Fe + 4HNO_3 = Fe(NO_3)_3 + NO + 2H_2O$$

$$3Co + 8HNO_3 = 3Co(NO_3)_2 + 2NO + 4H_2O$$

Nickel is resistant to the action of alkaline solutions in various concentrations, as well as in melts. Iron and cobalt are also stable in a basic medium. They interact with bases to form hydroxo complexes only when heated and at a high concentration of the base (more than 50%).

It should be noted that nickel has a high resistance to fluorine up to 500–600 °C due to the formation of a dense nonvolatile layer of nickel fluoride on its surface.

32.1.4 Chemical compounds

Iron forms compounds in all positive oxidation states from +2 to +6, cobalt – up to +5, and nickel – up to +4. The maximum oxidation state of the element corresponds to the sum of the s– and unpaired d-electrons in the valence shell.

Oxidation state +2 is the characteristic of all three elements. In this oxidation state, iron, cobalt and nickel form oxides, hydroxides and numerous salts. Oxides FeO, CoO, and NiO are compounds with a nonstoichiometric composition. They are obtained by heating hydroxides, carbonates, and oxalates in an inert atmosphere (oxidation is possible in air). Oxides do not react with water and alkalis, but easily interact with acids:

$$MO + 2H^+ = M^{2+} + H_2O$$

When salts of divalent iron, cobalt, and nickel react with bases, precipitates from the hydroxides $Me(OH)_2$ are released. Iron hydroxide is easily oxidized, so pure iron(II) hydroxide can be obtained only in the complete absence of oxygen in the solution:

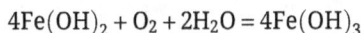

$$4Fe(OH)_2 + O_2 + 2H_2O = 4Fe(OH)_3$$

In the $Fe(OH)_2$–$Co(OH)_2$–$Ni(OH)_2$ series, stabilization of the divalent state is observed, and in the same order the reduction activity of hydroxides decreases. Cobalt dihydroxide oxidizes slowly during prolonged exposure to air. The process takes place more intensively when hydrogen peroxide is added to the solution:

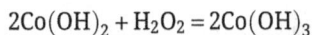

$$2Co(OH)_2 + H_2O_2 = 2Co(OH)_3$$

Spontaneous oxidation of $Ni(OH)_2$ by oxygen does not occur. Hydrogen peroxide also turns out to be an insufficiently strong oxidizing agent and the oxidation process of nickel(II) hydroxide becomes possible only when stronger oxidizing agents are used:

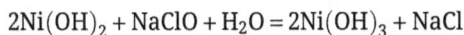

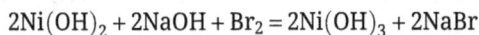

$$2Ni(OH)_2 + NaClO + H_2O = 2Ni(OH)_3 + NaCl$$
$$2Ni(OH)_2 + 2NaOH + Br_2 = 2Ni(OH)_3 + 2NaBr$$

The hydroxides of Fe(II), Co(II), and Ni(II) react with acids, which proves their basic character. The weak amphoteric character is obtained when reacting with highly concentrated solutions of bases, hydroxo complexes like $Na_2[Fe(OH)_4]$, $Na_2[Co(OH)_4]$ are formed, and in a melt of NiO with Na_2O, Na_2NiO_2 is obtained. A specific property of

Co(II) and Ni(II) hydroxides is their interaction with aqueous solutions of NH_3 with the formation of complexes:

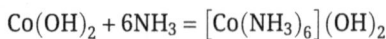

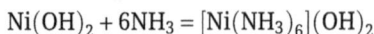

$$Co(OH)_2 + 6NH_3 = [Co(NH_3)_6](OH)_2$$

$$Ni(OH)_2 + 6NH_3 = [Ni(NH_3)_6](OH)_2$$

Among the numerous salts of Fe(II), Co(II), and Ni(II), sulfates, chlorides, and nitrates are the most commonly used. Iron(II) salts often appear as reducers in aqueous solution, for example:

$$10FeSO_4 + 2KMnO_4 + 8H_2SO_4 = 5Fe_2(SO_4)_3 + 2MnSO_4 + K_2SO_4 + 8H_2O$$

Their most stable complex compounds are cyanide compounds. They are formed when an excess of cyanides are added to solutions of their salts. The cyanide anion CN^- creates a strong electrostatic field around the complexing agent and therefore the complexes are low-spin and diamagnetic. The most common complex compound of this type is $K_4[Fe(CN)_6]$.

When it is added to solutions containing Fe^{3+} cations, a dark blue precipitate of Berlin blue (*Prussian blue*), $Fe_4[Fe(CN)_6]_3$, is released:

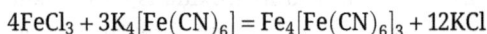

$$4FeCl_3 + 3K_4[Fe(CN)_6] = Fe_4[Fe(CN)_6]_3 + 12KCl$$

Unlike iron, cobalt, and nickel do not form stable oxides in its *oxidation state +3*. Fe(III) compounds are produced by oxidation of metallic iron or Fe(II) compounds. Brown-red iron(III) oxide Fe_2O_3 is a stable compound that exists in nature. Synthetic Fe_2O_3 is obtained by the decomposition of Fe(III) hydroxide and salts, for example, Fe$(NO_3)_3$. It easily interacts with acids:

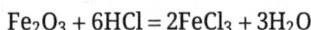

$$Fe_2O_3 + 6HCl = 2FeCl_3 + 3H_2O$$

Fe_2O_3 interacts with bases and carbonates of alkali metals in a melt, which indicates its amphoteric nature:

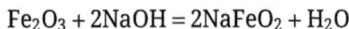

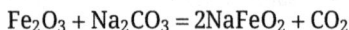

$$Fe_2O_3 + 2NaOH = 2NaFeO_2 + H_2O$$

$$Fe_2O_3 + Na_2CO_3 = 2NaFeO_2 + CO_2$$

The existence of double oxide $(FeO \cdot Fe_2O_3 \equiv Fe_3O_4 \equiv Fe(FeO_2)_2)$ is also an evidence of the amphoteric nature of Fe_2O_3.

The hydroxides $Fe(OH)_3$, $Co(OH)_3$, and $Ni(OH)_3$ are practically insoluble in water. They exhibit a basic character with weak amphoteric ratios. Their formulas $E(OH)_3$ are conditional, since in reality they are hydrated oxides $(E_2O_3 \cdot xH_2O)$ which when heated turn into monohydrates $(E_2O_3 \cdot H_2O \equiv EO(OH))$.

When dissolved in acids, $Co(OH)_3$ and $Ni(OH)_3$ exhibit strong-oxidizing properties and are reduced to Ni^{2+} and Co^{2+} ions:

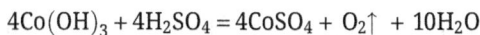

$$4Co(OH)_3 + 4H_2SO_4 = 4CoSO_4 + O_2\uparrow + 10H_2O$$

Iron(III) hydroxide, when heated with concentrated solutions of bases, forms ferrites – salts of ferric acid, which proves its amphoteric properties:

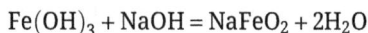

$$Fe(OH)_3 + NaOH = NaFeO_2 + 2H_2O$$

Simple (noncomplex) salts of Fe(III) are numerous. In solutions, they undergo hydrolysis with the formation of basic salts that color the solutions in a yellow-brown color. Fe^{3+} cations are medium-strength oxidizing agents:

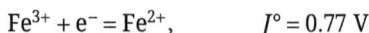

$$Fe^{3+} + e^- = Fe^{2+}, \qquad J° = 0.77 \text{ V}$$

That is why it is impossible to obtain Fe_2S_3, FeI_3 and some other salts in solution, because their formation proceeds according to the intramolecular redox type:

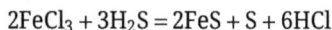

$$2FeCl_3 + 3H_2S = 2FeS + S + 6HCl$$

$$2FeCl_3 + 2KI = 2FeCl_2 + I_2 + 2KC1$$

Cobalt(III) salts are the strongest oxidizing agents:

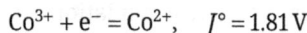

$$Co^{3+} + e^- = Co^{2+}, \quad J° = 1.81 \text{ V}$$

Although some of them exist in a solid state (CoF_3 and $CoCl_3$), when dissolved, they are reduced, oxidizing water:

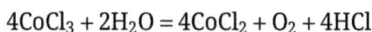

$$4CoCl_3 + 2H_2O = 4CoCl_2 + O_2 + 4HCl$$

Simple (noncomplex) nickel(III) salts practically does not exist.

Complexation increases the stability of iron, cobalt and nickel compounds in +3 oxidation state. Potassium hexacyanoferrate(III) $K_3[Fe(CN)_6]$ is quite stable and widely used. When it interacts with solutions of iron(II) salts, the so-called Turnbull's blue, $Fe_3[Fe(CN)_6]_2$, is formed, which quickly turns into Berlin blue – $Fe_4[Fe(CN)_6]_3$, since Fe^{2+} is oxidized to Fe^{3+} and reduces the hexacyanoferrate(III) ion to hexacyanoferrate(II):

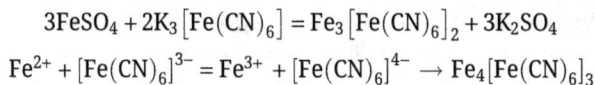

$$3FeSO_4 + 2K_3\left[Fe(CN)_6\right] = Fe_3\left[Fe(CN)_6\right]_2 + 3K_2SO_4$$

$$Fe^{2+} + [Fe(CN)_6]^{3-} = Fe^{3+} + [Fe(CN)_6]^{4-} \rightarrow Fe_4[Fe(CN)_6]_3$$

It was found that Berlin and Turnbull's blue have the same structure, because during precipitation there is an exchange of Fe^{3+} and Fe^{2+} between the inner and outer coordination spheres of the compound.

In *the oxidation state of +4*, ferrates(IV) of the type Na_2FeO_3 and Ba_2FeO_4, cobaltates(IV) of similar composition and nickelates(IV) of the type K_2NiO_3 and K_2NiF_6 are known.

The oxidation state of +5 appears to be the maximum in cobalt. There are compounds such as potassium cobaltate(V) (K_3CoO_4) and a similar compound in iron K_3FeO_4.

When Fe_2O_3 and $Fe(OH)_3$ react with strong oxidizing agents in concentrated solutions and melts of bases, iron compounds are formed in *an oxidation state of +6*, the so-called ferrates(VI):

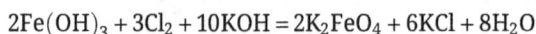

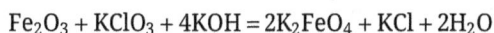

$$2Fe(OH)_3 + 3Cl_2 + 10KOH = 2K_2FeO_4 + 6KCl + 8H_2O$$

$$Fe_2O_3 + KClO_3 + 4KOH = 2K_2FeO_4 + KCl + 2H_2O$$

Ferrates(VI) are strong oxidizing agents stable in an alkaline environment. In other environments, they decompose in an intramolecular redox type:

$$4K_2FeO_4 + 10H_2O = 4Fe(OH)_3 + 8KOH + 3O_2$$

$$4BaFeO_4 + 4H_2SO_4 = 4BaSO_4 + 2Fe_2O_3 + 3O_2 + 4H_2O$$

For this reason, it is practically impossible to obtain ferric acid, H_2FeO_4.

Iron, cobalt, and nickel form carbonyls $Fe(CO)_5$, $Co_2(CO)_8$, and $Ni(CO)_4$, which are volatile liquids with a molecular structure. The metals in carbonyls are in a zero-oxidation state, as evidenced by their easy oxidation – the carbonyls in air spontaneously ignite:

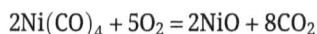

$$2Ni(CO)_4 + 5O_2 = 2NiO + 8CO_2$$

Carbonyls are formed when powdered metals react with carbon(II) oxide under pressure at ordinary temperature or under low heating (in an inert medium). The higher temperature leads to the decomposition of carbonyls to form finely dispersed pyrophoric metals.

Iron, cobalt, and nickel form complex compounds in which ligands are organic molecules or radicals. For example, for the analytical determination of Ni(II), its reaction with dimethylglyoxime is used. In the vital process of respiration in humans and animals, hemoglobin is involved – a complex compound of iron(II), which attaches oxygen and delivers it through the circulatory system to all organs. The complex compound ferrocene, $Fe(C_2H_5)_2$, is widely used (iron pentadienyl). These and other similar complex compounds are subjects of consideration in both inorganic and organic chemistry.

32.1.5 Usage and biological role

Iron in the form of cast iron and steel is the basis of modern technology. Nickel and cobalt are used as alloying additives to stainless steels, in heat-resistant alloys (nichrome, nimonic, inconel, and hastelloy), in magnetic alloys (permalloy) and in special alloys (invar, platinumite, nickeline, constantan, and monelmetal). Porous iron with additives is used as a catalyst in the synthesis of ammonia. Highly dispersed nickel (Reney nickel) is an active catalyst for the hydrogenation of organic compounds. It is prepared by melting nickel with aluminum with subsequent washing of aluminum. Cobalt is part of the catalyst desiccant, which accelerates the "drying" of oil paints. Iron oxides are used as pigments in paints (ochre and meerkat), from which ferrites are obtained, that are widely used in radio electronics and in audio cassettes. Nickel compounds are part of alkaline iron-nickel and cadmium-nickel batteries.

Iron is a necessary element for almost every form of vital activity, although it is poorly absorbed directly. On the other hand, an excess of iron in the body poses a "toxic risk." Difficulties in the absorption of iron are associated with the extremely low solubility of Fe(III) compounds contained in its minerals. Dissolved Fe(III) compounds are hydrolyzed at a slight increase in pH (up to ~2.0), which leads to the release of polymerized hydrated oxide, which is difficult for cells to access. The second problem is toxicity related to the ability of Fe(III) ions to catalyze the production of hydroxide radicals. Nature has created a system for the absorption, transport and accumulation of iron in vivo, based primarily on the ability of Fe(II) and Fe(III) ions to form highly stable complex compounds resistant to hydrolysis. Exhibiting oxidation states of +2, +3, +4 and coordination numbers 4 and 6, iron is a fairly mobile element in its compounds, easily moving from one type of coordination with the bioligand to another. Numerous organic iron compounds are involved in a number of biological processes in living organisms. Undoubtedly, the greatest importance is hemoglobin, which transports oxygen to living cells. Hemoglobin consists of the iron-containing active group "heme" and a protein part – "globin," connected to each other. Each iron ion is coordinated with five nitrogen atoms. Four of them are in the same plane and belong to the molecule of protoporphyrin, and the fifth is from the imidazole molecule and binds to the protein part of hemoglobin. The key moment of participation of hemoglobin in the respiratory cycle consists in the coordination of molecular O_2, the contraction of the hemoglobin fragment Fe . . . O^{2-} in the so-called imidazole pocket, and further distribution of O_2 through the bloodstream in the vessels. In the alveoli of the lung, hemoglobin attaches oxygen, turns into oxyhemoglobin, which spreads through the blood and supplies oxygen to all cells of the body. Iron is involved in the transport of oxygen in the respiratory cycle and in the form of another protein called myoglobin. It contains a complex of iron with porphyrin, which distributes oxygen to the cells of skeletal muscles and the muscles of the heart. Oxygen carriers in higher organisms – hemoglobin and myoglobin belong to the "heme" forms of iron in the body. The toxic effect of molecules and ions similar to the O_2 molecule, for example, CN^-, CO, etc., is determined by the formation of stronger bonds with iron ions. The absorption of iron from the external environment is carried out through specific ligands – siderophores. The transfer and circulation of iron in the biofluids of higher organisms occurs with the help of proteins transferrins, which, together with ferritins, are responsible for the accumulation of iron in the body.

Iron plays a key role in energy production as a constituent of some enzymes; including iron catalase, iron peroxidase, and cytochrome enzymes. It should be noted that in addition to the Fe-containing proteins – carriers of O_2, there are also proteins – electron carriers – cytochromes, which also contain some metals, including iron. Since iron forms an oxidation-reduction pair Fe^{2+}/Fe^{3+}, it is the potential of this pair that determines the transfer of electrons. In Fe-containing cytochromes, the coordination number of iron is usually equal to 6, amino acid fragments are always in the axial position (unlike hemoglobin), and iron is in a low-spin state in its two oxidation states. Electron carrier proteins also include iron-sulfur proteins containing a cluster fragment $(Fe-S)_n$ called ferrodoxin.

The body of an adult person contains about 3 g of iron, of which 75% is concentrated in the composition of hemoglobin. Iron stimulates bone marrow function and boosts the production of red blood cells. The daily need of the human body for iron is great. In case of anemia, weakness, exhaustion, and neurasthenia, iron is introduced into the body through various pharmaceutical preparations. Iron is better absorbed in the ionic state as iron(II) than as iron(III) and does not cause irritation of the lining of the stomach and intestines.

In small doses, cobalt is not toxic to humans, animals, and plants. It is one of the trace elements necessary for human life. In small quantities, cobalt salts act as catalysts for the biological synthesis of hemoglobin. It is used in the preparation of combined with iron. It is part of vitamin B_{12}, cobalamin, or coenzyme B_{12}. Vitamin B_{12} (cobalamin) is a group of related compounds – derivatives of corrin. It is an endogenous vitamin that is synthesized in the intestinal microflora. Cobalamin contains a macrocycle – a corrin ring linked to a nucleotide and dimethyl benzimidazole. The cavity in the center of the corrin ring is occupied by a Co atom with a coordination number of 5, and the sixth coordination site in coenzyme B_{12} is occupied by 5-deoxyadenosine linked to Co by the $-CH_2$ group. Due to this structure, coenzyme B_{12} is a rare example of a natural organometallic compound. If the sixth coordination site is occupied by other small ligands, aquacobalamin, hydroxocobalamin, and cyanocobalamin, known as vitamin B_{12}, are obtained. Cobalamin is vital for higher organisms (in humans, several mg per day are needed), although it is only made by microorganisms. But even a mild deficiency can cause anemia, chronic fatigue, and depression, and a long-term deficiency can be detrimental to cardiovascular health and cause permanent damage to the brain and central nervous system. Co^{2+} salts contribute to the accumulation of other vitamins: pyridoxine, nicotinamide, which have a positive effect on all types of metabolism: protein, mineral, and carbohydrate. An excess of Co suppresses the function of the thyroid gland, because it affects the iodine content in its hormone. This manifests itself as a disease, the so-called endemic goiter, common in regions with an increased content of cobalt in soil and drinking water. The radioactive isotope ^{60}Co is used in medicine.

It is interesting to note that the closest analogue of cobalt and nickel, is considered undesirable and even dangerous in the biosphere. The beneficial effect of nickel in hemoglobin synthesis has been proven when administered simultaneously with iron(II) salts. Nickel enhances the synthesis of sulfur-containing amino acids. Nickel (in an unprecedented oxidation state) enters the active center of the enzyme urease, which promotes the hydrolysis of urea.

Of the medicines containing elements of group VIIIB, iron preparations used for the treatment and prevention of iron deficiency anemia are widely known. They contain ferrous iron in the form of salts (ferrous sulfate, chloride, glutamate, gluconate, aspartate, etc.) and are mainly for oral administration. In patients with chronic blood loss, in whom oral treatment is not effective enough, iron preparations for parenteral administration are administered. They are in the form of trivalent iron. It is also known that sodium nitroprusside) $Na_2[Fe(CN)_5NO])$, which has been known since the 50 s of the twentieth century, serves as a means of lowering blood pressure, relaxing the muscles of the vessels. Complexes of Co and Ni with thiosemicarbazones and Schiff bases exhibit weak bactericidal properties.

32.2 Platinum metals

32.2.1 General characteristics

Under the general name of platinum metals, the d-elements of the eighth group, which are not included in the iron family, are united: ruthenium (Ru), rhodium (Rh), palladium (Pd), osmium (Os), iridium (Ir), and platinum (Pt). Their physical characteristics are listed in Table 32.2.

Table 32.2: Characteristics of atoms and physical features of elements of platinum subgroup.

Elements	Ru	Rh	Pd	Os	Ir	Pt
Radius of the atom (nm)	0.134	0.134	0.137	0.137	0.135	0.138
Ionization potential (eV)	7.4	7.5	8.3	8.5	9.1	8.9
Density (g/cm³)	12.4	12.4	12.0	22.5	22.4	21.5
T_m (°C)	2,250	1,960	1,554	3,030	2,450	1,770
Electrode potential (V)	0.45	0.80	1.00	0.70	1.00	1.20

The metals have very similar atomic radii and ionization potentials. This explains the similarity in their properties, their joint distribution in nature and the complexity of their separation. The density of these metals shows a strong horizontal similarity: 4d platinum metals have a density of about 12 g/cm^3, while the density of 5d metals is about 22 g/cm^3. The melting temperatures of platinum metals are also high, with values ranging from 1,500 °C to 3,000 °C.

Characteristic similarities of platinum metals are also found in their compounds, for example, fluoride with the general formula MF_5. In fact, all of these fluorides are isostructural (but not isoelectronic) tetramers M_4F_{20}, where M is Ru, Os, Rh, Ir, and Pt.

32.2.2 Occurrence and synthesis

Platinum metals are very rare elements and its total content in the earth's crust is about 10^{-6}%. These metals are found in nature mainly in a free state, most often united together in the so-called native platinum, which contains about 80% platinum, 10% of other platinum metals and 10% of other metals (Cu, Au, etc.). In the form of compounds (mainly sulfides and arsenides, e.g., $PtAs_2$), these "soft" metals are found in some Cu and Ni ores, so sludge from the electrolysis of Cu and Ni is an important source for the production of rare platinum metals.

32.2.3 Physical and chemical properties

Platinum metals, along with Au and Ag, are called precious metals because of their appearance and valuable properties: chemical inertness, high melting point and malleability.

In the row of relative activity, platinum metals are found after hydrogen. They are resistant to corrosion, do not oxidize to air, do not interact with individual acids, except palladium, which dissolves in hot HNO_3. Aqua regia oxidizes platinum and palladium at room temperature, while ruthenium, rhodium, and iridium oxidize when heated to form complex acids:

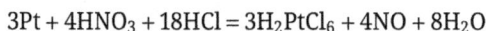

$$3Pt + 4HNO_3 + 18HCl = 3H_2PtCl_6 + 4NO + 8H_2O$$

Osmium reacts with aqua regia, which forms the oxide OsO_4.

Platinum metals interact with bases when heated in the presence of oxygen, since their oxides are amphoteric. Therefore, platinum vessels are not used for melting bases. For this purpose, nickel, silver or even iron vessels are used. All platinum metals exhibit extremely high catalytic activity. A large number of reactions are known to be accelerated by them.

All platinum metals absorb hydrogen in large quantities and form metal hydrides. Palladium is exceptional in this respect, which can absorb up to 900 volumes of hydrogen per volume of metal. A palladium membrane heated to 250 °C easily allows hydrogen to pass through, while other gases do not pass through it. The metallic conductivity slowly decreases as the absorbed hydrogen increases to a composition of $PdH_{0.5}$, whereby the substance becomes a semiconductor. The maximum absorption of hydrogen corresponds to the formula $PdH_{0.7}$. This ability of palladium is the reason why the palladium-hydrogen standard reference electrode is more preferred than the traditional platinum-hydrogen electrode.

32.2.4 Chemical compounds

In the compounds of platinum metals, different oxidation states are possible. Below are some of the most characteristic compounds that are more widely used in chemical practice.

32.2.4.1 Ruthenium and osmium

Although platinum metals have many similarities with each other, ruthenium and osmium in their chemical compounds rather continue the laws of the other transition metals. They are characterized by the maximum oxidation state +8 and d^0-electron configuration.

Ruthenium most often forms compounds in oxidation states +3 (halides RuX_3), +4 (RuO_2, RuS_2), +6 (ruthenates Me_2RuO_4) and +8 (RuO_4). Numerous complex compounds of ruthenium are known in oxidation states of +3 and +4.

Osmium resembles ruthenium in a variety of its valence states, but its compounds with high oxidation states are more stable. The most characteristic compound of osmium is OsO_4 – a pale-yellow crystalline substance with a molecular lattice. The OsO_4 molecule is tetrahedral with double osmium-oxygen bonds, isostructural and isoelectronic with tungstate WO_4^{2-} and perrenate ReO_4^- ions. The compound is easily obtained by heating osmium in an oxygen medium:

$$Os + 2O_2 \rightarrow OsO_4$$

Osmium(VIII) oxide is quite similar to xenon tetroxide, which is an example of similarity between the main (n) and secondary $(n + 10)$ groups. Osmium(VIII) oxide melts at a temperature of 40 °C and boils at 130 °C. Its vapors are highly poisonous even when inhaled at low concentrations. When OsO_4 is dissolved in water, the weak osmic acid H_4OsO_6 is formed. The acid is dibasic; therefore, its formula is $H_2[OsO_4(OH)_2]$. Simultaneously with the dissociation by acidic type $(K_1 = 10^{-12}, K_2 = 10^{-15})$, this compound also dissociates as a base $(K = 10^{-15})$. When H_4OsO_6 reacts with bases, the salts $Me_2[OsO_4(OH)_2]$ are formed. Osmates are salts of osmium of +6 oxidation state. In its complex compounds, Os occurs with oxidation states of +2, +3, and +4.

32.2.4.2 Rhodium and iridium

In platinum metals, the chemical properties of rhodium, iridium, palladium and platinum are much closer to each other than to those of ruthenium and osmium. For example, the highest oxidation state in them is +6 (instead of equal to the group number), and lower stable oxidation states are also observed.

Rhodium in its compounds is usually trivalent. The Rh_2O_3 oxide, $Rh(OH)_3$ hydroxide, RhX_3 halides, $Rh(NO_3)_3$ nitrate, $Rh_2(SO_4)_3$ sulfate, and $MeRh(SO_4)_2$ double salts are known.

Rhodium and iridium easily form complex compounds with an oxidation state of +3, which, like cobalt analogues, are kinetically inert. Similar to the 3d members of the group, rhodium and iridium form aquaions $[M(H_2O)_6]^{3+}$, where M = Rh or Ir. However, here too, there are noticeable differences between the 4d and 5d members; for example, an oxidation state of +4 is quite common for iridium. But in rhodium it is rare.

32.2.4.3 Palladium and platinum

The most common oxidation states for palladium and platinum are +2 and +4 (isoelectronic with +1 and +3 for rhodium and iridium, respectively). In oxidation state +2, the complexes are square planar, their aquaions are of composition $[M(H_2O)_4]^{2+}$, where M = Pd or Pt.

Palladium is usually found in the oxidation state of +2 and in its complex compounds in +4. The starting substance for the preparation of palladium compounds is usually $PdCl_2$ chloride, which is obtained by reacting chlorine with powdered palladium. A notable feature of $PdCl_2$ is its ability to react with carbon(II) oxide in aqueous solution:

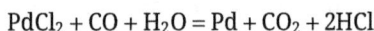

$$PdCl_2 + CO + H_2O = Pd + CO_2 + 2HCl$$

This reaction is used to detect CO because none of the other metals are reduced by CO at room temperature from solutions of their own salts.

In the 4d and 5d transition metals, the formulas do not always correspond to the true structure. For example, palladium forms fluoride with the empirical formula PdF_3, which is actually palladium(II) hexafluoropaladate(IV), $(Pd^{2+})[PdF_6]^{2-}$.

The most abundant element of this family, platinum, forms compounds mainly in the oxidation state of +4. These are PtO_2 dioxide, $Pt(OH)_4$ hydroxide, PtS_2 disulfide, PtX_4 halides and $Pt(SO_4)_2$ sulfate. Platinum(IV) oxide and hydroxide are amphoteric and interact with bases to form $Me_2[Pt(OH)_6]$ salts. Platinum forms numerous complex compounds in oxidation states +2 and +4.

When platinum reacts with aqua regia, hexachloroplatinic(IV) acid (H_2PtCl_6) is obtained. In solution it is a strong acid, and in a solid state it exists in the form of a crystal hydrate. At 360 °C it decomposes according to the equation:

$$H_2PtCl_6 \cdot 6H_2O = PtCl_4 + 2HCl + 6H_2O$$

Many salts of this acid were obtained. All of them, including those containing alkali metal cations and ammonium cation, are slightly soluble in water. Due to this, it is possible to separate platinum from other metals. The platinum precipitates in the form of ammonium hexachloroplatinate(IV), $(NH_4)_2[PtCl_6]$, whereby the other elements remain in the solution. When ammonium hexachloroplatinate(IV) is heated to 600 °C, a powdered metal is formed, which is further purified from impurities:

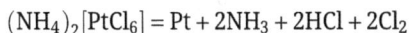

$$(NH_4)_2[PtCl_6] = Pt + 2NH_3 + 2HCl + 2Cl_2$$

Platinum(VI) fluoride is one of the strongest oxidizing agents, known to oxidize oxygen to O_2^+, and it itself is reduced to the more stable hexafluoroplatinate ion:

$$PtF_6 + O_2 \rightarrow (O_2^+)[PtF_6]^-$$

32.2.5 Usage and biological role

Although platinum metals are very expensive, they are widely used due to their high chemical inertness and heat resistance. In chemical laboratories, platinum crucibles and cups are used for analysis. Platinum is often used to make electrodes. Platinum is the only material that does not interact with molten glass, so it is used in the production of crucibles for melting pure optical glasses. The thinnest filament can be drawn from platinum – less than one micrometer thick, necessary for special physicochemical measurements.

Platinum and its alloys are used to make surgical instruments. Thermometers made of platinum and platinum-rhodium alloy are used to measure temperatures up to 1,400 °C, and from an alloy of iridium and tungsten – up to 2,300 °C. Platinum-rhodium alloy (hard and chemically resistant) is used to make crucibles for the growth of single crystals, which are necessary in quantum electronics. Platinum metals are widely used as catalysts and in the preparation of jewelry. Osmium(VIII) oxide is widely used in organic synthesis, especially in the oxidation of alkenes to diols. The compound is also widely used as a contrast agent in electron microscopy.

A particularly important role in recent decades has been acquired by the complex compounds of platinum used to treat cancer. The first preparation is the complex compound cisplatin with the formula $PtCl_2(NH_3)_2$ (*cis*-dichlorodiamineplatin), discovered by Rosenberg in 1969. The trans-isomer of cisplatin does not show biological activity. Since then, hundreds of complex compounds of platinum and other metals of VIIIB group have been synthesized and studied for cytotoxicity, but only in isolated cases has noticeable antitumor activity been observed. Platinum compounds are recognized as the best carcinostatics, which allow in 80–90% of cases to extend the life of patients. New platinum-based antitumor drugs have also been developed with proven lower toxicity than cisplatin, such as carboplatin and oxaloplatin, which are used to treat cisplatin-resistant tumors, as well as some that can be administered orally. The target of action of platinum preparations is DNA. It has been proven that the square-planar complexes of platinum are embedded between the helices of DNA, preventing unwanted division and growth of tumor cells. The complexes of Pt^{2+} (d^8 configuration) have a square-planar structure, and this is the configuration of the complexes of Pd (II), Au(III), Rh(I), and Ir(I), but these ions are either easily reduced or tend to pass into a nonplanar coordination polyhedron.

Ruthenium complexes are presently objects of great attention in medicinal chemistry, as anticancer agents with selective antimetastatic effects and low systemic toxicity. In addition, ruthenium is unique amongst the Pt group metals because it can exist in various oxidation states, the most common being the oxidation states +2, +3, and +4, which are accessible under biological conditions. Ru also finds application in photodynamic therapy. Many rhodium complexes, displaying low oxophylicity, broad functional-group tolerance, kinetical inertness and aqueous stability, have been studied as antineoplastic agents or enzyme inhibitors. Coordination compounds of Pd with heterocyclic N-containing ligands have found application as effective immunomodulators promoting cell recovery after radiation damage.

Chapter 33
Lanthanides and actinides

33.1 Lanthanides

33.1.1 General characteristics

In the sixth period, there are elements, the atoms of which contain electrons at the 4f-subshell. This subshell consists of seven orbitals, on which from one to fourteen electrons can be arranged. Therefore, the number of these elements is fourteen. They are located between lanthanum and hafnium, and are known by the common name of lanthanides ("lanthanum-like"). For example, lanthanum has an electron configuration of $[Xe]6s^25d^1$, while the next element cerium has a configuration of $[Xe]6s^24f^2$. Of interest is the electron configuration of gadolinium, which is $[Xe]6s^25d^1f^7$, in contrast to the predicted $[Xe]6s^24f^8$. Such electron transitions occur and are possible only in adjacent orbitals with similar energies. They also occur in the elements from actinium (Ac) to nobelium (No) due to the similarity in energies of the 7s, 6d, and 5f orbitals. The characteristics of atoms of lanthanides are listed in Table 33.1.

Table 33.1: Characteristics of atoms of lanthanides.

Elements	Z	Valence electrons	Atomic radius (nm)	Ionization potential (eV)	Electrode potential (V)	Oxidation states
Cerium (Ce)	58	$4f^26s^2$	0.183	5.50	−2.48	+3, +4
Praseodymium (Pr)	59	$4f^36s^2$	0.182	5.42	−2.46	+3, +4
Neodymium (Nd)	60	$4f^46s^2$	0.182	5.49	−2.43	+3
Promethium (Pm)	61	$4f^56s^2$	0.180	5.55	−2.42	+3
Samarium (Sm)	62	$4f^66s^2$	0.181	5.63	−2.41	+3, +2
Europium (Eu)	63	$4f^76s^2$	0.202	5.66	−2.40	+3, +2
Gadolinium (Gd)	64	$4f^75d^16s^2$	0.179	6.16	−2.40	+3
Terbium (Tb)	65	$4f^96s^2$	0.177	5.85	−2.39	+3, +4
Dysprosium (Dy)	66	$4f^{10}6s^2$	0.177	5.93	−2.35	+3
Holmium (Ho)	67	$4f^{11}6s^2$	0.176	6.02	−2.32	+3
Erbium (Er)	68	$4f^{12}6s^2$	0.175	6.10	−2.30	+3
Thulium (Tm)	69	$4f^{13}6s^2$	0.174	6.18	−2.28	+3
Iterbium (Yb)	70	$4f^{14}6s^2$	0.193	6.25	−2.27	+3, +2
Lutetium (Lu)	71	$4f^{14}5d^16s^2$	0.174	5.43	−2.25	+3

Lanthanides are very similar in their properties, because the appearance of new electrons in atoms occurs not on the outer or penultimate but on the third shell from the outside to the inside, with the same number of electrons in the outer shell ($6s^2$). Electrons from the 4f-subshell do not participate in the formation of chemical bonds. The difference in the energies of the 4f- and 5d-states is still very small. Thanks to this, one of the

https://doi.org/10.1515/9783111712246-033

4f-electrons is easily excited, moving to the 5d-subshell, where it becomes valent. Therefore, in most of their compounds, the lanthanides have an oxidation state of +3, which explains the proximity of their properties to the properties of IIIB group d-elements. That is why the lanthanides, together with lanthanum, yttrium, and scandium (IIIB) are united into one family – rare earth elements (REE), although some of them are not rare.

Small changes in the properties of the lanthanides (increase in ionization and decrease in electrode potential) are explained by the lanthanide contraction, namely the gradual reduction of their atomic and ionic radii. As a result of the lanthanide contraction, the sizes of the atoms and ions of the elements of the sixth period (Hf, Ta, W, etc.) are very close to the sizes of the atoms and ions of the corresponding elements of the fifth period (Zr, Nb, Mo, etc.).

33.1.2 Occurrence and synthesis

The REEs are found in nature together. Their total content is 0.016%. The most abundant among them are cerium and neodymium, and the rarest are thulium and lutetium. Promethium is not found in nature but it is produced artificially.

The main source for the production of REEs is monazite $(REE)PO_4$. The problem with the separation of REEs is complicated, but most often it is solved by the ion exchange method. In the form of simple substances, REEs are obtained by metallothermic reduction of their oxides, fluorides or chlorides with calcium.

33.1.3 Properties and chemical compounds

Lanthanides are active metals. In the row of relative activity, they are located close to magnesium. They displace hydrogen from acids and in dispersed state they are pyrophoric, and when heated they interact with oxygen, halogen elements, and sulfur.

In most of their compounds, the lanthanides are found in the *oxidation state of +3*. These are non-amphoteric thermodynamically stable oxides E_2O_3, slightly soluble nonamphoteric hydroxides $E(OH)_3$ and numerous salts.

Samarium, europium, and ytterbium form compounds in the *+2 oxidation state*, which are good reducers. The Eu(II) ion has a similar behavior to alkaline earth ions. For example, its carbonate, sulfate, and chromate are insoluble, as are those of the heavier alkali earth metals. The ionic radii of Eu(II) and Sr(II) are indeed very close in size, and as expected, many europium and strontium compounds are isomorphic.

Cerium, praseodymium, and terbium also form other compounds in the *oxidation state of +4*, which are strong oxidizing agents:

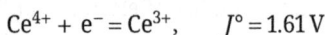

$$Ce^{4+} + e^- = Ce^{3+}, \qquad J° = 1.61 \, V$$

Cerium(III) sulfate is oxidized in an alkaline medium from chlorine to Ce(IV) hydroxide:

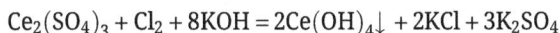

$$Ce_2(SO_4)_3 + Cl_2 + 8KOH = 2Ce(OH)_4\downarrow + 2KCl + 3K_2SO_4$$

While europium exhibits a lower-than-normal oxidation state, cerium exhibits in a higher-than-normal oxidation state of +4. The formation of the Ce(IV) ion corresponds to the configuration of a noble gas, which is evidenced by the fact that Ce has the lowest fourth ionization potential of all the lanthanides. The Ce(IV) ion behaves as zirconium(IV) and hafnium(IV) of IVB group and as thorium(IV) of the actinide group form insoluble fluorides and phosphates. There are particularly strong similarities in the chemistry of Ce(IV) and Th(IV). They form isomorphic nitrates $M(NO_3)_4 \cdot 5H_2O$, where M is Ce or Th, as well as hexanitrate complex ions $[M(NO_3)_6]^{2-}$. The main difference between the two elements in this oxidation state is that Th(IV) is the thermodynamically stable form of this element, while Ce(IV) is a strong oxidizing agent. In concentrated hydrochloric acid, $Ce(OH)_4$ is reduced to Ce(III):

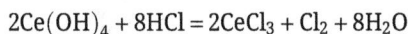

$$2Ce(OH)_4 + 8HCl = 2CeCl_3 + Cl_2 + 8H_2O$$

In a concentrated solution of NaOH, Ce(IV) hydroxide dissolves to form sodium cerate:

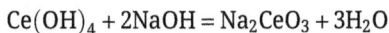

$$Ce(OH)_4 + 2NaOH = Na_2CeO_3 + 3H_2O$$

33.1.4 Usage

Lanthanides and their compounds are widely used. Cerium serves as an alloying additive to steel, used to make surgical instruments, as well as to obtain aluminum and tungsten alloys. The addition of cerium increases the electrical conductivity of aluminum, improves its mechanical properties, and also helps the rolling of tungsten. In addition, lanthanides are used to prepare various pyrophoric compositions. The alloy of cerium with iron is used to make pebbles for lighters. Strong and light alloys of lanthanides with magnesium are used in aircraft and rocket industry. Iron alloys with neodymium or samarium are super strong magnets. Oxides of cerium, neodymium, and praseodymium are part of optical and colored glass.

33.1.5 Biological role

REEs do not belong to the classical biogenic metals, but it is known that they are capable of exhibiting some biological activity. It has been observed that REEs accumulate in fern leaves, ginseng, as well as in some alkaloids isolated from plants. In living organisms, REEs are concentrated mainly in the skeleton (heavy REE), in the fatty tissue of the liver (mild REE), and also in the blood plasma. REEs can enter living organisms with drinking water and atmospheric pollution. Industrial waste also contains oxides of REEs.

The biological roles of REEs, their toxicological and pharmacological actions were not known for many years. The first mention of the pharmacological action of REE salts dates back to the end of the nineteenth century, when the action of cerium oxalate $Ce(C_2O_4)_2$ as an antiemetic agent (for prevention of vomiting). In 1897, the antimicrobial effect of REE salts in the treatment of tuberculosis and leprosy was reported. Only after the Second World War, some radioactive isotopes of REEs (^{141}Ce, ^{153}Sm, ^{153}Gd, ^{160}Tb, ^{170}Tm, ^{169}Yb, and ^{177}Lu) were discovered, which accumulate in certain tissues of the body and allow the diagnosis of various pathologies, and in many cases exhibit antitumor action, destroying malignant neoplasms because of their radiation.

In the 1920s, the influence of REE salts on blood clotting became known. It was clarified that the ability of REE ions to replace Ca ions can be used not only for diagnosis, but also for regulating the processes of Ca metabolism in the body, for example, during blood clotting. A key moment in the cascade process of blood clotting is the formation of Ca complex with prothrombin. REE preparations, applied in the form of soluble complexes with bioligands (amino acids, vitamins, etc.), displace Ca from the prothrombin complex, which prevents the formation of a blood clot. This is very important both for preventing the coagulation of donor blood during its storage and for preventing the formation of blood clots in vivo. These anticoagulant properties of REE compounds have not yet found full practical application due to the unfounded suspicion of toxic effect of REE during intravenous administration. In addition, the well-known anticoagulant heparin is much cheaper. Other directions in the pharmacological action of REE are also known. There are scientific reports that REEs lower blood pressure, lower cholesterol and glucose levels, suppress appetite, exhibit moderate anti-inflammatory effects, etc.

Biological utilization of lanthanides as complexing agents is very reasonable from a chemical viewpoint. The liability of Ln(III) complexes, very fast water exchange reaction, strong oxyphilicity, non-directionality of Ln-ligand bond and varying coordination numbers, all contribute towards Ln interaction with biomolecules. Smaller size of chelating bioligands can even suit larger Ln with lowered coordination number. Correspondingly, small lanthanides can expand their coordination number and can produce stable chelates with larger biomolecules. This can explain the different coordination power and different biological behavior of the respective lanthanides under different biological conditions.

So far, few medicines based on REE are known. These are ointments with samarium compounds, as well as gadolinium complexes with bulk ligands, for example, diethylenetriaminepentaacetic acid, used as contrast agents in medical tomography. Europium complexes are used in fluorescence immunoassay. Over the past 10 years, it has been reported that some compounds of REE (especially with Gd, Tb, Tm, Ce ions, but not La, Eu, and Lu) are antioxidants against the active oxygen-containing radicals $\bullet O_2^-$, $\bullet OH^-$, and H_2O_2.

As for the toxic effect of REE, it is usually greatly exaggerated. The REE ions themselves are not aggressive cations, because they do not exhibit pronounced oxidative properties, and in addition, they are significantly "closed" by their hydrate shells which further reduces their activity. Of course, the powdered oxides of REE have a strong toxic effect (especially when inhaled), and the salts of REE are embryotoxic, because they cause deformations in the skeleton.

33.2 Actinides

33.2.1 General characteristics

In the seventh period, the analogues of the lanthanides and d-elements of the third group are the actinides: thorium (Th), protactinium (Ra), uranium (U), neptunium (Np), plutonium (Pu), americium Am, curium (Cm), berkelium (Bk), californium (Cf), einsteinium (Es), fermium (Fm), mendelevium (Md), nobelium (No), and lawrencium (Lr). The elements of the actinide group are numbered from 90 (Th) to 103 (Lr).

It is curious that while the lanthanides have very small similarities with the elements outside their own series, some similarities with transition metals are observed in the actinides. In this regard, the similarities in the chemical structures and properties of the actinides (Th, Pa, and U) with the corresponding analogous transition metals can be noted. This resemblance is so strong that by 1944 the first five elements in the group (from thorium to plutonium) were assigned to the fourth row of the transition metals series (where Rf ($Z = 104$) and Db ($Z = 105$) are now located).

The regularities in actinides are much more complex. In accordance with Klechkowski's rule, the filling of atoms with electrons should take place at the 5f-subshell with a filled $7s^2$-subshell. However, the energies of the 5f- and 6d-states are so close that in the first half of the actinides (from Th to Cm) the electrons are located at both the 5f- and 6d-subshells, so these elements exhibit properties of both f-, and d-elements. In this range of nuclear charges ($Z = 90$–95), 6d and 5f electrons can take part in chemical interactions. So therefore, actinides exhibit a much wider range of oxidation states. For example, in neptunium, plutonium, and americium, are known compounds in which these elements are in a hexavalent state. For example, americium forms compounds in oxidation states +2, +3, +4, +5, +6, and +7 (similar to manganese).

Only in the elements after curium ($Z = 96$) the trivalent state becomes stable. As the atomic number increases, the binding energy of the 5f-electrons to the nucleus increases, and the binding energy of the 6d-electrons decreases. Consequently, the transition of 6d-electrons to the 5f-subshell becomes more advantageous. For this reason, the second half of the actinides (Bk–Lr) are similar in properties to typical f-elements.

Therefore, the properties of actinides differ significantly from those of the lanthanides, so the two families should not be considered similar. The actinide family ends with the element $Z = 103$ (lawrencium). A review of the chemical properties of Rf ($Z = 104$) and Db ($Z = 105$) shows that these elements should be analogous to hafnium and tantalum, respectively. That is why scientists suggest that after the family of actinides the systematic filling of the 6d-subshell begins in the atoms of the elements.

33.2.2 Occurrence and synthesis

Only three of the actinides (Th, Pa, and U) are found in the earth's crust. The rest do not occur in nature and were produced artificially. The nuclei of the atoms of all actinides are unstable and spontaneously decay into nuclei of atoms of lighter elements. As the atomic number increases, the instability of the nuclei increases and their half-life decreases. If Th has a half-life of 1.4×10^{10} years, uranium – 4.5×10^9 years, for lawrencium it is only 3 min.

ThO_2 was discovered by Berzelius in 1828, but metallic thorium was obtained relatively recently. The Th content in the earth's crust is $1.3 \times 10^{-3}\%$, and the uranium content is $4 \times 10^{-4}\%$.

Uranium is a very scattered element and is found in many minerals. Minerals in which uranium is contained in the form of UO_2 (uranite) and U_3O_8 are of practical importance. The latter is a compound formed by two oxides $UO_2 \cdot 2UO_3$.

Uranium is extracted from the ore with sulfuric or nitric acid:

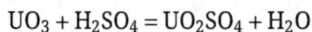

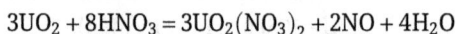

$$UO_3 + H_2SO_4 = UO_2SO_4 + H_2O$$

$$3UO_2 + 8HNO_3 = 3UO_2(NO_3)_2 + 2NO + 4H_2O$$

From the above solutions, when reacting with bases, the hydroxide $UO_2(OH)_2$ precipitates, which is reduced with hydrogen to dioxide:

$$UO_2(OH)_2 = UO_3 + H_2O$$

$$UO_3 + H_2 = UO_2 + H_2O$$

Uranium dioxide is converted into tetrafluoride, which is reduced with calcium to uranium:

$$UO_2 + 4HF = UF_4 + 2H_2O$$

$$UF_4 + 2Ca = 2CaF_2 + U$$

33.2.3 Properties and chemical compounds

Of all actinides, uranium is of the greatest practical importance. It is this element that is associated with significant success in the development of science and technology. During the study of this element, radioactivity was discovered, as well as two unknown chemical elements (radium and polonium). For the first time, a chain reaction of fission of atomic nuclei was carried out. After these discoveries, the development of the use of atomic energy for military and peaceful purposes began. Currently, nuclear power plants, where the fuel is uranium, produce nearly a fifth of the electricity, with 1 kg of uranium replacing more than 2,000 tons of coal.

Uranium is a heavy hard metal. In the row of relative activity, it is located before aluminum. In air, it is covered with an oxide layer, which hinders but does not stop oxidation. At 150 °C, uranium is rapidly oxidized to form U_3O_8. Hydrochloric acid interacts with uranium to form UCl_3 and UCl_4 chlorides, sulfuric acid to UO_2SO_4, and nitric acid to $UO_2(NO_3)_2$. With solutions and melts of bases, uranium interacts with the formation of uranates (Na_2UO_4) and diuranates ($Na_2U_2O_7$). In a highly dispersed state, uranium actively absorbs hydrogen with the formation of metal-like hydrides with a non-stoichiometric composition. When uranium reacts with an excess of fluorine, UF_6 hexafluoride is formed, which easily sublimates (at 56.5 °C). It is the only uranium compound existing in a gaseous state at low temperature.

Uranium can be compared in properties with the metals of VIB group. For example, it easily reaches +6 oxidation state as well as the members of VIB group. The most obvious similarity is in oxyanions, for example, the yellow diuranate ion ($U_2O_7^{2-}$) and the orange dichromate ion ($Cr_2O_7^{2-}$). Uranium forms uranyl chloride UO_2Cl_2, which is similar to chromyl and molybdenyl chlorides (CrO_2Cl_2 and MoO_2Cl_2). In general, as can be expected, uranium has the closest similarity to tungsten. For example, uranium and tungsten (but not Mo and Cr) form stable hexachlorides UCl_6 and WCl_6, respectively. Just as uranium resembles group VIB metals, so protactinium resembles group VB elements, and thorium resembles group IVB elements. The similarities in properties are the result of similarities in the external electron configuration of their atoms.

33.2.4 Usage and biological role

The main application of uranium is in nuclear energy. In nature, uranium is found in the form of a mixture mainly of two isotopes with atomic masses of 238 (99.3%) and 235 (0.7%). The isotope ^{235}U decays spontaneously. Therefore, uranium, used in reactors as nuclear fuel, is enriched in order to increase the content of the ^{235}U in it. In nuclear reactors, when absorbing neutrons, ^{235}U decays and releases a huge amount of energy. The separation of uranium isotopes is carried out by the method of gas diffusion of UF_6 using nickel porous membranes. Uranium compounds are also used as paints and dyes in the printing and silicate industries. Thorium dioxide is used in medicine, as well as in the preparation of some catalysts.

Plutonium was discovered in 1940, and the content of plutonium in the Earth's crust is negligible. It is obtained from the decay products of fuel in nuclear reactors and is used for the same purposes as uranium-235.

Novel lanthanide and actinide chelators have found widespread applications in MRI, PET imaging, targeted radionuclide therapy, and particularly in cancer chemotherapy.

Index

https://doi.org/10.1515/9783111712246-034

www.ingramcontent.com/pod-product-compliance
Lightning Source LLC
Chambersburg PA
CBHW080127220326
41598CB00032B/4981